Seismic Evaluation and Rehabilitation of Structures

GEOTECHNICAL, GEOLOGICAL AND EARTHQUAKE ENGINEERING

Volume 26

For further volumes:
http://www.springer.com/series/6011

Alper Ilki • Michael N. Fardis
Editors

Seismic Evaluation and Rehabilitation of Structures

Springer

Editors
Alper Ilki
Civil Engineering Faculty
Istanbul Technical University
Istanbul, Turkey

Michael N. Fardis
Department of Civil Engineering
University of Patras
Patras, Greece

ISSN 1573-6059
ISBN 978-3-319-00457-0 ISBN 978-3-319-00458-7 (eBook)
DOI 10.1007/978-3-319-00458-7
Springer Cham Heidelberg New York Dordrecht London

Library of Congress Control Number: 2013945282

Printed on acid-free paper

Springer is part of Springer Science+Business Media (www.springer.com)

Preface

In the past, old facilities which were considered to be at the end of their useful service life were demolished and replaced with new ones that better met the functional requirements of modern society, including new safety standards. However, humankind has recently recognised the threats to the environment and to our limited natural resources posed by the relentless determination to destroy the old and build anew. Awareness of these constraints and the increased emphasis on sustainability lead to the need to rehabilitate our existing structures and extend their service life as long as possible, regardless of economic or functional considerations which would have led in the past to their demolition and replacement. Indeed, while upgrading the well-known "Construction Products Directive" 89/106/CEE to the "Construction Products Regulation" 305/2011/EU, the European Union added a 7th "Essential Requirement": for sustainable use of resources. It is reminded in this regard that, according to the European Commission, the imperative to meet the first two "Essential Requirements", namely those for mechanical resistance and stability and for safety in case of fire, is the basis of the Eurocodes. It is not a coincidence, therefore, that the forthcoming revision of the set of ten EN-Eurocodes towards their second generation will include the new dimension of rehabilitating existing structures, after going through an evaluation of their structural condition and safety. Apart from the need to meet the new formulation of the "Essential Requirements" in Regulation 305/2011/EU, the EN-Eurocodes recognise with this extension of their scope the coming economic and societal realities and their impact on the Construction Sector, as retrofitting of existing facilities will increase at the expense of new construction.

The majority of the existing building stock and of civil infrastructures are seismically deficient. So, when the time comes for a decision to prolong their service life with the help of structural and architectural upgrading, the issue of seismic retrofitting arises. Besides, it is often decided to upgrade the earthquake resistance of facilities that still meet their functional requirements and fulfill their purpose from an architect's point of view, but are clearly unsafe in the event of an earthquake. It is indicative of the importance of seismic retrofitting for public safety that the only one in the set of 56 Parts in the first generation of EN-Eurocodes which deals with existing structures is Part 3 of Eurocode 8 on "(Seismic) Assessment and Retrofitting of Buildings". This Part will be one of the four focus areas in the upcoming revision of Eurocode 8 towards the second generation: the revision will encompass a major overhaul of the Normative part, including, among others, incorporation of specifics presently in the Informative Annexes, as well as extension of the scope to cover seismic assessment and retrofitting of bridges, presently a major void in Eurocode 8.

In order to decide how badly an individual structure needs seismic upgrading or to prioritise the rehabilitation of a population of structures, a seismic evaluation or assessment of the as-built structure is necessary. Such an evaluation may be carried out at various levels of approximation or sophistication, depending on the scale considered and the resources available. The outcome of the evaluation normally serves as a guide for the extent and details of the strengthening intervention, as well as for the technique to be used. Modern seismic codes or standards do not sufficiently cover the delicate phase of seismic evaluation nor the many potential technical options for seismic upgrading; therefore experimental and analytical research in these two areas is on-going and the state of the art is constantly evolving. All the more so as seismic evaluation and rehabilitation demand considerable expertise in order to make best use of the available safety margins in the existing structure, to adapt the engineering capabilities and techniques at hand to the particularities of a project, to minimise disruption of use or nuisance to occupants, etc. Further, as old structures are very diverse in terms of their materials, layout, etc., seismic retrofitting does not lend itself to straightforward codified procedures or cook-book approaches. As such seismic evaluation and rehabilitation need the best that the current state of the art can offer on all aspects of earthquake engineering: from the subsoil and its role in the event of an earthquake to the mechanical behavior of novel retrofitting materials and components.

In this volume, top seismic experts and researchers from around the world present the most recent outcomes of their work on seismic evaluation and retrofitting and closely related subjects. Many of the authors are partners in SERIES ("Seismic Engineering Research Infrastructures for European Synergies", www.series.upatras.gr, the largest research project in earthquake engineering in the European Union's 7th Framework Programme, FP7/2007–2013) supported under grant agreement n° 227887 of the Research Infrastructures Programme in FP7, or external researchers who have used Europe's seven largest and most advanced seismic testing facilities in the framework of SERIES.

The SERIES Workshop on the "Role of Research Infrastructures in Seismic Rehabilitation" took place in Istanbul in February 2012 and attracted a large audience to listen to renowned experts from around the world presenting close to 30 invited contributions. The Workshop itself and the publication and diffusion of these Proceedings are part of the Networking Activities of the SERIES project and have been made possible through European Community funding.

We gratefully acknowledge the support of Ms Vassia Vayenas for the preparation of these Proceedings. We are also thankful to Dr. Medine Ispir and Dr. Dionysis Biskinis for their contribution to the organisation of the SERIES Workshop in Istanbul.

Istanbul, Turkey
Patras, Greece
February 2013

Alper Ilki
Michael N. Fardis

Contents

1 **Surrealism in Facing the Earthquake Risk** 1
Mete A. Sözen

2 **Rapid Seismic Assessment Procedures for the Turkish Building Stock** 15
Ahmet Yakut, M. Altuğ Erberik, Alper Ilki, Haluk Sucuoğlu, and Sinan Akkar

3 **Post-Earthquake Risk-Based Decision Making Methodology for Turkish School Buildings** 37
Ufuk Yazgan and Reşat Atalay Oyguç

4 **Proposed Vulnerability Functions to Estimate the Real Damage State of RC Buildings After Major Turkish Earthquakes** 47
Ulgen Mert Tugsal and Beyza Taskin

5 **Probabilistic Path Finding Method for Post-Disaster Risk Estimation** 61
Florin Leon and Gabriela M. Atanasiu

6 **Seismic Behavior of Thin-Bed Layered Unreinforced Clay Masonry Shear Walls Including Soundproofing Elements** 77
Christophe Mordant, Matt S. Dietz, Colin A. Taylor, André Plumier, and Hervé Degée

7 **Assessing Seismic Vulnerability of Unreinforced Masonry Walls Using Elasto-Plastic Damage Model** 95
Basheer H. Al-Gohi, Cem Demir, Alper Ilki, Mohammed H. Baluch, and Muhammad K. Rahman

8 Implementation of Experimentally Developed Methodology for Seismic Strengthening and Repair of Historic Monuments 115
Veronika Shendova, Zoran T. Rakicevic, Mihail Garevski, Roberta Apostolska, and Zivko Bozinovski

9 Shaking Table Tests of a Full-Scale Two-Storey Pre-Damaged Natural Stone Building Retrofitted with the Multi-Axial Hybrid Textile System "Eq-Grid" 155
Lothar Stempniewski and Moritz Urban

10 Application of Mesh Reinforced Mortar for Performance Enhancement of Hollow Clay Tile Infill Walls 171
Pourang Ezzatfar, Barış Binici, Özgür Kurç, Erdem Canbay, Haluk Sucuoğlu, and Güney Özcebe

11 Shake Table Tests on Deficient RC Buildings Strengthened Using Post-Tensioned Metal Straps 187
Reyes Garcia, Iman Hajirasouliha, Kypros Pilakoutas, Yasser Helal, Yaser Jemaa, Maurizio Guadagnini, Mihail Petkovski, Philippe Mongabure, Mihaela Anca Ciupala, Nicholas Kyriakides, Christis Z. Chrysostomou, Alper Ilki, M. Saiid Saiidi, Lluis Torres, Nicolae Taranu, and Mihai Budescu

12 Bond Strength of Lap Splices in FRP and TRM Confined Concrete: Behavior and Design 203
Dionysios Bournas and Thanasis Triantafillou

13 Finite Element Modeling of Seismic Performance of Low Strength Concrete Exterior Beam-Column Joints 221
Danish Ahmed, Mohammed H. Baluch, Muhammad K. Rahman, and Alper Ilki

14 FRP Local Retrofit of Non-Conforming RC Beam-Column Joints 243
Andrea Prota, Marco Di Ludovico, Alberto Balsamo, Claudio Moroni, Mauro Dolce, and Gaetano Manfredi

15 Seismic Rehabilitation of Concrete Buildings by Converting Frame Bays into RC Walls 261
Michael N. Fardis, Antonis Schetakis, and Elias Strepelias

16 Pseudo-Dynamic Tests of 4-Storey Non-Ductile Frames with RC Infilling of the Bay 281
Elias Strepelias, Xenophon Palios, Stathis N. Bousias, and Michael N. Fardis

17 RC Infilling of Existing RC Structures for Seismic Retrofitting 303
Christis Z. Chrysostomou, Nicholas Kyriakides, Martin Poljanšek, Fabio Taucer, and Francisco Javier Molina

18 Hybrid Control of a 3-D Structure by Using Semi-Active Dampers 329
Gürsoy Turan

19 Substructure Pseudo-Dynamic Tests on Seismic Response Control of Soft-First-Story Buildings 341
Hideto Kanno, Tetsuya Nishida, and Jun Kobayashi

20 Towards Robust Behavioral Modeling of Reinforced Concrete Members 355
Kutay Orakcal

21 Earthquake Engineering Experimental Facility for Research and Public Outreach 379
Ece Eseller-Bayat, Seda Gokyer, and Mishac K. Yegian

22 Physical Modeling for the Evaluation of the Seismic Behavior of Square Tunnels 389
Grigorios Tsinidis, Charles Heron, Kyriazis Pitilakis, and Gopal Madabhushi

23 Susceptibility of Shallow Foundation to Rocking and Sliding Movements During Seismic Loading 407
Charles Heron, Stuart Haigh, and Gopal Madabhushi

24 Centrifuge Modeling of Liquefaction Effects on Shallow Foundations 425
Andreia Sofia Pedroso da Silva Marques, Paulo Alexandre Lopes de Figueiredo Coelho, Stuart Haigh, and Gopal Madabhushi

25 Stability Control of Rafted Pile Foundation Against Soil Liquefaction 441
Ahmed Mohammed Youssef Mohammed and Koichi Maekawa

26 Experimental Assessment of Seismic Pile-Soil Interaction 455
Armando L. Simonelli, Luigi Di Sarno, Maria Giovanna Durante, Stefania Sica, Subhamoy Bhattacharya, Matt S. Dietz, Luiza Dihoru, Colin A. Taylor, Roberto Cairo, Andrea Chidichimo, Giovanni Dente, Arezou Modaressi, Luìs A. Todo Bom, Amir M. Kaynia, George Anoyatis, and George Mylonakis

27 Experimental Investigation of Dynamic Behavior of Cantilever Retaining Walls 477
Panos Kloukinas, Augusto Penna, Anna Scotto di Santolo, Subhamoy Bhattacharya, Matt S. Dietz, Luiza Dihoru, Aldo Evangelista, Armando L. Simonelli, Colin A. Taylor, and George Mylonakis

Index 495

Contributors

Danish Ahmed Department of Civil Engineering, King Fahd University of Petroleum and Minerals, Dhahran, Saudi Arabia

Sinan Akkar Department of Civil Engineering, Earthquake Engineering Research Center, Middle East Technical University, Ankara, Turkey

Basheer H. Al-Gohi Department of Civil and Environmental Engineering, King Fahd University of Petroleum and Minerals, Dhahran, Saudi Arabia

George Anoyatis Department of Civil Engineering, Geotechnical Engineering Laboratory, University of Patras, Patras, Greece

Roberta Apostolska Institute of Earthquake Engineering and Engineering Seismology, IZIIS, SS Cyril and Methodius University, Skopje, Republic of Macedonia

Gabriela M. Atanasiu Multidisciplinary Center of Structural Engineering and Risk Management, "Gheorghe Asachi" Technical University of Iasi, Iasi, Romania

Alberto Balsamo Department of Structures for Engineering and Architecture, University of Naples Federico II, Naples, Italy

Mohammed H. Baluch Department of Civil and Environmental Engineering, King Fahd University of Petroleum and Minerals, Dhahran, Saudi Arabia

Subhamoy Bhattacharya Department of Civil Engineering, University of Bristol, Bristol, UK

Barış Binici Department of Civil Engineering, Middle East Technical University, Ankara, Turkey

Dionysios Bournas Department of Civil Engineering, The University of Nottingham, Nottingham, UK

Stathis N. Bousias Department of Civil Engineering, Structures Laboratory, University of Patras, Patras, Greece

Zivko Bozinovski Institute of Earthquake Engineering and Engineering Seismology, IZIIS, SS Cyril and Methodius University, Skopje, Republic of Macedonia

Mihai Budescu Faculty of Civil Engineering and Building Services, "Gheorghe Asachi" Technical University of Iasi, Iasi, Romania

Roberto Cairo Department of Civil Engineering, University of Calabria, Rende, Italy

Erdem Canbay Department of Civil Engineering, Middle East Technical University, Ankara, Turkey

Andrea Chidichimo Department of Civil Engineering, University of Calabria, Rende, Italy

Christis Z. Chrysostomou Department of Civil Engineering and Geomatics, Cyprus University of Technology, Limassol, Cyprus

Mihaela Anca Ciupala School of Computing, Information Technology and Engineering, University of East London, London, UK

Paulo Alexandre Lopes de Figueiredo Coelho Department of Civil Engineering, University of Coimbra, Coimbra, Portugal

Hervé Degée Department of Architecture, Geology, Environment & Constructions, University of Liege, Liege, Belgium

Cem Demir Department of Civil Engineering, Civil Engineering Faculty, Istanbul Technical University, Istanbul, Turkey

Giovanni Dente Department of Civil Engineering, University of Calabria, Rende, Italy

Marco Di Ludovico Department of Structures for Engineering and Architecture, University of Naples Federico II, Naples, Italy

Luigi Di Sarno Department of Engineering, University of Sannio, Benevento, Italy

Matt S. Dietz Department of Civil Engineering, University of Bristol, Bristol, UK

Luiza Dihoru Department of Civil Engineering, University of Bristol, Bristol, UK

Mauro Dolce Department of Structures for Engineering and Architecture, University of Naples Federico II, Naples, Italy

Department of Civil Protection, Presidency of Council of Ministers, Rome, Italy

Maria Giovanna Durante Department of Engineering, University of Sannio, Benevento, Italy

M. Altuğ Erberik Department of Civil Engineering, Middle East Technical University, Ankara, Turkey

Ece Eseller-Bayat Civil Engineering Faculty, Istanbul Technical University, Istanbul, Turkey

Aldo Evangelista Department of Civil and Environmental Engineering, University of Naples Federico II, Naples, Italy

Pourang Ezzatfar Department of Civil Engineering, Middle East Technical University, Ankara, Turkey

Michael N. Fardis Department of Civil Engineering, Structures Laboratory, University of Patras, Patras, Greece

Reyes Garcia Department of Civil and Structural Engineering, The University of Sheffield, Sheffield, UK

Mihail Garevski Institute of Earthquake Engineering and Engineering Seismology, IZIIS, SS Cyril and Methodius University, Skopje, Republic of Macedonia

Seda Gokyer Department of Civil and Environmental Engineering, Northeastern University, Boston, MA, USA

Maurizio Guadagnini Department of Civil and Structural Engineering, The University of Sheffield, Sheffield, UK

Stuart Haigh Department of Engineering, Schofield Centre, University of Cambridge, Cambridge, UK

Iman Hajirasouliha Department of Civil and Structural Engineering, The University of Sheffield, Sheffield, UK

Yasser Helal Department of Civil and Structural Engineering, The University of Sheffield, Sheffield, UK

Charles Heron Department of Engineering, Schofield Centre, University of Cambridge, Cambridge, UK

Alper Ilki Department of Civil Engineering, Civil Engineering Faculty, Istanbul Technical University, Istanbul, Turkey

Yaser Jemaa Department of Civil and Structural Engineering, The University of Sheffield, Sheffield, UK

Hideto Kanno Department of Architecture and Environment Systems, Faculty of Systems Science and Technology, Akita Prefectural University, Yurihonjo, Akita, Japan

Amir M. Kaynia Computational Geomechanics Section, Norwegian Geotechnical Institute, Oslo, Norway

Panos Kloukinas Department of Civil Engineering, Geotechnical Engineering Laboratory, University of Patras, Patras, Greece

Jun Kobayashi Department of Architecture and Environment Systems, Faculty of Systems Science and Technology, Akita Prefectural University, Yurihonjo, Akita, Japan

Özgür Kurç Department of Civil Engineering, Middle East Technical University, Ankara, Turkey

Nicholas Kyriakides Department of Civil Engineering and Geomatics, Cyprus University of Technology, Limassol, Cyprus

Florin Leon Multidisciplinary Center of Structural Engineering and Risk Management, "Gheorghe Asachi" Technical University of Iasi, Iasi, Romania

Gopal Madabhushi Department of Engineering, Schofield Centre, University of Cambridge, Cambridge, UK

Koichi Maekawa Civil Engineering Department, Faculty of Engineering, The University of Tokyo, Tokyo, Japan

Gaetano Manfredi Department of Structures for Engineering and Architecture, University of Naples Federico II, Naples, Italy

Andreia Sofia Pedroso da Silva Marques Department of Civil Engineering, University of Coimbra, Coimbra, Portugal

Arezou Modaressi Laboratoire MSS-Mat CNRS UMR 8579, Ecole Centrale Paris, Chatenay-Malabry, France

Ahmed Mohammed Youssef Mohammed Civil Engineering Department, Faculty of Engineering, The University of Tokyo, Tokyo, Japan

Francisco Javier Molina European Laboratory for Structural Assessment, Institute for the Protection and Security of the Citizen, Joint Research Centre, European Commission, Ispra, Italy

Philippe Mongabure CEA DEN/DANS/DM2S/SEMT/EMSI, Gif sur Yvette Cedex, Saclay, France

Christophe Mordant Department of Architecture, Geology, Environment & Constructions, University of Liege, Liege, Belgium

Claudio Moroni Department of Civil Protection, Presidency of Council of Ministers, Rome, Italy

George Mylonakis Department of Civil Engineering, Geotechnical Engineering Laboratory, University of Patras, Patras, Greece

Department of Civil Engineering, University of Bristol, Bristol, UK

Tetsuya Nishida Department of Architecture and Environment Systems, Faculty of Systems Science and Technology, Akita Prefectural University, Yurihonjo, Akita, Japan

Kutay Orakcal Department of Civil Engineering, Bogazici University, Istanbul, Turkey

Reşat Atalay Oyguç Newmark Civil Engineering Laboratory, University of Illinois at Urbana-Champaign, Urbana, IL, USA

Güney Özcebe Department of Civil Engineering, Middle East Technical University, Ankara, Turkey

Xenophon Palios Department of Civil Engineering, Structures Laboratory, University of Patras, Patras, Greece

Augusto Penna Department of Engineering, University of Sannio, Benevento, Italy

Mihail Petkovski Department of Civil and Structural Engineering, The University of Sheffield, Sheffield, UK

Kypros Pilakoutas Department of Civil and Structural Engineering, The University of Sheffield, Sheffield, UK

Kyriazis Pitilakis Department of Civil Engineering, Research Unit of Geotechnical Earthquake Engineering and Soil Dynamics, Aristotle University of Thessaloniki, Thessaloniki, Greece

André Plumier Department of Mechanics of Materials and Structures, University of Liege, Liege, Belgium

Martin Poljanšek European Laboratory for Structural Assessment, Institute for the Protection and Security of the Citizen, Joint Research Centre, European Commission, Ispra, Italy

Andrea Prota Department of Structures for Engineering and Architecture, University of Naples Federico II, Naples, Italy

Muhammad K. Rahman Center for Engineering Research, Research Institute, King Fahd University of Petroleum and Minerals, Dhahran, Saudi Arabia

Zoran T. Rakicevic Institute of Earthquake Engineering and Engineering Seismology, IZIIS, SS Cyril and Methodius University, Skopje, Republic of Macedonia

M. Saiid Saiidi Department of Civil and Environmental Engineering, University of Nevada, Reno, NV, USA

Antonis Schetakis Department of Civil Engineering, Structures Laboratory, University of Patras, Patras, Greece

Anna Scotto di Santolo Universita Telematica Pegaso, Naples, Italy

Veronika Shendova Institute of Earthquake Engineering and Engineering Seismology, IZIIS, SS Cyril and Methodius University, Skopje, Republic of Macedonia

Stefania Sica Department of Engineering, University of Sannio, Benevento, Italy

Armando L. Simonelli Department of Engineering, University of Sannio, Benevento, Italy

Mete A. Sözen School of Civil Engineering, Purdue University, West Lafayette, IN, USA

Lothar Stempniewski Department of Reinforced Concrete and Building Materials, Karlsruhe Institute of Technology, Karlsruhe, Germany

Elias Strepelias Department of Civil Engineering, Structures Laboratory, University of Patras, Patras, Greece

Haluk Sucuoğlu Department of Civil Engineering, Middle East Technical University, Ankara, Turkey

Nicolae Taranu Faculty of Civil Engineering and Building Services, "Gheorghe Asachi" Technical University of Iasi, Iasi, Romania

Beyza Taskin Department of Civil Engineering, Civil Engineering Faculty, Istanbul Technical University, Istanbul, Turkey

Fabio Taucer European Laboratory for Structural Assessment, Institute for the Protection and Security of the Citizen, Joint Research Centre, European Commission, Ispra, Italy

Colin A. Taylor Department of Civil Engineering, University of Bristol, Bristol, UK

Luìs A. Todo Bom Laboratoire MSS-Mat CNRS UMR 8579, Ecole Centrale Paris, Chatenay-Malabry, France

Lluis Torres Analysis and Advanced Materials for Structural Design, Polytechnic School, University of Girona, Girona, Spain

Thanasis Triantafillou Department of Civil Engineering, University of Patras, Patras, Greece

Grigorios Tsinidis Department of Civil Engineering, Research Unit of Geotechnical Earthquake Engineering and Soil Dynamics, Aristotle University of Thessaloniki, Thessaloniki, Greece

Ulgen Mert Tugsal Department of Civil Engineering, Civil Engineering Faculty, Istanbul Technical University, Istanbul, Turkey

Gürsoy Turan Department of Civil Engineering, Izmir Institute of Technology, Izmir, Turkey

Moritz Urban Department of Reinforced Concrete and Building Materials, Karlsruhe Institute of Technology, Karlsruhe, Germany

Ahmet Yakut Department of Civil Engineering, Middle East Technical University, Ankara, Turkey

Ufuk Yazgan Earthquake Engineering and Disaster Management Institute, Istanbul Technical University, Istanbul, Turkey

Mishac K. Yegian Department of Civil and Environmental Engineering, Northeastern University, Boston, MA, USA

Chapter 1
Surrealism in Facing the Earthquake Risk

Mete A. Sözen

Abstract The possibility of a violent ground motion in a population center poses multiple threats to the safety and continuity of society. It has to be met on multiple planes including the political and the economical. It is not an exaggeration to claim that the construction trades in many parts of the world understood the threat only in twentieth century. In many towns and cities in seismic zones, a strong ground motion can destroy 10 % or more of the existing buildings. To locate the vulnerable, it is necessary to investigate all. If this study is to be done in detail using codified criteria, it may involve an investment that would be unacceptable to a political system that seldom appreciates the risk. That condition constrains the engineering effort to the minimal and requires a procedure that has to be simple and transparent. Above all, data acquisition needs to be within the reach of workers without technical degrees. The paper investigates such a procedure, the Hassan Index, using available information from five earthquake events and finds it imperfect but useful.

1.1 Introductory Remarks

In planning for a battle in 1921 in which his chances of success were slim, Mustafa Kemal Atatürk made a declaration that may be translated freely as, "Our defense is not to be limited to a line. It is to include the entire country."[1] In two short sentences he was able to define the multi-dimensional nature of his defense strategy.

Defense of a region against the earthquake threat also involves more than one dimension. The threat may include any or even all of (a) unpredictable ground

[1] In Turkish, his statement was limited to six words.

M.A. Sözen (✉)
School of Civil Engineering, Purdue University, 550 Stadium Mall Drive, West Lafayette, IN 47907, USA
e-mail: sozen@ecn.purdue.edu

A. Ilki and M.N. Fardis (eds.), *Seismic Evaluation and Rehabilitation of Structures*, Geotechnical, Geological and Earthquake Engineering 26,
DOI 10.1007/978-3-319-00458-7_1,

motion, (b) disastrous landslides, (c) soil liquefaction, (d) an overwhelming tsunami followed by fire, and (e) social unrest. To boot, that what needs to be defended includes building structures of all types and ages, bridges, tunnels, roads, food-distribution system, health resources, and the communication infrastructure, all constructed under different requirements at different times. For a planner to focus on only one of these is, in a sense, declaration of defeat before engaging the threat.

Having said that, we continue with a limited objective, one aimed at safety of reinforced concrete building structures having seven or fewer stories. The focus is on planning the structural-risk assessment to be done before the earthquake occurs.

Proper assessment of the structural risk is not easy to define, because it is subject to the judgments of many professional groups. It has been said that the first material that needs to be purchased in starting construction is money. Thorstein Veblen has placed engineering between physics and banking. From the viewpoint of pure engineering concerns, "proper" is not difficult to define. But if the criteria are also controlled by banking interests, it becomes very difficult to define them, especially with respect to a threat that, unlike gravity, occurs infrequently.

In intellectual activities related to earthquake-risk assessment, the needed actions go against the grain of conventional engineering thought processes that begin with needs interpreted in terms of forces and space to produce a safe, serviceable, and durable artifact. The engineer's concern is to build the right thing in the right way. That, unfortunately, is not necessarily the way to success in earthquake-risk assessment. If the goal is to reduce the risk of severe damage or the time of recovery, there has to be a vision shared by all parties involved in the decision process. An unforgiving requirement of minimum safety, or an upper-bound estimate of the ground-motion demand is very likely to lead to inaction.

Because assessment involves large inventories, it is better if it is rule-based and best if the rules are simple. In the following section, we discuss a simple approach to assessment that is flexible enough to be modified to fit local engineering and banking goals.

For assessment of vulnerability to strong ground motion of individual structures, engineering seeks safety. The evaluation process requires robust data and detailed analysis. Only those structures that are unquestionably safe according to the driving criteria pass muster. The process may be called a high-pass filter. Its goal is to separate the safe from the questionable.

For risk assessment of earthquake vulnerability of building inventories, the constraints demand a different process. Large-scope decisions must be made based on inexpensively acquired data and evaluation algorithms that can be implemented rapidly. The goal is to separate the definitely vulnerable from the rest. This process may be deemed to be a low-pass filter. It is driven by the fear of failure rather than desire for success.

In regions with construction built without earthquake-related codes and construction control, the low-pass process is likely to be safer because explicit recognition of substantial need tends to lead to inaction.

The low-pass filter needs to be as simple as it can be. And it needs to have transparent rules for data acquisition that do not require an advanced degree in engineering to implement them. In fact, acquisition of the input data should be easy enough to be obtained by high school graduates after a week's training. It should be to engineering analysis what surrealist painting is to classical art. Even if the design of its overall concept requires talent, training and keen perception, it should be reproducible by almost any interested person.

Such a procedure is available in the Hassan Index (Hassan and Sözen 1994), a simple method that was inspired by the wall index proposed by Shiga et al. (1968). The Hassan index uses two indices to rank the relative safety of the structure. To organize data on damage states from earthquake events in different regions built at different times subjected to different earthquakes, and evaluated by different engineers using only two numbers, the column index and the wall index, in order to estimate vulnerability of existing buildings to earthquakes that have not yet happened, does appear to be something that cannot be done by anyone born of woman.[2] Within the scope of this paper, that hypothesis will be tested.

1.2 The Hassan Index

The Hassan Index has its roots in a brilliant insight contained in (Shiga et al. 1968). After studying the aftermath of the Tokachi-Oki Earthquake of 1968 in Northern Japan, they observed that buildings with reinforced concrete walls satisfying the following two simple criteria had survived the demands of the ground motion with light damage,

$$\frac{W}{\sum A_c + \sum A_w} \leq 1.2 \text{ MPa} \tag{1.1}$$

and

$$\frac{\sum A_w}{\sum A_f} \geq 0.3 \ \% \tag{1.2}$$

where:

W: Total weight of the building above base
$\sum A_c$: Total cross-sectional area of columns at base
$\sum A_w$: Total cross-sectional area of reinforced concrete walls in a given plan direction, continued over the building height, at base
$\sum A_f$: Sum of tributary floor areas above base

This simple and direct procedure turned out to be difficult to apply for evaluating the effects on buildings of the Erzincan 1992 earthquake, primarily because most of

[2] William Shakespeare, "Macbeth," Act 5, Scene 3.

the walls were masonry infill, not reinforced concrete, and the primary element of resistance was the column. A different set of measures were proposed: the Column Index (CI) and the Wall Index (WI), defined as:

$$CI = \frac{\sum A_c}{2 * \sum A_f} \times 100 \tag{1.3}$$

and

$$WI = \frac{\sum A_{wc} + \frac{\sum A_{mw}}{10}}{\sum A_f} \times 100 \tag{1.4}$$

where:

$\sum A_w$: Sum of cross-sectional areas of reinforced concrete walls in a given plan direction

$\sum A_{mw}$: Sum of cross-sectional areas of masonry infill walls in a given plan direction

Before getting into the data it is important to mention that the reported damage state of a building is not exactly a measurement that can be obtained with reasonable confidence, such as the weight of a concrete cylinder. The reported damage state depends usually on three factors: The first is the ground motion. Even if two identical buildings are subjected to the same earthquake event, and even if they are in the same district of the same town, they may be subjected to ground motions that may be different. The second factor is the toughness and strength of the structure. Geometrically identical structures, that would have similar CI and WI indices, may have different defects. An oversight in the detailing, or the presence of the well known captive column, in one and not in the other may make all the difference between success and failure. The third factor is likely not to affect evaluations of no damage, light damage, and collapse. But depending on the experiences and attitudes of two different observers, a damaged building may be classified as moderately damaged when it is severely damaged and vice versa.

In the next section CI and WI indices obtained after five different earthquakes are reported. The earthquakes occurred in Erzincan 1992, Düzce 1999, Bingöl 2003, Wenchuan 2008, and Haiti 2010. It must be emphasized that (1) the damage statistics included only a fraction of the buildings at each site, and (2) structural and architectural properties of the buildings varied from site to site, as did the earthquake demand. The Wenchuan data were limited to a university campus. No collapses occurred. Collapses occurred in Haiti, but they were not reported, because of lack of information on the buildings that collapsed.

Earthquake magnitudes, M_w, and their approximate epicentral distances to the site of the damage are listed in Table 1.1.

Table 1.1 Earthquake magnitudes and their approximate epicentral distances to the site of the damage

Earthquake	Magnitude, M_w	Approx. distance to epicenter, km
Erzincan 1992	6.7	17
Düzce 1999	7.2	10
Bingöl 2003	6.4	15
Wenchuan 2008	7.9	70
Haiti 2010	7.0	25

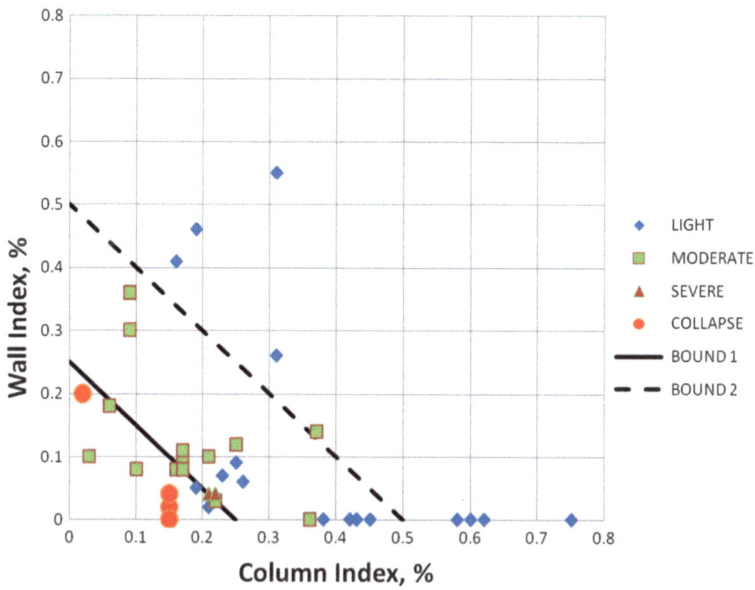

Fig. 1.1 Data from Erzincan 1992

1.3 Evaluation of Recorded CI and WI Indices

Figure 1.1 contains data obtained by investigation teams from Middle East Technical University, Ankara, after the Erzincan earthquake of 1992 reported in http://www.anatolianquake.org. The diagonal lines in the plot, Boundaries 1 and 2, represent priority indices, PI, determined after the data were plotted. Index PI is defined as

$$PI = CI + WI \tag{1.5}$$

such that Boundary 1 defines the locus of points with CI + WI = 0.25 and Boundary 2 defines the locus of points with CI + WI = 0.5.

Data available from five earthquake events are summarized in terms of the column and wall indices in Figs. 1.1, 1.2, 1.3, 1.4 and 1.5. Before discussing the data from the earthquakes, we should recognize that not all the data from a single

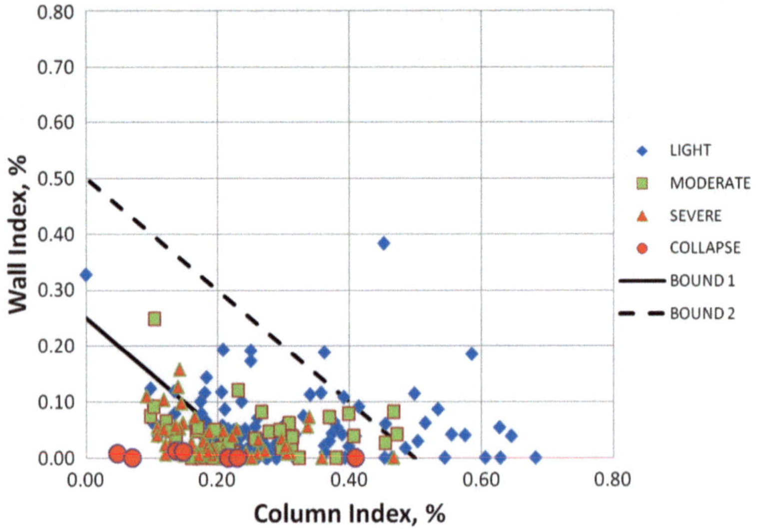

Fig. 1.2 Data from Düzce 1999

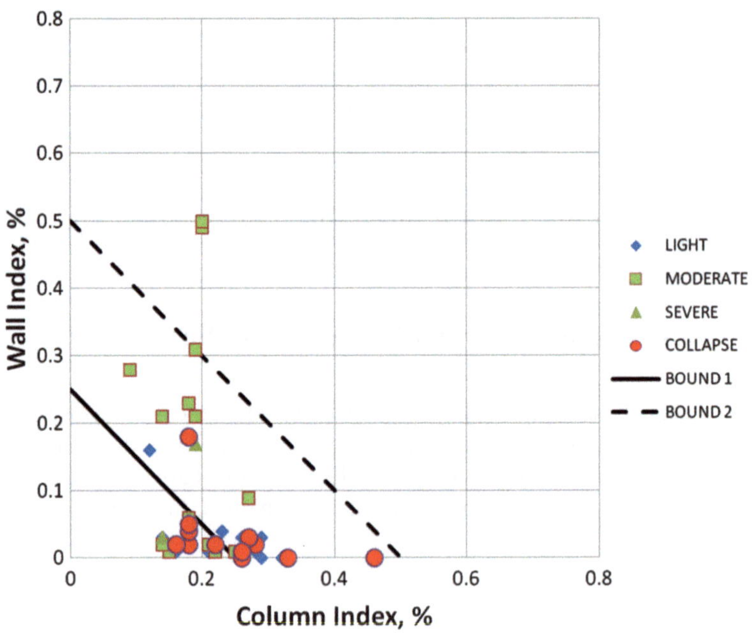

Fig. 1.3 Data from Bingöl 2001

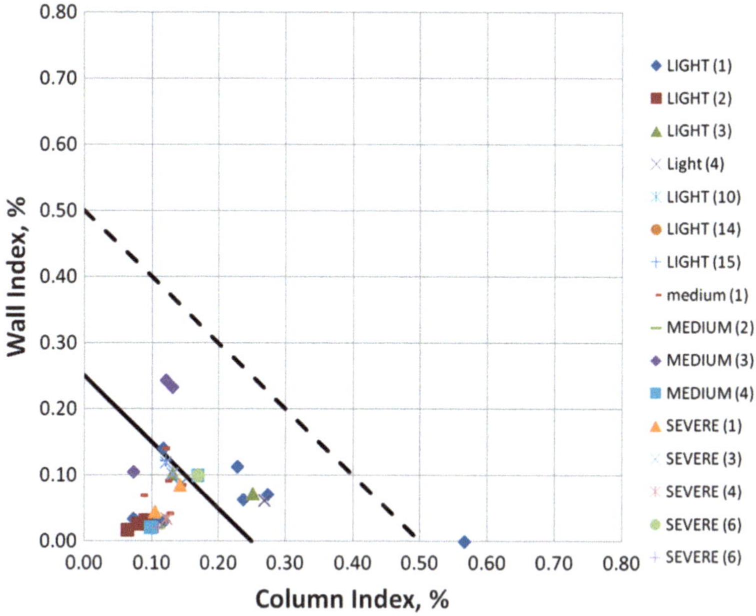

Fig. 1.4 Data from Wenchuan 2008

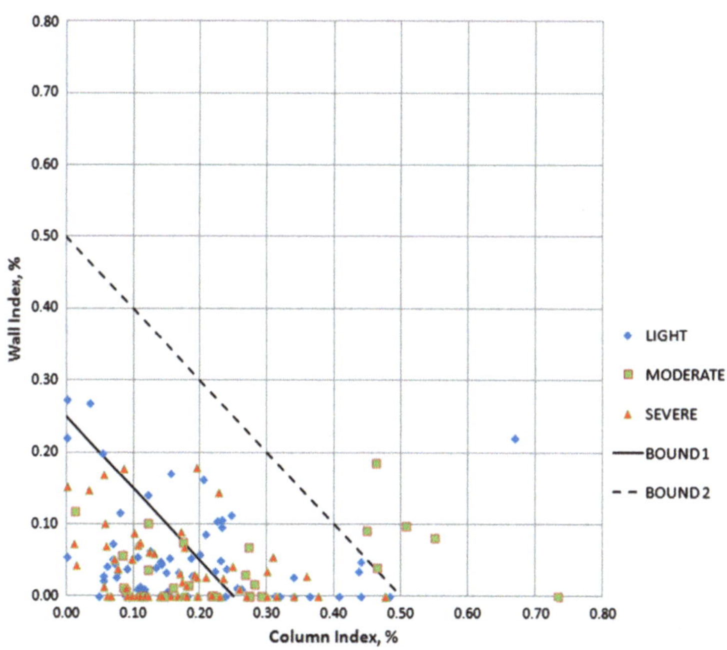

Fig. 1.5 Data from Haiti 2010

Table 1.2 Cases of light, moderate, severe damage and collapse within Bound 1

	Erzincan 1992	Düzce 2003	Bingöl 1999	Wenchuan 2008	Haiti 2010
Total cases documented with Light Damage	28	125	11	66	130
Cases with light damage included within bound 1	2	30	7	43	43
Total cases documented with moderate damage	16	40	21	25	33
Cases with moderate damage within Bound 1	6	21	8	7	13
Total cases documented with severe damage	2	55	6	24	67
Cases with severe damage within Bound 1	1	36	5	6	49
Total collapse cases	4	8	12	None	None reported
Collapse cases within Bound 1	4	7	6	None	None reported
Total cases for each event	50	228	50	115	230
Grand Total				673	

event were reported in a format suitable for evaluating the WI and CI indices. We are not dealing with complete data but only with a partial view. In fact, it may be readily apparent that, if a building collapsed, one could not always obtain the needed dimensions unless plans were available.

The data are categorized in four broad damage groups: Light, moderate, severe and collapse. The category "light" includes apparently undamaged buildings, as well as those with flexural cracks not exceeding approximately 0.3 mm in thickness at the level of the tensile reinforcement. "Moderate" refers to those with flexural and inclined (combined bending and shear) cracks and without permanent drift exceeding approximately 0.25 %. "Severe" refers to a standing building with one or more structural elements having localized failure.

Figure 1.1 contains data from the Erzincan earthquake of 1992. It also includes two additional data items from the Skopje, 1963, and San Fernando, 1971, examples that had been used as a spot check of the overall bounds. The dominant trend to be inferred from the plot is that the closer the CI and WI coordinates place a building with respect to the origin, the more vulnerable it is to earthquake demand. Trajectory of Bound 1 was selected to contain the four collapse cases. Bound 2 was selected to contain the cases with moderate damage. That the bound trajectories connect to the same values on each axis is not dictated by mechanics. It is a choice made to enable convenient comparison of the sum of the CI and WI indices with the particular bound that is being used as a criterion. Otherwise, the bounds can be set in any shape or distance from the origin depending on the local conditions and demands. As summarized in Table 1.2, two cases of light and six cases of moderate damage were also within Bound 1, showing that the bound is not as good as a litmus test.

All cases with moderate damage were within Bound 2. That was the reason for setting Bound 2 with PI = 0.25.

The trajectories of the bounds refer to constant values of the Priority Index, PI = CI + WI. For Bound 1 PI is 0.25 and for Bound 2 it is equal to 0.5.

It goes without saying that if decisions about strengthening had been made before the earthquake using Bound 1, resources would be wasted on two buildings that did not require strengthening. On the other hand, lives in four buildings would have been saved. Admittedly, this is a light conjecture because the bounds were determined after the data were obtained.

In Fig. 1.2, the data acquired in Düzce, Bolu, and Kaynaşlı after the 1999 Düzce earthquake are compared with the two bounds established after the Erzincan 1992 earthquake (Boğazici University and other university teams 1999, Dönmez and Pujol 2005). On the positive side, we note five collapses that would have been identified. On the negative side, the data indicate that the CI/WI coordinates of 30 out of 125 cases sustaining no or light damage (about 25 % of the total) placed the buildings within Bound 1. Furthermore, 21 or approximately one half of the 40 buildings that sustained moderate damage, also would have been identified as vulnerable. On the positive side, it was observed that two-thirds of the severely damaged buildings would have been identified (36 buildings out of 55).

Figure 1.3 refers to Bingöl (Ozcebe et al. 2004). Five cases of collapse would have been identified by Bound 1. Unfortunately there are seven cases of collapse that would not have been identified by Bound 1. The recurring cause of collapse in Bingöl was the "captive column." It could be that an experienced observer would have identified these buildings as vulnerable before the earthquake; but it has to be accepted that such a person might not have been available for the inspection. It has to be accepted that these tragic collapses would have not been identified by a routine application of the CI/WI indices with Bound 1 defining the limit for serious vulnerability.

Figure 1.4 is based on studies on buildings belonging to a university in Chengdu, Sichuan, after the 2008 Wenchuan earthquake (Zhou et al. 2011). Bound 1 would cover all cases of severe damage. Out of 66 cases with light or no damage, 43 or two-thirds were within Bound 1 and would have been diagnosed as vulnerable. Only one-fourth (6 out of 24) of the severe damage cases reported were within Bound 1.

In Haiti (Fig. 1.5), of the 132 buildings reported to have light damage, 43 of approximately 1/3 had CI/WI coordinates within Bound 1. Approximately 40 % or 13 of the 33 buildings reported to be moderately damaged were within Bound 1. A total of 67 buildings were reported to have severe damage. Of these, 49 or nearly ¾ of total had indices that placed them within Bound 1. Even though many buildings collapsed in Port au Prince, there was no relevant report because information on dimensions of the structural elements was lacking (O'Brien et al. 2011).

All of the 673 observed damage states are plotted in Figs. 1.6, 1.7, 1.8, and 1.9. Each figure contains observations for a single level of reported damage. It is to be noted that the data combined were obtained from sites in three different countries and five different cities that experienced different earthquakes. Uniformity is not to be expected.

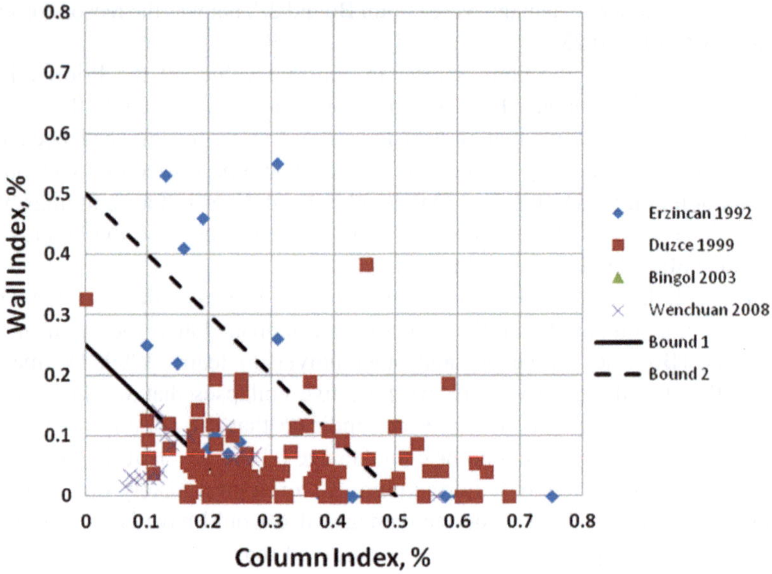

Fig. 1.6 Cases with light or no damage

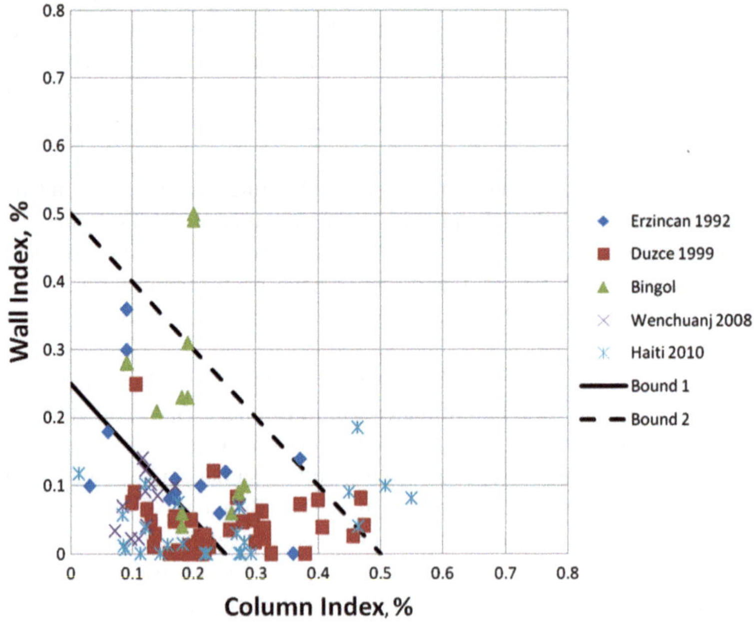

Fig. 1.7 Cases with moderate damage

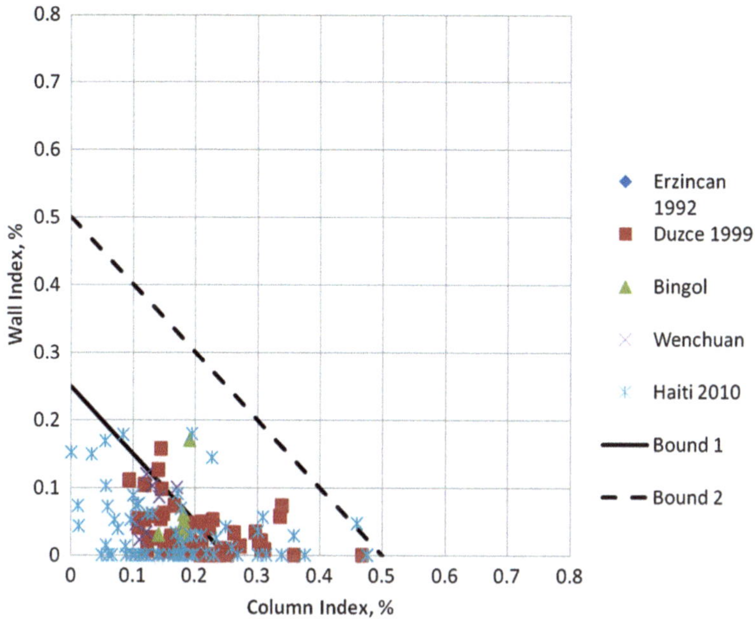

Fig. 1.8 Cases with severe damage

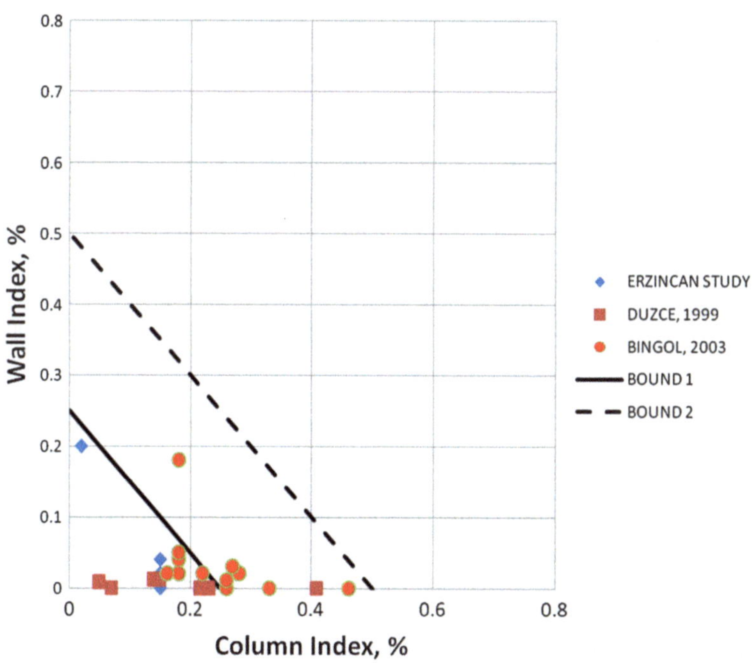

Fig. 1.9 Reported collapses

One definite conclusion from the four figures is that the CI/WI coordinates of no building that collapsed or had severe damage was observed to be out of Bound 2. Strong as this conclusion is, it is of little practical value, because identifying every building that is within Bound 2 would make remedial action practically impossible. There is at least one earthquake-threatened city with large districts in which over 95 % of the low-rise reinforced concrete building inventory would fall within Bound 2.

"Prediction" of collapses is certainly flawed. The index does not by itself identify factors such as captive columns or inadequate splices.

1.4 Concluding Discussion

The Hassan Index has its flaws, but it can be used as a first filter in making decisions about which low-rise reinforced concrete buildings of a city inventory ought to be subjected to a detailed vulnerability assessment.

If the building inventory of the city is comparable to that included in the combined set, using Boundary 2 for selection of the buildings to be investigated would be appropriately conservative, but would mean that approximately 95 % of the building inventory would be involved in evaluation. It is highly unlikely that such a plan would be accepted by either the political or the banking communities.

If Bound 1 is used as the criterion for selecting the buildings to be evaluated, approximately 40 % of the inventory would be included in the study. It is probably doable, provided that economic and engineering resources are available and the political leadership is willing to assign those resources to the cause of limiting human and economic losses.

If the evaluations are made properly in this sample of building quality, 3 % of the total inventory in low-rise reinforced concrete might be identified as susceptible to collapse and would be strengthened or demolished. Strengthening or demolishing would also demand sizeable expense.

Approximately 1 % of the same inventory would be susceptible to collapse, but would not be identified unless the study was expanded beyond Bound 1. Considering that not all collapse cases were included in the statistical sample, 1 % may be optimistic.

The above statements, besides being arguable, are easy to make. But the conclusions take on a sinister form if they are projected to a particular city (Griffiths et al. 2007).

Consider a major city that is likely to be subjected to a strong ground motion in the near future, although it is difficult to state how near. Assume that the reinforced concrete low-rise building inventory within reach of the energy from the fault is approximately 500,000 and is similar in quality to those included in the five sites considered. If 3 % is found to need strengthening or demolition, the target may be approximately 15,000 buildings. The task is expensive, but within reach of a rich community. If 1 % "miss" of buildings susceptible to collapse is correct, we should expect approximately 5,000 or even more collapses; this disastrous result will occur after having strengthened or demolished approximately 15,000 buildings.

The vision that is based on observed damage is gloomy if its projection, in terms of the quality of building stock in the five sites, is plausible.

Whatever the conditions, it is not mandatory but wise to assess earthquake risk of the existing building stock in a city threatened by earthquake. As an accurate indicator of vulnerability, the Hassan Index is a failure. It is, after all, surrealistic to expect success using two simple indices to identify every failure in an environment where details such as height or strength of an architectural wall or an oversight in proper spacing of a few ties can lead to disaster. However, a timely pre-selection for assessment using the Hassan Index or a similar vetted and simple criterion is likely to help the cause of reducing human and economic losses in an earthquake.

References

Boğazici University and other university teams (1999) Building data. http://www.anatolianquake.org

Dönmez C, Pujol S (2005) Spatial distribution of damage caused by the 1999 earthquakes in Turkey. Earthq Spectra 21:53–69

Griffiths JHP, Irfanoglu A, Pujol S (2007) Istanbul at the threshold: an evaluation of the seismic risk in Istanbul. Earthq Spectra 23(1):63–75

Hassan AF, Sözen MA (1994) Seismic vulnerability assessment of low-rise buildings in regions with infrequent earthquakes. ACI Struct J 94:31–39

O'Brien P et al (2011) Measures of the seismic vulnerability of reinforced concrete buildings in Haiti. Earthq Spectra 27(S1):373–386

Ozcebe G et al (2004) 1 May 2003 Bingöl earthquake engineering report. Tübitak Structural Engineering Research Unit. Report no 2004/1, Ankara

Shiga T, Shibata A, Takahashi T (1968) Earthquake damage and wall index of reinforced concrete buildings. In: Proceedings of Tohoku district symposium. Architectural Institute of Japan, vol 12, pp 29–32

Zhou W et al (2011) Seismic vulnerability of reinforced concrete structures affected by the 2008 Wenchuan earthquake. https://nees.org/warehouse. p 21

Chapter 2
Rapid Seismic Assessment Procedures for the Turkish Building Stock

Ahmet Yakut, M. Altuğ Erberik, Alper Ilki, Haluk Sucuoğlu, and Sinan Akkar

Abstract Seismic performance assessment procedures based on street survey have been developed for low- to mid-rise reinforced concrete and masonry buildings in Turkey. These procedures rely on data that can be collected through visual examination of each building. The data is collected through the forms designed for that purpose. The attributes that are believed to affect seismic performance have been determined and used to evaluate seismic vulnerability. It is important to note that these procedures are not appropriate for determination of seismic vulnerability of individual buildings but to rank a population of buildings according to their relative vulnerability. The primary parameters used for RC buildings include seismic hazard, structural system, number of stories, irregularities in plan and elevation, architectural features and building adjacency. Similar parameters are

A. Yakut (✉)
Department of Civil Engineering, Middle East Technical University,
K2/209, Dumlupinar Bulvari 1, Cankaya, Ankara 06800, Turkey
e-mail: ayakut@metu.edu.tr

M.A. Erberik
Department of Civil Engineering, Middle East Technical University,
K7/103, Dumlupinar Bulvari 1, Cankaya, Ankara 06800, Turkey
e-mail: altug@metu.edu.tr

A. Ilki
Department of Civil Engineering, Civil Engineering Faculty, Istanbul Technical University, Maslak, Istanbul 34469, Turkey
e-mail: ailki@itu.edu.tr

H. Sucuoğlu
Department of Civil Engineering, Middle East Technical University,
K7/101, Dumlupinar Bulvari 1, Ankara 06800, Turkey
e-mail: sucuoglu@metu.edu.tr

S. Akkar
Department of Civil Engineering, Earthquake Engineering Research Center,
Middle East Technical University, K6/109, Dumlupinar Bulvari 1, Ankara 06800, Turkey
e-mail: sakkar@metu.edu.tr

A. Ilki and M.N. Fardis (eds.), *Seismic Evaluation and Rehabilitation of Structures*, Geotechnical, Geological and Earthquake Engineering 26,
DOI 10.1007/978-3-319-00458-7_2,

used for masonry buildings. The procedures developed use the selected attributes to determine seismic vulnerability scores for each building. These scores include a base score that is modified for each attribute to get the vulnerability score. The procedures proposed for rapid seismic assessment were validated and calibrated based on field data and results of detailed assessment procedures.

2.1 Introduction

Observations from recent earthquakes in Turkey reveal that a remarkable number of existing buildings have poor seismic performance. In order to reduce the seismic risk in future earthquakes, seismic assessment procedures that are applicable to Turkish buildings need to be developed. Reinforced concrete and masonry buildings are the two dominant construction types in Turkey. Given the large population of building stock to be assessed, quick assessment procedures are preferred over detailed ones as the first level to classify buildings with high risk. In this context, quick seismic assessment procedures were developed to rank seismic risk of RC and masonry buildings in Turkey. The procedures are generally based on visual examination of each building, from outside and limited entry to the building, to identify attributes that are considered to influence seismic performance. These procedures take into account not only building attributes but also the seismicity of the location based on the current seismic zone map of the country.

Since RC buildings are very common both in Turkey and around the world, a number of previous studies have focused on their seismic performance assessment. Several rapid or preliminary assessment procedures have been proposed: FEMA 154 (2002), Japanese Seismic Index (Kaminosono 2002), DURTES (Temür 2006), P25 (Bal et al. 2007), Sextos et al. (2008), İlki et al. (2003), Yakut (2004). Among these, FEMA 154 is a rapid assessment procedure that has been developed for US buildings, thus reflects practice there. Other procedures require more detailed data than those used in the proposed methodology.

Masonry buildings in Turkey can be classified into two major groups: urban-type and rural-type. Rural-type masonry buildings reside in small residential areas, towns and villages. They are generally low-rise dwellings built of adobe, brick or stone units. They are also called "non-engineered buildings" since they have been constructed in a traditional manner without the intervention of an engineer or architect. On the other hand, urban-type masonry buildings are low-rise or mid-rise dwellings with a larger floor area than rural counterparts, sometimes having irregular plan geometry with many projections. Generally, brick units and concrete blocks are used in their construction and they have an "engineering touch" as opposed to rural-types. The major earthquakes that have occurred in Turkey in the last three or four decades revealed the fact that both rural- and urban-type masonry buildings suffered damages to different extents, while a considerable portion were severely damaged or collapsed. This is also the case in many earthquake-prone regions of the world, especially in developing countries. Hence it is not surprising to encounter various rapid screening procedures developed for masonry buildings.

The most popular rapid visual screening methodology, which was developed by the Applied Technology Council (ATC) for the Federal Emergency Management Agency (FEMA) in 1998 and was updated in 2002 (ATC 2002), includes also the seismic safety assessment of masonry buildings among various construction types. But it should be stated that some of the specific structural parameters regarding masonry buildings are not included in the FEMA evaluation form since the methodology had been designed to evaluate 15 different types of structures with one single form. In another study that has been inspired by the FEMA approach, rapid visual screening of existing buildings, including reinforced and unreinforced masonry construction, has been carried out for the province of Quebec in Canada (Karbassi and Nollet 2008). The researchers modified the base scores of the FEMA methodology in order to represent the characteristics of the building stock in Quebec. Arya (2007) also worked on a rapid screening methodology similar to the FEMA approach, but he added some more structural parameters regarding masonry structures in the proposed data collection form.

There also exist rapid visual screening methodologies developed in Turkey for the seismic safety assessment of unreinforced masonry buildings in urban and rural regions. One such a method was developed under the project “Earthquake Master Plan for Istanbul (EMPI)” as the first-stage evaluation of a multi-stage approach in order to prevent or mitigate seismic risk, and prepare emergency rescue and restoration plans for the earthquake prone areas identified in the city of Istanbul (Sucuoğlu et al. 2007). In the first-stage evaluation procedure, also referred to as the “sidewalk survey”, the masonry buildings under inspection were examined from the street by considering their basic structural parameters that can be determined without entering the building. Then the results of the first-stage evaluation were used to distinguish those buildings with high damage risk, to be examined in detail in the second stage. The details of the methodology and its application to sub-provinces of Istanbul can be found elsewhere (Erberik 2008, 2010).

Another method was developed by the technical teams from the (formerly known as) Ministry of Public Works and Settlement in order to assess the seismic safety of existing buildings in the province of Denizli (Kocaman et al. 2009). Within the context of this study, 4,151 masonry buildings were examined from the street and the obtained information was gathered in a database with the help of data collection forms. Then this information was used in order to rank the buildings in a relative manner according to their seismic vulnerabilities.

2.2 Proposed Procedure for RC Buildings

The procedure developed for low- to mid-rise (1–7 stories) reinforced concrete buildings is an extension of a previously developed procedure (Sucuoğlu and Yazgan 2003; Sucuoğlu et al. 2007) incorporating more parameters and treating seismic hazard differently. The parameters considered, the form developed for data collection and the methodology are described in the following sections.

2.2.1 Parameters Employed

The structural system, number of stories, existing condition and apparent quality, soft story, vertical irregularity, heavy overhangs, plan irregularity, short column, building adjacency, topography, seismic hazard and soil type are the primary parameters considered. These parameters have been determined based on examination and analyses of observations from recent earthquakes, previous studies on the vulnerability of Turkish buildings, peculiarities and practice for the local building stock. A short description of these parameters is given below.

1. The number of unrestrained (free) stories (n_s): This is one of the most important parameters for the seismic vulnerability of RC buildings in Turkey (Sucuoglu and Yazgan 2003; Ozcebe et al. 2004). This parameter shows the number of stories above the ground level for buildings without basements. It represents the number of unrestrained stories for buildings with basements.
2. Type of structural system: Reinforced concrete frame buildings (RCF), RC frame buildings with shear walls (RCFS) and RC shear wall buildings (RCS) are among the RC systems considered.
3. Apparent quality and existing condition: This parameter is expected to reflect the quality of workmanship and material. Apparent quality of the building is assigned considering three classifications as good, moderate and poor.
4. Soft story: This attribute is determined visually based on the height difference as well as a significant difference in the rigidities of stories, including the potential contribution of the infill walls, if possible.
5. Vertical irregularity: Discontinuous frames, columns and shear walls along the height are reflected through this parameter.
6. Heavy overhangs: It represents the difference between the floor area of the ground floor and floors above. It causes irregularity in elevation by shifting the frames and beam lines at the stories above and has a significant negative influence on seismic performance of RC buildings in Turkey.
7. Irregularity in plan torsion: Unsymmetrical plan and distribution of vertical elements may cause torsion. If this effect can be determined visually then it should be taken into account.
8. Short column: This attribute can easily be observed from street survey.
9. Building adjacency/pounding: Due to inadequate gap between the adjacent buildings pounding may result in damage. Buildings located at corner of blocks are affected the most. Additionally, if floor levels of adjacent buildings are different then the effect of pounding becomes more significant. All these cases can easily be determined during street survey.
10. Topographic effect: This attribute is taken into consideration for buildings that are located at hills having a significant slope.
11. Seismicity and soil type: The level of seismicity is considered, based on the seismic zone map and soil types in the Turkish seismic code (TEC 2007).

2.2.2 *Data Collection Form*

The parameters explained above are determined based on data collected through a form developed for this purpose. This form also helps develop a building inventory. The form shown in Fig. 2.1 has two sides; the front side contains necessary information for determining the parameters and the back side gives some examples and explanations for identifying certain features.

2.2.3 *Performance Assessment Methodology*

The main objective of the proposed procedure is to make a risk prioritization using the parameters that can be determined through a walk down survey with limited entry to the building. Based on the analysis of the parameters and considering their relative influence on the performance, a quantitative evaluation is carried out to calculate a performance score for each building. These performance scores are first determined for a reference seismicity and are then modified to reflect the effect of the building's hazard and soil conditions. The performance scores of all buildings are used to obtain a risk prioritization.

a **RAPID ASSESSMENT DATA COLLECTION FORM FOR RC**

DATE :

FORM 1 BUILDING ID **No:..........**

REGION NO				
MAHALLE				
STREET				
APT NAME AND NUMBER				
PAFTA / ADA / PARSEL				
CITY INFO. SYST.NO				
BUILDING AGE				
COORDINATES (GPS) (E/ N)				
FORM 2 BUILDING DATA				
STRUCTURAL SYSTEM	☐ RC FRAME	☐ RC WALL	☐ RC FRAME AND WALL	
NUMBER OF FREE STORIES	....			
CORROSION LEVEL	☐ NO	☐ LIGHT	☐ MODERATE ☐ HEAVY	
MEMBER STRUCTURAL DAMAGE	☐ LIGHT		☐ MODERATE ☐ HEAVY	
PLAN WIDTH	............m			
PLAN LENGTH	............m			
ADJACENCY	☐ SEPARATED	☐ ADJACENT	☐ ADJACENT CORNER	
FLOOR LEVEL	☐ SAME	☐ DIFFERENT		
ADDITIONAL STORY	☐ YES(........)	☐ NO	☐ UNKNOWN	
HEAVY OVERHANGS	☐ YES	☐ NO		
WEAK / SOFT STORY	☐ YES	☐ NO		
SHORT COLUMN	☐ YES	☐ NO		
VERTICAL IRREGULARITY	☐ YES	☐ NO		
PLAN IRREGULARITY	☐ YES	☐ NO		
SOIL CLASS	☐ Z1	☐ Z2	☐ Z3	☐ Z4

Fig. 2.1 (**a**) Data collection form for RC (*front side*). (**b**) Data collection form for RC (*back side*)

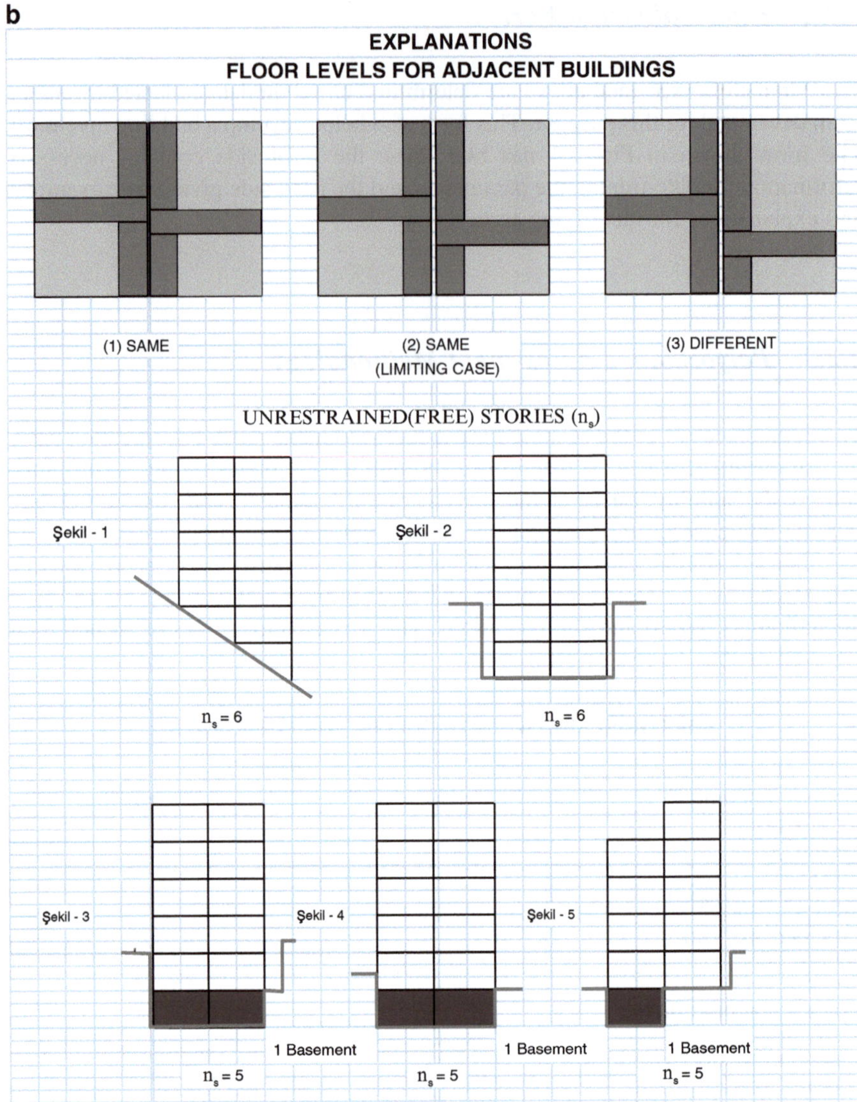

Fig. 2.1 (continued)

2.2.4 *Calculation of Performance Score*

Performance scores of buildings are calculated based on the seismicity of the building's location and the parameters reflecting its properties. Turkey is divided into four hazard zones determined based on the current seismic zone map and soil types defined in the Turkish seismic code (TEC 2007). A base score (BS) is

Table 2.1 Base and structural system type scores

Number of unrestrained stories	Base score (BS) Seismic hazard zone I	II	III	IV	Positive scores (POS) Structural system RCF	RCFS	RCS
1 or 2	90	120	160	195	0	100	200
3	80	100	140	170	0	85	170
4	70	90	130	160	0	75	150
5	60	80	110	135	0	65	130
6, 7	50	65	90	110	0	55	110

assigned to the building according to its hazard zone and the number of unrestrained stories. The base scores have been determined taking reinforced concrete frame buildings (RCF) as reference. For RCFS and RCS additional positive scores (POS) are assigned to take into account the structural system type. Apparent quality, soft story, heavy overhang, pounding effect, irregularities, short column and topographic effect are included through penalty scores (PESi).

Tables 2.1 and 2.2 give the base scores and penalty scores for each parameter, respectively. The relationship between the seismic hazard zones, number of stories and base scores is given in Table 2.1, and the current seismic zone map and soil types in the code in Table 2.3. The performance score is calculated via Eq. (2.1). Here, "yes" or "no" type of assessment is made for all parameters having negative influence, except for the apparent quality and building adjacency. The influence indices (Ii) for these parameters take a value of 0 for "no" and 1 for "yes". If the apparent quality assessment is "good", the influence index is 0, for "moderate" it is 1 and for "poor" 2. Table 2.4 gives influence indices for each parameter.

$$\mathrm{PS} = \mathrm{BS} + \sum \mathrm{I_i} * \mathrm{PES_i} + \mathrm{POS} \tag{2.1}$$

2.2.5 *Calibration of Proposed Procedure*

The scores assigned to the parameters used in the performance assessment procedure explained above are based on observations from past earthquakes and their relative influence on structural performance. Accuracy and validity of these scores are assessed using two databases: 1. Building damage database compiled after the 1999 Düzce earthquake; 2. Buildings whose seismic performance was assessed using detailed procedures.

Table 2.2 Penalty scores for parameters employed

Number of unrestrained stories	Penalty scores (PES)										
				Floor level/building adjacency							
	Soft story	Apparent quality	Heavy overhang	Same Middle	Same Edge	Different Middle	Different Edge	Vertical irregularity	Plan irregularity/ torsion	Short column	Topographic effect
1, 2	−10	−10	−10	0	−10	−5	−15	−5	−5	−5	−3
3	−20	−10	−20	0	−10	−5	−15	−10	−10	−5	−3
4	−30	−15	−30	0	−10	−5	−15	−15	−10	−5	−3
5	−30	−25	−30	0	−10	−5	−15	−15	−10	−5	−3
6 or 7	−30	−30	−30	0	−10	−5	−15	−15	−10	−5	−3

Table 2.3 Seismic hazard zones according to current seismic zones and soil types in the code

Seismic hazard zone (PGV, m/s)	Seismic zones (TEC 2007)	Soil type (TEC 2007)
I (0.60–0.80)	1	Z3/Z4
II (0.30–0.60)	1	Z1/Z2
	2	Z3/Z4
III (0.15–0.30)	2	Z1/Z2
	3	Z3/Z4
IV (0–0.15)	3	Z1/Z2
	4	All soils

Table 2.4 Influence indices for parameters

Parameter	Influence indices (I_i)
Soft story	No (0); Yes (1)
Heavy overhang	No (0); Yes (1)
Apparent quality	Good (0); Moderate (1); Poor (2)
Short column	No (0); Yes (1)
Topographic effect	No (0); Yes (1)
Plan irregularity	No (0); Yes (1)
Vertical irregularity	No (0); Yes (1)

Table 2.5 Damage distribution of buildings in Düzce database

	Observed damage state				
No. of stories	None	Light	Moderate	Heavy/collapsed	Total
3	18	62	29	15	124
4	17	43	60	27	147
5 or 6	18	30	60	75	183
Total	53	135	149	117	454

2.2.5.1 Application to Düzce Database

After the 1999 earthquakes, a comprehensive post earthquake damage assessment that contained 454, 3–6 story buildings in Düzce was carried out (Ozcebe et al. 2004; SERU 2003). The buildings were classified into four damage states; none, light, moderate, and heavy or collapsed. Results of damage assessment are summarized in Table 2.5.

In order to check the predictions of the proposed procedure, performance scores have been calculated (HPS) for each building in the Düzce database. A separate score to reflect observed performance (GPS) has been assigned to each building based on its damage state. The relation between observed damage score and the damage state is shown in Table 2.6. A threshold score value (SD) is needed to classify the buildings according to their HPS. Therefore, if HPS is less than SD and the GPS score for this building classifies it as "high risk" then the building can be considered to be correctly classified. Similarly, if HPS is greater than SD and the GPS score classifies the building as "low risk" then this building is also classified

Table 2.6 Observed performance scores

Observed damage	Corresponding risk level	Observed performance score (GPS)
None	Low	100
Light	Low	80
Moderate	Low	50
Heavy/Collapse	High	0

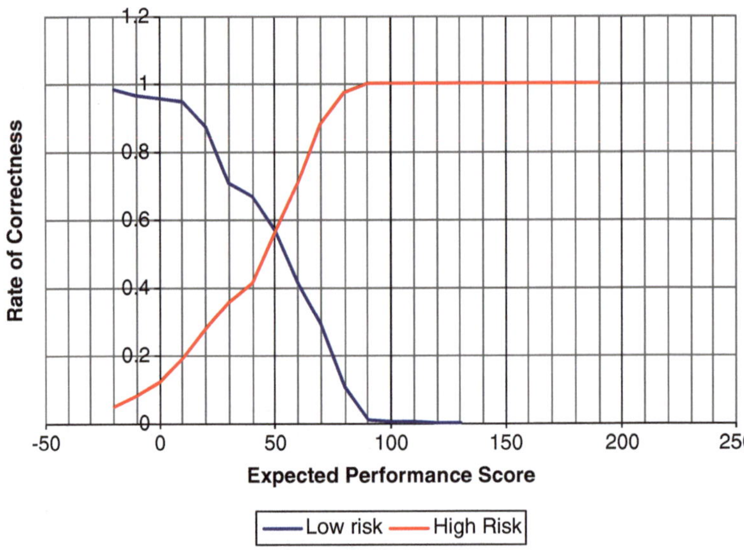

Fig. 2.2 Correctness ratio for Düzce database

correctly. Consequently, a correctness ratio that shows the ratio of correctly classified buildings in the Düzce database to the buildings in the same classification according to their observed behavior (i.e. GPS) can be defined. Figure 2.2 shows the correctness ratio for a selected SD for low and high risk buildings. If SD is taken as 52 (i.e., at the intersection of two curves) then 60 % of both low risk and high risk buildings would be correctly classified. If SD is selected as 65, then 80 % of high risk buildings and 40 % of low risk buildings would be correctly classified. Figure 2.3 shows comparison of calculated and observed performance scores through cumulative distribution function. A reasonably good match is observed.

2.2.5.2 Application to Existing Building Database

131 RC buildings located in different parts of Turkey were assessed using detailed assessment based mainly on the 1997 earthquake code (TEC 1997). Final decisions for the buildings after the detailed assessment revealed that 36 buildings were

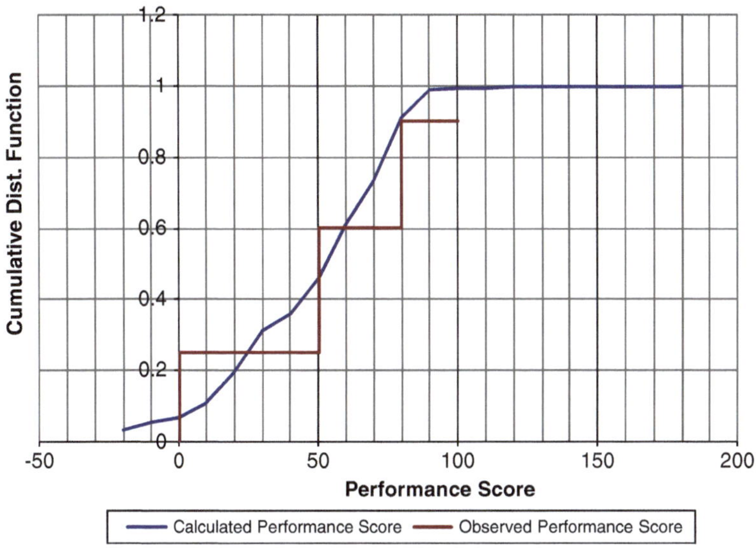

Fig. 2.3 Comparison of calculated and observed scores for Düzce database

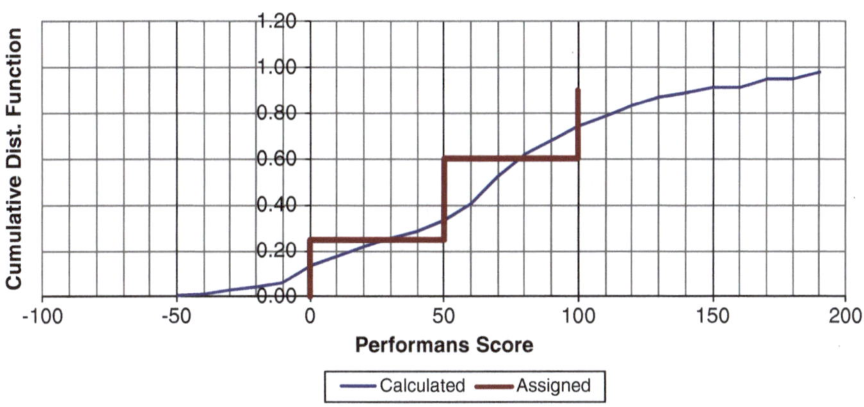

Fig. 2.4 Comparison of calculated and observed scores for the existing building database

sufficiently safe, 7 buildings were too weak to consider strengthening, and the remaining 88 buildings needed strengthening intervention to reach an acceptable safety level. Similar to observed performance scores for the Düzce database, GPS scores were also assigned to these buildings; GPS of 100 is assigned to safe buildings, 50 to buildings that needed strengthening, and 0 to the ones to be demolished (weakest 7 buildings). HPS scores have also been calculated for these buildings and are compared with GPS scores as shown in Fig. 2.4. As can be seen, the calculated and observed scores are in good agreement.

2.3 Proposed Procedure for Masonry Buildings

2.3.1 Data Collection Form and Methodology

The data collection form and seismic performance assessment methodology developed for masonry buildings is very similar to the one developed for reinforced concrete buildings. The structural parameters in the form, which provide the input to calculate the performance score of the surveyed building, have been selected in such a way that they can be obtained simply in a short period of time by street survey and without entering the building. There also exist some parameters in the form that cannot be obtained without entering the building, but these parameters are optional and the absence of these parameters does not prevent the calculation of the performance score of the building. The form is composed of two pages (see Fig. 2.5). The first page contains the structural parameters collected from the building. The second page acts as a commentary; providing simple and illustrative explanations for some of the parameters in the first page.

The first page of the form is composed of five sections: building ID info, type of masonry construction, observations from outside the building, observations from inside the building and general observations. The first section contains the parameters related with ID code, address, coordinates and construction year of the building. In the second section, type of masonry construction for the inspected building should be provided since it is an important parameter for the seismic safety evaluation of masonry buildings. There exist four options: unreinforced masonry, confined masonry, reinforced masonry and hybrid construction (masonry wall + RC frame). In the third section regarding the observations from outside the building, the parameters included can be listed as number of stories, plan geometry, vertical irregularity, position of the building (adjacency) and previous damage (if any). Section four is related to the observations from inside the building and it is optional. The parameters included are as follows: story height, wall thickness, unrestrained wall length, length of wall segment between openings, etc. The final section is composed of general parameters such as masonry unit type, mortar type, workmanship quality, floor type, presence of horizontal and vertical bond beams, weak/soft story, wall-to-wall and wall-to-floor connections.

In order to calculate the performance score (PS) of the inspected building some of the major parameters in the form are employed. First a base score (BS) is determined in accordance with the seismic zone and number of stories. Then the base score is reduced by the penalty scores (Pi) that reflect the structural deficiencies of the inspected building, as:

$$\mathrm{PS} = \mathrm{BS} + \sum \mathrm{w_i} * \mathrm{P_i} + \mathrm{R} \tag{2.2}$$

a SURVEY FORM FOR SEISMIC SAFETY ASSESSMENT OF MASONRY BUILDINGS

BUILDING ID INFO

BUILDING ID	
DATE OF SURVEY	
BUILDING ADRESS	
GPS COORDINATES (E/N)	
CONSTRUCTION YEAR	
TECHNICAL PERSON	

Photo of the building

CONSTRUCTION TYPE (See -1-)

☐ UNREINFORCED	☐ CONFINED
☐ REINFORCED	☐ HYBRID (URM + RC)

OBSERVATIONS OUTSIDE THE BUILDING (See -2-)

NUMBER OF STORIES	..... (NUMBER)		
HIGH SLOPE ?	NO ( )	YES ( )	
BASEMENT FLOOR	NO ( )	YES ( )	N/A ( )
PLAN GEOMETRY	REGULAR ( )	IRREGULAR ( )	
FAÇADE LENGTH (FRONT) ..... Meters	CRITICAL STORY OPENING LENGTH (FRONT) ..... Metre		
FAÇADE LENGTH (SIDE) ..... Meters	CRITICAL STORY OPENING LENGTH (SIDE) ..... Metre		
VERTICAL OPENING LAYOUT	REGULAR ( )	F. REGULAR ( )	IRREGULAR ( )
LOCATION OF BUILDING	SEPARATED ( )	ADJ. MIDDLE ( )	ADJ. CORNER ( )
BUILDING HEIGHT DIFFERENCE	NO ( )	YES ( )	
FLOOR ELEVATION DIFFERENCE	NO ( )	YES ( )	
PREVIOUS DAMAGE	NO ( )	YES ( )	
ADJACENT TO HISTORICAL BUILDING	NO ( )	YES ( )	

OBSERVATIONS INSIDE THE BUILDING (See -3-)

TYPICAL STORY HEIGHT	..... Meters	
TYPICAL WALL THICKNESS	..... Meters	
UNCONSTRAINED WALL LENGTH (L_m) > 5.0 m ?	YES ( ) ..... TIMES	NO ( )
WALL LENGTH BTW TWO OPENINGS (L_b) < 1.0 m ?	YES ( ) ..... TIMES	NO ( )
WALL LENGTH BTW CORNER & OPENING (L_k) < 1.5 m ?	YES ( ) ..... TIMES	NO ( )

GENERAL OBSERVATIONS (See -4-)

MASONRY WALL TYPE	SOLID BRICK ( )	HOLLOW BRICK ( )	SOLID CMU ( )	
	HOLLOW CMU ( )	AAC ( )	CUT STONE ( )	
	RUBBLE STONE ( )	ADOBE ( )		
MORTAR TYPE	CEMENT ( )	LIME ( )	MUD ( )	NO ( )
WORKMANSHIP	GOOD ( )	MODERATE ( )	POOR ( )	
FLOOR TYPE	RC ( )	WOODEN ( )	ARCHED ( )	
HORIZONTAL BOND BEAM ?	OVER WINDOW ( )	FLOOR LEVEL ( )	NO ( )	
VERTICAL BOND BEAM ?	YES ( ) ..... metre interval	NO ( )		
LINTEL ?	YES ( )	NO ( )		
LINTEL/BEAM MATERIAL?	RC ( )	WOODEN ( )		
ROOF TYPE	FLAT ( )	SHED ( )	GABLE ( )	HIPPED ( )
ROOF MATERIAL	TILE ( )	RC ( )	METAL SHEET ( )	EARTHEN ( )
CONNECTIONS	GOOD ( )	POOR ( )		
SOFT/WEAK STORY	YES ( )	NO ( )		

Fig. 2.5 (**a**) First page of the proposed data collection form for masonry buildings. (**b**) Second page of the proposed data collection form for masonry buildings

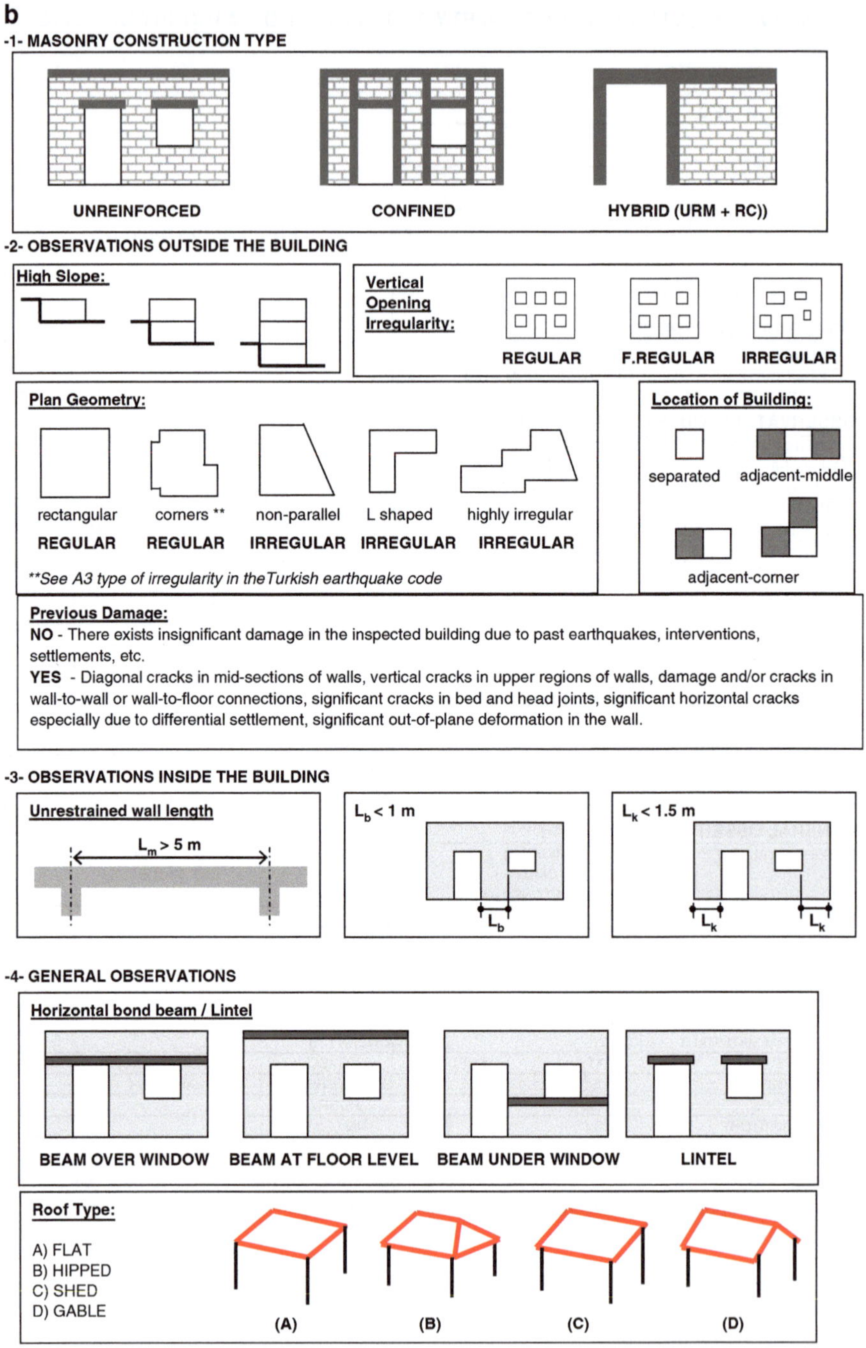

Fig. 2.5 (continued)

Table 2.7 Base scores for masonry buildings other than adobe

No. of stories	Zone-I (PGA≥0.4g)	Zone-II (0.2g ≤PGA<0.4g)	Zone-III (PGA<0.2g)
1	110	120	130
2	100	110	120
3	90	100	110
4	80	90	100
5	70	80	90

In Eq. (2.2), w_i stands for the weighing factor multiplied by the negative penalty score in order to reflect the negative effect of the corresponding structural deficiency on the potential performance of the inspected building and can take values of 0, 1, or 2. Parameter R represents the type of masonry construction and takes positive values in the case of confined (30 points) and reinforced masonry (60 points) buildings.

The base score is a function of seismic zone and number of stories. Masonry buildings are evaluated in three seismic zones in accordance with the peak ground acceleration (PGA) limits as presented in Table 2.7. The base scores are calculated by considering the maximum number of stories permitted by the code for masonry buildings according to the seismic zone that the buildings reside. For instance, the cells in gray color in Table 2.7 represent the maximum allowable number of stories for a specific seismic zone. The base score for these cases is considered as 100. Each additional story means violation of the code requirement and hence punished by a penalty score of 10 per story. On the other hand, if the number of stories of the inspected building is less than the number of stories allowed by the code, then the base score is increased by 10 per story. However, it should be noted that the scores in Table 2.7 are for masonry structures other than adobe. According to the earthquake code, adobe buildings are allowed to be constructed as a single story regardless of the seismic zone, since they are the most vulnerable type of masonry construction. Hence the scores for adobe construction are adjusted in such way that all scores in Table 2.7 are reduced by 30.

Penalty scores, which represent the structural deficiencies of masonry buildings, can be grouped as follows:

- Existing condition and apparent quality of the building
- Irregularities in plan
- Irregularities in elevation
- Location of the building (adjacency)
- Out-of-plane vulnerability of the masonry walls

Penalty scores for the parameters related to the existing condition and apparent quality of the building are presented in Table 2.8. These include the material quality (both masonry units and mortar), workmanship quality and existence of previous

Table 2.8 Penalty scores and weighing factors for the existing condition and apparent quality

No. of stories	Material quality (0/1/2)	Workmanship (0/1/2)	Previous damage (0/1)
1–5	−10	−5	−5

Table 2.9 Penalty scores for irregularities in plan and elevation

	Irregularities in plan			Irregularities in elevation		
No. of stories	Geometry (0/1)	Wall length (0/1/2)	Bond beams (0/1)	Openings (0/1/2)	High slope (0/1)	Soft/weak story (0/1)
1	−5	−5	−5	0	−5	0
2	−10	−5	−5	−5	−5	−5
3	−10	−10	−5	−5	−5	−5
4	−15	−10	−5	−10	−5	−10
5	−20	−15	−5	−10	−5	−10

damage or not. Penalty scores are assumed to be independent of the number of stories for the given parameters. Weighing factors for material quality are 0, 1 and 2, representing good, moderate and poor quality, respectively. The same weighing factors are also valid for the workmanship quality. For the previous damage parameter, the weighing factors take values of 0 or 1 for absence and presence of any significant damage.

The second and third groups of penalty scores include the parameters related to the irregularities in plan and elevation for masonry buildings. These parameters can be listed as plan geometry, load-bearing wall length, existence of horizontal and vertical bond beams, vertical arrangement of openings in walls, high slope and existence of soft/weak story. The penalty scores and the weighing factors of these two groups are presented in Table 2.9. Since the structural deficiencies in elevation become more pronounced with an increase in the number of stories, this should also be reflected in the penalty scores as given in Table 2.9. The most influential parameters seem to be the plan geometry and the total wall length, which is consistent with the observations during field studies on damaged buildings. In this methodology, plan geometry is classified as "regular" or "irregular" with weighing factors of 0 and 1, respectively. The definition of plan irregularity is based on the requirements of the current earthquake code. Accordingly, a building is irregular in plan if the projections (ax, ay in Fig. 2.6) beyond the re-entrant corners in both of the two principal directions exceed the total plan dimensions of the building (Lx, Ly in Fig. 2.6) in the respective plan directions by more than 20 %. In addition, buildings with L-shaped or U-shaped plans are also deemed as irregular.

The total length of the load-bearing walls in the critical story (generally the ground story) of the building in both principal directions can be roughly determined during the street survey. Accordingly, if the total length of the openings (door and window) is less than 1/3 of the total length of the walls on façade under consideration, the total length is considered as "adequate" with a weighing factor of 0. On the other hand, if the total length of the openings is more than 2/3 of the façade length, it is considered as "inadequate" with a weighing factor of 2. In between, the total length is considered as "fairly adequate" with a weighing factor of 1. If it is

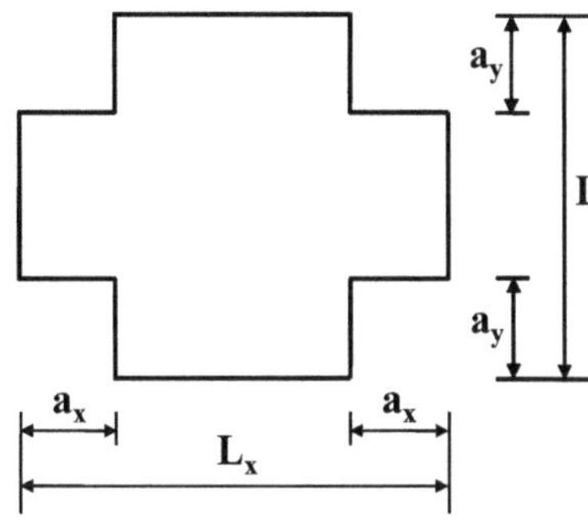

Fig. 2.6 The definition of irregularity in plan according to the Turkish earthquake code

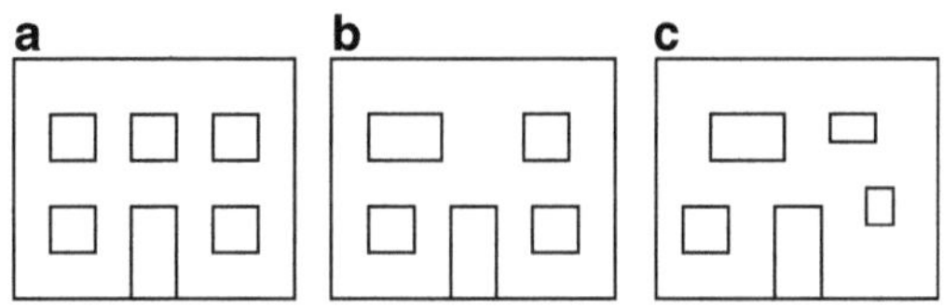

Fig. 2.7 The illustrations for classification of masonry buildings according to vertical arrangements of openings, (**a**) regular, (**b**) fairly regular, (**c**) irregular

possible to enter the building, a more precise evaluation can be performed regarding the total length of the load-bearing walls. The last parameter related with the irregularities in plan is the presence or absence of the horizontal and vertical bond beams with weighing factors of 0 and 1, respectively.

The buildings are classified according to the vertical arrangement of the openings in walls as "regular", "fairly regular" and "irregular" with weighing factors of 0, 1 and 2, respectively. The definitions used for this classification are illustrated in Fig. 2.7. Buildings on high slopes and having soft/weak stories are also considered as seen in Table 2.9 with weighing factors of 0 (absence) and 1 (presence).

The location of the building with respect to the adjacent buildings is also an important parameter. For adjacent buildings, if the floor levels are different, it is possible that the buildings get some damage during an earthquake because of the pounding effect. In this study, five different cases are considered for the location of the building and the floor levels of adjacent buildings:

1. Case A: Separated building
2. Case B: Adjacent building in the middle – floor levels at the same elevation
3. Case C: Adjacent building in the middle – floor levels at different elevations
4. Case D: Adjacent building in the corner – floor levels at the same elevation
5. Case E: Adjacent building in the corner – floor levels at different elevations

The penalty scores for these five different cases are listed in Table 2.10. The most critical case seems to be Case E, representing a building at the corner of a block with floor levels at different elevations with respect to the adjacent building.

Table 2.10 Penalty scores for location of the building and floor levels of adjacent buildings

No. of stories	Case A	Case B	Case C	Case D	Case E
1–5	0	0	−5	−5	−10

There exist some minor parameters which affect the seismic performance of masonry buildings so they should also be included into Eq. (2.1) as penalty scores. The first one is the existence of heavy earthen roofs, which is generally encountered in rural-type masonry construction. Masonry buildings with heavy earthen roofs usually become death traps during an earthquake and in most of the cases it is not possible to escape outside during an earthquake. This structural deficiency increases the final death toll significantly as experienced recently during the 2010 Elazığ (Turkey) earthquake (Akkar et al. 2011; Celep et al. 2011). Hence a penalty score of 10 is assigned to masonry buildings with earthen roof. Another important parameter in the seismic performance of masonry buildings is the out-of-plane capacity of load-bearing walls. In some cases, out-of-plane damage precedes in-plane damage in masonry buildings and causes the partial or total collapse of the building. The structural deficiencies that cause out-of-plane vulnerability can be listed as poor wall-to-wall and wall-to-floor connections, flexible floor diaphragm, having poor or no mortar (dry joint) in walls, existing out-of-plane deformations in the walls and unrestrained gable end walls. These are also the parameters that exist in the data collection form. If at least three of the aforementioned deficiencies are present in the inspected building, then a penalty score of 10 is assigned to that building.

2.3.2 Evaluation of Performance Scores Using Field Data

The scores in this methodology have been determined by considering previous experience from past earthquakes, expert opinion and the results of the analytical fragility functions for different types of masonry construction (Erberik 2008). However the reliability of the proposed methodology should be checked by using actual damage data. For this purpose, the damaged building data obtained after the 1995 Dinar (Turkey) earthquake has been employed. This includes 102 masonry buildings, for which the damage levels were determined by using the damage assessment form of the Ministry of Public Works and Settlement (METU-EERC 1996). According to this form, damage scores are assigned for the walls of the most damaged story of the masonry structure as well as for the stairs and roof. Damage is classified as undamaged (0–1), minor (1–3), moderate (4–6), severe (7–9) or collapse (>9). The numbers in parenthesis represent the score range for each damage state. The comparison of the inverse of the damage score with the performance score obtained by using the proposed methodology is presented in Fig. 2.8. There seems to be an increasing trend line for the scattered data. In other words, as expected, buildings with serious damage after the earthquake receive low

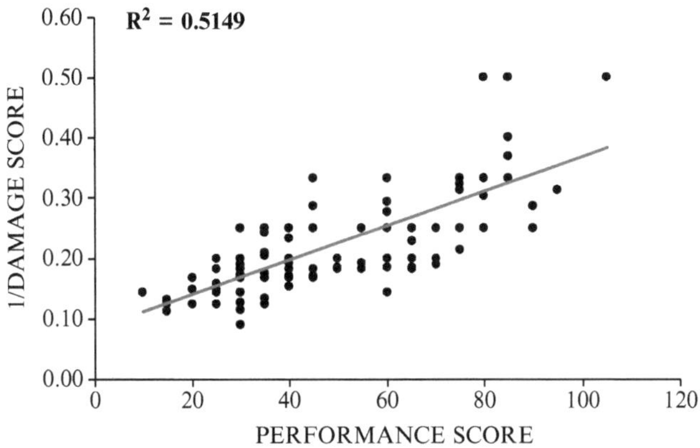

Fig. 2.8 Comparison of performance score with the inverse of the damage score obtained after the 1995 Dinar earthquake

performance points. The R^2 value is not sufficiently high since the damage score obtained after the earthquake does not contain many of the parameters used in calculating the performance score by the proposed methodology, so it is not possible to make a one-to-one comparison between these two measures.

Then the proposed methodology is applied to two different existing masonry building databases for comparison with the existing procedures. The first one is the Zeytinburnu (Istanbul) database, which is composed of 69 masonry buildings. The seismic safety of buildings in this database were examined by using a performance measure named as the "weighted shear strength factor (WSSF)", details of which are provided in the technical report by Sucuoğlu and Erberik (2005). The WSSF can take positive values, for which high values mean high seismic risk. The comparison of the inverse of the WSSF with the performance score is provided in Fig. 2.9a. The second database considered is the Fatih (Istanbul) database, composed of 9,457 masonry buildings. The seismic safety evaluation study carried out by using this database makes use of a vulnerability measure named as the "vulnerability score (VS)". This score is obtained by using analytical fragility curves and takes values between 0 (high risk) and 1 (low risk). The details of the procedure and the VS can be found elsewhere (Erberik 2008, 2010). The comparison of the VS with the performance score is shown in Fig. 2.9b.

The statistical comparisons in Fig. 2.9 reveal that there is a consistent trend but a low R^2 value in both cases. The dispersion in the results arises from the fact that each methodology uses different approaches with different parameters. The assumptions used are also different, so it is not possible to make a one-to-one comparison between different methodologies. Overall, the proposed procedure for masonry buildings is promising but it has to be calibrated further by using field data, which is not easy to find.

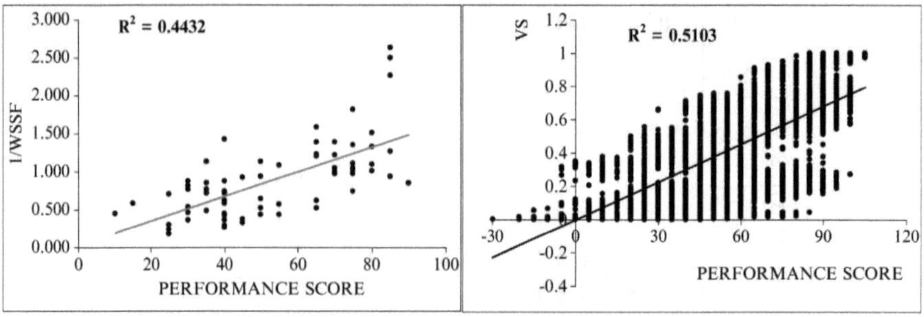

Fig. 2.9 Comparison of the performance measures obtained previously for two different building databases with the performance score developed in this study

2.4 Conclusions

Rapid assessment procedures have been developed for the two most common construction types in Turkey. The procedures reflect peculiarities of Turkish building stock and are believed to reflect their performance observed in recent earthquakes in Turkey. These procedures are applicable to a population of buildings to determine their seismic risk prioritization, but are not appropriate to determine seismic vulnerability of an individual building. Their validity has been tested using field data of existing building databases as well as past earthquake damage databases. The proposed procedures are expected to be used officially in Turkey for both risk prioritization and obtaining building inventory.

References

Akkar S, Aldemir A, Askan A, Bakır S, Canbay E, Demirel O, Erberik MA, Gülerce Z, Gülkan P, Kalkan E, Prakash S, Sandikkaya MA, Sevilgen V, Ugurhan B, Yenier E (2011) 8 March 2010 Elazığ-Kovancılar (Turkey) earthquake: observations on ground motions and building damage. Seismol Res Lett 82(1):42–58

Applied Technology Council (ATC) (2002) Rapid visual screening of buildings for potential seismic hazards: a handbook, 2nd edn, FEMA 154. Federal Emergency Management Agency, Washington, DC

Arya AS (2007) Rapid visual screening of masonry buildings. Technical report prepared under GoI – UNDP Disaster Risk Management Programme. National Disaster Management Division, New Delhi

Bal İE, Tezcan S, Gülay FG (2007) Betonarme binaların göçme riskinin belirlenmesi için P25 hızlı değerlendirme yöntemi. Altıncı Ulusal Deprem Mühendisliği Konferansı, 16–20 Ekim 2007, İstanbul

Celep Z, Erken A, Taskin B, Ilki A (2011) Failures of masonry and concrete buildings during the March 8, 2010 Kovancilar and Palu (Elazig) earthquakes in Turkey. Eng Fail Anal 18(3):868–889

Erberik MA (2008) Generation of fragility curves for Turkish masonry buildings considering in-plane failure modes. Earthq Eng Struct Dyn 37:387–405

Erberik MA (2010) Seismic risk assessment of masonry buildings in Istanbul for effective risk mitigation. Earthq Spectra 26:967–982

FEMA 154 (2002) Rapid visual screening of buildings for potential seismic hazards: a handbook, 2nd edn. Federal Emergency Management Agency; FEMA 500 C Street, SW Washington, DC

İlki A, Boduroğlu H, Özdemir P, Baysan F, Demir C, and Şirin S (2003) Mevcut ve güçlendirilmiş yapılar için sismik indeks yöntemi ve yapısal çözümleme sonuçlarının karşılaştırılması. Beşinci Ulusal Deprem Mühendisliği Konferansı, İstanbul

Kaminosono T (2002) Evaluation method for seismic capacity of existing reinforced concrete buildings in Japan. In: Proceedings of the international conference on earthquake loss estimation and risk reduction, Bucharest

Karbassi A, Nollet MJ (2008) Development of an index assignment procedure compatible with the regional seismicity in the province of Quebec for the rapid visual screening of existing buildings. Can J Civ Eng 35:925–937

Kocaman C, Kuran F, Demirok E, Avşar Ö (2009) Seismic safety evaluation studies in the province of Denizli. Technical report, Ministry of Public Works and Settlement, General Directorate of Disaster Affairs, Earthquake Research Division, Ankara

METU-EERC (1996) Repair and rehabilitation project for moderately damaged masonry structures after 1995 Dinar Earthquake. Earthquake Engineering Research Center, Civil Engineering Department, Middle East Technical University, Ankara

Ozcebe G, Yucemen MS, Aydogan V (2004) Statistical seismic vulnerability assessment of existing reinforced concrete buildings in Turkey on a regional scale. J Earthq Eng 8(5):749–773

SERU [Structural Engineering Research Unit] (2003) Middle East Technical University, Ankara, Turkey. http://www.seru.metu.edu.tr/archives/databases/Duzce_Database/

Sextos AG, Kappos AJ, Stylianidis KC (2008) Computer-aided pre- and post-earthquake assessment of buildings involving database compilation, GIS visualization, and mobile data transmission. Comput-Aided Civ Infrastruct Eng 23:59–73

Sucuoğlu H, Erberik MA (2005) Earthquake master plan for Istanbul, pilot study in Zeytinburnu sub-province: seismic safety evaluation of masonry and hybrid buildings. Technical report, Middle East Technical University, Ankara

Sucuoğlu H, Yazgan U (2003) Simple survey procedures for seismic risk assessment in urban building stocks. In: Wasti ST, Ozcebe G (eds) Seismic assessment and rehabilitation of existing buildings, vol 29, Earth and environmental sciences. Kluwer Academic, London, pp 97–118

Sucuoğlu H, Yazgan U, Yakut A (2007) A screening procedure for seismic risk assessment in urban building stocks. Earthq Spectra 23:441–458

Temür R (2006) Hızlı Durum Tespit (DURTES) yöntemi ve bilgisayar programının geliştirilmesi. Yüksek lisans tezi, İstanbul Üniversitesi, İstanbul

Turkish Earthquake Code (TEC) (1997, 2007) Specification for buildings to be built in seismic areas. Ministry of Public Works and Settlement, Ankara

Yakut A (2004) Preliminary seismic performance assessment procedure for existing RC buildings. Eng Struct 26(10):1447–1461

[illegible]

Chapter 3
Post-Earthquake Risk-Based Decision Making Methodology for Turkish School Buildings

Ufuk Yazgan and Reşat Atalay Oyguç

Abstract Safety assessment of damaged school buildings is one of the challenging tasks after damaging earthquakes. In this study, a new post-earthquake safety assessment method is proposed for school buildings. The method enables researchers to evaluate the level of risk and to judge the safety of the school building by taking into account the observed level of structural and non-structural damage. First, the vulnerability of the building is identified by considering the relevant damage indicators. Subsequently, the likely consequences of the various performance levels of the school building are identified. The likelihood of a specific consequence is evaluated by jointly considering all failure mechanisms that can lead to the considered consequence. Critical decisions regarding the school are made based on the level of risk and the available resources for risk mitigation. The methodology is expected to be a useful supporting tool for the post-earthquake decision-making process. Using the proposed method, critical decisions, such as continued use, strengthening or decommissioning school buildings can be handled in a rational and consistent way.

3.1 Introduction

Post-earthquake evaluation of public buildings that shelter large groups of people is a critical task. Particularly in the case of school buildings, high occupant density imposes further challenges. Making decisions related to repair or decommissioning

U. Yazgan (✉)
Earthquake Engineering and Disaster Management Institute,
Istanbul Technical University, Maslak, Istanbul 34469, Turkey
e-mail: ufukyazgan@itu.edu.tr

R.A. Oyguç
Newmark Civil Engineering Laboratory, University of Illinois at Urbana-Champaign,
205 North Mathews Avenue, Urbana, IL 61801, USA
e-mail: oyguc@illinois.edu

A. Ilki and M.N. Fardis (eds.), *Seismic Evaluation and Rehabilitation of Structures*, Geotechnical, Geological and Earthquake Engineering 26,
DOI 10.1007/978-3-319-00458-7_3, © Springer International Publishing Switzerland 2014

of a damaged school building has a strong influence on the recovery of the affected region. Schoolchildren in the affected area can only continue their education if sufficient classroom space can be provided in safe buildings. In addition, school buildings are spread over a wide area in rural regions. This imposes additional difficulties related to providing necessary resources and properly inspecting the repair efforts. All the above make post-earthquake evaluation of school buildings a challenging task after damaging earthquakes.

Existing approaches to post-earthquake evaluation of school buildings in Turkey are based on the evaluation of the extent of structural damage. Columns, structural walls, beams, – and, whenever suitable, foundation system – of the affected structure are inspected. During this inspection, the severity of damage to individual components and the location of the damage are identified. The safety of the building is judged by assessing the degree of change in the safety level compared to its pre-earthquake one. If the damage is judged to be insignificant, the building is repaired and put back into service. On the other hand, if the damage is found to be so severe that the anticipated repair cost almost exceeds the rebuilding cost, the building is rebuilt with the available resources. For the school buildings which can be cost-effectively retrofitted, seismic performance of the building structure is improved using common techniques so that its expected seismic performance complies with the code regulations. Safety decisions in current practice are made only based on the aim of limiting the probability of a structural failure during a possible future earthquake.

A new post-earthquake decision-making scheme is introduced here for the post-earthquake safety evaluation of school buildings. Unlike current practice, the proposed scheme is based on assessing the likely consequences due to structural as well as non-structural failure during potential future earthquakes. The consequences are evaluated in terms of casualties, injuries and economic losses. Each potential damage mechanism is evaluated in terms of the potential losses that mechanism may result in. In this evaluation, failure of non-structural components is taken into account as well as failure of structural components. The damage mechanisms that are expected to result in severe consequences are prioritized against other failures that are expected to lead to less important consequences. As a result of this prioritization, the available resources for seismic improvement of schools can be allocated to avert unacceptable consequences and reduce the seismic risk more effectively.

3.2 Characteristics of School Buildings

Public school buildings in Turkey share some common characteristics related to their structural and architectural layout. These buildings are designed and constructed based on a set of templates. These standardization measures are introduced by public administration with the aim of improving the efficiency of the construction and maintenance works. Making use of these design templates, application of the post-earthquake safety assessment method outlined in this investigation can be standardized as well.

The school buildings typically have regular structural systems. Most of the multistory school buildings have regular reinforced concrete structural systems consisting of regular frames and structural walls. The design loads assumed for school buildings in the Turkish Earthquake Code (TEC) are higher compared to ordinary buildings with the same structural properties. TEC (1975) enforced design loads for school buildings 50 % higher compared to ordinary buildings. Later, this difference was reduced to 40 % in TEC (1998, 2007). Unfortunately the special care taken in setting the of design loads is not always reflected in the construction of the school buildings. The quality of construction materials and detailing practice for school buildings are known to have significant flaws (Gülkan 2004; Yüzügüllü et al. 2004; Gür et al. 2009).

After damaging earthquakes, architectural details (e.g. claddings and suspended ceilings) may prevent detection of cracks and other damage indicators related to structural components. In the case of public school buildings, such architectural details are not common and the detection of damage to structural components is relatively simpler compared to other types of buildings. This is very beneficial in the post-earthquake safety assessment of the buildings.

In the majority of rural settlements in Turkey, the school building is the only multistory public building which is designed according to engineering principles. That is to say, school buildings are often the only accessible public building in rural regions. As a result, they may directly serve as temporary shelters for the earthquake victims provided that the building safety can be evaluated reliably after damaging earthquakes.

3.3 Proposed Safety Assessment Method

The safety assessment approach proposed in this study aims at evaluating the estimated consequences of potential damage mechanisms. As a result, post-earthquake safety of a school building is not judged based only on the level of increase of the seismic vulnerability of the school, but also on the basis of the estimated risk associated with the building and its occupants. Towards this aim, the likely consequences of a potential failure of both structural and non-structural components are taken into account.

The probability of a specific consequence of interest C_i that may occur due to a failure mechanism F_j can be estimated using the following triple integral:

$$\Pr(C_{i,j}) = \int_F \int_{EDP} \int_{IM} F_{C_i|F_j}(w|x)\, f_{F_j|EDP}(x|y)\, f_{EDP|IM}(y|z)\, f_{IM}(z)\, dx\, dy\, dz \qquad (3.1)$$

In the equation above, $f_{IM}(z)$ is the mean annual rate of a given peak intensity measure *IM* (e.g. $S_a(T)$, *PGA*, *PGV*) exceeding a threshold value z at the site, $f_{EDP|IM}(y|z)$ is the conditional probability density function of the considered engineering demand parameter *EDP* (e.g. peak roof drift, peak inter-story drift) given the ground motion intensity, $f_{F_j|EDP}(x|y)$ is the conditional probability of the considered failure mechanism F_j taking place given the *EDP* level, $F_{C_i|F_j}(x|y)$ is the conditional cumulative probability of consequence $C_{i,j}$ occurring given that failure mechanism F_j has occurred. Note that the triple integral presented in Eq. 3.1 is a slightly modified version of the Pacific Earthquake Engineering Research (PEER) Center framework equation (Cornell and Krawinkler 2000). The likelihood associated with all consequences C_i and the failure modes F of interest can be estimated using Eq. 3.1. The conventional method to estimate $f_{IM}(z)$ is the probabilistic seismic hazard assessment method proposed by Cornell (1968). Conditional probability distributions $f_{F_j|EDP}(x|y)$ and $F_{C_i|F_j}(x|y)$ may be estimated using any of the suitable methods available in the literature (Aslani and Miranda 2005). Conditional probability distribution $f_{EDP|IM}(y|z)$ can be estimated using the procedure presented in the next section.

As an example, the loss of lives due to failure of non-structural brick gable walls that are attached to the roof is considered here. Such a gable wall failed during October 23rd, 2011 M7.2 Van Earthquake. The part of the gable wall which failed during the *Abdurrahman Gazi School for the Hearing Impaired* and was repaired afterwards is indicated in Fig. 3.1. As seen in this figure, the gable wall is directly located above the entrance of the building.

Potential consequences due to failure of a gable wall are directly related to the location of the wall. If the gable wall is located above a small protected area with limited access, the failure of the wall has minor consequences. On the other hand, if the wall is located directly above the entrance (Fig. 3.1), it may fall over building occupants that are evacuating the building during the earthquake. As a result, failure of the wall may lead to casualties. It is important to note that the potential consequences of non-structural failure are not taken into account directly in current practice.

The likelihood of each consequence (e.g. casualties, economic losses, interruption of occupancy) related to a given failure mechanism can be evaluated using Eq. 3.1. However, a specific consequence C_i may be triggered by a number of different failure modes (i.e. F_1, F_2, … F_{nf}). In such cases, the consequence likelihood associated with each mechanism may be evaluated separately using Eq. 3.1 and the results can be combined as follows:

$$\Pr\left(C_i^*\right) \cong 1 - \prod_{j=1}^{nf} \left(1 - \Pr\left(C_{i,j}\right)\right) \tag{3.2}$$

In the equation above, $\Pr\left(C_i^*\right)$ is the estimated probability for consequence C_i^* obtained by considering the whole set of different failure modes, and *nf* is the number of failure modes. The resulting probability estimated for consequence C_i^* may be adopted as the basis for evaluating the safety of the damaged structure.

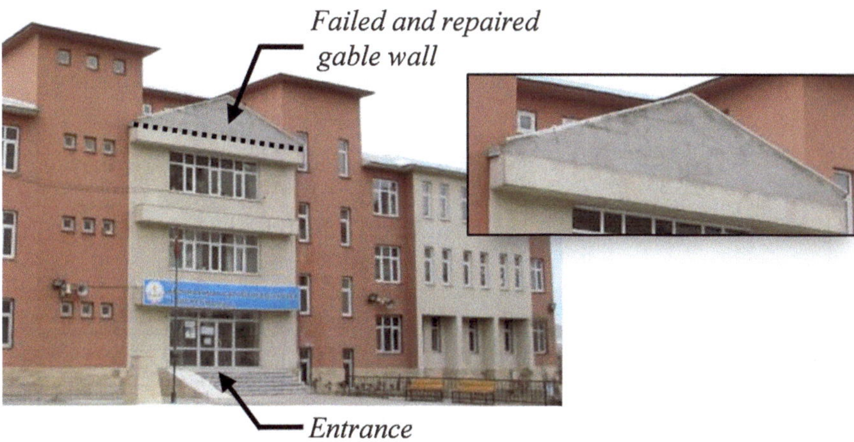

Fig. 3.1 Failed and repaired gable wall in a school building in Van (Photo taken after October 23rd, 2011 M7.2 Earthquake)

3.4 Considering Post-Earthquake Indicators in the Evaluation of Safety

The damage observations made during the post-inspection of the school building are considered to improve the accuracy of the probability distribution $f_{EDP|IM}(y|z)$ estimated for the *EDP* given *IM*. Detection of damage indicators like cracks or spalled cover concrete in columns, structural walls and beams provide information related to key dynamic characteristics of the structure. For this purpose, a method to estimate dynamic characteristics of structures based on damage indicators was proposed by Yazgan and Dazio (2012). This procedure can be extended to enable taking into account damage evidence related to non-structural components and the building's contents.

For example, damage evidence such as detection of a tipped-over bookshelf is considered here. Such tipped over bookshelves were observed in school buildings after the recent October 23rd, 2011 Van Earthquake (Fig. 3.2a, b). The peak horizontal acceleration level that would result in a rigid body tipping over can be estimated by considering the equilibrium of forces acting on the system (Fig. 3.2c). For a rectangular general rigid body with center of mass located in the middle, the horizontal acceleration threshold a_{tip} that would cause tipping over can be calculated as follows:

$$a_{tip} = \frac{b \cdot g}{h} \tag{3.3}$$

where b and h are the width and height of the rigid body, respectively and g is the gravitational acceleration. Due to the vertical component of earthquake excitations and to the sliding occurring at the base of the rigid body, there is some uncertainty

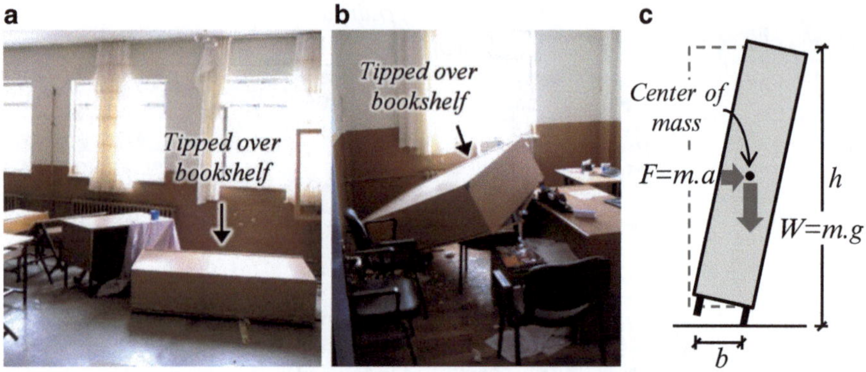

Fig. 3.2 Bookshelves that had tipped over during 2011 M7.2 Van Earthquake: (**a**) in a typical classroom, and (**b**) in the director's office, and (**c**) forces acting on a bookshelf subjected to horizontal acceleration

in the actual acceleration level that may have caused the tipping over. The approach proposed by Konstantinidis and Makris (2009) may be adopted to take such effects into account and to evaluate the uncertainty related to a_{tip}. Based on the estimated a_{tip}, an upper bound to the peak absolute acceleration that had taken place at the position of the bookshelf can be calculated. Detection of a tipped over bookshelf may be used for establishing a lower bound for the accelerations that had taken place at that location during the earthquake. On the other hand, an upper bound is obtained for the absolute acceleration if a tipped over bookshelf is detected.

Post-earthquake damage indicators can be taken into account in the evaluation of the likelihood of consequences using Eq. 3.1. A convenient way to achieve this is to evaluate the integral by discretizing it and conditioning the estimates to the observed damage evidence, E. Accordingly, conditional probability $\Pr(C_i|E)$ for consequence C_i given E can be evaluated as follows:

$$\Pr(C_i|E) \cong \sum_j \sum_k \sum_l \Pr(C_i|F_j)\ \Pr(F_j|EDP_k)\ \Pr(EDP_k|IM_l, E)\ \Pr(IM_l) \quad (3.4)$$

where $\Pr(C_i|F_j)$, $\Pr(F_j|EDP_k)$ and $\Pr(IM_l)$ are the probabilities obtained by discretizing the continuous probability models $F_{C_i|F_j}(x|y)$, $f_{F_j|EDP}(x|y)$, and $f_{IM}(z)$ in Eq. 3.1 by considering specific intervals of the random variables. The term $\Pr(EDP_k|IM_l, E)$ is the new term that represents the conditional probability of an engineering demand parameter being in the kth interval EDP_k given the ground motion intensity IM_l and observed damage evidence E. This latter conditional probability establishes the key link between the post-earthquake damage inspection results and the estimated likelihood of each consequence.

The most common approach to evaluating the probabilistic character of an engineering demand parameter given an intensity measure level is to perform a series of numerical simulations (e.g. Kunnath 2007). In these simulations, the variability of strong ground excitation and uncertain model parameters (e.g. material properties and damping) are modeled by simulating the response by assuming a set of randomly generated parameter values in each simulation. Typically, the statistical distribution of *EDP* is evaluated at the various *IM* levels by assuming an equal likelihood for each simulation.

In case of post-earthquake safety evaluation, the simulation likelihood can be adjusted by considering damage evidence detected during the inspection. Accordingly, the conditional probability $\Pr(EDP_k|IM_l, E)$ in Eq. 3.4 can be obtained by marginalizing the distribution as follows:

$$\Pr(EDP_k|IM_l, E) = \sum_{m=1}^{ns} \Pr(EDP|IM_l, S_m)\Pr(S_m|E) \tag{3.5}$$

where $\Pr(EDP_k|IM_l, S_m)$ is the conditional probability distribution of *EDP* resulting from the response simulation-*m* S_m, $\Pr(S_m|E)$ is the conditional probability distribution for S_m given the damage evidence E and ns is the total number of simulations. In brief, the expression in Eq. 3.5 is the marginalization of the individual conditional probabilities for each *EDP* interval given each simulation, over the likelihood of each simulation, given the damage observed (damage evidence, *E*).

The likelihood assumed of each simulation may be updated based on the damage observations using Bayes Theorem as follows:

$$\Pr(S_m|E) = \frac{\Pr(E|S_m)\Pr(S_m)}{\Pr(E)} \tag{3.6a}$$

where

$$\Pr(E) = \sum_{k=1}^{ns} \Pr(E|S_k)\Pr(S_k) \text{ and } \Pr(S_m) \cong 1/ns \tag{3.6b}$$

In the equations above, $\Pr(S_m)$ is the prior probability assumed for S_m and $\Pr(E|S_m)$ is the likelihood of observing the considered evidence given simulation-*m* S_m. The conditional likelihood $\Pr(E|S_m)$ for observing a damage evidence E can be estimated using the simulated response parameters from simulation-*m* S_m and the fragility model related to the considered damage evidence E. For the case of tipping over behavior of a bookshelf, the response is primarily controlled by peak absolute story acceleration. Thus, likelihood of each response simulation S_m can be updated based on the observation of a tipped over bookshelf through the simulated peak absolute acceleration (Fig. 3.3).

The proposed approach may be adopted to obtain simulation estimates conditioned on the damage evidence observed in the post-earthquake inspection of a school building. The conditional probabilities estimated using Eq. 3.5 are then substituted

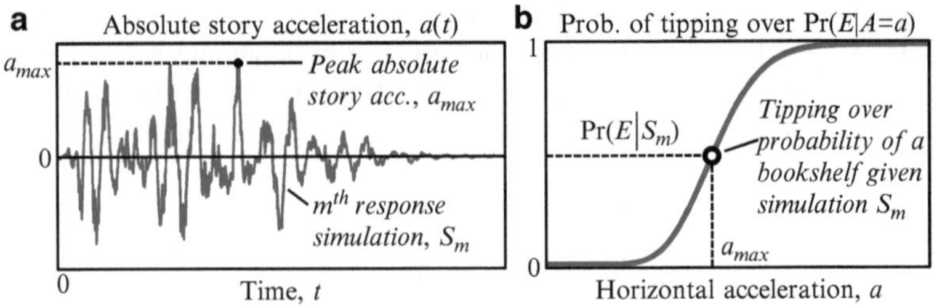

Fig. 3.3 Evaluation of the conditional probability of observing a tipped over bookshelf conditioned on the response-history simulation S_m: (**a**) simulated absolute story acceleration, and (**b**) likelihood of tipping over estimated as a function of peak acceleration

into Eq. 3.4 in order to obtain the likelihood of each consequence C_i conditioned on the observed set of damage evidence E. The likelihood obtained using the proposed approach should be compared with the threshold levels that define the acceptable level of risk. The safety of the school is judged based on this comparison.

3.5 Summary and Conclusions

The overview of a new post-earthquake safety assessment method for school buildings is presented. In the proposed method, the seismic safety of the considered building is evaluated objectively by estimating the likelihood of a set of consequences of interest. Post-earthquake damage observations are utilized to estimate the likelihood of each consequence. The proposed method enables researchers to take into account both structural and non-structural damage in order to estimate more accurately the likelihood of each consequence. Unlike current practice, post-earthquake safety of the building is not judged based only on the estimated structural performance. The safety is judged directly by evaluating the seismic risk associated with the occupants of the school and the building itself due to structural as well as non-structural failures.

Tipped over bookshelves are considered as an example of damage evidence. Such bookshelves were frequently observed in school buildings excited by strong ground motions during past earthquakes. The procedure for taking into account this non-structural damage observation in post-earthquake safety evaluation is presented. The means to update the likelihood of a specific consequence based on the observation of a tipped over bookshelf is explained.

Acknowledgments The authors acknowledge the devoted educators in Van who guided the authors in the region after the 2011 earthquake disaster for their helpful support. In addition, the support provided by the Ministry of Education and Istanbul Technical University, which enabled the authors to observe the post-earthquake states of the affected schools, is acknowledged.

References

Aslani H, Miranda E (2005) Probability-based seismic response analysis. Eng Struct 27:1151–1163. doi:10.1016/j.engstruct.2005.02.015

Cornell CA (1968) Engineering seismic risk analysis. Bull Seismol Soc Am 58:1583–1606

Cornell CA, Krawinkler H (2000) Progress and challenges in seismic performance assessment. PEER Center News. http://peer.berkeley.edu/news/2000spring/index.html. Accessed 15 Sept 2012

Gülkan P (2004) Obstacles to improving seismic safety of school buildings in Turkey. In: Yelland H, Tucker B (eds) Keeping schools safe in earthquakes. OECD, Paris

Gür T, Pay AC, Ramirez JA, Sözen MA, Johnson AM, İrfanoğlu A, Bobet A (2009) Performance of school buildings in Turkey during the 1999 Düzce and the 2003 Bingöl earthquakes. Earthq Spectra 25:239–256

Konstantinidis D, Makris A (2009) Experimental and analytical studies on the response of freestanding laboratory equipment to earthquake shaking. Earthq Eng Struct Dyn 38:827–848. doi:10.1002/eqe.871

Kunnath SK (2007) Application of the PEER PBEE methodology to the I-880 viaduct. PEER 2006/10, Pacific Earthquake Engineering Research Center, University of California, Berkeley

Turkish Earthquake Code (1975) Specification for structures to be built in disaster areas. Ministry of Public Works and Settlement, Government of Republic of Turkey, Ankara

Turkish Earthquake Code (1998) Specification for structures to be built in disaster areas. Ministry of Public Works and Settlement, Government of Republic of Turkey, Ankara

Turkish Earthquake Code (2007) Specification for structures to be built in disaster areas. Ministry of Public Works and Settlement, Government of Republic of Turkey, Ankara

Yazgan U, Dazio A (2012) Post-earthquake damage assessment using residual displacements. Earthq Eng Struct Dyn 41:1257–1276. doi:10.1002/eqe.1184

Yüzügüllü Ö, Barbarosoglu G, Erdik M (2004) Seismic risk mitigation practices of school buildings in Istanbul. In: Yelland H, Tucker B (eds) Keeping schools safe in earthquakes. OECD, Paris

Chapter 4
Proposed Vulnerability Functions to Estimate the Real Damage State of RC Buildings After Major Turkish Earthquakes

Ulgen Mert Tugsal and Beyza Taskin

Abstract The Chapter focuses on the applicability of fragility relationships, to predict the seismic vulnerability of existing structures. Since these relationships offer the probability of exceeding a predefined structural response limit in terms of a ground motion intensity parameter, fragility functions are very practical tools to be employed during urban renewal of metropolitan cities with high seismicity. A building ensemble which experienced various damage levels after major Turkish earthquakes is considered herein. Planar structural models for each building are established utilizing DRAIN-2DX computer program and nonlinear dynamic analyses are carried out. The demand parameters are obtained and the capacity is determined in terms of limit states. Finally, fragility relationships recently proposed by various researchers are employed for the building set and compared with the analytical results by means of reflecting the most reliable actual damage state.

4.1 Introduction

Seismic damage to buildings during the recent major earthquakes in Turkey has emphasized the need for risk assessment of the existing building stock, specifically low- and mid-rise reinforced concrete (RC) structures, to estimate the potential damage during future events. The need to predict the vulnerability of these structures has become a challenge among researchers concerned with the development of practical, applicable and handy methods. Fragility relationships are used to estimate the probable damage for an individual building as well as building stocks. Henceforth, considering the difficulty in making a decision about the whole building stock, fragility curves are employed in various software and risk analyses carried out in

U.M. Tugsal (✉) • B. Taskin
Department of Civil Engineering, Civil Engineering Faculty, Istanbul Technical University, Maslak, Istanbul 34469, Turkey
e-mail: tugsal@itu.edu.tr; btaskin@itu.edu.tr

A. Ilki and M.N. Fardis (eds.), *Seismic Evaluation and Rehabilitation of Structures*, Geotechnical, Geological and Earthquake Engineering 26,
DOI 10.1007/978-3-319-00458-7_4,

metropolitan cities subjected to high seismicity. Although a fragility curve offers the probability of reaching or exceeding a level of damage under a ground motion intensity parameter, it is observed that the level of damage state can be related to the structural capacity and demand. Other than the structural system properties, however, the post-earthquake damage state of a building is dependent upon many parameters, including local site conditions, structural material qualities, adequacy of the workmanship, and detailing of reinforcement.

Approximately 75 % of the total building stock in Turkey consists of low- and mid-rise RC frame structures which are generally non-engineered and have not been adequately designed according to the earthquake code regulations. Destructive earthquakes within the last decades proved that structures, especially those constructed prior to the Turkish Earthquake Resistant Design Code of Turkey (TERDC) of 1975, have experienced severe damage and partial or even total collapse, mainly depending on poor structural material quality, inadequate reinforcement detailing, lack of confinement zones, heavy and large-span cantilevers and indirect supporting preventing the formation of regular structural frames. Therefore, the damage to buildings has urged the need for risk assessment of the existing building stock to estimate the potential damage from future earthquakes.

Fragility curves offer the probability of exceeding a predefined structural dam-age limit due to earthquakes in terms of a ground motion intensity parameter: e.g., peak ground acceleration or velocity (PGA or PGV), elastic spectral acceleration or displacement (Sa or Sd). They are scientific and practical tools in cases of urban renewal, damage or loss estimation and retrofitting decisions. Existing vulnerability functions can be classified into the four generic groups: empirical, judgmental, analytical and hybrid according to whether the damage data used in their generation derives mainly from observed post-earthquake surveys, expert opinion, analytical simulations or combinations of these (Rossetto and Elnashai 2003). Analytical fragility relationships can be established either by employing a set of nonlinear dynamic or pushover analyses.

Only few attempts to estimate the seismic vulnerability of Turkish RC buildings by the use of fragility curves have been made (Akkar et al. 2005; Kircil and Polat 2006). In order to investigate the applicability of these proposed vulnerability functions for a typical building stock in Turkey, nonlinear dynamic analyses are carried out for a selected set of low- and mid-rise RC frame structures which have suffered structural damage during the major Turkish earthquakes settling on firm soil in seismic zone-1. Time-history analyses are carried out utilizing the DRAIN-2DX (Prakash et al. 1993) computer program. Planar structural models are established, including the contribution of non-structural infill walls. Seven simulated ground motions compatible with the design spectrum for an effective acceleration coefficient of 0.40g defined for seismic zone-1 and Z2 soil class defined for firm soil in the TERDC-2007 are employed during the analyses. For an adequate comparison with the proposed vulnerability curves, the peak ground acceleration of each simulated ground motion in the ensemble is scaled from lower values up to 0.8g with 0.1g increment. The demand parameters in terms of maximum inter-story drift ratio, story shear forces and base shear ratio are obtained for the earthquake ensemble from artificial records and the capacity is determined in terms of limit states. Finally, fragility relationships recently proposed by various researchers, as in DiPasquale and Cakmak (1987), Erberik and Elnashai (2004), Kircil and Polat (2006) are employed to the building set and compared with each other and to the analytical results.

4.2 Investigated RC Buildings and Earthquake Ensemble

4.2.1 Structural Properties of Building Stock

Most of the damaged buildings during earthquake shakes in Turkey are observed to be designed brittle, weak in stiffness and strength and poor in material qualities due to the lack of proper amount of engineering service during the design and site-control stages. Therefore, it is a vital issue in case of urban renewal to replace these buildings with an earthquake resistant building stock using a rapid risk assessment method by means of vulnerability functions.

A set of 23 existing residential RC buildings, which have experienced different levels of damage during major earthquakes in Turkey are considered herein. Damage levels of the selected ensemble are specified as no-damage (ND), moderately damaged (MD) and heavily damaged (HD) and building codes are given accordingly with the damage rank as given in Table 4.1. The number of stories for the buildings vary from 2-story to 8-story, therefore the set generally consists of low- and mid-rise structures. All buildings with frame structural systems are investigated in detail in terms of structural material quality, reinforcement amount, rebar detailing and local site conditions. Concrete class is found to vary from C8 ~ C25 (characteristic compressive strength $f_{ck} = 8.0 \sim 25.0$ MPa), while the reinforcing steel is S220 (characteristic yield strength $f_{yk} = 220$ MPa) class. Table 4.1 tabulates some characteristics of the buildings, where T_0 is the dominant vibration period for the virgin state for both directions and W is the total weight of the building.

4.2.2 Strong Ground Motion Ensemble

An ensemble of seven acceleration-time histories are generated utilizing TARSCTHS computer program (Papageorgiou et al. 2000), which is capable of simulating earthquake motions compatible with a defined spectrum. According to the TERDC-2007, if a set of seven strong motions or more is used, then the mean values of the structural responses may be used to represent the structural demands. Either recorded or simulated earthquake motions should satisfy the following conditions:

(i) The duration of the strong motion acceleration record shall neither be less than five times the first natural vibration period of the building nor less than 15 s.

Table 4.1 Structural characteristics of the inspected buildings

Building code	Direction	No of stories	h_{total} (m)	f_{ck} (MPa)	W (kN)	$T_{0(x\text{-}x/y\text{-}y)}$ (s)
01-ND-ST4	y-y	4	12.00	14.0	17100.0	0.469
02-MD-ST3	x-x/y-y	3	8.25	16.0	3950.7	0.324/0.351
03-MD-ST5	x-x/y-y	6	14.5	16.5	10750.7	0.502/0.544
04-MD-ST5	x-x/y-y	6	13.75	13.3	8104.4	0.442/0.445
05-MD-ST3	x-x/y-y	3	8.85	8.6	7676.8	0.439/0.319
06-MD-ST4	x-x/y-y	4	12.1	10.0	6382.8	0.532/0.490
07-MD-ST3	x-x/y-y	3	8.7	8.3	3872.7	0.356/0.379
08-MD-ST5	x-x/y-y	5	11.00	11.4	7859.3	0.196/0.519
09-MD-ST4	x-x/y-y	4	11.05	12.0	5890.4	0.539/0.557
10-MD-ST3	x-x/y-y	3	8.55	11.9	4009.2	0.302/0.285
11-MD-ST3	x-x/y-y	3	8.10	9.4	3556.9	0.393/0.421
12-MD-ST4	x-x/y-y	4	12.25	13.1	5731.8	0.527/0.496
13-MD-ST5	x-x/y-y	6	13.75	11.5	8104.4	0.451/0.454
14-MD-ST5	x-x/y-y	5	14.25	8.2	10587.2	0.921/0.604
15-MD-ST5	x-x/y-y	5	15.3	11.7	7383.9	0.855/0.545
16-MD-ST2	x-x/y-y	2	5.60	12.0	4,594	0.222/0.233
17-HD-ST3	x-x/y-y	3	7.95	12.0	5,789	0.341/0.258
18-ND-ST2	x-x/y-y	2	6.3	25.0	4,238	0.272/0.253
19-ND-ST8	x-x/y-y	8	22.4	14.0	11154.7	0.821/0.638
20-ND-ST6	x-x/y-y	6	17.55	14.0	9,214	0.742/0.608
21-ND-ST2	x-x/y-y	2	6.4	14.0	10086.7	0.309/0.275
22-ND-ST2	x-x/y-y	2	6.4	14.0	7400.3	0.339/0.273
23-MD-ST6	x-x/y-y	6	15.2	8.3	8208.1	0.571/0.56

(ii) Mean value of spectral acceleration for each recorded or simulated acceleration record with 5 % damping ratio at zero period (T = 0) shall not be less than the spectral acceleration $A_0 \times g$; where A_0 is the effective ground acceleration coefficient and g is the acceleration of gravity.

(iii) Mean of the spectral acceleration values computed for each recorded or simulated acceleration record with 5 % damping ratio within a period range of $0.2\ T_1 \sim 2\ T_1$, where T_1 is the dominant vibration period of the structure, shall not be less than 90 % of the elastic spectral acceleration $S_{ae}(T)$.

Considering locations and the local site conditions of the investigated buildings, the effective ground acceleration of $A_0 = 0.40g$ for seismic zone-1 and a local soil class *Z2* with characteristic periods of $T_A = 0.15$ s and $T_B = 0.40$ s are taken into account. Seven simulated motions, each with duration of 25 s, are generated compatible with the design spectrum defined in the TERDC, which has a probability of exceedance of 10 % within 50 years. Table 4.2 summarizes the engineering intensities such as peak ground acceleration *PGA*; strong motion duration t_{SM} is defined as the duration for accelerations greater than 5 % of g; effective duration t_{eff} is calculated as the time interval from 5 to 95 % of the Arias Intensity curve.

Among the values given in Table 4.2, the duration of the strong motion is calculated as 18.54 s averagely, which is consistent with condition *(i)* above.

Table 4.2 Characteristics of the earthquake ensemble

		Simulated motions							
		SimEQ-1	SimEQ-2	SimEQ-3	SimEQ-4	SimEQ-5	SimEQ-6	SimEQ-7	Mean
PGA	(gal)	420.7	435.0	412.8	406.2	413.2	405.3	427.0	417.2
t_{SM}	(s)	17.33	19.79	18.08	18.67	17.69	18.36	19.83	18.54
t_{eff}	(s)	11.27	11.80	12.02	11.91	11.82	11.69	11.95	11.78

Mean of the PGA is computed as 417.2 cm/s^2. It is well known that this value corresponds to the zero period spectral acceleration which satisfies the above condition *(ii)* with a design spectral acceleration of 0.40g (=392.3 cm/s^2) for seismic zone-1. Condition *(iii)* is also satisfied depending on the dominant vibration periods of each building which will be shown in the following. Figure 4.1 shows the acceleration response spectra for the earthquake ensemble compared with the 'target' design spectrum for a 5 % damping ratio. The mean tripartite elastic response spectrum, as well as the corresponding acceleration, velocity and displacement-sensitive spectral regions and separating periods with values $T_c = 0.46$ s for acceleration to velocity-sensitive regions and $T_d = 3.67$ s for velocity to displacement-sensitive regions, are also shown in the same figure.

4.2.3 *Analytical Modeling of the Structural Systems*

Buildings are modeled utilizing DRAIN-2DX, however the modified version by Ascheim (2005) is preferred, which is capable of handling the stiffness and strength degradation through Takeda hysteretic model. Nonlinear behavior of columns and beams is assumed to be represented with a stiffness degrading hysteresis model. A single earth-retaining basement RC wall in one of the buildings is modeled considering the pinching effects. For the entire building stock, a structural damping of 5 % and a strain-hardening of 3 % are taken into account.

Assigning element type-7 for stiffness-degrading elements and type-9 for the non-structural and reinforced concrete walls, the analytical model for DRAIN-2DX computer program is established. For each state of the building, the structural system is modeled as planar frames connected to each other with elastic tension/compression link elements representing rigid diaphragm effect. The structural model for non-structural walls is established as in Al-Chaar and Lamb (2002), where walls are modeled as two diagonal compression struts having a width of a_w defined by the following Eq. 4.1:

$$a_w = 0.175 r_w \left\{ H \left(\frac{E_w t \sin 2\theta}{4 E_c I_c h_w} \right)^{\frac{1}{4}} \right\}^{-0.4} \tag{4.1}$$

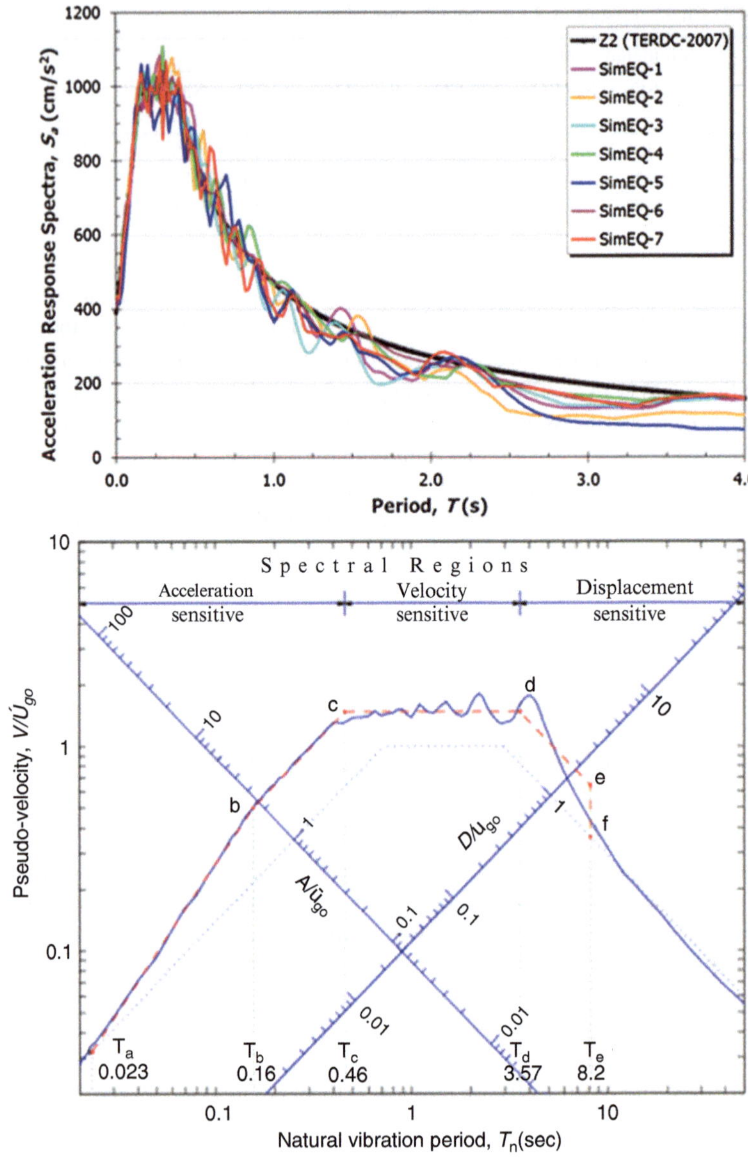

Fig. 4.1 Elastic response spectra of ground motions and design spectra in TERDC-2007 (*top*); Tripartite elastic response spectra of the earthquake ensemble (*bottom*)

Here t is the thickness, r_w is the diagonal length and θ is the diagonal slope angle of the wall. Considering the wall height as h_w, story height as H, the moment of inertia of the neighboring columns as I_c, the modulus of elasticity for infill walls $E_w = 1{,}000$ MPa and that of concrete, E_c, between 24,000 and 27,000 MPa, strut widths a_w are computed for each non-structural wall (Fig. 4.2).

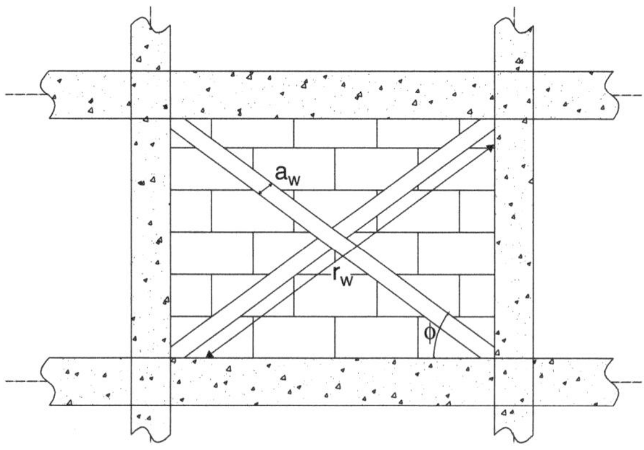

Fig. 4.2 Modeling of infill walls

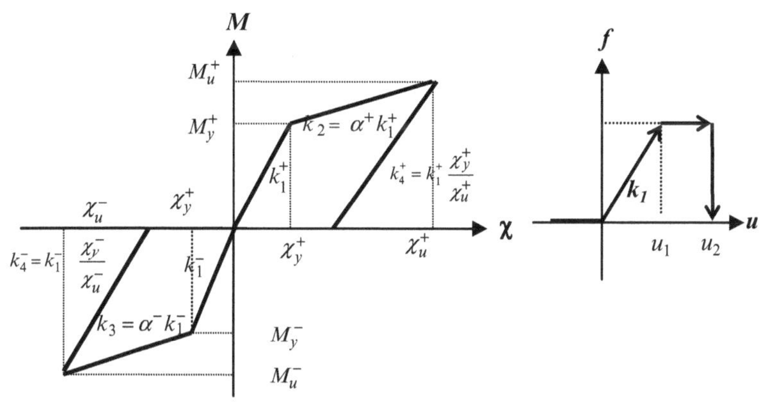

Fig. 4.3 Hysteresis curves for structural (*left*) and non-structural elements (*right*)

Figure 4.3 shows the backbone curves for the hysteretic relations employed to the structural and non-structural elements during the nonlinear dynamic analysis. In the left figure, M_y and M_u represent the yield and ultimate moments and χ_y and χ_u are the corresponding curvatures. k_1 is the initial stiffness; k_2 and k_3 are the post-yield stiffnesses in the positive and negative directions where α is the strain-hardening ratio and k_4 is the unloading stiffness. In the model for the infill walls, k_1 represents the axial stiffness of the uncracked section, u_1 and u_2 are the cracking and ultimate deformations, respectively.

4.3 Nonlinear Dynamic Analyses Results

Initially, modal analyses are carried out for each building and natural vibration periods for both directions are calculated. Substituting the dominant vibration period of each building, a comparison of the acceleration response spectra of the simulated motions with the design spectrum given in Fig. 4.1 is made concerning satisfaction of condition (*iii*) in Sect. 4.2.2.

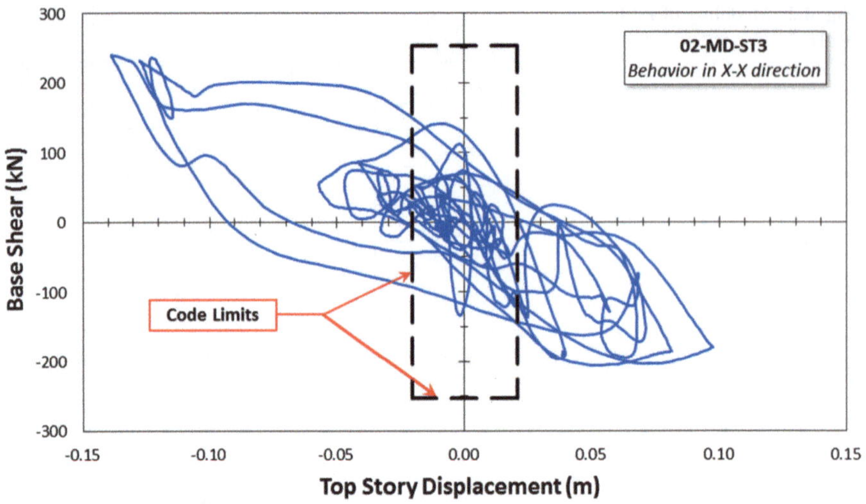

Fig. 4.4 Top story displacement-base shear variation of 02-MD-ST3 building with code limits

Next, nonlinear dynamic analyses under the effect of earthquake ensemble are carried out for the entire building stock. The top story displacements (U_{top}) versus base shear (V_{b}) demands for each building in each direction are obtained. As an example, 02-MD-ST3 coded building in X-X direction subjected to SimEQ-4 ground motion is given in Fig. 4.4. The code limit for story drift Δ_{i} and base shear demand V_{b} are also shown in the figure with dashed lines. In most of the buildings, base shear demands slightly exceeded the code limits. Furthermore, for the entire building set, it should be emphasized that the top story displacements significantly exceeded the code limits. These results can be attributed to weakness in structural stiffness and poor concrete quality causing increased displacement demands but decreased base shear results from low structural resistance.

Subjected to the simulated ground motion ensemble, a number of 315 time history analyses are carried out basically for a *PGA* of 0.4g and for each scaled ground motion these calculations are repeated. Envelopes for story displacement and drifts are obtained for the entire levels of scaled motions and the mean of the structural demands are obtained for each strong motion ensemble.

Table 4.3 shows the computed demands for story drifts Δ_i, base shear V_b and the overturning moment M_0 for each direction of each building on the ensemble for a *PGA* of 0.4*g*.

4.4 Comparison of the Results with Proposed Functions

Although various researchers focused on establishing vulnerability functions in order to estimate the probability of structural damage due to earthquakes, only few attempts to estimate the seismic vulnerability of Turkish RC buildings by use of

Table 4.3 Structural characteristics of the inspected buildings

Code name	W, kN	$U_{top(x\text{-}x)}$, m	$U_{top(y\text{-}y)}$, m	$V_{base(x\text{-}x)}$, kN	$V_{base(y\text{-}y)}$, kN	$M_{0(x\text{-}x)}$, kNm	$M_{0(y\text{-}y)}$, kNm
01-ND-ST4	17,100	–	0.124	–	1649.4	–	19574.1
02-MD-ST3	3,951	0.131	0.101	234.5	242.4	2966.4	3695.3
03-MD-ST5	10,751	0.126	0.139	1206.3	1239.8	19080.4	17809.6
04-MD-ST5	8,104	0.099	0.098	1059.2	1004.8	12875.3	12279.9
05-MD-ST3	7,677	0.124	0.076	733.3	1034.8	6889.7	10174.1
06-MD-ST4	6,383	0.166	0.146	636.2	527.9	5515.7	6359.1
07-MD-ST3	3,873	0.102	0.117	310	315	3539.8	3333.9
08-MD-ST5	7,859	0.031	0.15	380.8	335.7	5077.6	6370.4
09-MD-ST4	5,890	0.152	0.16	428.8	374.1	5345.2	4976.1
10-MD-ST3	4,009	0.066	0.063	32.3	35.2	24451.9	5766.9
11-MD-ST3	3,557	0.11	0.143	205.2	144.4	3233.7	2404.8
12-MD-ST4	5,732	0.147	0.142	429.9	512.3	5094.2	5918.8
13-MD-ST5	8,104	0.104	0.103	957.8	930.4	12,133	11781.3
14-MD-ST5	10,587	0.238	0.175	543	807.7	6142.6	9882.3
15-MD-ST5	7,384	0.195	0.145	223.8	367.1	4499.1	7791.9
16-MD-ST2	4,594	0.044	0.041	480.5	485.6	7007.8	5916.2
17-HD-ST3	5,789	0.063	0.047	1144.8	1161.4	8426.8	9741.5
18-ND-ST2	4,238	0.089	0.066	888.8	672.3	5208.6	4602.1
19-ND-ST8	11,155	0.234	0.155	792.3	1430.8	10,395	22065.7
20-ND-ST6	9,214	0.257	0.19	804.4	826.7	8412.3	8522.7
21-ND-ST2	10,087	0.083	0.071	1319.1	1651.3	9766.9	12836.9
22-ND-ST2	7,400	0.093	0.072	898.1	1128.4	6697.3	15,729
23-MD-ST6	8,208	0.159	0.147	1121.3	1109.7	11,428	11704.4

fragility curves have been made. In order to investigate the applicability of the proposed vulnerability functions to the typical building stock in Turkey, some of these fragility curves are employed for the building set and compared with the analytical results.

4.4.1 Damage Indices Defined by DiPasquale and Cakmak (1987)

In their study that primarily concerns damage assessment of buildings, DiPasquale and Cakmak mentioned several indices that can be proposed as measures of structural damage which are functions of the fundamental periods, $(T_O)_i$, estimated during an earthquake, as well as of the initial fundamental period $(T_O)_{initial}$ and the final one $(T_O)_{final}$.

The functional form of the indices may depend upon phenomenological aspects of damage at the local level, upon analytical considerations and upon the analysis of data recorded from damaged structures. They proved the possibility to compute averages of the local stiffness degradation using the vibrational parameters of the body, in particular $(T_O)_{final}$. According to their study by comparing the fundamental period $(T_O)_{initial}$ and $(T_O)_{final}$ before and after the earthquake, a measure of the structure's global stiffness degradation can be obtained. The index thus defined is called final softening and indicated with δ_f as given in Eq. 4.2:

$$\delta_f = 1 - \frac{(T_0)^2{}_{initial}}{(T_0)^2{}_{final}} \tag{4.2}$$

With the aim to investigate the applicability of this criterion in the ultimate state evaluation, modal analyses are renewed for the damaged state of the entire building set at the end of each nonlinear dynamic analysis and vibration periods are recalculated. Table 4.4 summarizes the results for the building set for each direction. Depending on the site-observed damage of the related structures and the damage indices values, it can be assumed that the degradation of the low-amplitude characteristics of existing structures can in fact be quantified by monitoring the periods of the vibration.

4.4.2 Empirical Functions Proposed by Rossetto and Elnashai (2003)

Empirical fragility curves for European-type RC building populations are derived in their study based on a data bank of 99 post-earthquake damage distributions observed in 19 earthquakes and concerning a total of 340,000 RC structures. The heterogeneous observational data are reinterpreted in terms of a damage scale,

Table 4.4 Damage index of the inspected buildings

Code Name	$T_{0,in,x\text{-}x}$ (s)	$T_{0,in,y\text{-}y}$ (s)	$T_{0,fin,x\text{-}x}$ (s)	$T_{0,fin,y\text{-}y}$ (s)	$\delta_{f,x\text{-}x}$	$\delta_{f,y\text{-}y}$
01-ND-ST4	–	0.469	–	1.753	–	0.928
02-MD-ST3	0.324	0.351	2.389	2.517	0.982	0.981
03-MD-ST5	0.502	0.544	1.7	1.788	0.913	0.907
04-MD-ST5	0.442	0.445	1.233	1.22	0.871	0.867
05-MD-ST3	0.439	0.319	1.721	1.248	0.935	0.935
06-MD-ST4	0.532	0.49	2.096	2.075	0.936	0.944
07-MD-ST3	0.356	0.379	1.673	1.729	0.955	0.952
08-MD-ST5	0.196	0.519	0.697	2.166	0.921	0.943
09-MD-ST4	0.539	0.557	2.255	2.61	0.943	0.954
10-MD-ST3	0.302	0.285	1.382	1.244	0.952	0.948
11-MD-ST3	0.393	0.421	2.542	2.899	0.976	0.979
12-MD-ST4	0.527	0.496	2.484	2.085	0.955	0.943
13-MD-ST5	0.451	0.454	1.351	1.298	0.889	0.878
14-MD-ST5	0.921	0.604	4.136	2.68	0.950	0.949
15-MD-ST5	0.855	0.545	3.082	2.379	0.923	0.948
16-MD-ST2	0.222	0.233	1.241	1.133	0.968	0.958
17-HD-ST3	0.341	0.258	0.897	0.806	0.855	0.898
18-ND-ST2	0.272	0.253	0.987	1.031	0.924	0.940
19-ND-ST8	0.821	0.638	5.878	2.006	0.980	0.899
20-ND-ST6	0.742	0.608	5.911	5.028	0.984	0.985
21-ND-ST2	0.309	0.275	1.412	1.206	0.952	0.948
22-ND-ST2	0.339	0.273	1.493	1.242	0.948	0.952
23-MD-ST6	0.571	0.56	1.959	1.886	0.915	0.912

namely homogenized reinforced concrete (HRC-scale), which is calibrated experimentally. The feasibility of using observation-based data for the generation of vulnerability functions for different strong ground motion parameters is investigated.

The functions most commonly used in existing relationships are cumulative normal and log-normal distributions. Rossetto and Elnashai (2003), after considering many alternatives, proposed a relationship in the form of Eq. 4.3, which is found to yield the optimum fit for all considered ground motion parameters.

$$P(d \geq DI_{HRC}|GM) = 1 - \exp\left(-\alpha.GM^{\beta}\right) \tag{4.3}$$

Using the parameters α and β, which define the mean curves according to the equation, empirical fragility relations were developed for six different HRC damage states for a ground motion parameter, PGA. Figure 4.5 illustrates the comparison of the building set results with the proposed vulnerability functions. It can be observed from the figure that highly vulnerable existing structures to seismic actions in Turkey experience higher damage under the effect of average-values of PGA.

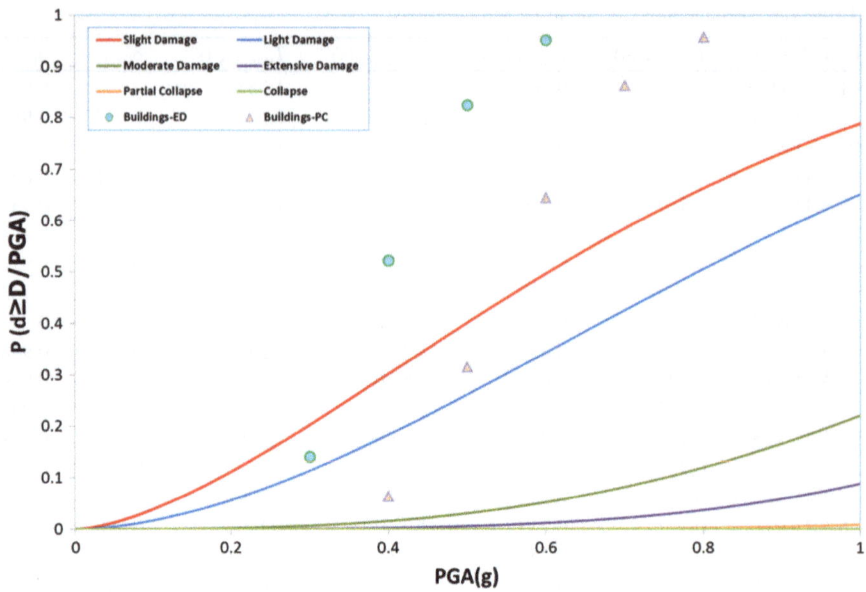

Fig. 4.5 Comparison of the analyses results with the proposed curves for different damage states

4.4.3 Analytical Functions Proposed by Kircil and Polat (2006)

Analytical fragility curves for mid-rise RC frame buildings in Istanbul, which have been designed according to the 1975 version of the TERDC, are developed based on numerical simulation with respect to the number of stories of the buildings. For this purpose, 3-, 4-, 5- and 6-storey representative buildings are designed according to the above-mentioned design code. Incremental dynamic analyses are performed using 12 artificial ground motions and the yielding and collapse capacities of each sample building are determined. Based on those capacities, vulnerability functions are developed in terms of different ground motion parameters such as *PGA*, S_a, S_d and inter-story drift ratio (*ISD %*) with lognormal distribution assumption.

Figure 4.6 depicts the comparison of analytically derived fragility functions for different story numbers and nonlinear dynamic analyses results of buildings having similar heights. It can be clearly seen that, the distribution of the analyses results for each building in the ensemble significantly differ from the proposed functions for the buildings.

4.5 Conclusion

Existing vulnerability relationships developed for RC structures are investigated in this study by means of their adequacy to reflect the most reliable actual damage state for existing buildings of Turkey. A number of 23 frame buildings which suffered

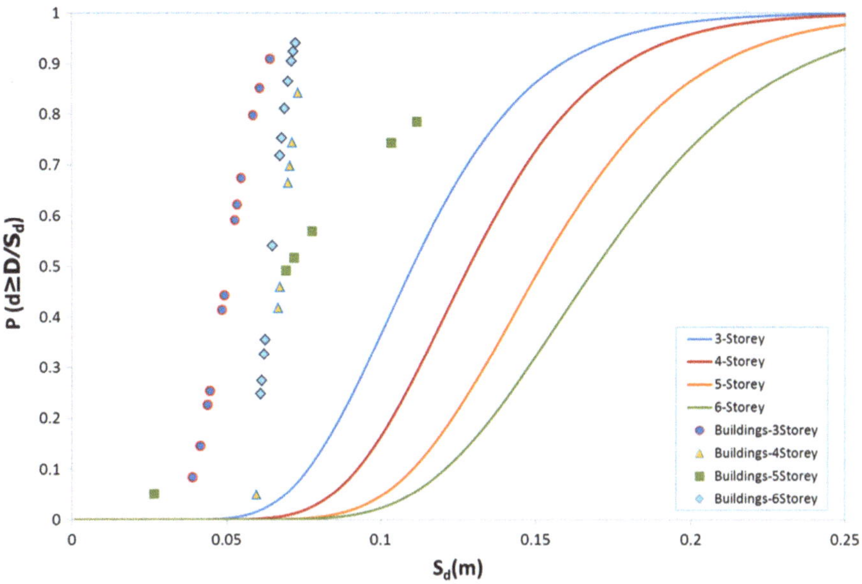

Fig. 4.6 Comparison of the analyses results with the proposed curves for different story numbers

various levels of structural damage during major destructive Turkish earthquakes are selected considering similarities in seismic zone and local site conditions. Structural demands are computed using nonlinear dynamic analyses under the effect of an earthquake ensemble consisting of seven simulated strong motions. Later, vulnerability functions proposed by various researchers are investigated and three of the fragility functions are selected and compared with the structural demands and responses calculated by the nonlinear dynamic analysis solutions.

One of the main findings of this research is that existing buildings with the general characteristics of common structure types in Turkey have significantly high values of displacement demands due to their weaknesses in stiffness, but low values of base shear demand, depending on their strength deficiencies and how poor concrete quality is. A second finding is the poor agreement between those fragility functions and the analytical results of the building set. To investigate further this bias, it seems necessary to increase the amount of buildings in the analysis set in future studies.

The results of the investigation also showed that estimating vulnerability curves by taking into account different structural parameters, such as base shear resistance capacity, is necessary, because the building stock in Turkey is deemed unsafe. Its deficiencies stem from the lack of control mechanism during construction for most of the structures constructed prior to the National Construction Control and Supervising Law.

References

Akkar S, Sucuoglu H, Yakut A (2005) Displacement based fragility functions for low- and mid-rise ordinary concrete buildings. Earthq Spectra 21(4):901–927

Al-Chaar G, Lamb G (2002) Design of fiber-reinforced polymer materials for seismic rehabilitation of in-filled concrete structures. ERDC/CERL TR-02-33, US Army Corps of Engineers

Ascheim M (2005) http://nisee.berkeley.edu/software/drain2dx

DiPasquale E, Cakmak AS (1987) Detection and assessment of seismic structural damage. Technical report NCEER-87-0015, State University of New York, Buffalo

Erberik MA, Elnashai AS (2004) Fragility analysis of flat-slab structures. Eng Struct 26(7):937–948

Kircil MS, Polat Z (2006) Fragility analysis of mid-rise R/C frame buildings. Eng Struct 28(9):1335–1345

Papageorgiou A, Halldorsson B, Dong G (2000) Target acceleration spectra compatible time histories –TARSCTHS user manual. V:1.0, State University of New York-Buffalo, Engineering Seismology Laboratory report, New York

Prakash V, Powell GH, Campbell S (1993) DRAIN-2DX Base program description and user guide, SEMM report, UCB/SEMM-93/17, University of California, Berkeley

Rossetto T, Elnashai AS (2003) Derivation of vulnerability functions for European-type RC structures based on observational data. Eng Struct 25(10):1241–1263

TERDC (2007) Turkish earthquake resistant design code. Ministry of Public works and Settlement, Ankara

Chapter 5
Probabilistic Path Finding Method for Post-Disaster Risk Estimation

Florin Leon and Gabriela M. Atanasiu

Abstract One of the main problems in an immediate post-disaster situation is to find the shortest path between command centres or/and important life care institutions and affected areas, respectively. In this paper, a modified A* search algorithm is proposed, that can find the shortest path between any two points on a map, weighted by the damage probability of the existing infrastructure situated in different locations of an urban area. The proposed approach combines the risk probabilities given by the fragility curves of some relevant constructions located in the disaster area with the deterministic search algorithm. In this case, the real costs provided to the A* algorithm are replaced with expected costs, which are estimated in a stochastic framework. The concept is exemplified on the case study of an urban sample identified in Iaşi, a city of around 300,000 inhabitants, located in the North Eastern region of Romania, exposed to repetitive earthquakes with a recurrence period of around 35–40 years. Thus, probabilistic scenarios can be created for emergency interventions, based on previously recorded local values of Peak Ground Acceleration (PGA). Moreover, the proposed routes for the emergency intervention teams in the post-disaster stage can be visualized on a GIS map, shown actually in our case study. In case of a real extreme event, the information about the proposed routes can be updated in real time, as new data are collected in the field and transmitted to the decision centre.

F. Leon • G.M. Atanasiu (✉)
Multidisciplinary Center of Structural Engineering and Risk Management,
"Gheorghe Asachi" Technical University of Iasi, 43 Bdul. Dimitrie Mangeron,
Iasi 700050, Romania
e-mail: florinleon@gmail.com; gabriela.atanasiu@gmail.com

A. Ilki and M.N. Fardis (eds.), *Seismic Evaluation and Rehabilitation of Structures*, Geotechnical, Geological and Earthquake Engineering 26,
DOI 10.1007/978-3-319-00458-7_5,

5.1 Introduction

Disaster risk reduction measures are designed to protect the lives and assets of communities and individuals from the impact of hazards by mitigation and preparedness (Concern 2007). The meaning of mitigation is to make the consequences of disasters less severe. Mitigation measures aim at reducing the frequency, scale, intensity and impact of hazards, and are typically thought of as being physical in nature, for example strengthening and retrofitting buildings. Preparedness plans can be considered contingency plans for when a hazard overwhelms the capacity of a community and any mitigation measures that may have been put in place. They are usually knowledge-based and include early warning systems that monitor and predict the occurrence of hazards, and plans for effective response and recovery.

In a post-event situation, one of the most important problems that intervention teams face is finding not necessarily the shortest path, but the quickest or the safest path between locations, for example between the command centres such as the Prefecture or the Fire Department headquarters and the affected areas, and from there to the hospitals or to other similar facilities. In the immediate circumstances after an earthquake, there are not enough data to precisely know the damage state of buildings or lifeline systems from an existing infrastructure, but one can estimate the damage probabilities based on fragility curves.

Given the fact that immediately in a post-event situation there could be additional damage because of partial or total collapse of some buildings, the methodology presented in the paper can be applied to improve the rescue planning and thus to reduce the risk for people and other infrastructure.

This chapter is organized as follows. Sections 5.2 and 5.3 concentrate on the process of risk mitigation, with a focus on the short period of time after an earthquake. Section 5.4 describes the modification of an artificial intelligence algorithm, A* search, in order to find safe post-disaster routes for the intervention teams. Section 5.5 presents the results of the case studies and Sect. 5.6 contains the conclusions of our work.

5.2 Related Work

The need to help the intervention or evacuation teams by creating computational models is an actual research direction.

Several studies have focused on methods to improve the planning and operational aspects of the evacuation process to maximize the utility of the existing transportation network (Han et al. 2006). Stepanov and Smith (2009) also review a series of evacuation models and suggest that modeling techniques can be grouped by computational techniques into analytical and simulation techniques. One class uses analytical optimization methods to offer routing policies, and then these are evaluated with traffic simulation models. The other one defines a set of optimal routes and evaluates their performance measures simultaneously.

Sari and Kubat (2012) describe a model that provides an index of intervention designed to save maximum life within shortest term and with limited economic resources in case of an earthquake. In the first phase of the model, the definition of the vulnerability and calculations of the number and locations of people in danger are provided. The second phase of the process is the evaluation of the blockage risks of the roads within a network because of the collapsed buildings. Afterwards, predictions of the major routes that people use frequently to reach the city main road network and major destinations in the urban configuration are estimated with space syntax theory. The outputs of the model are the "road risk" and the "index of intervention". Each value provides information for total risk assessment and intervention priorities against the earthquake risk. The model is applied to a case study from Istanbul, Turkey.

Campos et al. (2012) present a heuristic algorithm that can be applied for defining some independent routes for developing the evacuation plans. The basic idea is to use interactively an algorithm for finding the two independent paths in the network from the disaster area to each destination, with the least travel time, associating to each path an index of capacity/time that facilitates the evaluation of each route in comparison to the others.

To collect data from the affected areas more efficiently, some works have studied hybrid networks for data collection in disaster situations. These systems employ cellular systems and sensor networks in parallel to achieve superior performance, such as high speed, high capacity and wide area coverage. Liu et al. (2011) propose an idea which utilizes the surviving time interval of sensor nodes, namely the duration in which wireless sensor networks still function after the disaster, to transmit vital data to the sensor nodes in the safe zone. It relies on the observation that sometimes the buildings or local resources do not get damaged or destroyed at the beginning of most disasters, and the deployed sensor network can keep working for a while before it becomes paralyzed. This period can be used to transfer vital data gathered by the networks.

Besides technical issues, the psychological and emotional traumas associated with disasters are also important and have been studied, e.g. by Ehrenreich (2001).

5.3 Fragility Curves

Fragility curves of different structures or structural elements, also known as damage functions, are used to approximate structural damage from natural hazards. They represent a measure of vulnerability or an estimate of overall seismic risk, developed so far by using different methods such as heuristic, empirical, analytical ones or a combination of these.

Immediately after an earthquake, there could be not enough data to help the intervention teams, therefore the probabilities from the fragility curves can be used

Table 5.1 Acceptable values for the φ coefficient

	Solution quality		
Damage	*Standard error*	*Correlation coefficient*	Coefficient φ
Minor	0.0086499	0.9997215	0.007836079
Moderate	0.0102153	0.9996280	0.020117283
Major	0.0100495	0.9996191	0.037445433
Collapse	0.0125385	0.9994262	0.10592203

to make initial estimates. Of course, afterwards, as actual information comes from the field, the state of the problem can be updated and the path finding results can be revised in real time.

In a previous work (Leon and Atanasiu 2007), we noticed that many fragility curves have a typical shape, similar to a sigmoid function. Through regression, we estimated them with this simplified analytical model which takes into account only one coefficient:

$$y(x) = \frac{x^{\mathrm{e}}}{\varphi + x^{\mathrm{e}}}. \tag{5.1}$$

Because the model is very flexible, we can increase or decrease the coefficient φ, and cause the fragility curves to move to the left or to the right, and thus the probabilities themselves change. Also, this model can account for dynamical changes during the lifetime of a building, caused by damages or repairs, which increase or decrease this coefficient. The corresponding values of the coefficient for the four damage levels of the fragility curves as suggested by the HAZUS methodology (FEMA 1999) are displayed in Table 5.1.

In our case, we consider that each building has its own value for the coefficient. Thus, each building behaves differently during the simulation.

We used this model in order to provide the probabilities for the buildings in the case studies, in which we used a sample area from Iaşi, Romania with almost 600 buildings. Also, a GIS based software, namely *NetSET* (Data Invest 2012), was used, which has the advantage of having access to all the buildings in the city, the lifelines and so on. For example, in *Google Maps* or *OpenStreetView*, the streets of Iaşi are present, but not all the buildings, and therefore *NetSET* provides a full model.

5.4 A Search Algorithm for Safe Route Finding in Emergency Situations

In this section, a modified A* search algorithm will be presented, which can find the shortest path between any two points on a map, weighted by the damage probability of the existing infrastructure situated on different locations of the urban area.

One of the simplest search strategies is to minimize the estimated cost to reach the goal. That is, the node in the search graph whose state appears to be the closest to the goal state is always expanded first. Using a good heuristic function, the space and time complexity can be substantially reduced, and the amount of reduction especially depends on the quality of the heuristic function. The distinctive feature of A* (Hart et al. 1968) is that it uses the sum between the cost of the path travelled so far from the initial state and the heuristic estimation of the cost of the path from the current position to the goal to guide the search process. It is proven that A* is complete in the sense that it will always find a solution if there is one; this should be an optimal one because it finds the lowest-cost solution. It is also optimally efficient, that is, no other optimal algorithm is guaranteed to expand fewer nodes than A* (Russell and Norvig 2009).

The goal of the A* algorithm is to find the shortest route between two points. In our case, on the street graph, the distances are known, but the graph is so complex that there are exponentially many possibilities of traversing the graph. What we need to do is to find a way to simplify this process of finding the optimal path. This is where heuristic estimates are used to guide the process, so that the best, most promising options can be tried first.

For route-finding, the nodes of the graph are the states of the problem, i.e. the intersections of the roads. The arcs are the actual road segments, and each arc has a cost, for example the length of a road segment.

The main idea of the algorithm is to sort the nodes that are to be expanded at a certain moment in decreasing order of the f function, such as for a node n:

$$f(n) = g(n) + h(n), \tag{5.2}$$

where $g(n)$ is the known, best cost so far for node n, and $h(n)$ is the estimated cost of getting from node n to the goal.

Therefore, at each step, the next node to be selected for expansion is the one with the minimum value of the f function.

A* is a general search algorithm. It is useful for route finding, which is our case, but it can be used in any search problem, for example solving the Rubik cube, booking flight or train tickets, any kind of situation when we can have some information about the distance between a state and the goal. In this general setting, there is one property that the heuristic functions must have: to be admissible, or optimistic, their values should always be smaller than the actual cost. In our case, when we use a straight-line distance we can be sure that it is less than the real distance because a road segment can also have curves and thus it is guaranteed that the straight-line distance is always smaller.

The proposed approach is combining the risk probabilities given by the fragility curves with the deterministic search algorithm. In this case, the real costs provided to the A* algorithm are replaced with costs estimated in a stochastic framework. Therefore, by applying the concept on the case study of an urban sample identified in Iaşi, we can create probabilistic scenarios for emergency interventions, based on previously recorded local values of Peak Ground Acceleration (PGA). Moreover, the proposed routes for the emergency intervention teams in the post-disaster stage can be visualized on a GIS map. In case of a real event, the information about the proposed routes can be updated in real time, as new data are collected in the field and transmitted to the decision centre.

By using geospatial data, one can develop useful scenarios to improve the knowledge on the structural vulnerability of the urban built infrastructure. Through the analysis of urban seismic risk for different classes of seismic scenarios, the city stakeholders can create a risk management plan for defining the actions for rehabilitation needs in order to improve the seismic safety of constructions and important infrastructure lifelines, thus improving the institutional response and the preparedness of the population in case of a possible earthquake.

5.4.1 Deterministic Scenarios

In order to find an optimal path between two locations, we modified the problem of A* by introducing the so-called *induced costs*. The costs are the real distances in A*, but when a road segment is affected by an earthquake, it should appear to increase its length, in order to make it less attractive for the algorithm. For example, if the road is completely blocked, its distance should go to infinity. We define a vicinity, a neighborhood of a collapsed building and if it intersects a road segment, it will affect that part of the road. We must emphasize the fact that this simulation is pixel-based, not road-segment-based. So we know that a road segment is affected, but we also know where and to what degree. This fact increases the number of states, and in the simulations there are almost 2,000 states, yielding quite a large search graph.

Let c be the actual cost, as used by the classical A* algorithm. Let d be the distance from a building to the road. We denote by $I(c,d)$ the induced cost which will be used by the algorithm instead of simple distances. When defining the I function, it is clear that it must obey the following requirements:

- when $d = 0$ the damage must have a maximum effect, therefore $I(c,0) = c \cdot n$;
- when $d = d_{max}$, the damage must have no effect, therefore $I(c,d_{max}) = c$.

We consider that the collapse of the building will affect the road up to a certain distance, d_{max}. Beyond that distance we consider that there is no effect and closer to the building the effect is greater, with a linear relationship. Of course, this can be changed, to make it quadratic or any other type of relationship. The equation is a sum because if there are for example two buildings on opposite sides of the road, they can affect the road together.

Since t buildings can affect a road segment, the equation used to compute the induced costs, i.e. the modified distances, is the following:

$$I(c,\mathbf{d}) = c \cdot \sum_{i=1}^{t} ((d_{max} - d_i) \cdot n), \qquad (5.3)$$

where **d** is the vector of all the distances from the collapsed buildings to the road and n is a scaling factor, which was empirically considered in our case to be 5.

Under these assumptions, each scenario can generate a different route.

5.4.2 *Probabilistic Scenarios*

In order to apply the probabilistic A*, one must take into account the *expected costs* (or distances) instead of the real ones.

Since the main goal is to minimize the time to arrive to the point of intervention, we consider that the expected distance will increase as the buildings in the vicinity are affected, because the actual speed of the car will decrease. If the road is completely blocked, the distance of that road segment will increase to infinity.

If a building has some minor or moderate damage, it is considered that it will not have much influence of the intervention vehicles using the roads nearby, however in some cases it may affect them. If the buildings have some major damage or have collapsed, debris can block part of the road and this may pose problems for the intervention teams.

We propose a statistical approach in order to give an estimate of the building vulnerability, unique for a specific PGA, using fragility curves. We follow the HAZUS methodology (FEMA 1999) with 4 levels of damage: minor, moderate, major and collapse, corresponding to 5 probabilities P_i, the 4 above and 1 for no damage:

$$\sum_{i=1}^{5} P_i = 1. \tag{5.4}$$

We developed the model with the induced costs by using these weighted probabilities. The d part is the same as in the deterministic case, and we also use the p part with probabilities. The weighted probability is a number from 0, when a building is not affected, up to 1 (but it will never get to 1), because the sum of probabilities is 1 and they are weighted by coefficients which are less than 1.

We are mainly interested in the last two; therefore we compute the *expected probability of a damage level* with the relationship given as follows:

$$E(P_{d_i}) = w_i \cdot (P_i - P_{i+1}), \tag{5.5}$$

where the weight w_i is the weight of a certain state, with the following empirical values considered in increasing order of damage effect: $\mathbf{w} = \{0,\ 0.1,\ 0.2,\ 0.5,\ 1\}$.

In Eq. 5.5, $P_i - P_{i+1}$ is the probability of the building being in a certain state, and we weigh this probability. For example, the collapse state is the most important; therefore its weight is 1. The major damage state has a coefficient of 0.5. And so on, for the moderate and minor damage states, and finally no damage. Of course, these weights are arbitrary and can be changed.

One can notice that:

$$\sum_{j=1}^{5} P_{d_j} < 1. \tag{5.6}$$

In a similar way to the deterministic case, we should assume that when $d = 0$, the damage must have a maximum effect, and when $d = d_{max}$, there should no longer be an effect.

Correspondingly, when $E(P_d) \approx 0$, the damage should have no effect, and when $E(P_d) \approx 1$, it should have its maximum effect.

We therefore define the *expected cost* as:

$$E(c, \mathbf{d}, \mathbf{P}) = c \cdot \sum_{i=1}^{t} ((d_{max} - d_i) \cdot n \cdot E(P_{d_i}) \cdot m), \tag{5.7}$$

where n and m are scaling factors, empirically chosen as $m \cdot n = 10$ in the following case studies.

5.5 Case Studies

For the deterministic scenarios, we considered three simulated artificial earthquakes with different Peak Ground Accelerations. Based on different levels of the PGA, we can make different seismic scenarios regarding the degree of damage of the buildings in a selected area of the city, presented in Fig. 5.1.

Landmark buildings (e.g. the Palace of Culture), churches, patrimony buildings, other public-use structures were not considered in the simulation. They may have special anti-seismic designs or they may behave in a different way from residential buildings. These buildings are displayed in light gray.

5.5.1 Deterministic Scenarios

In order to analyze the performance of the route finding algorithm, we considered different seismic scenarios, where each building has a damage state associated with it. In almost white we present the buildings not affected, in light gray the buildings with minor damage, in gray buildings with moderate damage, in dark gray buildings with major damage and in black the collapse state.

Figure 5.2 presents the first scenario of an earthquake with a PGA of 0.2g.

It is important to note that in these scenarios, the probabilities of each building are computed independently. Therefore, the damage states are just samples of the probability distribution that define the overall behavior of the building.

Another scenario based on the same PGA is presented in Fig. 5.3.

Fig. 5.1 The selected city area used to develop emergency scenarios

Fig. 5.2 The first scenario with a PGA value of 0.2g

Fig. 5.3 The second scenario with a PGA value of 0.2g

For the same value of PGA (in this case 0.2g), one can see that the results are different. The overall aspect of the two trials is similar: most buildings are not affected. However, there are some changes: buildings that are not affected in one trial can have minor or moderate damage in the other trial, depending on the fragility curves.

When the PGA value increases, the percent of affected building also increases. Figures 5.4 and 5.5 present two different scenarios for a PGA of 0.35g. In this case, one can see that more buildings are affected, and most of them are in at least a minor damage state.

When the PGA increases, e.g. 0.5g, several buildings reach the point of collapse, as shown in Figs. 5.6 and 5.7.

Regarding the actual path-finding, the results of the algorithm are the same as in the deterministic case for a small earthquake, as shown in Fig. 5.8.

For a larger earthquake in terms of accelerations and PGA, the building near the first entry collapses. The first route went through that area. The corresponding road segment is affected and it would be safer for the intervention team to go the other way. Therefore, the route is changed, as displayed in Fig. 5.9.

5.5.2 Probabilistic Scenarios

As stated above, the previous case studies were based on trials, and in each scenario the results were different. Therefore the deterministic approach is useful to devise

Fig. 5.4 The first scenario with a PGA value of 0.35g

Fig. 5.5 The second scenario with a PGA value of 0.35g

Fig. 5.6 The first scenario with a PGA value of 0.5g

Fig. 5.7 The second scenario with a PGA value of 0.5g

Fig. 5.8 The results of the path-finding algorithm for a small earthquake with a PGA value of 0.2g

Fig. 5.9 The results of the path-finding algorithm for a larger earthquake with a PGA value of 0.5g

Fig. 5.10 The results of the path-finding algorithm taking into account expected fragility for a hypothetical major earthquake with a PGA value of 1.2g

different scenarios as drills of intervention in the preparedness phase. In case of an actual earthquake, this approach cannot give much information about which parts of the city would be particularly damaged. If we repeated the trials over and over again, they would give us a statistical distribution. However, this statistical distribution is embedded in the fragility curves. So we can compute the expected level of damage where we weight the probabilities in the fragility curves.

Thus, we can redraw the map of the city, as shown in Fig. 5.10. In this final case study, the earthquake is supposed to be very strong, with a PGA of 1.2g. In this figure, we also represent the expected fragility, weighted by the coefficients above. This representation is unique for a given PGA and fragility curves of buildings. Since this presents a probabilistic approach, the meaning of the colors is different. For each building we have a number associated with it between 0 and 1. We divided this interval into 5: from 0 to 0.2 the buildings are almost white, from 0.2 to 0.4 they are light gray, then gray, dark gray and finally black. Of course, these boundaries can be changed. By applying the safe path finding algorithm, it can be seen that the route changes once again.

5.6 Conclusions

The paper presents an application of an artificial intelligence algorithm for the mitigation of seismic risk in urban dense areas, in the early stages of a post-event emergency situation. We presented two cases, the deterministic and the probabilistic methods. The deterministic evaluation is based on scenarios for three simulated earthquakes. The probabilities given by fragility curves can be weighted to compute the expected level of damage. The underlying model was based on the idea of increasing the distance of a road segment affected by the nearby buildings. It was shown that routes can change depending on the scenario. The results can be displayed on a GIS map and be used to increase the quality of decision making in post-event situation or for making informed decisions about the rehabilitation of critical buildings and for improving the emergency planning scenarios.

Regarding the future development, actual data about the buildings can be integrated into the model in order for the methodology to be useful in a real situation. Since the approach is very flexible, the damage of the lifelines can also be included into the present model, taking into account the gas stations or other industrial facilities and critical infrastructure.

Acknowledgments This research leading to these results was supported in part by the European Community's Seventh Framework Programme [FP7/2007–2013] under grant agreement no. 227887 [SERIES].

References

Campos V, Bandeira R, Bandeira A (2012) A method for evacuation route planning in disaster situations. In: Proceedings of the 15th edition of the Euro Working Group on Transportation. http://www.lvmt.fr/ewgt2012/compendium_107.pdf. Accessed 7 Sept 2012

Concern Worldwide US Inc (2007) Approaches to disaster risk reduction. Emergency Unit, September 2005. http://www.concernusa.org/media/pdf/2007/10/Concern_ApproachestoDRR%2520paper%2520-%2520final.pdf. Accessed 7 Sept 2012

Data Invest Iaşi (2012) http://www.datainvest.ro. Accessed 7 Sept 2012

Ehrenreich JH (2001) Coping with disasters: a guidebook to psychosocial intervention, rev edn. http://www.toolkitsportdevelopment.org/html/resources/7B/7BB3B250-3EB8-44C6-AA8E-CC6592C53550/CopingWithDisaster.pdf. Accessed 7 Sept 2012

FEMA (1999) HAZUS, Earthquake loss estimation methodology. Technical Manual, National Institute of Building Sciences for the Federal Emergency Management Agency

Han L, Yuan F, Chin S, Hwang H (2006) Global optimization of emergency evacuation assignments. Interfaces 36(6):502–513

Hart PE, Nilsson NJ, Raphael B (1968) A formal basis for the heuristic determination of minimum cost paths. IEEE Trans Syst Sci Cybern 4(2):100–107. doi:10.1109/TSSC.1968.300136

Leon F, Atanasiu GM (2007) Estimating fragility curves of buildings using genetic algorithms. In: Proceedings of the 9th international conference on the application of artificial intelligence to civil, structural and environmental engineering. Civil-Comp Press, Stirlingshire

Liu M, Gong HG, Wen YG, Chen GH, Cao JN (2011) The last minute: efficient data evacuation strategy for sensor networks in post-disaster applications. In: Proceedings of the 2011 I.E.

conference on computer communication (INFOCOM 2011), Shanghai. http://www.ntu.edu.sg/home/ygwen/Paper/LGWCC-Infocom-11.pdf. Accessed 7 Sept 2012
Russell S, Norvig P (2009) Artificial intelligence: a modern approach, 3rd edn. Prentice Hall, Upper Saddle River
Sari F, Kubat AS (2012) Syntactic properties of evacuation and access routes in earthquake vulnerable settlements. In: Greene M, Reyes J, Castro A (eds) Proceedings of the eighth international space syntax symposium, Santiago de Chile PUC. http://www.sss8.cl/media/upload/paginas/seccion/8038.pdf. Accessed 7 Sept 2012
Stepanov A, Smith J (2009) Multi-objective evacuation routing in transportation networks. Eur J Oper Res 1498:435–446

Chapter 6
Seismic Behavior of Thin-Bed Layered Unreinforced Clay Masonry Shear Walls Including Soundproofing Elements

Christophe Mordant, Matt S. Dietz, Colin A. Taylor, André Plumier, and Hervé Degée

Abstract According to the current standards, unreinforced masonry may only be used in regions of low seismicity as the material for the lateral-load resisting system. This requirement may be too safe-sided and leading to not cost-effective solutions for moderately seismic regions. This chapter presents overview of experimental results from shake table tests on unreinforced masonry shear walls carried out in the EQUALS Laboratory of Bristol University, in order to assess, and possibly enhance, the current seismic design rules. The study also includes as additional parameter the presence of soundproofing devices required in buildings with numerous dwellings, in order to achieve the acoustic isolation recommended by recent standards. In practice the required level of acoustic isolation is obtained by locating horizontal rubber layers in the wall. These layers are likely to influence significantly the dynamic response of the wall and hence of the whole structure under seismic actions. Tests are performed on walls realized with masonry units and construction methods typical of North-Western Europe.

C. Mordant (✉) • H. Degée
Department of Architecture, Geology, Environment & Constructions, University of Liege, Chemin des Chevreuils, 1, Liege 4000, Belgium
e-mail: cmordant@ulg.ac.be; h.degee@ulg.ac.be

M.S. Dietz • C.A. Taylor
Department of Civil Engineering, University of Bristol, Queen's Building, University Walk, Bristol BS8 1TR, UK
e-mail: m.dietz@bristol.ac.uk; colin.taylor@bristol.ac.uk

A. Plumier
Department of Mechanics of Materials and Structures, University of Liege, Chemin des Chevreuils, 1, Liege 4000, Belgium
e-mail: a.plumier@ulg.ac.be

A. Ilki and M.N. Fardis (eds.), *Seismic Evaluation and Rehabilitation of Structures*, Geotechnical, Geological and Earthquake Engineering 26,
DOI 10.1007/978-3-319-00458-7_6, © Springer International Publishing Switzerland 2014

6.1 Introduction

For centuries masonry is a traditional construction material for several types of buildings, like private dwellings, public buildings, churches, etc. Historically, the design of masonry structures has been relying on good practice. The contribution of engineers is often limited, usually leaving the task of designing the buildings to architects and builders. However, the contribution of structural engineers is nowadays increasingly necessary. Indeed, masonry structures are more and more used at the limit of the material capacity, for example with compression stresses under service loads higher than in the recent past (e.g. through the increased use of multi-storey apartment buildings or of lightweight concrete blocks in houses). Moreover, ecological and economic considerations require a more efficient use of materials. All these considerations are behind "Eurocode 6 – Design of masonry structures", where modern design methodologies and verification rules are proposed.

The consideration of earthquakes and their impact requires an additional understanding of the structural behavior under horizontal dynamic actions. The basic principles and the consequent analysis and design methodologies are presented in (Tomazevic 1999). These general considerations need however to be specifically transposed to each particular type of masonry structure. In this perspective, several research works have been carried out in the past 15 years (Calìo et al. 2012; Nakagawa et al. 2012; Milani et al. 2009). None of the identified references, however, deals especially with one the most common type of masonry structural elements in North-Western European areas, i.e. relatively thin bearing walls (from 140 to 200 mm) with high strength units (compression resistance up to 15 N/mm^2 or more) working at a very high compression ratio under service loads, implemented more and more with horizontal thin-bed layered joints and open "tongue and groove" vertical joints for construction efficiency.

Moreover, the spreading use of unreinforced masonry for load bearing in multistorey apartment buildings gives rise to new regulations for the acoustic and thermal performance of buildings. Some of these rules can have an important impact on the structural behavior; so, their consequences need to be characterized and quantified. Consequently, tests on unreinforced masonry walls have been carried out in the Earthquake and Large Structures Laboratory (EQUALS) at the University of Bristol. The present test series investigates the dynamic response of four simple unreinforced clay masonry walls, two of them including soundproofing devices ("*rubber layers*").

The experimental study has a double aim. On one hand, it is expected to provide a better understanding of the general behavior of single walls in dynamic conditions, in order to calibrate theoretical models and extend the conclusions to the global structural response of the entire building. On the other hand, the comparison between the structural response of walls with the same overall geometry, with and without soundproofing rubber devices, will assess the consequences of this type of device on seismic behavior.

This chapter presents the direct observations derived from the experimental measurements and the resulting conclusions.

6.2 Description of Test Specimens

6.2.1 Mechanical and Geometrical Characteristics

The units used to realize the walls are "*Zonnebeke POROTHERM*" (Wienerberger 2012). The walls are built with thin-bed layered masonry and "*PORO+*" glue-mortar. Vertical joints are empty tongue and groove systems (see Fig. 6.1). The unit size is: $(Length \times Width \times Heigth \equiv 300\ mm \times 138\ mm \times 188\ mm)$. Mechanical characteristics of the units and masonry are:

- Normalised compressive strength of units per EN 772-1, Annex A: $f_b = 13.0\ N/mm^2$
- Measured characteristic masonry compressive strength per EN 1052-1: $f_k = 5.6\ N/mm^2$
- Characteristic compressive strength per EN 1996-1-1: $f_k = 4.2\ N/mm^2$
- Characteristic compressive strength per NBN-EN 1996-1-1: $f_k = 3.9\ N/mm^2$

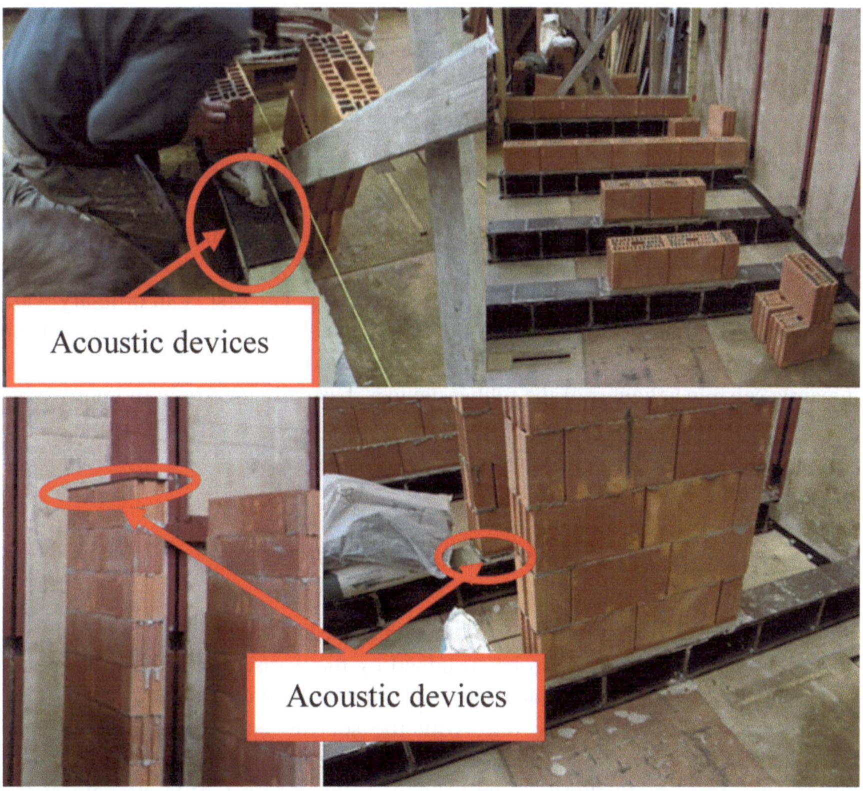

Fig. 6.1 Wall specimens with soundproofing devices at top and bottom

No specific characterization has been carried out for shear behavior. Usual standard values will thus be considered for further assessment:

- Initial shear strength per NBN-EN 1996-1-1: $f_{vk0} = 0.3\ N/mm^2$
- Characteristic shear strength per NBN-EN 1996-1-1: $f_{vk} = 0.5f_{vk0} + 0.4\sigma_d \leq 0.045f_b (= 0.585\ N/mm^2)$

Characteristics of the rubber (layer of 10 mm) can be found in (Wienerberger 2012) in terms of dimensions, acoustic and mechanical properties. Different densities of the rubber are available, depending on the acoustic performance targeted. It is herein equal to 810 kg/m^3.

6.2.2 Description of the Walls

The geometry of the walls is chosen in such a way to focus on different failure modes. The first mode is a failure in shear and corresponds to a wall with an aspect ratio close to 1 ("*Long wall*"). The second one is a failure in bending. It corresponds to a wall with an aspect ratio close to 0.4 ("*Short wall*"). Exact dimensions of the two generic types of walls are the following:

- *Long wall*: Length × Height × Width = 2.1 m × 1.8 m × 0.14 m
- *Short wall*: Length × Height × Width = 0.72 m × 1.8 m × 0.14 m

The structural floor load is simulated by an additional mass of 5 tons on top of the wall. The resulting average compressive stress in the wall is about 0.5 MPa for the short wall and about 0.17 MPa for the long one. These values are in the usual range of the compression level for this type of masonry.

Two different walls are built for each aspect ratio. The first is an ordinary wall, while the second one includes soundproofing devices (rubber layers) at the bottom and the top (Fig. 6.1). This will allow the comparative study of the influence on dynamic seismic behavior of the rubber devices, basically used only to improve the acoustic performance.

6.2.3 Preliminary Assessment of the Design

Prior to the tests, a theoretical assessment has been carried out to estimate the maximum acceleration that can be sustained by the walls, based on the structural models commonly used for structural seismic assessment of masonry elements.

The model is a very basic one, with the seismic action represented by an equivalent static horizontal load V. The procedure consists in the calculation of the compressive length through equilibrium, assuming no tensile strength at the interface between the base of the wall and the foundation. A linear distribution of compressive stresses is assumed at the interface. Once the compressive length is

Table 6.1 Properties of test specimens

Wall length [m]	Compressive load [kg]	Mean compressive stress [MPa]	$V_{\text{Rd,max}}$ [kN]	$a_{g,\text{ max}}$ [m/s^2]	Stiffness [N/m]	Frequency [Hz]
2.1	5,000	0.17	26.09	2.15	58,690,000	17.24
0.72	5,000	0.5	8.97	0.72	3,920,000	4.46

known, consequent verifications are made to check the level of shear stress, compression of the units and global overturning of the wall. This procedure results in an estimate of the maximum allowable equivalent horizontal load, $V_{\text{Rd,max}}$. The maximum allowed acceleration is then derived considering the wall as a Single-Degree-of-Freedom (SDOF) system: the value of the maximum allowed horizontal load is divided by the mass applied at the wall top and an amplification factor of 2.5, according to the 5 %-damped acceleration response spectrum of Eurocode 8 in the constant spectral acceleration range, where the fundamental period normally falls for masonry walls of usual stiffness, Eq. (6.1).

$$a_{g,\max} = \frac{1}{2.5} \frac{V_{Rd\max}}{Mass} \tag{6.1}$$

The value of the natural frequency f can be estimated from Eq. (6.2), based on the expression of the wall stiffness K in Eq. (6.3).

$$f = \frac{1}{2\pi} \sqrt{\frac{K}{Mass}} \tag{6.2}$$

$$K = \frac{1}{\frac{H^3}{3E\frac{L^3 t}{12}} + \frac{H}{G\frac{5}{6}L\,t}} \tag{6.3}$$

The short-term secant modulus of elasticity is taken equal to $500\,f_k$ according to the suggestions of Eurocodes 6 and 8. The data and the results are summarized in Table 6.1.

6.3 Test Description

6.3.1 Axis Convention

For all information below regarding instrumentation and results, the following convention is followed according to the usual practice at EQUALS laboratory:

- The X-axis is positive from the near to the far end of the table;
- The Y- axis is positive from left to right;
- The Z-axis is positive upwards

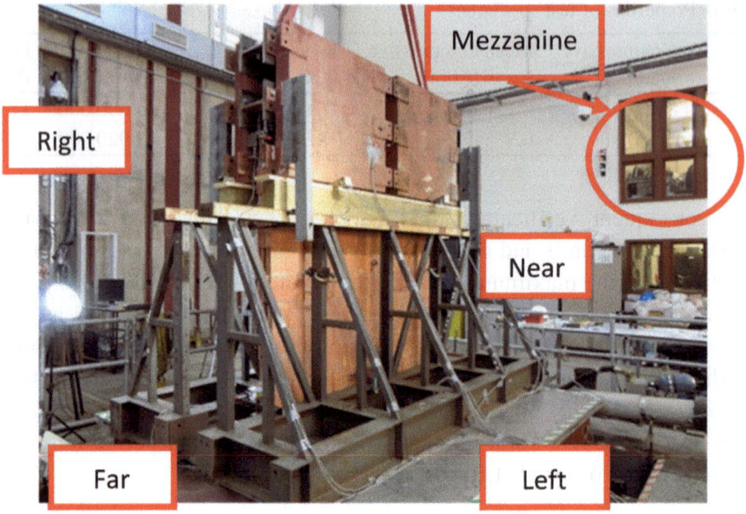

Fig. 6.2 Test specimen and set-up and nomenclature

Fig. 6.3 Symbols for the instrumentation in Fig. 6.4

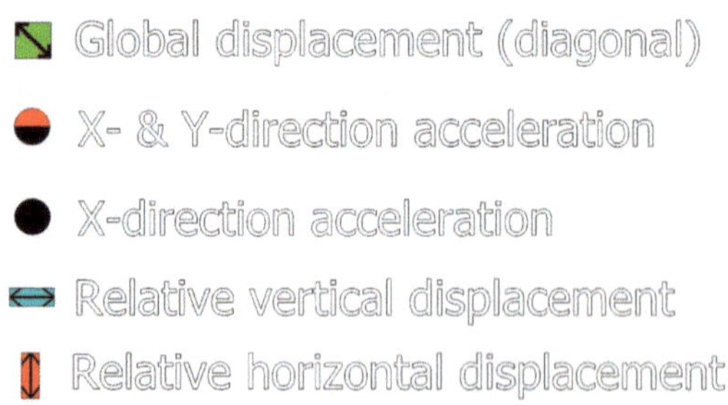

The reference for "near end", "far end", "left" and "right" of the shaking table is according to the control mezzanine, as shown in Fig. 6.2.

6.3.2 Instrumentation of Specimens

Two sets of devices are used for the instrumentation of the walls:

- SETRA type 141A accelerometers for acceleration measurements.
- Displacements of the test specimen relative to the table or relative displacements between parts of the specimen are monitored either with Celesco draw wire displacement transducers or with Linear Variable Differential Transformers (LVDT), whichever is most convenient.

The instrumentation layout is described in Figs. 6.3 and 6.4. Two complementary accelerometers are used on the table.

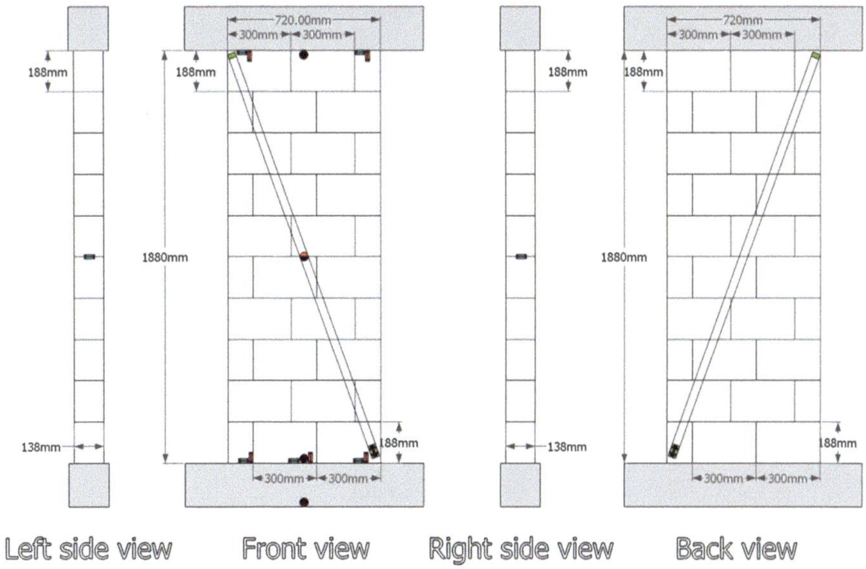

Fig. 6.4 Specimen instrumentation

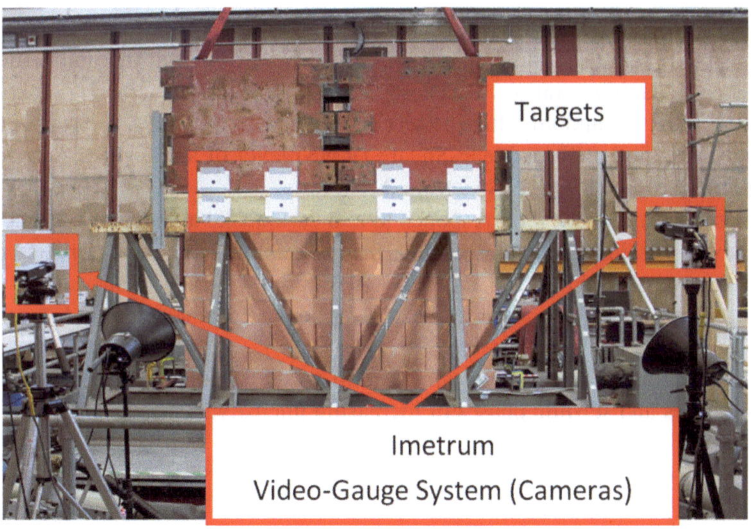

Fig. 6.5 Imetrum Video-Gauge System to monitor displacements through targets at the wall top

In addition to these devices, eight targets are located at the top of the wall (four on the loading mass lying on the wall and four on the steelwork fixed to table – see Fig. 6.5) in order to monitor the corresponding displacements (Imetrum Video-Gauge System).

The same instrumentation layout was used for all four walls.

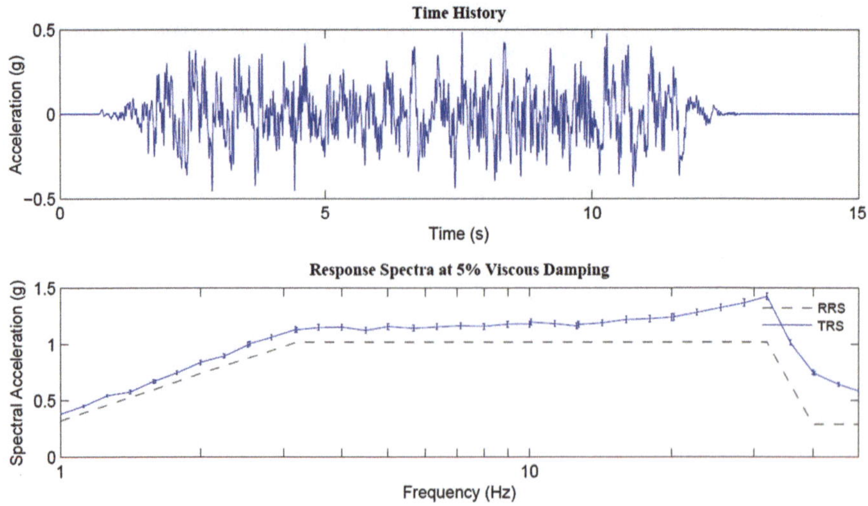

Fig. 6.6 Input motion and corresponding response spectrum (target vs. actual)

6.3.3 *Testing Procedure*

The testing procedure comprises two types of tests. The first one is a low level white noise test to obtain the main dynamic characteristics of the wall (natural modes and frequencies, damping ratio). This test lasts about 3 min, to reach a stationary vibration state. A random white noise excitation with frequency content between 1 and 100 Hz and at Root-Mean-Square (RMS) level of about 0.1 g is generated using an Advantest R9211C Spectrum Analyser.

The second type of test corresponds to real earthquake action using an acceleration waveform simulating the seismic ground motion. The characteristics of the waveform are given in Sect. 6.3.4. Successive input signals with increasing values of the peak ground acceleration (PGA) are applied to the table. Section 6.4 gives the acceleration levels.

In practice, the specimens are subjected to an alternation of white noise identification phases and seismic phases of increasing intensity.

6.3.4 *Excitation Waveforms for Seismic Tests*

Specimens have been pre-assessed according to Eurocodes 6 and 8. The waveform of the input signal used for the seismic phases of the testing procedure is chosen consistently as an artificial input compatible with the Eurocode 8 Type 2 spectrum. The excitation waveform and the corresponding response spectrum are given in Fig. 6.6, where TRS is the target spectrum and RRS the spectrum corresponding to the generated signal.

6.4 Test Results

The main qualitative observation from the tests is that all specimens exhibit a significant rocking behavior, independent of presence of rubber, even for long walls for which the preliminary assessment predicts failure in shear and not rocking or overturning. Moreover, thanks to the rocking phenomenon, the specimens are able to sustain an acceleration level much higher than foreseen by the equivalent static theoretical models.

Quantitative results are provided in the following sections in terms of natural frequencies, modal properties, damping level and seismic response of the system. Detailed information on the test results is available in Mordant (2012).

6.4.1 *Frequency Drop and Evolution of Damping*

Modal characterisation is carried out under low-level white noise excitation, during which the 'input' and 'response' channels of a spectrum analyser are connected to appropriate instrumentation (i.e. table X-accelerometer and top mass X-accelerometer, respectively). The spectrum analyser acquires 32 segments of time data, converts them to the frequency domain and averages the results to produce a transfer function. Natural frequency and damping values are determined for all coherent resonances, using a curve-fitting algorithm running directly on the spectrum analyser. Identification is carried out before and after each seismic shake. This allows the characterisation of a possible deterioration of the walls. The first and second resonant frequencies are given in Figs. 6.7 and 6.8, respectively, for the long and short walls, with and without rubber.

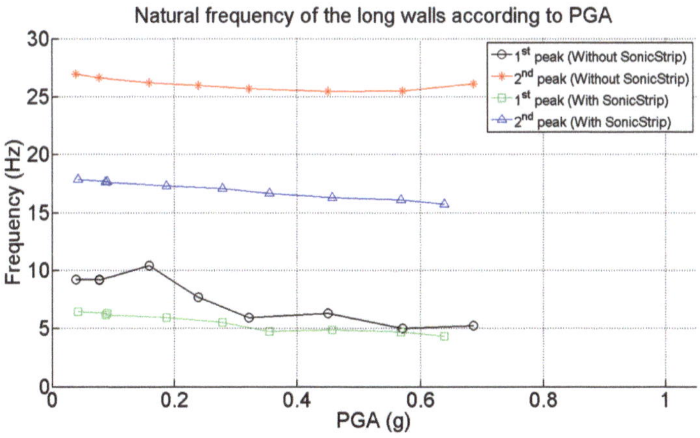

Fig. 6.7 Reduction of first and second natural frequencies with excitation level in long walls

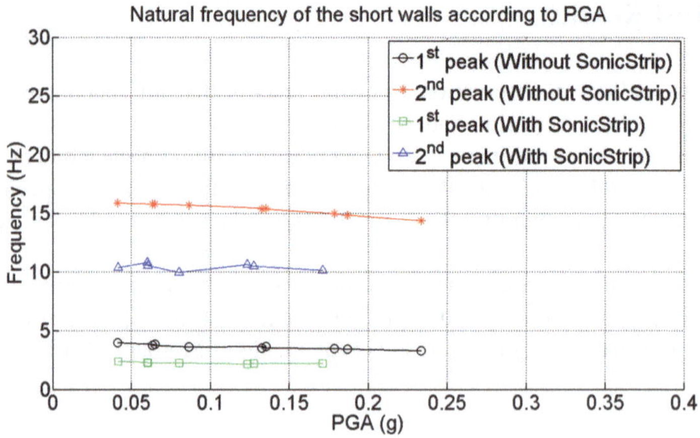

Fig. 6.8 Reduction of first and second natural frequencies with excitation level in short walls

It can be observed that the walls without acoustic isolation devices are characterized by higher frequencies than walls of the same dimensions but with such devices. Their difference is about 30–40 % in the undamaged (virgin) situation. A decrease is then observed after each seismic test, meaning that some damage occurs in the structural system. The frequency drop is more important for the walls without acoustic devices, showing a higher level of damage at similar ground acceleration level (see Figs. 6.7 and 6.8 and Tables 6.2 and 6.3).

These conclusions are in line with observations in (Michel et al. 2010) for non-isolated walls.

A progressive increase of the damping ratio is also observed when the acceleration level of the seismic test increases. In Tables 6.2 and 6.3, the value of damping particularly increases for the first natural frequency in the walls without acoustic devices. The increase is more striking for the long wall, even in presence of acoustic devices. Note that the accuracy of the procedure used for identifying the damping ratio may be questioned, in view of the abnormally high values obtained for very much damaged states (more than 100 %).

It is finally interesting to notice that seismic tests have been doubled at selected acceleration levels, as shown in Tables 6.2 and 6.3, and that the degradation rate is similar for these repetitions as for successive tests with increasing amplitude.

Although the seismic action has an effect on the wall frequency and damping for all cases, the positive influence of the rubber elements for improving the acoustic performance is clearly visible. These devices are indeed mitigating both the frequency drop and the damping increase. This can be related to a significant change in the behavior due to the presence of rubber elements. The rocking motion is indeed changing from a classical rocking phenomenon, where the wall behaves as a rigid body rotating about its toe, to a wall on elastic foundation. The deterioration caused by the loss of contact between the wall and the foundation and by the vertical impact when the wall is coming back to its upright position, as classically observed

Table 6.2 First and second natural frequencies and corresponding damping values after shaking at different PGA levels – long walls with or without rubber

	Long wall without rubber				Long wall with rubber			
After test at PGA	First natural frequency [Hz]	First mode damping [%]	Second natural frequency [Hz]	Second mode damping [%]	First natural frequency [Hz]	First mode damping [%]	Second natural frequency [Hz]	Second mode damping [%]
Virgin	9.01	96.49	25.50	2.76	5.35	44.88	17.51	8.60
0.05g	9.23	8.94	26.94	1.56	6.45	8.33	17.86	5.88
0.10g	9.19	23.71	26.61	1.74	6.29	14.30	17.65	5.78
0.10g	9.17	28.46	26.62	1.82	6.16	13.90	17.62	5.93
0.19g	10.41	93.77	26.21	2.21	5.91	28.16	17.30	6.23
0.29g	7.67	82.94	25.97	2.42	5.53	40.43	17.05	6.59
0.39g	5.93	126.33	25.69	2.50	4.75	26.29	16.62	7.18
0.49g	6.29	132.20	25.45	2.50	4.87	42.54	16.29	6.81
0.58g	5.02	161.90	25.52	2.65	4.68	36.83	16.08	7.80
0.68g	5.26	95.75	26.11	2.41	4.31	31.93	15.70	8.37

Table 6.3 First and second natural frequencies and corresponding damping values after shaking at different PGA levels – short walls with or without rubber

	Short wall without rubber				Short wall with rubber			
After test at PGA	First natural frequency [Hz]	First mode damping [%]	Second natural frequency [Hz]	Second mode damping [%]	First natural frequency [Hz]	First mode damping [%]	Second natural frequency [Hz]	Second mode damping [%]
Virgin	4.04	17.05	16.04	2.02	2.53	3.51	11.14	2.70
0.05g	3.99	3.86	15.88	1.30	2.38	9.14	11.37	4.01
0.08g	3.86	7.27	15.81	1.45	2.30	6.53	10.83	3.76
0.08g	3.77	14.97	15.81	1.45	2.26	6.70	10.54	3.87
0.10g	3.63	10.87	15.71	1.56	2.24	6.52	10.79	3.99
0.15g	3.67	15.06	15.44	1.91	2.17	9.41	10.61	2.74
0.15g	3.54	15.54	15.38	1.74	2.21	8.45	10.49	4.29
0.19g	3.50	17.44	15.00	2.05	2.22	9.19	10.16	5.22
0.19g	3.45	19.60	14.87	2.00	–	–	–	–
0.24g	3.30	17.84	14.38	2.32	–	–	–	–

Table 6.4 Measured first natural frequency in virgin state vs. value on the basis of the material properties per EC 6 for walls without acoustic devices

	Measured frequency [Hz]	Frequency based on EC6 [Hz]
Long wall	9.01	15.59
Short wall	3.81	4.20

Table 6.5 Contributions of flexure and shear to flexibility per Eq. (6.3) for $G = 0.1E$ and resulting first natural frequency in virgin state – walls without acoustic devices

	Flexural flexibility [m/N]	Shear flexibility [m/N]	Stiffness [N/m]	Frequency [Hz]
Long wall	9.38e-9	3.83e-8	20.97e6	10.31
Short wall	2.33e-7	1.12e-7	2.90e6	3.84

in the usual rocking motion, is thus clearly reduced. This behavior will be quantified in a following section in terms of contact length between the wall and its foundation.

The comparison between the measured first natural frequency in the virgin state and the value obtained on the basis of the material properties according to Eurocode 6 (EC 6) is given in Table 6.4 for the walls without acoustic devices. The values are in fairly good agreement in the case of the short wall, where the bending stiffness controls the response, but not for the long wall. The origin of the difference seems to be the assumption on the shear modulus value (let's recall that Eurocode 6 suggests a value of G equal to $0.4E$, E being the elastic modulus in compression) and hence the shear contribution to the overall transverse stiffness per Eq. (6.3). A better fitting is obtained if the shear modulus is taken as $0.1E$, as shown in Table 6.5. It can therefore be concluded that the recommended value of shear modulus in EC 6 is too high for the type of masonry considered in the present study, where the vertical joints are not filled with mortar.

6.4.2 *Eigenmodes of the Walls*

Natural vibration modes have been identified from the low-level white noise tests, via the squared matrix frequency-dependent of cross PSD of the wall, represented by a 6-Degree-of-Freedom (DOF) structure, each DOF corresponding to the location of one acceleration measurement point. Details of the identification procedure are given in (Mordant 2012). The objective of this post-processing is to identify the shape of the vibration mode of the wall, the vibration of the added mass with respect to the wall and the influence of the rubber layers on modal shape. Note that, for such a low level of ground acceleration, no significant rocking is induced in the structure and the system behaves as fixed to the foundation. The analysis is focused on the first two vibration modes of the structure.

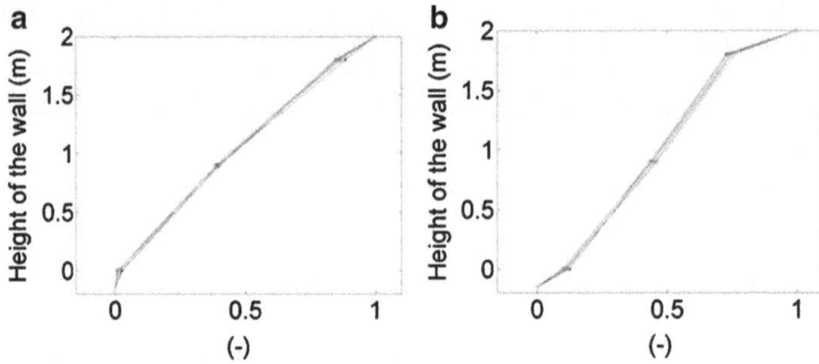

Fig. 6.9 Shape of first eigenmode extracted from the white-noise tests: (**a**) walls without acoustic devices; (**b**) walls with acoustic devices

The identified modal shapes show that the first mode is close to a straight line and corresponds to the classical triangular modal shape (Fig. 6.9a); the second mode (not plotted herein) corresponds to a vibration of the added mass in opposite phase with respect to the wall. The rubber layers result in deformable zones at top and bottom of the wall, translated in the identified vibration modes by a break in the general slope of the first mode (Fig. 6.9b).

6.4.3 Seismic Behavior

As already pointed out above, the main observation on the tests is the generalised rocking behavior of all walls. Rocking is a typical dynamic feature and can hardly be observed in tests where the seismic action is simulated by a static cyclic load (Degée and Lascar 2011a, b).

A deeper insight in the present dynamic test results allows an interesting focus on two important aspects. The first one concerns the estimation of the contact length at the interface between the wall and its foundation, while the second one deals with the transfer mechanism of the horizontal seismic forces from the added mass through the wall to its foundation.

The contact length at the interface between the base of the wall and its foundation has been evaluated from the measurement of the relative displacement at the edges and in the middle of the wall, assuming that the baseline remains straight, as detailed in the study of Mordant (2012). The resulting values, also expressed in percentage of the total wall length, are given in Fig. 6.10 and Table 6.6. When referring to the classical design methodology, suggested for instance by Eurocode 6, this contact length is understood as the so-called compressive length. The comparison of the theoretical compressive length with the experimental contact length for the same combination of normal force and bending moment at the base of the wall shows that, for a low PGA level (and hence for a low base bending moment), the measured compressive length is between the theoretical values obtained assuming a linear or a constant stress distribution over this length. For higher PGA levels, the theoretical

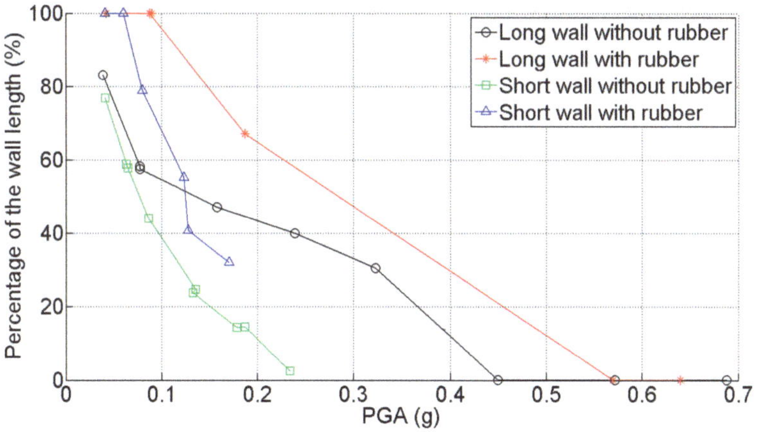

Fig. 6.10 Contact length of the base of the wall and the foundation

Table 6.6 Measured versus predicted compressive length at the wall base

Specimen	PGA [g]	Measured compressive length [%]	[mm]	Compressive length predicted for assumed linear stress distribution [%]	[mm]	Compressive length predicted for assumed constant stress distribution [%]	[mm]
Long wall without rubber	0.04	83.20	1747.2	100.00	2,100	82.86	1,740
	0.15	47.15	990.15	57.23	1201.9	35.71	750
	0.66	0.03	0.5	0	0	0	0
Long wall with rubber	0.04	100.00	2100.0	–	–	–	–
	0.19	67.20	1411.2	–	–	–	–
	0.64	0.05	1.1	–	–	–	–
Short wall without rubber	0.01	77.01	554.5	100.00	720	87.50	630
	0.02	58.70	422.6	100.00	720	75.00	540
	0.15	14.37	103.5	0	0	0	0
Short wall with rubber	0.04	100.00	720.0	–	–	–	–
	0.06	100.00	720.0	–	–	–	–
	0.17	32.13	231.3	–	–	–	–

model underestimates the compressive length whatever the stress distribution considered is. This comparison was only made for walls without rubber devices. Note that the estimation of the compressive length is an important parameter in the conventional verification of an unreinforced masonry wall subjected to combined compression and in-plan horizontal loading, since collapse is assumed to be reached either when the compressive length becomes zero or when the average shear stress over that length exceeds the conventional shear resistance. It is also important to note that, in practice, reaching a zero value of the compressive length does not necessary mean collapse of the wall, but initiation of the pure dynamic rocking phase during which the wall may remain stable, as shown for example in the tests with large PGA. The soundproofing rubber element is beneficial for the rocking behavior. Indeed, the measured compressive length increases at the same PGA level, reducing thus the

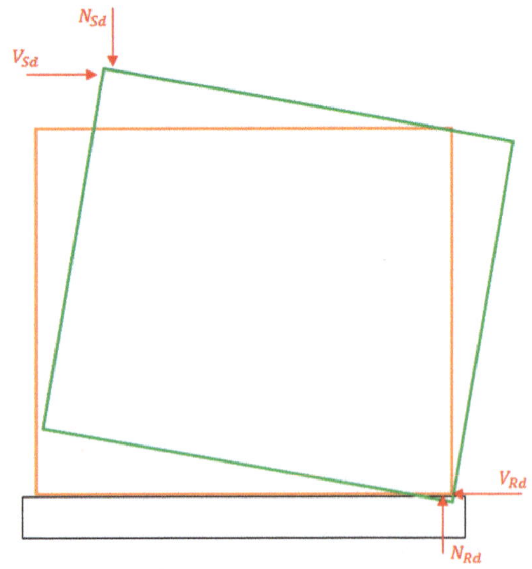

Fig. 6.11 Uplift response of tested walls

stress level. The resulting global horizontal displacements are however larger, because of the reduced overall stiffness of the system.

In the common equivalent static models, the horizontal seismic force is assumed to be transferred to the foundation through a global shear mechanism in the wall. The global resistance verification is then carried out by comparing this shear level to the applicable conventional shear resistance. This is for instance the condition that theoretically triggers the collapse for the geometry and loading corresponding to the tested long wall. However, when the rocking mechanism develops, the load transfer clearly shifts to a resistance governed by the compression in the diagonal strut, for which no specific design rules are proposed (Fig. 6.11). This can be seen as one of the main reasons for the significant underestimation of the maximum acceleration that can be sustained by the walls: for all four tests, a maximum acceleration of about three times the theoretically assessed value was reached and the tests were stopped essentially because of the large displacements and not because of a resistance problem. This mechanism is also clearly influenced by the presence of the rubber layers, although the effect is difficult to quantify, because of a current lack of information on the resistance of the rubber layer under combined nonuniform compression and shear.

6.5 Conclusions

This chapter presents the results of shake table tests carried out with the prime objective of characterising the dynamic rocking behavior of high strength clay masonry walls with glued horizontal joints and empty vertical joints subjected to seismic action. The main observation is that rocking occurs for all walls, even for situations where equivalent static models predict an anticipated shear failure.

Detailed evaluation of the load transfer mechanism and of the contact length was also presented. The identification of the natural vibration modes under low level white noise excitation raises the question of the accuracy of the values of the shear modulus usually suggested in the main reference documents, this modulus seeming to be significantly overestimated for the studied type of masonry.

The influence of soundproofing rubber layers was also investigated. Their effect on the overall stiffness, and hence on natural frequencies, was quantified. Their presence results in an increase of the compressive length for a given input acceleration level, a higher initial damping level, a limited progression of the damage, but much larger displacements, compared to similar geometrical situations without rubber layers. A first conclusion may be that these elements in general influence positively the behavior, but increase the displacements with consequences that need to be investigated in the context of entire buildings.

Acknowledgments The research leading to these results has received funding from the European Union Seventh Framework Programme [FP7/2007–2013] for access to the Bristol Laboratory for Advanced Dynamics Engineering (BLADE), University of Bristol, UK under grant agreement n° 227887 [SERIES]. H. Degée also acknowledges the direct support received from F.R.S.-FNRS (Belgian Fund for Research).

References

Caliò I, Marletta M, Pantò B (2012) A new discrete element model for evaluation of the seismic behaviour of unreinforced masonry buildings. Eng Struct 40:327–338

Degée H, Lascar L (2011a) Cyclic shear behaviour of clay masonry walls – Part 1 : walls including acoustic devices or with a door opening. Research report, Departement Argenco, Univerisity of Liege

Degée H, Lascar L (2011b) Cyclic shear behaviour of clay masonry walls – Part 2 : cyclic shear test on prefabricated clay masonry walls with T-shaped. Research report, Departement Argenco, Univerisity of Liege

Michel C, Zapica B et al (2010) Quantification of fundamental frequency drop for unreinforced masonry buildings from dynamic tests. Earthq Eng Struct Dyn 40:1283–1296

Milani G, Beyer K, Dazio A (2009) Upper bound limit analysis of meso-mechanical spandrel models for the pushover analysis of 2D masonry frames. Eng Struct 31:2696–2710

Mordant C (2012) Contribution to experimental tests on the seismic behaviour of masonry structural elements. Master dissertation, University of Liège, Liège

Nakagawa T, Narafu T, Imai H et al (2012) Collapse behaviour of a brick masonry house using a shaking table and numerical simulation based on the extended distinct element method. Bull Earthq Eng 10:269–283

Tomazevic M (1999) Earthquake-resistant design of masonry buildings. Imperial College Press, London

Wienerberger (2012) Caractéristiques rubber devices. http://www.wienerberger.be/fr/fiche-technique-sonicstrip-tapis-isolant-acoustiques-en-caoutchouc.html. Accessed 5 Dec 2012

Chapter 7
Assessing Seismic Vulnerability of Unreinforced Masonry Walls Using Elasto-Plastic Damage Model

Basheer H. Al-Gohi, Cem Demir, Alper Ilki, Mohammed H. Baluch, and Muhammad K. Rahman

Abstract In this paper, the seismic vulnerability of unreinforced masonry walls is assessed by conducting a numerical study on the interaction of axial and lateral resistance. The walls are modeled in an ABAQUS environment, using a plastic damage model originally developed by Lubliner et al. (Int J Solids Struct 25(3):299–326, 1989) and further extended by Lee and Fenves (J Eng Mech ASCE 124(8):892–900, 1998). The model yields interesting interactive collapse mechanisms that occur as the axial loading on the wall is increased. The different modes of failure identified as the axial load is increased include (i) rocking mode, (ii) sliding mode, (iii) staggered head/bed joint failure, (iv) diagonal cracks through wall blocks accompanied by staggered head/bed joint cracking, and (v) crushing of wall blocks or bricks.

7.1 Introduction

Unreinforced masonry (URM) construction is one of the oldest types of structures in human history. This type of construction is still being used up to the present day in many urban areas as bearing walls, and in other areas in heritage type of construction.

B.H. Al-Gohi • M.H. Baluch
Department of Civil and Environmental Engineering, King Fahd University of Petroleum and Minerals, Dhahran 31261, Saudi Arabia
e-mail: algohi@kfupm.edu.sa; mhbaluch@kfupm.edu.sa

C. Demir • A. Ilki
Department of Civil Engineering, Civil Engineering Faculty, Istanbul Technical University, Maslak, Istanbul 34469, Turkey
e-mail: demirce@itu.edu.tr; ailki@itu.edu.tr

M.K. Rahman (✉)
Center for Engineering Research, Research Institute, King Fahd University of Petroleum and Minerals, Dhahran 31261, Saudi Arabia
e-mail: mkrahman@kfupm.edu.sa

A. Ilki and M.N. Fardis (eds.), *Seismic Evaluation and Rehabilitation of Structures*, Geotechnical, Geological and Earthquake Engineering 26,
DOI 10.1007/978-3-319-00458-7_7,

Masonry components are also used in modern buildings, especially as infill walls. According to the work done by Abrams (2001) and Filiatrault et al. (2001), URM has been recognized, following nearly every medium to strong earthquake, as one of the most vulnerable forms of construction. From the past experience of earthquakes, well-engineered and constructed URM buildings possess good characteristics and can withstand such earthquakes.

7.2 Literature Review

Masonry walls subjected to in-plane cyclic loading have been investigated quite extensively. Several researchers have conducted experimental and numerical studies aiming at better understanding of the response of masonry walls to seismic loading. Demir (2012) studied the effect of cyclic loading on a heritage masonry wall representing the walls of Ottoman period monumental structures in Istanbul. In his work, Demir (2012) tested eight multi-leaf masonry walls that had dimensions of 1.2×1.2 m and a thickness of 0.3 m. He found that walls exhibit different behavior, mode of failure, and lateral strength according to the level of axial stress acting on them. Demir performed a series of finite element analyses using ABAQUS/Explicit in which the results showed good agreement between the experiments and FEM analysis. However, the range of axial load used in the FEM analysis was restricted to that used in the experiments reported in Demir (2012). It should be noted that the numerical analyses presented in the current study are based on the modeling approach proposed in Demir (2012).

Li et al. (2005) investigated the behavior of masonry concrete walls subjected to diagonal loading. They also investigated the effect of NSM (near surface mounted) reinforcement using GFRP and stainless steel bars on the strength of masonry walls. The authors discussed their results in the framework of a mechanistic model. Vasconcelos (2005) studied the seismic performance of masonry walls made of granite stone, through an extensive experimental campaign, which included tests on three types of masonry walls. All three types of walls were subjected to pre-compression stresses of 0.5, 0.87 and 1.25 MPa. Vasconcelos reported that the failure modes depend clearly on the level of axial load and on the textural arrangement. He also reported that masonry walls subjected to high levels of pre-compression exhibit a brittle failure.

In this paper, a finite element study using an elasto-plastic damage model was conducted to relate the level of axial load to both lateral resistance and type of failure mode. The simulation was conducted for walls tested by Demir (2012) and by Li et al. (2005). The loading types adopted in this study are axial and lateral loading, in which the effect would be also similar to that of diagonal loading adopted by Li et al. (2005). Finite element simulation has been conducted using a number of initial axial load intensities, as high as the full axial capacity of the walls, so that the interaction diagram in the axial and lateral load space can be determined. Results show a universal behavior in the interaction between axial force and lateral resistance and a confluence of failure modes for both sets of walls.

7.3 Mode of Failure of Unreinforced Masonry Walls

As mentioned before, masonry walls subjected to in-plane loading exhibit different mechanistic response based on the intensity of axial loading applied. These modes of failures have been studied by Li et al. (2005) using mechanistic modeling. Based on the level of axial force applied on the wall, three equations have been proposed by Li et al. (2005) for calculating the lateral strength of the wall. These equations are as follows:

- For low applied axial force:

$$V_{m,1} = \tau \times A_n = A_n(\tau_0 + \mu\sigma_n) \tag{7.1}$$

- For moderate applied axial force:

$$V_{m,2} = \left(\frac{f_{tb}'}{2.3}\sqrt{1 + \frac{\sigma_n}{f'_{tb}}}\right)A_n \tag{7.2}$$

- For high applied axial force:

$$V_{m,3} = \frac{(f_m' - \sigma_n)(2d)}{3b}A_n \tag{7.3}$$

where τ_0 is the shear bond strength, μ is the global friction coefficient, σ_n is the normal compressive stress on the wall, A_n is the sliding shear area, $\tilde{f}_{tb}$ is the tensile strength of the masonry, $\tilde{f}_m$ is the compressive strength of the masonry wall, d is the block length, and b is the block depth. These equations have been plotted in Fig. 7.1.

Figure 7.1 shows that the lateral strength of walls increases with increase in the level of axial stress applied to the wall up to a certain limit, after which the lateral strength of the wall sharply reduces. As shown in Fig. 7.1, the curve is composed of three line segments with clear and sharp boundaries between the segments.

7.4 Review of Plastic Damage Model

Concrete Damage Plasticity (CDP) is one of the constitutive models that can efficiently and accurately describe the behavior of concrete type materials under different conditions of loading. Several laboratory tests are necessary to identify the CDP model, as introduced and further developed by Kachanov (1986), Rabotnov (1969) and others. The elastic–plastic response of the concrete damaged plasticity model is described in terms of the effective stress and the hardening variables with

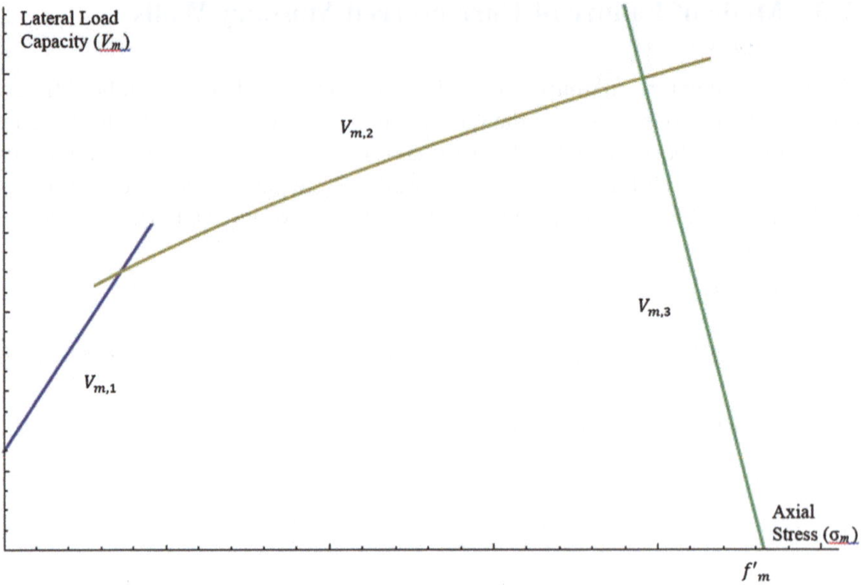

Fig. 7.1 Shear-axial interaction diagram for URM walls

scalar isotropic damage (Eqs. 7.4, 7.5 and 7.6). The Cauchy stress is calculated in terms of the stiffness degradation variable, *d*, and the effective stress (Eq. 7.7).

$$\bar{\sigma} = D_0^{el} : \left(\varepsilon - \varepsilon^{pl}\right) \in \{\bar{\sigma}\,|F(\bar{\sigma}, \bar{\varepsilon}^{pl}) \leq 0\} \tag{7.4}$$

$$\dot{\bar{\varepsilon}}^{pl} = h(\bar{\sigma}, \bar{\varepsilon}^{pl}) \cdot \hat{\dot{\varepsilon}}^{pl} \tag{7.5}$$

$$\dot{\varepsilon}^{pl} = \frac{\dot{\lambda}(\partial G(\bar{\sigma}))}{\partial \bar{\sigma}} \tag{7.6}$$

$$\sigma = (1 - d)\bar{\sigma} \tag{7.7}$$

7.4.1 Yield Function

The yield function *F* represents a surface in effective stress space, which determines the states of failure or damage. For the inviscid plastic-damage model the yield function can be expressed as:

$$F(\bar{\sigma}, \tilde{\varepsilon}^{pl}) = \frac{1}{1 - a}(\bar{q} - 3\alpha\bar{p} + \beta(\tilde{\varepsilon}^{pl})\langle\hat{\bar{\sigma}}_{\max}\rangle - \gamma\langle-\hat{\bar{\sigma}}_{\max}\rangle) - \bar{\sigma}_c(\tilde{\varepsilon}_c^{pl}) \leq 0 \tag{7.8}$$

where α, γ are dimensionless material constants; $\bar{p}$ is the effective hydrostatic pressure; $\bar{q}$ is the Mises equivalent effective stress; $\bar{\sigma}, \hat{\bar{\sigma}}$ is the algebraically maximum eigenvalue of $\bar{\sigma}$. The function β is given as

$$\beta(\tilde{\varepsilon}^{pl}) = \frac{\bar{\sigma}_c(\tilde{\varepsilon}_c^{pl})}{\bar{\sigma}_t(\tilde{\varepsilon}_c^{pl})}(1-a) - (1+a) \tag{7.9}$$

where e $\bar{\sigma}_c, \bar{\sigma}_t$ are the effective tensile and compressive cohesion stresses, respectively, and obtained from 1-D tests in uniaxial compression and uniaxial tension (stress-plastic strain data). The coefficient α can be determined by direct application of yield criteria as given by Eq. 7.8 for equal biaxial compression and for uniaxial compression (for both cases, $\hat{\bar{\sigma}}_{\max} = 0$). Solving for α, one obtains:

$$\alpha = \frac{\left(\frac{\sigma_{b0}}{\sigma_{c0}}\right) - 1}{2\left(\frac{\sigma_{b0}}{\sigma_{c0}}\right) - 1} \tag{7.10}$$

where σ_{co} is the uniaxial compression strength of concrete and σ_{bo} is the biaxial compression strength of concrete. Typical experimental values of the ratio σ_{bo}/σ_{co} for concrete are in the range from 1.10 to 1.16, yielding values of α between 0.08 and 0.12 (Lubliner et al. 1989). The coefficient γ enters the yield function only for stress states of triaxial compression, when $\hat{\bar{\sigma}}_{\max} \leq 0$. This coefficient can be determined by comparing the yield conditions along the tensile and compressive meridians. By definition, the *tensile meridian* (TM) is the locus of stress states satisfying the condition $\hat{\bar{\sigma}}_{\max} = \hat{\bar{\sigma}}_1 \geq \hat{\bar{\sigma}}_2 = \hat{\bar{\sigma}}_3$, and the *compressive meridian* (CM) is the locus of stress states such that $\hat{\bar{\sigma}}_{\max} = \hat{\bar{\sigma}}_1 = \hat{\bar{\sigma}}_2 \geq \hat{\bar{\sigma}}_3$, where, $\hat{\bar{\sigma}}_1, \hat{\bar{\sigma}}_2$, and $\hat{\bar{\sigma}}_3$ are the eigenvalues of the effective stress tensor. It can be shown in general that:

$$\bar{s}_1 = \frac{2}{\sqrt{3}}\sqrt{\bar{J}_2}\cos\theta \tag{7.11}$$

$$\bar{s}_2 = \frac{2}{\sqrt{3}}\sqrt{\bar{J}_2}\cos\left(\frac{2\pi}{3} - \theta\right) \tag{7.12}$$

$$\bar{s}_3 = \frac{2}{\sqrt{3}}\sqrt{\bar{J}_2}\cos\left(\frac{2\pi}{3} + \theta\right) \tag{7.13}$$

where $\bar{s}_i$ are the principal values of the effective deviatoric stress tensor $\bar{s}_{ij}$, and θ is the angle of similarity measured from the projection of $\bar{\sigma}_1$ in the deviatoric plane. For the tensile meridian, $\theta = 0$, whereas for the compression meridian, $\theta = 60$. Substituting the appropriate values of θ in expressions for $\bar{s}_i$ and expressing $\bar{\sigma}_1$ in terms of $\bar{s}_1$.

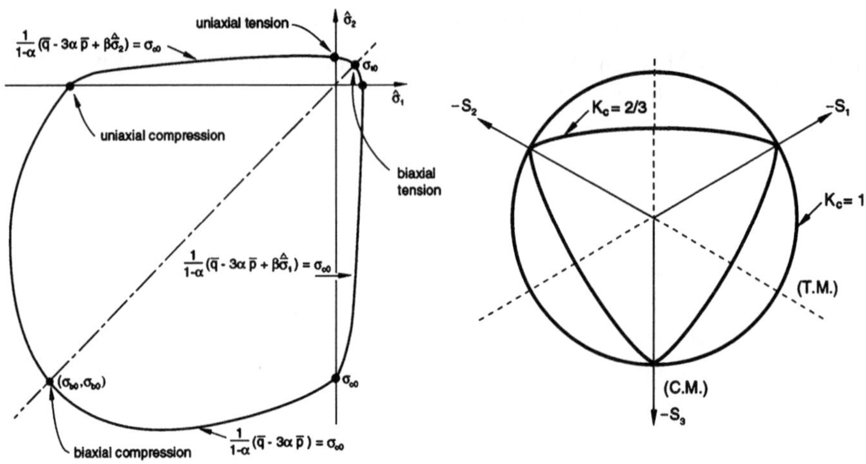

Fig. 7.2 Yield surfaces in deviatoric and in plane (different values of K_c)

$$(\hat{\bar{\sigma}}_{\max})_{TM} = \bar{\sigma}_1 = \bar{s}_1 - \bar{p} = \frac{2}{\sqrt{3}}\sqrt{\bar{J}_2} - \bar{p} = \frac{2\bar{q}}{3} - \bar{p} \tag{7.14}$$

$$(\hat{\bar{\sigma}}_{\max})_{CM} = \bar{\sigma}_1 = \bar{s}_1 - \bar{p} = \frac{1}{\sqrt{3}}\sqrt{\bar{J}_2} - \bar{p} = \frac{\bar{q}}{3} - \bar{p} \tag{7.15}$$

With $\hat{\bar{\sigma}}_{\max} \leq 0$, the corresponding yield conditions are:

$$\left(\frac{2}{3}\gamma + 1\right)\bar{q} - (\gamma + 3\alpha)\bar{p} = (1 - \alpha)\bar{\sigma}_c(TM) \tag{7.16}$$

$$(\frac{1}{3}\gamma + 1)\bar{q} - (\gamma + 3\alpha)\bar{p} = (1 - \alpha)\bar{\sigma}_c(CM) \tag{7.17}$$

Let $K_c = \bar{q}_{(\mathrm{TM})}/\bar{q}_{(\mathrm{CM})}$ for any given value of the hydrostatic pressure $\bar{p}$ with $\hat{\bar{\sigma}}_{\max} \leq 0$; then:

$$K_c = \frac{\gamma + 3}{2\gamma + 3} \tag{7.18}$$

The fact that K_c is constant does not seem to be contradicted by experimental evidence (Lubliner et al. 1989). The coefficient γ is, therefore, evaluated as

$$\gamma = \frac{3(1 - K_c)}{2K_c - 1} \tag{7.19}$$

It was suggested by Lubliner et al. (1989) that the value of K_c ranges from 0.66 to 0.8. Typical yield surfaces are shown in Fig. 7.2 in the deviatoric plane for $\hat{\bar{\sigma}}_{\max} \leq 0$ and for plane-stress conditions.

7.4.2 Flow Rule

Plastic flow is governed by a flow potential function $G(\bar{\sigma})$ according to the nonassociative flow rule:

$$G = \sqrt{(\in \sigma_{t0} \tan \psi)^2 + \bar{q}^2} - \bar{p} \tan \psi \tag{7.20}$$

where ψ is the dilation angle measured in the p–q plane at high confining pressure; σ_{to} is the uniaxial tensile stress at failure; and ϵ is a parameter referred to as eccentricity. More details can be found in the ABAQUS Manual (version 6.12).

7.5 Computational Modeling

7.5.1 Computational Simulation of Demir (2012) Walls

As mentioned in Sect. 7.2, Demir (2012) conducted a study to examine the in-plane seismic response of walls representing the walls of Ottoman period structures in Istanbul. As shown in Fig. 7.3, dimensions of the walls were 1.2 by 1.2 m with thickness of 300 mm. External leaves were constructed with finely cut stone units with dry joints.

The walls were subjected to pre-compression axial force prior to the application of cyclic lateral load. The variables are summarized in Table 7.1. Demir (2012) has reported different modes of failure of the walls according to the level of axial force. In the work presented here, a finite element simulation using micro-analysis

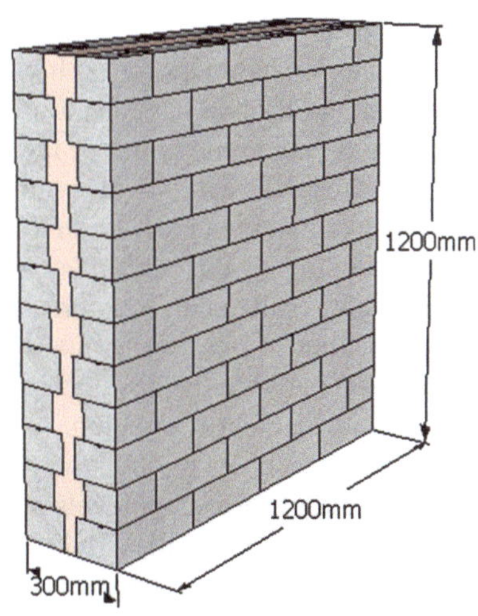

Fig. 7.3 Geometry of Demir walls

Table 7.1 Variables in Demir's experiments

Wall sample	Blocks clamped or not	Axial stress magnitude (MPa)
M-25-C	Yes	0.25
M-50-C	Yes	0.50
M-75-C	Yes	0.75
M-100-C	Yes	1.00
M-50	No	0.50

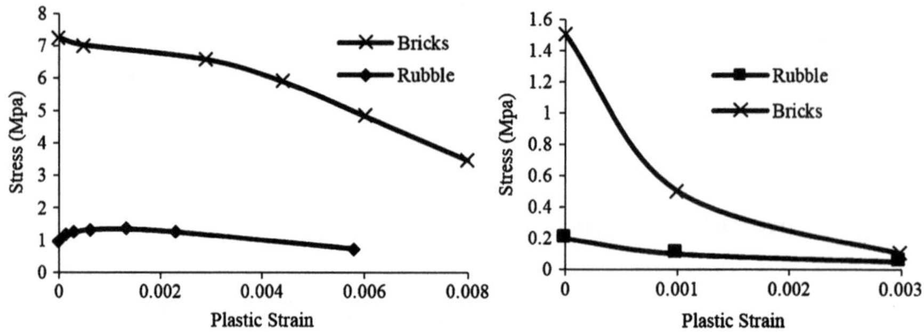

Fig. 7.4 Plastic strain–stress in compression and tension (Demir 2012)

Table 7.2 Parameters used in Plastic Damage Model

Mass density (tone/mm^3)	Young's modulus (MPa)	Poisson's ratio	Dilation angle ψ (degree)	Eccentricity ϵ	f_{bo}/f_{co}	K	Viscosity parameter
2.4E-009	3,200	0.18	36	0.1	1.16	0.67	0

approach has been conducted, taking the full range of axial stress starting from zero up to the full axial capacity of the wall. In the FEM simulations, pinned support (Ux = 0, Uy = 0, and Uz = 0) was assumed at the base of the walls. Uniaxial stress-plastic stain data for both external leaves and rubble core in uniaxial compression and tension have been used in the plastic damage model incorporated in ABAQUS (Fig. 7.4). Some of the parameters in the Plastic Damage model were assumed to be the default values and some others were based on actual experiments. These parameters are shown in Table 7.2. A total of 20 cases were investigated, corresponding to different axial stress level intensities. A comparison between cyclic test results and monotonic FEM simulation results for a wall subjected to axial loading of 0.5 MPa is made in Fig. 7.5, clearly showing that there is good agreement between the envelope of the experimental cyclic loading and the monotonic finite element simulation.

Figure 7.6 shows the relation between the wall lateral displacement and lateral resistance for all different levels of axial load. From the curves shown in Fig. 7.6, it can be seen that the walls exhibit high ductility when the axial pre-compression is small. This ductility decreases as the pre-compression stress increases and the walls

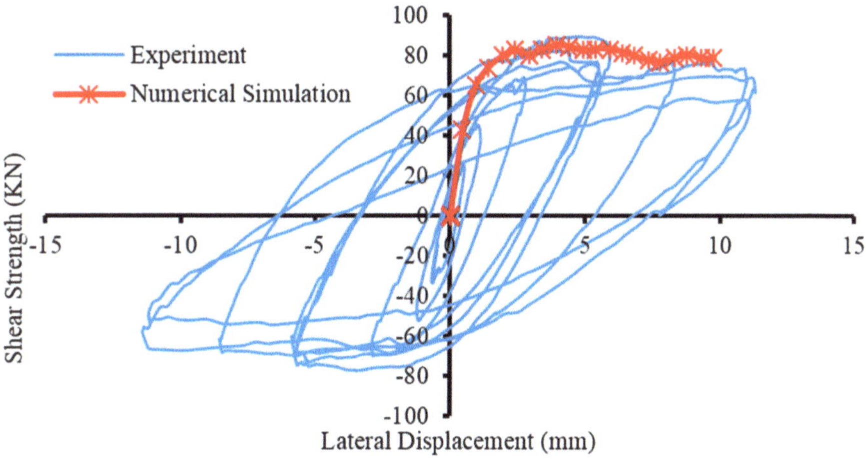

Fig. 7.5 Lateral strength-lateral displacement (axial stress = 0.5 MPa)

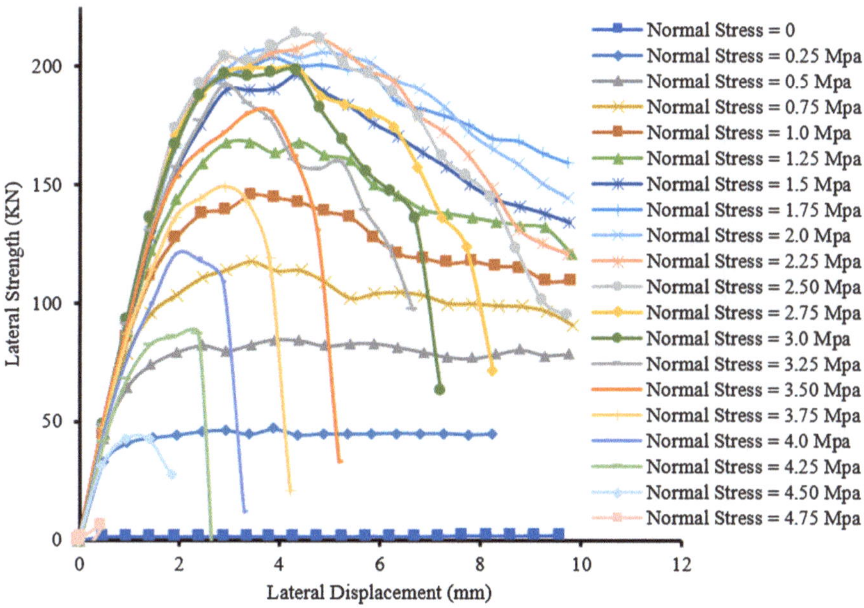

Fig. 7.6 Lateral strength-lateral displacement

tend to exhibit brittle failure. It is also clear from the curves that the lateral strength increases as the axial stress increases. Initially, this trend is true up to a certain level of axial load, after which the walls begin to weaken and lateral resistance decreases as the axial stress increases. The axial forces corresponding to the lateral strengths of the wall are presented in Table 7.3.

The data in Table 7.3 are shown in Fig. 7.7.

Table 7.3 FEM results for normal stress and force and lateral force for Demir walls

Normal stress (MPa)	Normal force (kN)	Lateral force-Exp (Demir 2012) (kN)	Lateral force-FEM (kN)	% Diff	Normal stress (MPa)	Normal force (kN)	Lateral force-FEM (kN)
0	0	–	1.9	–	2.75	990	199.3
0.25	90	54	46.8	13 %	3	1,080	198.2
0.5	180	90	84.8	6 %	3.25	1,170	192.5
0.75	270	122.4	117.6	4 %	3.5	1,260	180.8
1	360	136.8	145.5	6 %	3.75	1,350	149.3
1.25	450	–	168.2	–	4	1,440	120.9
1.5	540	–	196.9	–	4.25	1,530	87.1
1.75	630	–	203.8	–	4.5	1,620	42.8
2	720	–	207.3	–	4.75	1,710	5.5
2.25	810	–	211.2	–	4.85	1,745	0.0
2.5	900	–	213.7	–	–	–	–

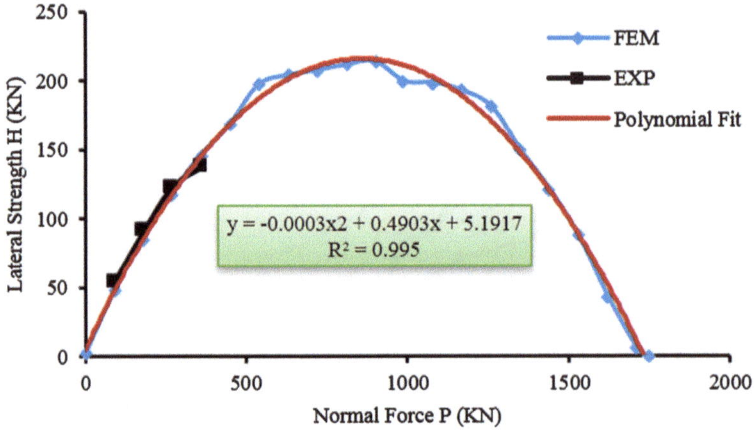

Fig. 7.7 Interaction relation between lateral strength H and normal force P

A regression fit of the lateral force at collapse of the wall H (kN) on the applied level of pre-compression P (kN), is parabolic in P. A similar trend has been reported by Voon (2007). The resulting regression is:

$$H = -0.0003P^2 + 0.4903P + 5.1917 \tag{7.21}$$

This relationship may be normalized by taking the ratio of lateral strength to the maximum lateral force attained ($H_{max} = 213.7$ kN) and the ratio of axial force applied to the maximum applied axial force ($P_{max} = 1{,}745$ kN), as in Fig. 7.8. The regression analysis of this curve results in the following equation:

$$y = -4.0635x^2 + 4.0033x + 0.0243 \tag{7.22}$$

where $y = H/H_{max}, x = P/P_{max}$.

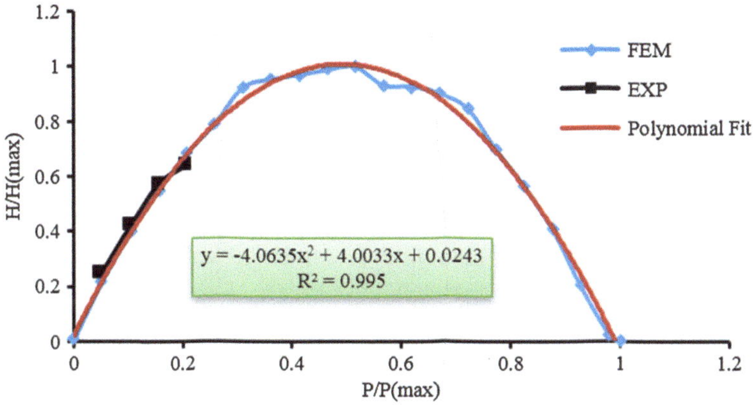

Fig. 7.8 Interaction relation between normalized lateral force and normalized axial force

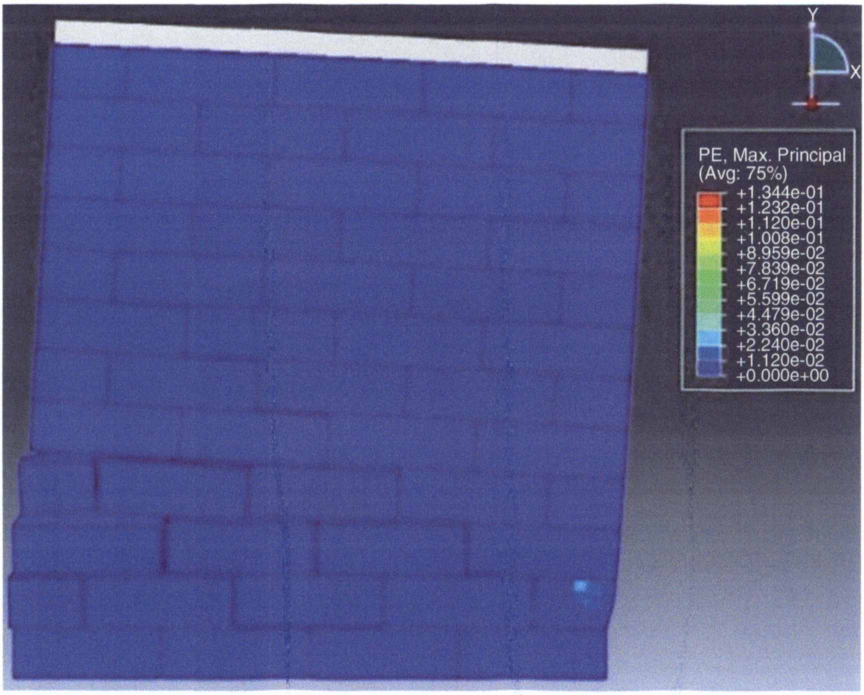

Fig. 7.9 Plastic strain and failure mode (0 axial stress) (Wall rocking)

Five failure modes were observed, namely sliding, rocking, head/bed joints opening, cracking in bricks, and crashing of wall. Figures 7.9, 7.10, 7.11, 7.12, 7.13, and 7.14, show these failure modes.

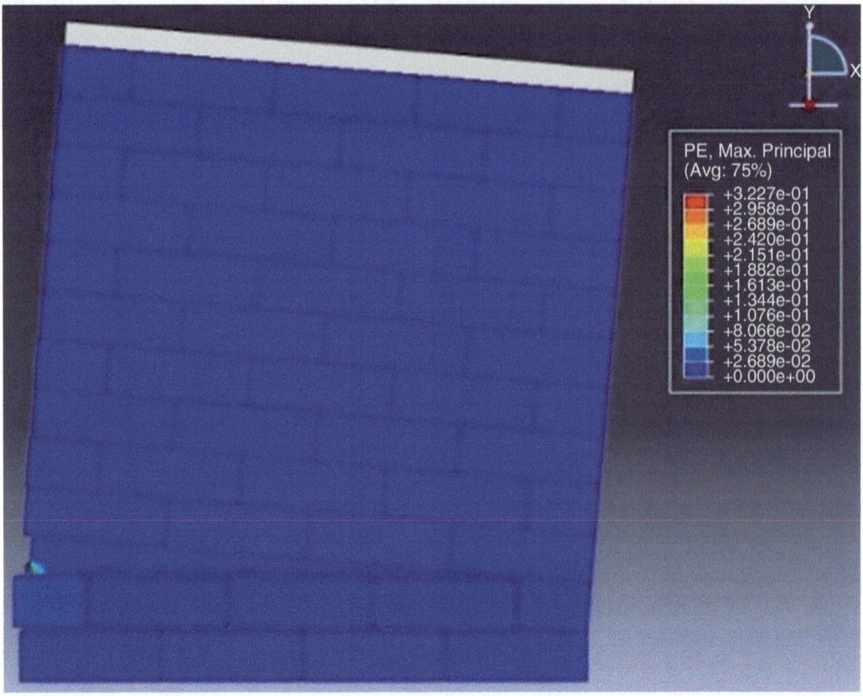

Fig. 7.10 Plastic strain and failure mode (0.25 MPa axial stress) (Wall sliding and rocking)

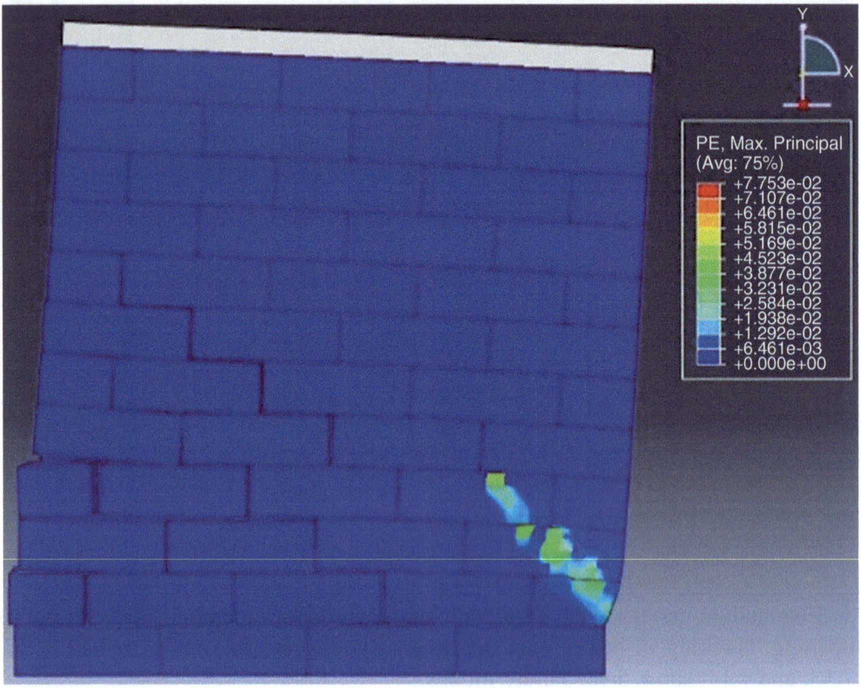

Fig. 7.11 Plastic strain and failure mode (0.5 MPa axial stress) (Wall sliding, rocking and toe crashing)

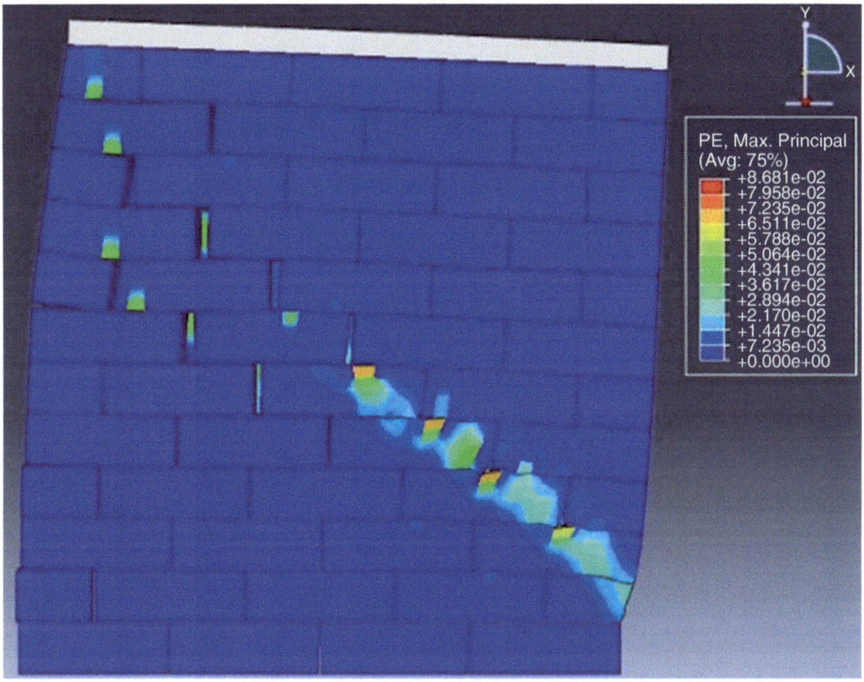

Fig. 7.12 Plastic strain and failure mode (1.0 MPa axial stress) (Wall head/bed joints opening and cracking in bricks)

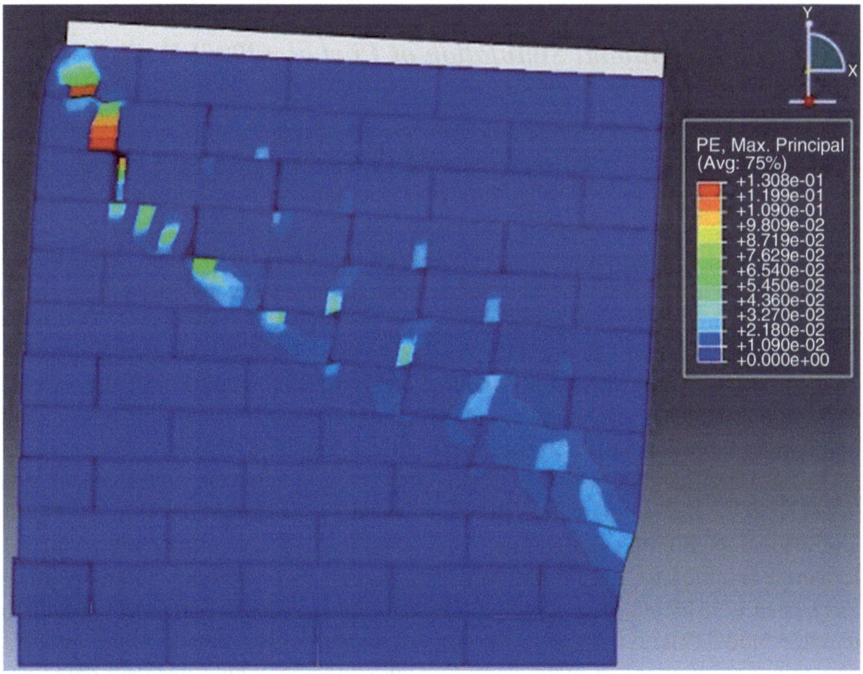

Fig. 7.13 Plastic strain and failure mode (2.0 MPa axial stress) (Wall head/bed joints opening and cracking in bricks)

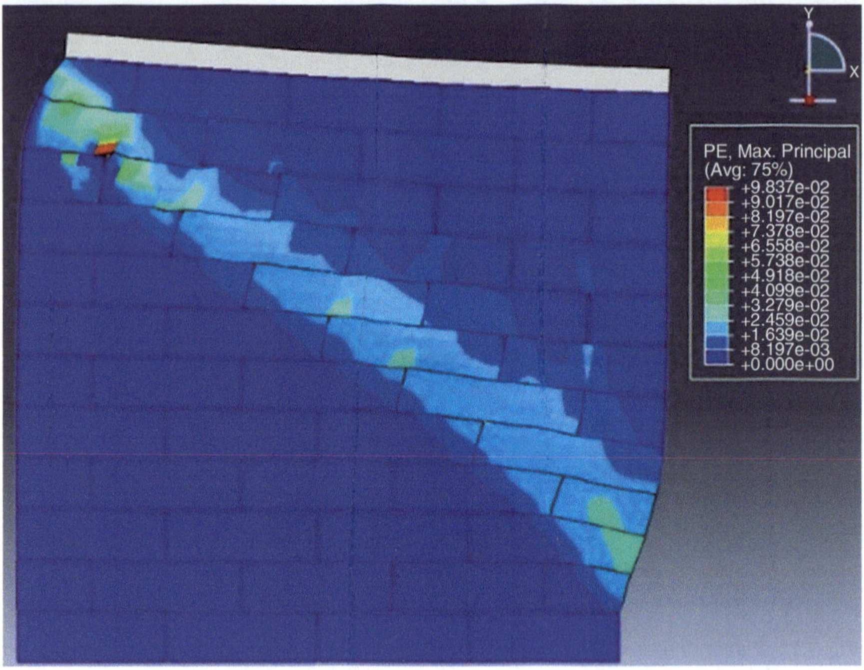

Fig. 7.14 Plastic strain and failure mode (3.0 MPa axial stress) (Cracking and crashing in bricks)

7.5.2 Computational Simulation of Li et al. (2005) Walls

The second finite element simulation has been carried out for the wall tested by Li et al. (2005). In this study, finite element simulations have been conducted on concrete masonry walls. The loading types adopted in this study are axial and lateral loading in which the effect is similar to that when diagonal loading is used. Finite element simulation has been conducted for this type of walls using all the range of axial loading starting from zero up to the full axial capacity of the wall, so that the interaction between axial and lateral loading can be extracted. The walls geometry adopted by Li et al. (2005) was 1,625 mm × 1,625 mm × 152 mm. The concrete masonry units had dimensions of 406 mm × 203 mm × 152 mm. A type N mortar (ASTM C270) was used to build the walls. The dimensions and configurations of the units and wall are shown in Fig. 7.15.

In the FEM simulations, pinned support (Ux = 0, Uy = 0, and Uz = 0) was assumed at the base of the walls. The Plastic Damage model available in ABAQUS was also adopted in the simulations. The materials used in the walls were concrete (concrete brick units) and mortar (head/bed joints between bricks). Both materials have been considered as a continuum media and simulated using the continuum damage model. The only material properties available in (Li et al. 2005) were the concrete and mortar ultimate compression strengths. Other needed data was extracted using ACI formulas. Some of the parameters needed in the Plastic

Table 7.4 Parameters used in plastic damage model

Mass density (tone/mm^3)	Young's modulus (MPa)	Poisson's ratio	Dilation angle ψ (degree)	Eccentricity ϵ	f_{bo}/f_{co}	K	Viscosity parameter
2.4E-009	19,264	0.2	36	0.1	1.16	0.67	0

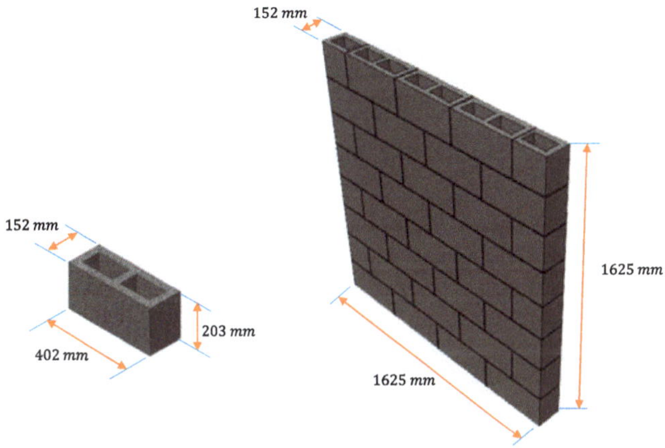

Fig. 7.15 Geometry and configuration of Li et al. (2005) walls

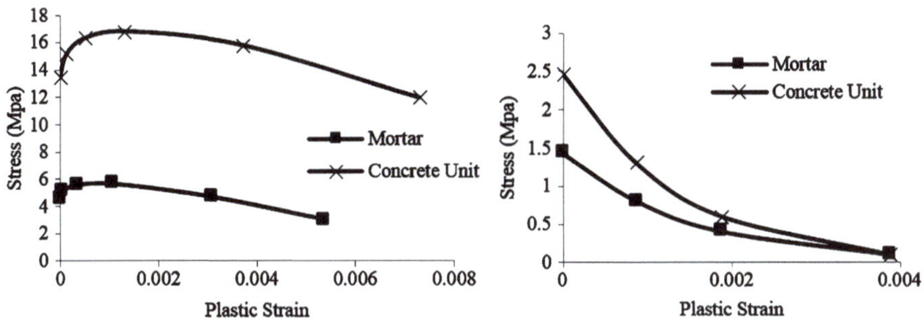

Fig. 7.16 Plastic strain–stress in compression and tension

Damage model were assumed to be the default values and some others were calculated based on ACI (Table 7.4). The stress plastic strain values needed in the damage model incorporated in ABAQUS are shown in Fig. 7.16.

A monotonic displacement loading was adopted in the simulations with 10 mm maximum lateral displacement. In the cases associated with low axial stresses, FEM converged up to the end of maximum lateral displacement of 10 mm. However, when axial loads were higher, the lateral displacements, for cases in which the FEM converged, were less than 10 mm. Figure 7.17 shows the relation between the wall lateral displacement and lateral resistance for all different levels of axial load.

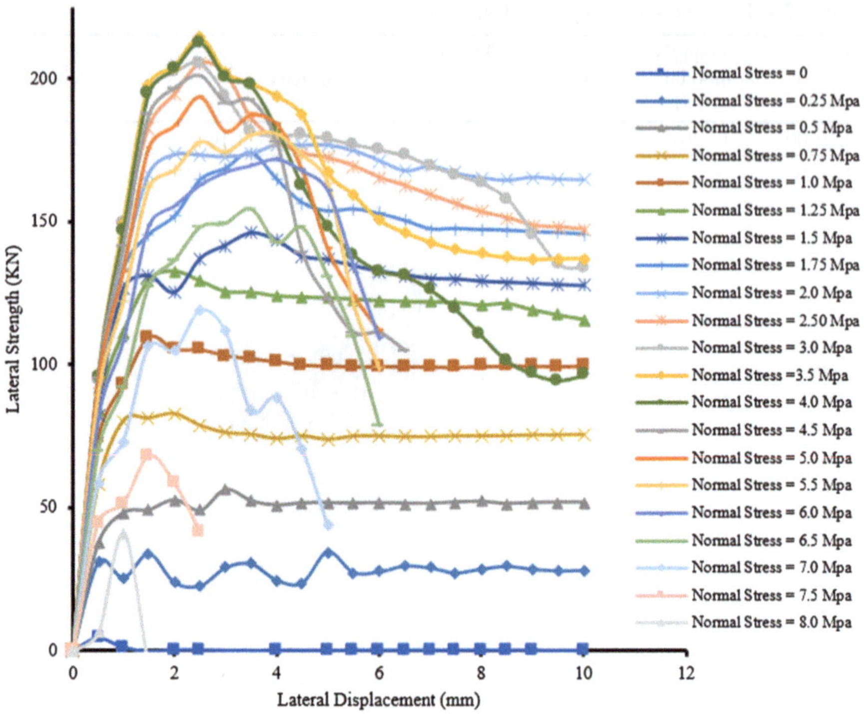

Fig. 7.17 Lateral strength-lateral displacement of the Li et al. (2005) walls

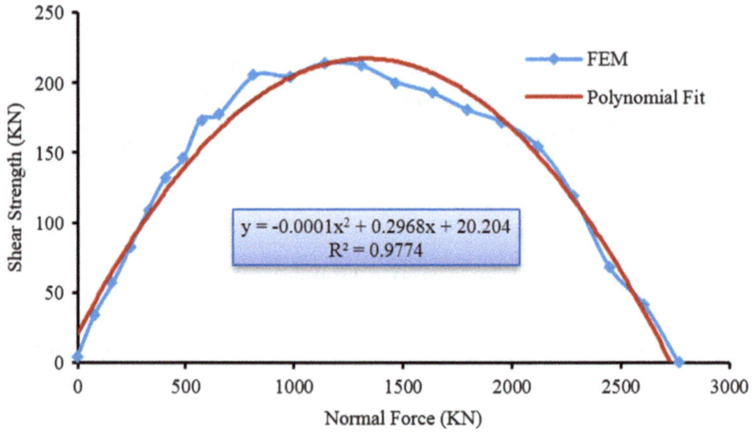

Fig. 7.18 Interaction relation between lateral strength H and normal force P

The behavior of the concrete masonry wall is noted to be very similar to that observed in the brick masonry wall (Fig. 7.6). The results for maximum lateral resistance associated with applied normal force are plotted in Fig. 7.18. A regression fit of the lateral force at collapse of the wall H (kN) on the applied level of

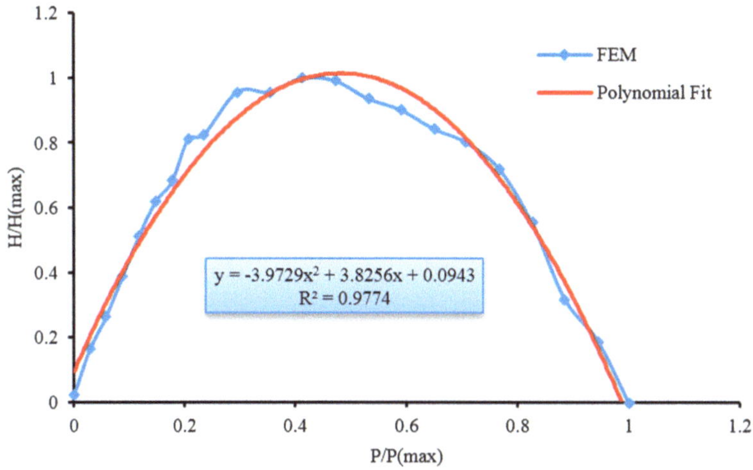

Fig. 7.19 Interaction relation between normalized lateral force and normalized axial force

pre-compression P (kN), is parabolic in P. A similar trend has been reported by Voon (2007). The resulting regression is:

$$H = -0.0001P^2 + 0.2968P + 20.204 \tag{7.23}$$

This relationship may be normalized by taking the ratio of lateral strength to maximum lateral force attained ($H_{max} = 214.3$ kN) and the ratio of axial force to maximum applied axial force ($P_{max} = 2{,}763$ kN). Figure 7.19 shows the normalized curve. The regression analysis of this curve results in the following equation:

$$y = -3.9729x^2 + 3.8256x + 0.0943 \tag{7.24}$$

where $y = H/H_{max}, x = P/P_{max}$.

When investigating the results of FEM simulation, five failure modes were observed, namely sliding, rocking, head/bead joints opening, cracking in units, and crashing of wall. Failure mode in concrete masonry walls per Li et al. (2005) were observed as those in stone masonry walls per Demir (2012).

7.6 Universal Behavior of URM Subjected to In-Plane Loading

From the FEM analysis of Demir (2012) and Li et al. (2005) walls, it is clear that masonry walls subjected to in-plane loading and an aspect ratio close to 1.0 follow a universal behavior. Figure 7.20 shows the normalized relationship between lateral strength and axial force for the FEM simulation of both walls.

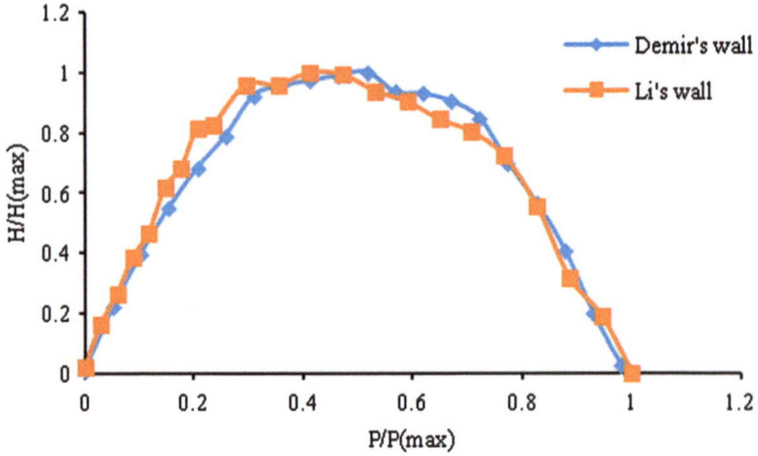

Fig. 7.20 Interaction relation between normalized lateral force and normalized axial force (Demir (2012) and Li et al. (2005) walls)

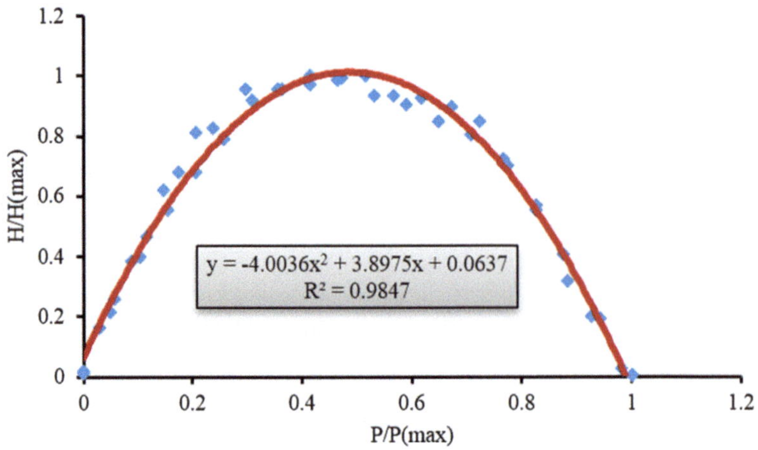

Fig. 7.21 Interaction relation between normalized lateral force and normalized axial force (Demir (2012) and Li et al. (2005) walls)

Regression analysis was carried out for the data of both simulations together. Data from FEM simulation of both types of walls and fitted curve are shown in Fig. 7.21.

The regression analysis of this curve results in the following equation:

$$y = -4.0036x^2 + 3.897x + 0.0637 \tag{7.25}$$

where $y = H/H_{max}, x = P/P_{max}$.

It is clear that Eqs. 7.22, 7.24, and 7.25 are almost identical. This means that URM follows a universal behavior when relating the normalized axial applied force to the normalized lateral strength of the wall.

7.7 Conclusions

This Chapter has demonstrated the simulation of unreinforced masonry walls using an elasto-plastic damage model developed by Lubliner et al. (1989) and further extended by Lee and Fenves (1998), available in the ABAQUS environment. The use of this model requires certain material parameters, including the stress-plastic strain data for hardening and softening in uniaxial compression and uniaxial tension. The wall has been treated as a strong/weak material combination, with blocks being represented as the strong material and the mortar as the weak material.

The results from the interaction of the wall lateral strength to the axial pre-compression show that masonry walls behave in a systematic manner to in-plane loading, regardless of wall size, patterns, and wall materials. When axial stress is low, the wall tends to fail in rigid type modes of sliding and rocking. However, when axial stress is slightly higher, the wall tends to fail by head/bed joints opening and mortar cracking. This mode of failure is common when dry contact system is adopted or when the mortar strength is weak.

For walls with moderate to high axial stress (more than 30 % of the wall axial capacity), cracks start to initiate within the bricks themselves due to diagonal tension in the bricks exceeding the tensile strength of the bricks. These cracks are also often complemented by staggered step cracking in head and bed joints, especially with wet mortar construction.

In cases where the axial load becomes excessive, the lateral strength of the walls is severely compromised; the wall is pre-damaged due to the presence of cracks as a consequence of high pre-compression, and failure of the wall under lateral loading is primarily due to extension of existing cracks in shear-compression.

Acknowledgments The authors wish to acknowledge the support of KFUPM Research Project No. DSR-IN101016 (2011–2014) funded by the Deanship of Scientific Research, King Fahd University of Petroleum and Minerals, Dhahran 31261, Saudi Arabia.

References

ABAQUS Theory Manual, version 6.12, Hibbitt Karlsson & Sorensen, Inc 2012

Abrams D (2001) Performance-based engineering concepts for unreinforced masonry building structures. Prog Struct Eng Mater J 3:48–56

Demir C (2012) Seismic behaviour of historical stone masonry multi-leaf walls. PhD dissertation, Istanbul Technical University (ITU), Istanbul

Filiatrault A, Uang C, Folz B, Chrstopoulos C, and Gatto K (2001) Reconnaissance report of February 28, 2001 Nisqually (Seattle-Olympia) earthquake. Report No SSRP-200102, Structural Systems Research Project, Department of Structural Engineering, University of California, San Diego

Kachanov LM (1986) Introduction to continuum damage mechanics. Springer, New York

Lee J, Fenves GL (1998) Plastic-damage model for cyclic loading of concrete structures. J Eng Mech ASCE 124(8):892–900

Li T, Galati N, Tumialan G, Nanni A (2005) Analysis of unreinforced masonry concrete walls strengthened with glass fiber-reinforced polymer bars. Struct J 102(4):569–577

Lubliner J, Oliver J, Oller S, Oñate E (1989) A plastic-damage model for concrete. Int J Solids Struct 25(3):299–326

Rabotnov YN (1969) Creep problems in structural members. North-Holland, Amsterdam

Vasconcelos GF (2005) Experimental investigations on the mechanics of stone masonry: characterization of granites and behaviour of ancient masonry shear walls. PhD dissertation, University of Minho, Minho. Available at http://www.civil.uminho.pt/masonry/

Voon KC (2007) In-plane seismic design of concrete masonry structures. PhD thesis, The University of Auckland, Auckland

Chapter 8
Implementation of Experimentally Developed Methodology for Seismic Strengthening and Repair of Historic Monuments

Veronika Shendova, Zoran T. Rakicevic, Mihail Garevski, Roberta Apostolska, and Zivko Bozinovski

Abstract The problem of earthquake protection of historic structures and monuments is radically different from that of other structures, due to the priority given to the preservation of the aesthetic, architectonic and historic value, instead of keeping the structure operational. In their effort to protect these structures with the least intervention and the greatest care to preserve authenticity, experts are challenged by the fast development and improved performance of new materials and techniques. However, the implementation of a particular strengthening methodology depends on the extent it has been investigated. Proving the effectiveness of a selected consolidation system can be successfully overcome by design assisted by testing. This Chapter presents the implementation of an original, experimentally verified strengthening methodology in the reconstruction of three very important cultural-historic monuments.

8.1 Introduction

Historic buildings and monuments provide the most tangible legacy of our past civilization; in some cases they speak clearer than any surviving manuscripts. They are usually severely damaged in strong earthquakes due to their stiff and brittle structural components and their lack of ductility that prevents them from being able to sustain the displacements and distortions caused by severe earthquakes. The goal should then be to strengthen these structures in a manner that requires the least

V. Shendova (✉) • Z.T. Rakicevic • M. Garevski
R. Apostolska • Z. Bozinovski
Institute of Earthquake Engineering and Engineering Seismology, IZIIS,
SS Cyril and Methodius University, Salvador Aljende 73,
P.O. BOX 101, Skopje 1000, Republic of Macedonia
e-mail: veronika@pluto.iziis.ukim.edu.mk; zoran_r@pluto.iziis.ukim.edu.mk;
garevski@pluto.iziis.ukim.edu.mk; beti@pluto.iziis.ukim.edu.mk;
zivko@pluto.iziis.ukim.edu.mk

A. Ilki and M.N. Fardis (eds.), *Seismic Evaluation and Rehabilitation of Structures*, Geotechnical, Geological and Earthquake Engineering 26,
DOI 10.1007/978-3-319-00458-7_8,

intervention and greatest care to preserve authenticity (Feilden 1982, 1987; Kelley and Crowe 1995; Tassios 2010; Oliveira and Costa 2010).

The repair and/or strengthening of historic monuments depends on the earthquakes to which they have been exposed in the past and the ground motions to which they will be exposed in the future, as well as the materials and methods used in their construction. It is therefore of great importance that repair and strengthening, as part of the preservation, restoration and conservation of historic monuments in seismic regions, be planned based on detailed studies on the expected seismic hazard, the local soil conditions, the dynamic characteristics of the structure, the strength and deformability characteristics of the structural elements and materials and the dynamic response of structures to expected earthquake ground motions. On the other hand, repair and strengthening should enable economically justified and technically consistent seismic protection, by providing the necessary load-bearing and deformation capacity for an acceptable level of damage in future earthquakes.

The key for selecting materials and techniques is the classification of repair and strengthening techniques into two main categories: reversible and irreversible. In selecting materials to be used in reversible interventions, there are usually only a few limitations. The materials used in irreversible interventions do impose two additional limitations: compatibility of new with old materials and durability.

To define an adequate concept of repair and strengthening, it is necessary to carry out a detailed analysis of the existing structure, the type and physical-mechanical characteristics of masonry, the dynamic properties of the structure, the criteria and the expected seismic action. If this analysis proves that the structure has sufficient load-bearing and deformation capacity, measures for its repair shall be sufficient. Otherwise, depending on the vulnerability level, strengthening should increase the strength of the existing structure or/and its deformability.

The specific nature of the structural systems and the limited possibilities for higher resistance and deformability of the main materials constituting the historic monuments prove the need for a strengthening methodology that will comply with the specific characteristics of these monuments (Danieli et al. 2008; Martelli 2009; Tassios et al. 2007; Carydis et al. 1996; De Canio et al. 2008).

8.2 Development of Methodology for Seismic Strengthening and Repair of Byzantine Churches

Macedonia is known for the large number of historic monuments among which are churches dating from the Byzantine period, representative of topmost architectonic creation with an extraordinary collection of highly valuable frescos. Being located in one of the most seismically active regions in Europe, these buildings have been exposed not only to climatic and man-made damaging factors but also to earthquake effects during the long period of their existence.

Due to the need for development of an appropriate methodology for the repair, strengthening, conservation and restoration of Byzantine churches, ample field,

Fig. 8.1 The church of St. Nikita in Banjani

analytical and experimental studies were performed within the framework of the research project "Study for Seismic Strengthening, Conservation and Restoration of Churches Dating from the Byzantine period (ninth–fourteenth century) in Macedonia", realized by IZIIS in Skopje, the Republic Institute for Protection of Cultural Monuments (RZZSK) in Skopje and the Getty Conservation Institute (GCI) in the US (Gavrilovic et al. 1995, 1999; Shendova 1998).

So far 154 medieval churches and monasteries, 54 of which are architectural creations with an extraordinarily rich collection of invaluable fresco-paintings dating from the period referred to as the golden age of medieval Byzantine art (ninth–fourteenth century), have been registered and put under the protection of Law in the Republic of Macedonia. Based on certain criteria (typology, existing state, interventions and authenticity), four representative churches were selected for investigation, which involved field and analytical tests in order to define seismic parameters and main dynamic characteristics, and to analyse seismic resistance. The church of St. Nikita in the village of Banjani has been selected as a prototype church representative of the Byzantine churches in Macedonia, Fig. 8.1.

8.2.1 Investigations of the Prototype Church of St. Nikita

Ample field and analytical studies, in situ and lab tests were carried out for the existing structure of the St. Nikita church, so as to define the physical-mechanical

Table 8.1 Mechanical characteristics of prototype material

Trial sample	Specific gravity, kN/m^3	Compressive strength, MPa	Shear strength, MPa	Bending strength, MPa	Dimensions of sample, cm
Stone	19.0	13.95	0.145		a/b/H = 5/5/5
Brick	15.30	31.00		5.00	a/b/H = 4/4/4
Mortar	18.10	1.34	0.102		f/H = 5/6

Table 8.2 Maximum ground accelerations and seismic intensities for St. Nikita church

Seismic influence excited by:			Return period of tp (years)						
			25	50	100	200	500	1,000	10,000
Local	Skopje	A_{max}	0.064	0.097	0.142	0.198	0.292	0.340	0.360
Foci		I	6.19	6.79	7.34	7.82	8.38	8.60	8.68
R < 40 km	Skopje +	A_{max}	0.089	0.117	0.146	0.198	0.292	0.340	0.360
	Kacanik	I	6.66	7.06	7.36	7.82	8.38	8.60	8.68
Adjacent		A_{max}	0.060	0.076	0.086	0.090	0.097	0.100	0.100
Foci	R > 40 km	I	6.09	6.43	6.61	6.68	6.79	6.83	6.83
All		A_{max}	0.093	0.117	0.146	0.198	0.292	0.340	0.360
Foci		I	6.73	7.06	7.36	7.82	8.38	8.60	8.68

and chemical characteristics of the built-in materials, the dynamic characteristics of the structure and the seismicity of the terrain (Tables 8.1 and 8.2).

Detailed investigations of the main dynamic characteristics have been performed for the prototype church in both orthogonal directions by applying the ambient vibration technique. Defined in this way were the natural frequencies 4.8 and 6.0 Hz for the N-S and E-W directions respectively.

A preliminary analysis of the seismic stability of the existing structure points to insufficient ultimate load-bearing and deformation capacity according to the design criteria on seismic safety. The proposed concept of strengthening consists of placing horizontal and vertical steel ties and filling the area around them with an appropriate material, to improve the deficient capacity of the existing structure.

8.2.2 Experimental Investigations of the Church Model

To experimentally verify the proposed methodology for repair and seismic strengthening of Byzantine churches, a 1:2.75 scale model of the prototype church of St. Nikita (M-SN-EXIST) was constructed and tested on the seismic shaking table of the IZIIS Dynamic Testing Laboratory, Fig. 8.2. The geometrical scale of the model church was selected on the basis of the characteristics of the seismic shaking table and the precisely defined objectives of testing, i.e. realistic reproduction of nonlinear behavior and failure mechanisms. Satisfying these criteria and adopting the "gravity forces neglected" modelling principle, the following main

Fig. 8.2 Models on shaking table

scales were adopted: geometrical scale $L_r = 1{:}2.75$, scale for the bulk density of the material $\rho_r = 1$, scale for the stresses $E_r = 1$.

The main targets of the experimental tests performed on the church model can be summarized into two groups: (i) assessment of the vulnerability of structures of interest, and (ii) selection of the most appropriate procedures for repair and strengthening of damaged structures in post-earthquake protection. To that effect, a programme of experimental tests was adopted by gradually increasing the intensity of input earthquake excitations and monitoring the progressive development of cracks and failure mechanism, the modification of the dynamic characteristics as well as the phases of dynamic behavior of the model, i.e., defining the elasticity limit (occurrence of first cracks).

The investigation was performed by simulating two main types of earthquake: the 1976 Friuli (IT) earthquake (Breginj record) as a local one, and the 1979 Montenegro (Petrovac record) and 1940 El Centro earthquakes as distant ones. From the general behavior of the model, it was concluded that it behaved as a rigid body in the elastic range, but when the first larger cracks occurred there was separation of the bearing walls and damage up to a state close to complete failure. This was proved by the decrease in natural frequency from 11. to 6.6 Hz.

Both from the analysis of the seismic stability of the existing prototype structure and the experimental investigations of model M-SN-EXIST, it was observed that it did not possess the necessary stability and resistance to intensities corresponding to the design and maximum earthquake. Therefore, after repairing the model by injection of lime-based mixture, it was structurally strengthened in accordance with the proposed methodology. Strengthening elements were the horizontal and the vertical belt courses formed by incorporating steel reinforcement into the wall mass and by then filling the area around them with an appropriate injection mixture that enabled contact with the surrounding existing masonry, Fig. 8.3.

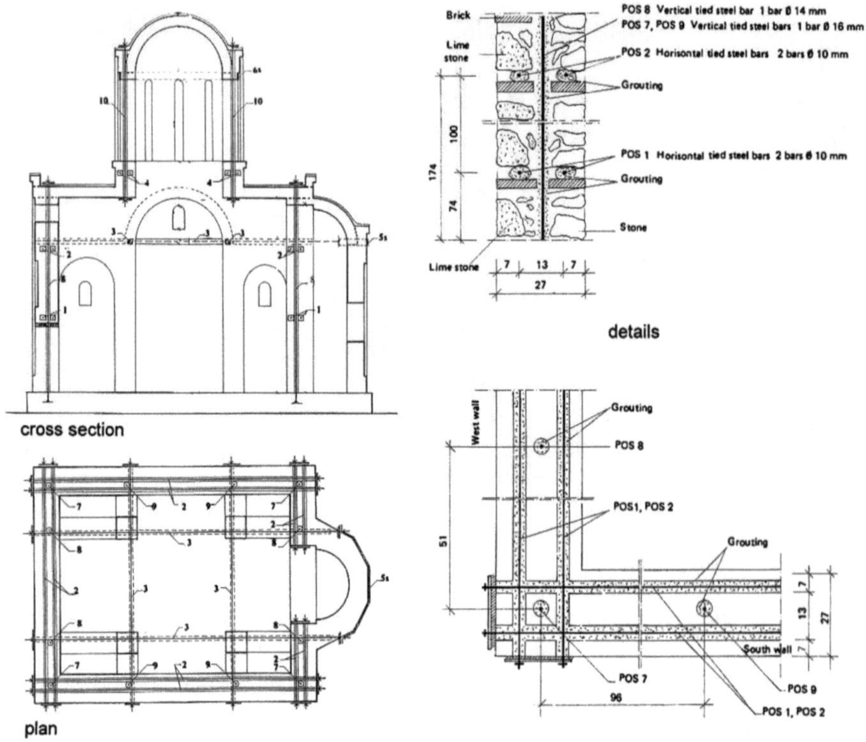

Fig. 8.3 Applied method for strengthening

Table 8.3 Comparison between the experimental results of models

		Output acceleration (in g) for the church models			
		M-SN-EXIST		M-SN-STR	
Earthquake	Input acc (g)	Level 1	Level 2	Level 1	Level 2
El Centro	0.17	0.29	0.55	0.20	0.47
El Centro	0.30	–	–	0.65	1.10
El Centro	0.49	–	–	0.91	1.59
El Centro	0.54	–	–	0.77	1.41
El Centro	0.63	–	–	–	–
Petrovac	0.19	0.39	0.76	0.27	0.48
Petrovac	0.40	–	–	0.77	1.36
Breginj	0.17	0.22	0.52	0.30	0.55
Breginj	0.28	–	–	0.20	0.40
Breginj	0.38	–	–	0.34	0.79

Level 1 base of the tambour, *Level 2* top of the dome

The repaired and strengthened model (M-SN-STR, Fig. 8.2), was subjected to the same series of dynamic tests for the purpose of proving the efficiency of the applied method of strengthening. However, due to the high resistance of the strengthened model, the tests were continued under higher intensities (Table 8.3).

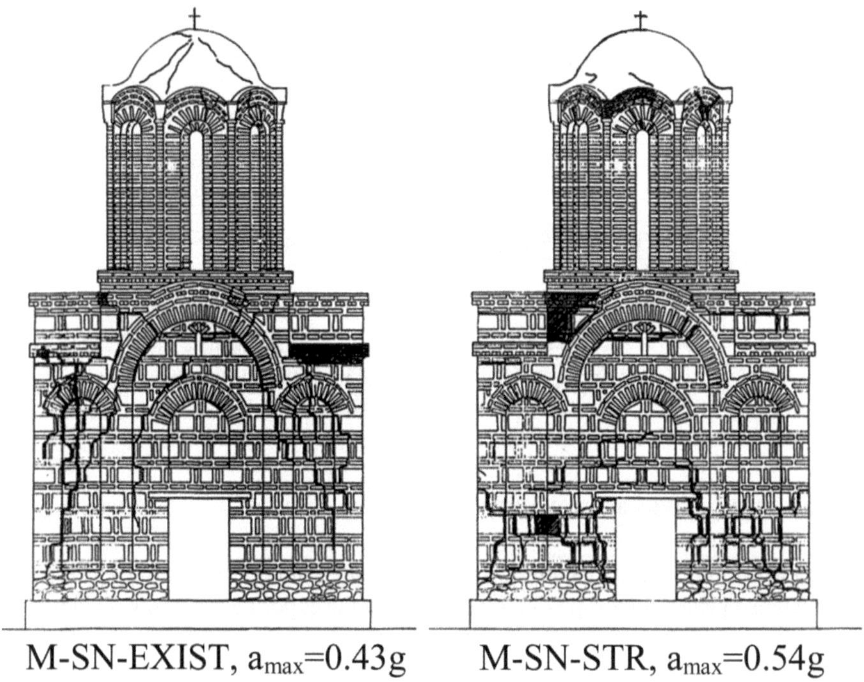

Fig. 8.4 Failure mechanisms (west facade)

The response of the strengthened model was considerably different from that of the original model. The increased elasticity limit and reduction of displacements at the top were characteristic. Although there was a considerable deterioration in the load-bearing capacity under maximum seismic effect, the complete stability of the model structure was not disturbed due to the presence of ductile elements, while the damage was such that it was repairable. The structure of the strengthened model was in a state of deep nonlinearity, but still far from failure.

Comparative analysis of experimentally obtained results shows that:

- The response of the structures to input acceleration of up to $a_{max} = 0.2g$ entails first cracks in the original model that remain open, whereas the strengthened model remains evidently completely elastic.
- Under $a_{max} \approx 0.4g$, the original model suffers severe nonlinear damage close to failure, whereas the same excitation level is the elasticity limit and beginning of nonlinearity for M-SN-STR.
- A qualitatively different type of failure mechanism has also been observed. The strengthened model does not suffer separation of bearing walls and vertical crack, but the failure is transferred to the lower zone and results in occurrence of diagonal cracks (Fig. 8.4).
- The applied methodology for repair and strengthening that is in compliance with the principle of "minimum interventions – maximum protection" increases the

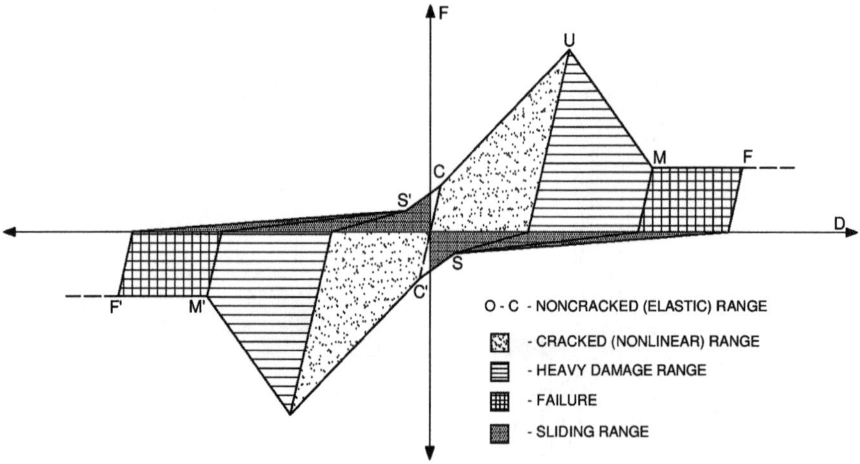

Fig. 8.5 Tri-linear IZIIS hysteretic model

load-bearing capacity and deformability of the structure up to the level of the designed protection, as proved by comparing the experimentally obtained results for both models.

8.2.3 Analytical Investigations of the Model Dynamic Response

As part of the analytical investigations of the dynamic response of the original and the strengthened model, (equivalent) static and dynamic analyses were performed according to the type of external effects, whereas according to the adopted discretization level, the Finite Element Method (FEM) and modelling by masses concentrated at two characteristic levels of the model structure were applied. Two ranges of dynamic response are distinguished for the M-SN-EXIST model: *elastic* ($a_{max}{}^{input}$ = 0.1–0.12g) and *nonlinear* ($a_{max}{}^{input}$ = 0.12–0.2g) range. For model M-SN-STR, three ranges are evident: *elastic* ($a_{max}{}^{input}$ < 0.2g), *nonlinear* ($a_{max}{}^{input}$ = 0.2–0.4g) and *ultimate range* ($a_{max}{}^{input}$ > 0.4g with evident stiffness degradation and large nonlinear deformations). The FE analysis gives possibilities in elastic static or dynamic analysis and determination of the main dynamic characteristics. Depending on the refinement of the discretization, a complete and precise definition of the stress and strain state and detection of weak points of the structure is possible.

Using a model with lumped masses at two characteristic levels (top and level 1), a nonlinear dynamic analysis was performed with a complex hysteretic (IZIIS) model (Fig. 8.5, Shendova 1998). To calibrate the computations in defining the capacity degradation in hysteretic models, the results from seismic shaking table testing of the model were used. With this, an attempt was made to model the dynamic response in a simple way, suitable for everyday analyses, resulting however in satisfactory final results on the behavior at individual levels.

Table 8.4 Input parameters for dynamic analysis using the IZIIS model

	Existing structure		Strengthened structure	
	Force Q	Displacement D	Force Q	Displacement D
Point C	$K_C = K$, $Q_C = (25–30)\%Q$	$D_C = 0.2‰H$	$K_C = K$, $Q_C = (25–30)\%Q$	$D_C = (0.2–0.3)‰H$
Point U	$Q_U = Q$	$D_U = (1.5–2)‰H$	$Q_U = Q$	$D_U = (1.5–2)‰H$
Point M	$Q_M = (20–30)\%Q$	$D_M = (2–3.5)‰H$	$Q_M = Q$	$D_M = (3–3.5)‰H$
Point F	$Q_F = (10–15)\%Q$	$D_F = (4–4.5)‰H$	$Q_F = (10–15)\%Q$	$D_F = (5–5.5)‰H$
Point S	$Q_S = (5–10)\%Q$	$D_S = (.05–0.1)‰H$	$Q_S = (5–10)\%Q$	$D_S = (.05–0.1)‰H$

With the IZIIS model the behavior of the church model is successfully described in all the characteristic phases. In the linear dynamic analysis of both models, the initial stiffness was defined based on the experimental main frequencies. In the nonlinear dynamic response all the necessary input parameters (stiffness, displacements, and damping) for the IZIIS model were defined on the basis of identification of the analytical and the recorded response. The response of model M-SN-STRN under $a_{max}^{input} > 0.5g$, was analysed by including the post-ultimate range of the three-linear hysteretic IZIIS model. The results obtained by applying the idealized model with two masses and the IZIIS hysteretic diagram correlate well with the recorded values, the shape of the time histories and the frequency content of accelerations for all characteristic ranges of behavior of the models, the correlation being greater with greater regularity of response.

The shaking table testing of a large scale model of a historic monument presented herein is a unique example in world practice. The knowledge gained, therefore, is valuable and necessary for the seismic strengthening of important cultural-historic structures, particularly in cases where the effect of interventions upon the authenticity of the monument is considered a priority.

Table 8.4 summarizes the knowledge gained in the form of input parameters defining the envelope points from the hysteretic model as empirical relationships between storey stiffness (K), storey height (H) and ultimate storey shear force (Q). In this way, knowing the values of K and Q, it is possible to analyse the dynamic response of an actual structure without previous experimental testing.

8.3 IZIIS' Integrated Approach in Seismic Protection of Historic Buildings and Monuments

Within the framework of IZIIS' research activities, which also cover the seismic design of modern structures, the experience gathered in the field of protection of structures pertaining to the cultural historic heritage is particularly noteworthy.

With an experience of over 30 years in this field, the Institute has realized important scientific research projects involving experimental and analytical research, field surveys and application of earthquake protection knowledge to important cultural historic structures and monuments.

Extensive research activities have been performed by IZIIS with the aim to evaluate procedures for repair and strengthening of valuable historic monuments. An integrated approach to the seismic protection of important historic structures has been adopted by the Institute and used in the process of reconstruction of the three monuments that are the subject of this chapter (Gavrilovic et al. 2003, 2004; Shendova et al. 2006). This approach, which complies with the restoration and conservation requirements as well as legislative regulations for high importance structures, encompasses the following:

- Definition of expected seismic hazard;
- Definition of soil conditions and dynamic behavior of soil media;
- Determination of structural characteristics and load-bearing and deformation capacity of existing structures;
- Definition of criteria and selection of concept for repair and/or strengthening;
- Definition of structural methods, techniques, materials and types of excitation;
- Analysis of dynamic response of repaired and/or strengthened structures and verification of their seismic stability;
- Definition of field works, execution and inspection.

Although the above seems to be the "normal procedure", it is the only way to provide high quality protection of cultural heritage. This task is much more than just listing of what is to be done, since it requires a lot of knowledge and efforts.

The main purpose of detailed geophysical surveys for the definition of the geotechnical and geodynamic models of the site is to define the seismic parameters for evaluation of the seismic stability of the structure. Namely, for structures of extraordinary importance according to the applicable regulations, it is necessary to define the seismic input for the site, in order to perform correctly the dynamic analysis. The investigations have been carried out in compliance with the latest achievements in the field of earthquake engineering. The main concept of the applied procedure is to consider the expected earthquake effect through a probabilistic approach, including also the local soil effects through nonlinear dynamic analysis of a representative geotechnical model.

Presented further in this chapter are the methods that have been used for the analysis under gravity and seismic load of three real monument structures, for which the developed strengthening methodology has been implemented.

8.3.1 Analysis of Load-Bearing and Deformation Capacity

The load-bearing and deformation capacity is the main parameter defining the behavior of the structure and of the individual structural elements. Presented in the following is the methodology for defining the load-bearing capacity of the church in the form of ultimate storey shear force that, compared to the equivalent seismic force, yields the factor of safety against failure (a procedure which is widely applied in equivalent static analyses of masonry structures at IZIIS).

Structural Treatment: The principal historic masonry structure consists of walls in two orthogonal directions, so that modelling for seismic effects is reduced to the modelling of the shear walls. In the analysis, the walls are usually treated as vertical cantilevers fixed at the base and interconnected by floor structures that are assumed to have the role of rigid diaphragms that transfer the horizontal effect of the individual walls proportionally to their rigidity in the considered direction. The assumption about the vertical cantilevers holds in the case when the individual walls are interconnected by flexible floor structures (timber ones) and when the openings are of minor proportions so that the simple cantilever mechanism of the behavior of the integral wall is not disturbed. In the cases where there are horizontal belt courses and rigid floor structures, the wall element is modelled as fixed at both ends. If the openings in the walls are of greater proportions, the weak point of the structure under seismic effect are the columns between the windows or the lintel beams so that one cannot speak about modelling of the integral wall but modelling of the individual wall elements.

In the historic structures, there are systems of arches, vaults and pendentives that bridge horizontal areas as floor structures. These are assumed to have the role of transferring the horizontal loads to the walls depending on their rigidity. The facade massive walls most frequently have openings in the form of doors, monophoriums, biphoriums or weakening of the walls in the form of niches and pilasters. The interior walls are always with large openings that form, in that way, the central columns supporting the upper part of the structure. The individual walls (external) or wall elements (internal) behave in different ways which aggravate the mechanism of behavior of the system to such an extent that it is practically impossible to perform precise mathematical modelling and analysis, even with the present level of knowledge of the problem.

The mathematical model that is applied in further analyses is based on modelling of the facade walls and the individual elements of the interior walls as vertical cantilevers fixed at the base. If there are data on the existence of horizontal belt courses, the walls can be modelled as fixed at both ends.

Stiffness and Deformability: In principle, the problem of formulation of a theoretical mathematical model of an actual structure to represent the effect of seismic forces in general and separately on masonry structures, is reduced to determining the factors that define the deformation of the structure in the horizontal direction and defining the distribution of masses along the height of the structure. The deformability and stiffness characteristics depend on the geometry of the structure and the material. Generally, the total deformation D of the wall loaded by a horizontal force consists of the following components:

$$D = D_s + D_b + D_x + D_\varphi \tag{8.1}$$

where:

D_s – shear deformation
D_b – bending deformation
D_x – horizontal deformation, displacement in the direction of the force
D_φ – deformation due to rotation of the base around the neutral axis

Assuming a cantilever wall with height h, length l and thickness t is acted on by a unit horizontal force, the following is obtained:

$$D = k_i h/GA + h^3/3EJ + 1/k_x + h^2/k_\varphi \tag{8.2}$$

where:

k_i – coefficient of shape of the cross-section (5/6 for rectangular cross-sections)
G, E – shear and elasticity moduli
A, J – surface area and moment of inertia of cross-section
k_x – soil translational stiffness
k_φ – soil rotational stiffness at the base

In strip foundations with depth less than the width k_x and k_φ are large, so their contribution to the total deformation basis in Eq. (8.2) may be neglected. The stiffness, K = 1/D, of a fixed element or a cantilever including shear effects is:

$$K_{fixed} = 12EJ/h^3(1+\alpha),\ K_{cantilever} = 3EJ/h^3(1+\alpha/4), \alpha = 12EJk_i/GA \tag{8.3}$$

For a rectangular cross-section, $J = Al^2/12$, and the following is obtained

$$\begin{aligned} K_{fixed} &= GA/1.2h\left[1 + 0.83Gh^2/El^2\right], \\ K_{cantilever} &= GA/1.2h\left[1 + 3.33Gh^2/El^2\right] \end{aligned} \tag{8.4}$$

Bending and shear bearing capacity: Bending and shear bearing capacity is expressed as the maximum (ultimate) horizontal force that can be developed in the wall during bending or shear failure; it is given in a tabular form (Table 8.5).

Computation of load-bearing and deformation capacity: While calculating the load-bearing and deformation capacity of a masonry structure, it is necessary to:

1. Compute masses lumped at individual characteristic structural levels; define the geometry of the bearing walls with the corresponding total vertical loads;
2. Adopt physical-mechanical characteristics of the masonry (elastic modulus E, shear modulus G, ultimate strength in compression, f_c, and tension, f_t);
3. Define the total seismic force at the base, S, and the storey horizontal forces S_i in both orthogonal directions, compliant with the applicable seismic codes:
4. Adopt adequate mathematical models and compute the initial stiffness k_j of the individual walls in the analysed direction;

Table 8.5 Bending and shear bearing capacity

	Bending capacity (cantilever: $\alpha = 1$, fixed: $\alpha = 1/2$)	Shear capacity
Plain masonry	$Q_u^{bending} = M_u/(\alpha h)$	$Q_u^{shear} = A\tau_u$
	$M_u = 0.5\sigma_0 tl^2(1 - \sigma_0/f_c)$	$\tau_u = f_t/b\ (\sigma_0/f_t + 1)^{1/2}$
Confined masonry	$Q_u^{bending} = M_u/(\alpha h)$	$Q_u^{shear} = A\tau_u$
	$M_u = 0.5\sigma_0 tl^2(1 - \sigma_0/f_c) + 2A_sF_{sy}(1/2 - 1')$	$\tau_u = f_t\left\{h/2l + \left[(h/2l)^2 + N/A/f_t + 1\right]^{1/2}\right\}$
Jacketed masonry	$Q_u^{bending} = M_u/(\alpha h)$	$Q_u^{shear} = A\tau_u$
	$M_u = 0.5\sigma_0 t_{ekv} l^2(1 - \sigma_0/f_c) + 0.85\, F_a\sigma_y l$	$\tau_u = f_t(\sigma_y/f_z + 1)^{1/2} + 0.2F_{av}\sigma_y + F_{ah}\sigma_y$

5. Define the associated seismic forces S_j by distributing the storey seismic force, S_i, in proportion to the stiffness k_j, as $S_j = S_ik_j/\sum k_j$, where S_j is the shear equivalent force in the j-th wall and k_j is the initial stiffness of the j-th wall
6. Define the load-bearing capacity of the horizontal cross-section as well as the bending ($Qu^{bending}$) and shear capacity (Qu^{shear}) of each individual wall, adopting the lower values as competent ultimate force Qu_j.
7. Compare the competent ultimate shear force Qu_j to the associated horizontal seismic force for each wall S_j and express their relationship in the form of factor of safety against failure of the wall ($F_j = Qu_j/S_j$);
8. Define the storey Q-D diagram by summing up the Q_j-D_j diagrams for all the walls in the considered direction;
9. Compare the storey ultimate force Qu_i to the storey force S_i and obtain the factor of safety against failure Fu_i of the considered storey i, ($Fu_i = Qu_i/\ S_i$).

The interpretation of the results obtained in this way makes it possible to consider the behavior of each wall separately, as well as the behavior of the entire structure. For the described procedure, the computer programme has been elaborated; input parameters include the geometry of the walls, characteristics of the material and the seismic forces, whereas the output data are the factors of safety against occurrence of the first cracks and the factors of safety against failure for each of the analysed walls and for the structure as a whole.

Through this procedure, the individual structural units of the three monuments have been analysed and the strengthening elements proportioned. The results for the three monuments are given in Sects. 8.4.1, 8.4.2 and 8.4.3.

8.3.2 *Analysis of Stress-Deformation State of the Structure by Use of the Finite Element Method*

Based on the defined structural systems of the three individual monuments and the defined strengthening structural elements, a static and equivalent seismic analysis has been performed by using the finite element method and the software SAP 2000 (Structural Analysis Programme, UC Berkeley, California). Taking into account the complexity and the specific nature of the structural system and the built-in materials of the structure on the one hand and the possibilities offered by the programme package on the other, an attempt has been made to define a FE model that most appropriately represents the structure. A moderately dense mesh has been adopted, involving the global geometrical characteristics of the model without paying attention to the inhomogeneity of the material.

The bearing massive walls and the tambour walls have been modelled by the SOLID 3D finite element with eight nodes. The steel vertical and horizontal strengthening elements were modelled by 3D-FRAME, i.e., 3D-TRUSS elements. All the vaults and the domes have been modelled by SHELL elements.

A static analysis has been performed for dead weight and equivalent seismic forces computed according to the valid technical regulations. The input geometrical and physical-mechanical characteristics of the corresponding elements as well as the output results on the stress–strain state for each of the analysed monument structures are given in Sects. 8.4.1, 8.4.2 and 8.4.3 respectively.

8.3.3 Dynamic Analysis of the Structure

A nonlinear dynamic analysis has been performed for the three monuments (Sect. 8.2.3), with the masses lumped at characteristic levels and applying a corresponding storey hysteretic model obtained by summing up the elastoplastic characteristics of each of the bearing walls, with the load-bearing capacity of each of them limited to the bending and shear capacity, whichever is less.

The dynamic analysis has been performed for the seismic parameters, defined by geotechnical investigation as different types of earthquake with defined maximum expected input acceleration for the corresponding return period. Obtained as results from the dynamic analysis are the storey displacement and the ductility, required by the earthquake, that has to comply with the design seismic safety criteria (Sect. 8.3.4). The results from this analysis are given in Sects. 8.4.1, 8.4.2 and 8.4.3 for each of the analysed monument structures, respectively.

8.3.4 Design Seismic Safety Criteria

The seismic safety criteria and the method of strengthening of a particular historic structure are defined based on the structural response to expected earthquakes and its seismic stability, taking into account its main characteristics and artistic value.

As in design of new structures, in repair and strengthening of damaged historic monuments or preventive seismic strengthening of structures of vital importance, it is necessary to define the design seismic criteria as follows:

Level I: Under earthquakes of a lower intensity and shorter return period, the dynamic behavior of the structure may not lead to vibrations and induce damage to both structural and secondary, non-structural elements (the behavior should be completely in the elastic range with required ductility of $\mu < 1$);

Level II: Under earthquakes of a higher intensity, i.e., under the so-called design earthquakes, the structure should generally remain in the linear range of behavior, with possible limited nonlinear deformations of individual elements of the system, which means limited stiffness deterioration and energy dissipation (initial nonlinear behavior with required ductility of $\mu < 1.5$)

Level III: Under maximum expected earthquake effects, the structural and non-structural elements of the structure are deeply in the nonlinear range of behavior, while the stiffness and the resistance of the structure are considerably reduced. However, such earthquakes must also not disturb completely the stability of the bearing structure, i.e., the damages induced should be repairable (nonlinear behavior with required ductility of $\mu < 2$).

In addition to this, in each intervention to be carried out for such type of structures, certain principles and rules must be respected. Among these, as mentioned before, the main principle is to provide maximum safety with minimal intervention. For each historic monument, the design criteria are defined on the basis of special conditions that depend on the historic, architectural and artistic value of the structure, the seismic hazard and the possibility of application of a corresponding measure for repair and strengthening.

The above design criteria have been used in the process of reconstruction and strengthening of three monuments as described in the following.

8.4 Reconstruction of Monuments Implementing the Developed Strengthening Methodology

After the realization of the project on Byzantine churches, IZIIS became partner of the National Conservation Center in R. Macedonia, which enabled direct application of the unique knowledge gained in actual conditions and for specific historic monuments. Presented in the following are the three most characteristic examples of application of the developed methodology.

8.4.1 Consolidation, Reconstruction and Seismic Strengthening of St. Clement's Church, St. Panteleymon, Plaoshnik, Ohrid

The church of St. Panteleymon was situated in Plaoshnik, between Lake Ohrid and the ramparts of Samuil's fortress, which date back some 1,100 years. The Plaoshnik compound is protected by the national Law for Protection of Cultural Monuments.

Fig. 8.6 Existing state of the church (excavated material 2001)

The archaeological investigations and excavations in Plaoshnik in 1942, 1965, 1999 and particularly in 2000 enabled complete definition of the phases of construction of St. Cement's church, dating back to the ninth–fourteenth century (Fig. 8.6).

In 2001, on the initiative of the Organizational Board for Restoration of St. Clement's church "St. Panteleymon" in Plaoshnik, Ohrid, the idea to reconstruct the church based on the original remains of the church walls without damaging the excavated fresco fragments was put forward. A professional body consisting of eminent architects, structural engineers, archaeologists, art historians and conservators was constituted, with the aim to find the most appropriate solution for the consolidation, rebuilding and strengthening of the church.

Based on the archaeological investigations, architectonic documentation was elaborated by the Institute for Protection of Cultural Monuments and National Museum – Ohrid. IZIIS contributed the design of the structural system, technical solutions for the consolidation of existing foundations and walls, as well as the methodology for repair and strengthening of the church (Gavrilovic et al. 2001a, b; Necevska-Cvetanovska and Apostolska 2008; Shendova and Gavrilovic 2004). The abovementioned activities were a complex task because of the historic, architectonic, artistic and national value of the structure. Therefore, according to IZIIS' integrated approach, detailed investigations were carried out to define the following parameters:

- Physical-mechanical characteristics of materials (stone, brick, lime mortar) that, besides the strength-deformation characteristics, had to meet the conservation requirements concerning the authenticity of the built-in material,
- Seismic design parameters,
- Geomechanical investigations,
- Dynamic and strength-deformability characteristics of the structural system,
- Dynamic response of the structure under expected ground motion.

Since seismic microzonation for definition of the effects of the local soil conditions was not performed for the site, approximate values were obtained using the results from detailed geological and geotechnical investigations done at a nearby location. The basic design criteria and requirements for the structure were:

- To be built from traditional materials, like in the original church, and on the existing foundations.
- To meet the criteria for seismic stability and safety under the specified ground acceleration of $a_{max} = 0.36g$, to safeguard the global integrity of the structure.

Presented further herein are parts of the results from the static and dynamic analyses performed according to the procedure described in Sect. 8.3, as well as the characteristic details of repair and strengthening of the structural system and consolidation of the existing foundations and walls.

8.4.1.1 Load-Bearing and Deformation Capacity of the Structure

In accordance with the procedure and methodology developed at IZIIS (Sect. 8.3.1), two variants of the structural system of the church of St. Panteleymon were analyzed. In the first, the church structure was analyzed only with bearing walls in both orthogonal directions, i.e. plain masonry. The total horizontal force at the base was computed in accordance with the valid national regulations for the construction of structures under seismic conditions. The following physical–mechanical characteristics of masonry are input to the analysis:

- Modulus of elasticity 1,200 MPa
- Shear modulus 720 MPa
- Ultimate compressive strength 1 MPa
- Ultimate tensile strength 0.1 MPa

However, since the safety factor on occurrence of the first cracks was less than unity, i.e., the structure did not have sufficient load-bearing and deformation capacity, strengthening of the principal structural system is needed.

Strengthening of the structure was done by incorporation of horizontal and vertical ties in accordance with the originally developed and experimentally verified methodology (Sect. 8.2). As second variant, the walls were treated as confined masonry at three story levels with horizontal and vertical ties. The results for the strengthened structure showed that the story safety factor against failure in both directions is greater than unity (Tables 8.6 and 8.7), i.e., that the walls had a sufficient load-bearing and deformation capacity.

8.4.1.2 Analysis of the Structure with the Finite Element Method (FEM)

3-D static and linear dynamic analysis of the structural model with FEM (SAP 2000) enabled to obtain and interpret the stress state under static effect and to calculate the main dynamic characteristics (Sect. 8.3.2). The bearing massive walls

Table 8.6 Load-bearing and deformation capacity – longitudinal direction

Storey	Weight, kN	Q_b per code, kN	Q_y, kN	Q_u, kN	D_y, cm	D_u, cm	F_y	F_u
3	892	91	1,181	2,108	0.13	0.54	1.29	2.30
2	5,136	4,534	4,983	5,531	0.43	0.51	1.10	1.22
1	12,040	8,131	9,252	10,717	0.28	0.34	1.14	1.32

Q_y – force at first cracking
Q_u – ultimate bearing capacity
D_y – displacement at first cracking
D_u – ultimate displacement
F_y – safety factor against first cracking
F_u – safety factor against failure

Table 8.7 Load-bearing and deformation capacity – transverse direction

Storey	Weight, kN	Q_b per code, kN	Q_y, kN	Q_u, kN	D_y, cm	D_u, cm	F_y	F_u
3	892	91	1,713	1,864	0.22	0.26	1.82	1.98
2	5,136	4,534	4,603	5,861	0.49	0.66	1.01	1.29
1	12,040	8,131	6,976	8,065	0.41	0.49	0.86	0.99

of the church were modeled by a three dimensional finite element with eight nodes of the SOLID type, i.e., a total of 4,083 elements. The steel vertical and horizontal ties were modeled by 3D-FRAME i.e., a total of 1,061 3D-TRUSS elements. The domes of the church bell tower, the central dome and the altar were modeled by 265 SHELL elements (Fig. 8.7). The above discretization was adopted to encompass the global geometrical characteristics of the model without paying particular attention to the inhomogeneity of the material.

A static analysis has been performed to compute the effect of dead loads and equivalent seismic forces according to the code. This analysis may give insight into the design values of axial stresses at the base of the model under dead weight, i.e., where cracks are most likely to occur (wherever the principal tensile stress exceeds the masonry tensile strength). The maximum calculated compressive stresses under vertical loads are less than the compression masonry strength.

8.4.1.3 Dynamic Analysis of the Structure

The analytical investigations involved definition of the seismic response of the structure in the nonlinear range of behavior. To model the behavior of the structural system in the elastic phase, at the beginning of nonlinearity (occurrence of cracks) and deep nonlinearity, the IZIIS three-linear hysteretic model with stiffness degradation and "pinching" effect was used (Fig. 8.5).

To obtain the dynamic response (according to Sect. 8.3.3) of the structure, four different types of earthquakes (El Centro 1940, Parkfield 2004, Ulcinj 1979, Petrovac 1979) with a maximum input acceleration of 0.36g have been applied. This set of records was chosen in order to investigate nonlinear structural response to excitations with different frequency content and duration. The results (Tables 8.8 and 8.9) show that, under acceleration of 0.36, the structure behaves in accordance with the designed seismic safety criteria (Sect. 8.3.4).

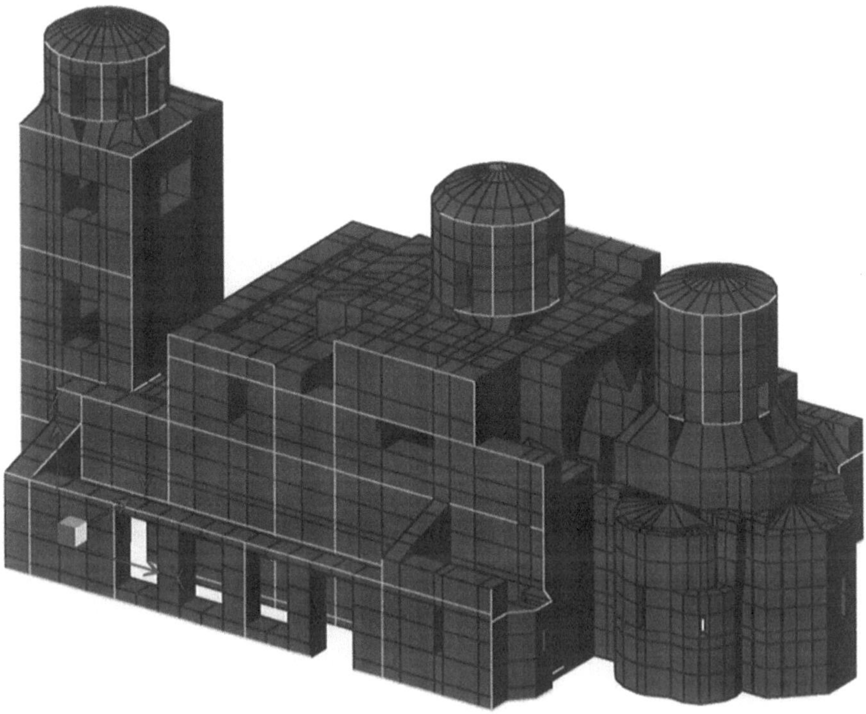

Fig. 8.7 3D view of the model of the structure

8.4.1.4 Consolidation and Reconstruction of the Church

Based on the performed analyses the structural system of the church was defined in accordance with the developed methodology (Sect. 8.2), the horizontal and vertical steel ties were proportioned and a solution for the foundation of the structure was given.

- The principal structural system consists of massive stone and brick masonry with incorporated horizontal and vertical steel ties. The ties of the external walls have size 50 mm × 50 mm × 3 mm, whereas those of the internal walls are 60 mm × 60 mm × 4 mm. The distribution and the distance between the steel ties can be seen on the plan and cross-section of the structure in Fig. 8.8.
- The stone and brick masonry were constructed in lime mortar with characteristics selected through static and dynamic analyses.
- The existing walls below the floor level were systematically injected with cement emulsion. The injection of the walls over the floor level with remains of fresco paintings was done by use of emulsions that did not contain cement.
- With the injection of foundation walls moisture penetration must be eliminated.
- The contact between existing and rebuilt walls was solved along certain levels, depending on the level of the existing walls.

Table 8.8 Results from dynamic analysis – longitudinal direction

	Storey	Required displacement (cm)	Capacity displacement (cm)	Required ductility
Petrovac, N-S, 0.36g	3	0.186	0.640	0.372
	2	0.391	0.511	0.891
	1	0.327	0.340	1.088
Ulcinj, N-S, 0.36g	3	0.282	0.640	0.563
	2	0.574	0.511	1.309
	1	0.397	0.340	1.318
El-Centro, 0.36g	3	0.254	0.640	0.508
	2	0.462	0.511	1.082
	1	0.288	0.340	0.957
Parkfield, 0.36g	3	0.137	0.640	0.274
	2	0.261	0.511	0.594
	1	0.215	0.340	0.713

Table 8.9 Results from dynamic analysis – transverse direction

	Storey	Required displacement (cm)	Capacity displacement (cm)	Required ductility
Petrovac, N-S, 0.36g	3	0.119	0.260	0.534
	2	0.607	0.660	1.090
	1	0.545	0.487	1.334
Ulcinj, N-S, 0.36g	3	0.131	0.260	0.590
	2	0.687	0.660	1.233
	1	0.548	0.487	1.341
El-Centro, 0.36g	3	0.112	0.260	0.505
	2	0.563	0.660	1.010
	1	0.472	0.487	1.155
Parkfield, 0.36g	3	0.063	0.260	0.286
	2	0.328	0.660	0.589
	1	0.358	0.487	0.876

- For the strengthening and consolidation of the existing foundation walls up to level 0 (up to the slab), a reinforced concrete (RC) belt course was formed from the inner and outer side of the existing walls, running under the existing wall along about 0.5 m and along the height of the existing wall with a thickness of 0.2 m up to the floor slab. The belt course was appropriately reinforced and connected to a RC slab with thickness of 0.2 m (Fig. 8.9).
- The sub-floor plate was constructed as a reinforced concrete one and was made watertight by special additives according to elaborated research on mortars, injection masses, insulation and the protection against corrosion.

Note that, incorporation of steel ties in the masonry is necessary for seismic stability, as it provides ductility. Regarding conservation, the same elements are present in the old architecture but are replaced with new materials – steel.

The church was reconstructed on the original remains of the church walls without inflicting damage to the excavated fresco fragments and after removing the parts of the walls constructed during the conservation of the structure in 1965. It was done by using traditional materials (stone, brick and lime mortar) and

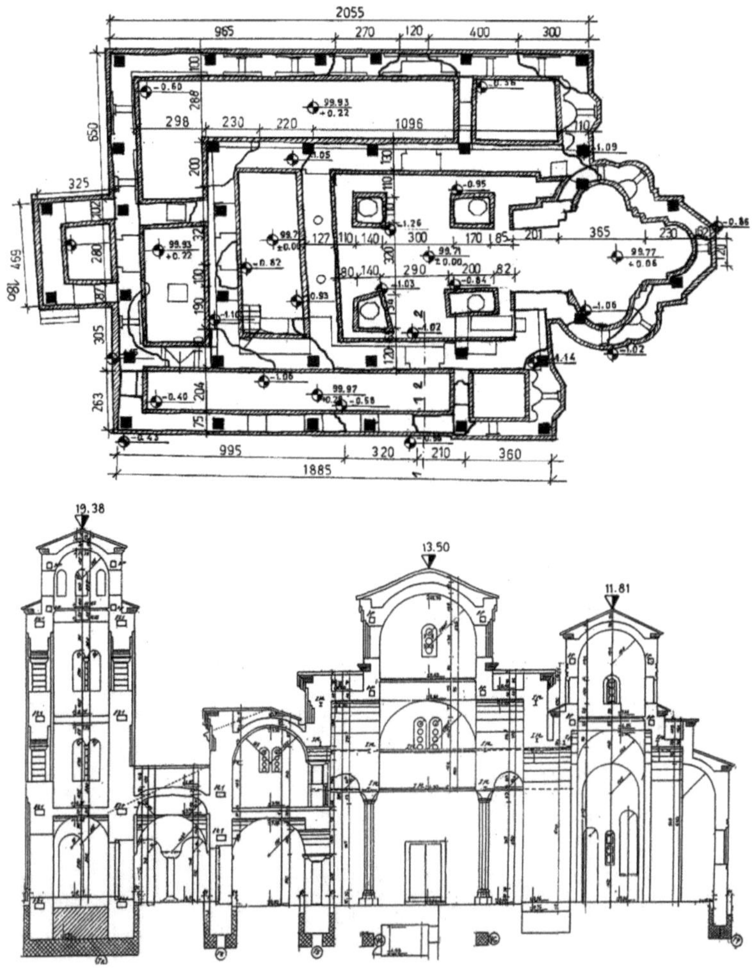

Fig. 8.8 Plan at level −0.22 m and cross section of the structure

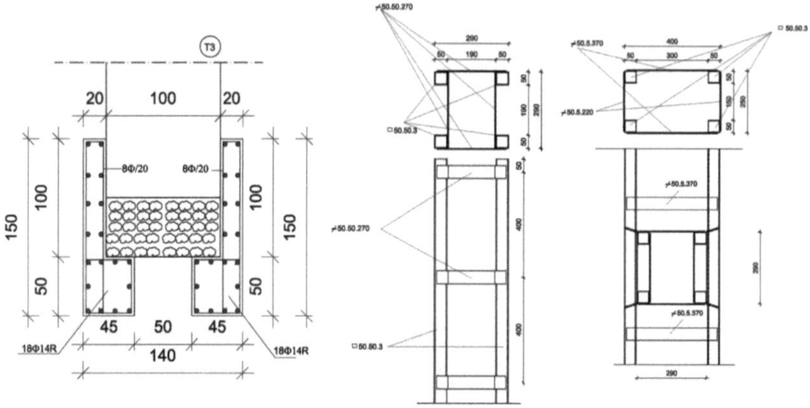

Fig. 8.9 Details of the foundation and of the horizontal and vertical steel ties

Fig. 8.10 The church during reconstruction

Fig. 8.11 The rebuilt St. Panteleymon church

preserving at the same time the original walling pattern of each of the chronologically different periods of construction of the structure. The church was reconstructed in the course of 2001 (Figs. 8.10 and 8.11).

8.4.2 Reconstruction, Seismic Strengthening and Repair of St. Athanasius Church in Leshok

The church of St. Athanasius is situated within the monastic compound of St. Bogoroditsa in the village of Leshok, Tetovo. It was constructed in the 1930s and is protected by the national Law on Protection of Cultural Monuments. It is a three-apse structure with an elongated narthex and belfries (Fig. 8.12). The walls were constructed of stone masonry in lime mortar while the vaults and the domes were constructed of brick masonry. The massive, circular (from the inside) and polygonal (from the outside) apses served to support the dome and the tambour through a system of spherical triangles – pendentives. The tambour was octagonal from the outside and circular from the inside. To the left of the dome (toward the narthex) and to the right of the dome (toward the altar), the massive facade walls are vaulted by parts of semi-circular vaults.

On August 21, 2001, during the armed conflict in R. Macedonia, the monastic church of St. Athanasius in Leshok experienced a strong detonation, resulting in its almost complete collapse (Figs. 8.13 and 8.14). The vaulted structure over the gallery area was completely torn down. The only remains of the floor were the timber floor beams, with visible deformation. In the part of the two preserved belfries, there were visible large cracks (over 20 mm wide) along the height of the bearing walls, in the staircase area, the walls of the tambours and the domes.

The foundation has been fully identified from the geotechnical surveys. It consists of strip foundation, about 1.2 m wise, at a depth of 4.2 m from the floor.

Fig. 8.12 Northwest view of the church of St. Athanasius

Fig. 8.13 The church after detonation

Fig. 8.14 Damage to the floor structure

It was concluded that the existing foundation walls are in good condition, which justified placing the foundation of the structure on the existing foundation walls.

For the needs of the repair and strengthening design of the existing damaged part as well as the reconstruction of the ruined part of the structure, it was necessary to test the strength and mechanical characteristics of the built-in materials, particularly the construction mortar and the sandstone used for the facades in order to define their role in the bearing wall. The tests were performed on mortar samples taken from the relatively sound stone masonry of the existing north wall, i.e., prisms with prescribed proportions of samples from the facade sandstone. The obtained mean values of compressive and tensile strength of the facade stone are $f_c = 25.5$ MPa and $f_t = 10$ MPa, respectively. The obtained mean value of elasticity modulus is $E = 14{,}690$ MPa, whereas that of the compressive strength of the mortar used in the construction is $f_c = 0.8$ MPa.

8.4.2.1 Concept for Repair, Strengthening and Reconstruction

Two approaches were followed to renovate and reconstruct the structure (Shendova and Stojanoski 2004, 2008). For the existing damaged part, *repair and structural strengthening* was selected; the demolished part was thoroughly *reconstructed* with elements of structural strengthening. For both individual structural units, a developed methodology for repair and strengthening was proposed (Fig. 8.15):

1. For the damaged, still standing part of the structure, *repair and structural strengthening* up to the necessary level of seismic safety has been adopted considering the (architectural-conservatory) preservation of the structure. The solution for repair and structural strengthening anticipates (i) injection of all the cracks and (ii) incorporation of strengthening elements (vertical RC jackets along the inner side of the walls of the staircase core and the columns of the tambours, horizontal RC belt courses at the level of the floor structure and at the base of the domes, RC slab below the floor level, as well as steel ties besides the timber beams in the floor structure of the gallery).
2. For the demolished part of the structure, *complete reconstruction* with maximum possible use of selected material has been adopted; elements for structural strengthening so as to provide the necessary level of seismic safety have also been chosen: (i) adding a RC belt course below the floor level, in the existing foundation walls, below the massive walls for the purpose of connection with the vertical strengthening elements, (ii) incorporation of vertical strengthening steel elements at the necessary height, at the ends of the massive walls and around the openings, (iii) incorporation of vertical strengthening steel elements into the tambour columns composed of deformed reinforcement, (iv) incorporation of horizontal steel elements along the massive walls, in the base of the tambour and in the base of the dome.
3. Due to the different treatment of the structural units constituting the integral structure, an *expansion joint* between them is anticipated to be constructed and dictate the concentration of damage during future earthquakes.

8.4.2.2 Analysis of the Structure

An analysis has been carried out for both structural units in accordance with the applicable regulations and European pre-standards. Two methods were used:

1. Analysis of load-bearing and deformation capacity of the structure and nonlinear dynamic analysis for maximum expected actual earthquake effects with intensity of $a_{max} = 0.24g$ with a return period of 1,000 years.
2. Static and equivalent seismic analysis of the structure in 3D with SAP 2000.

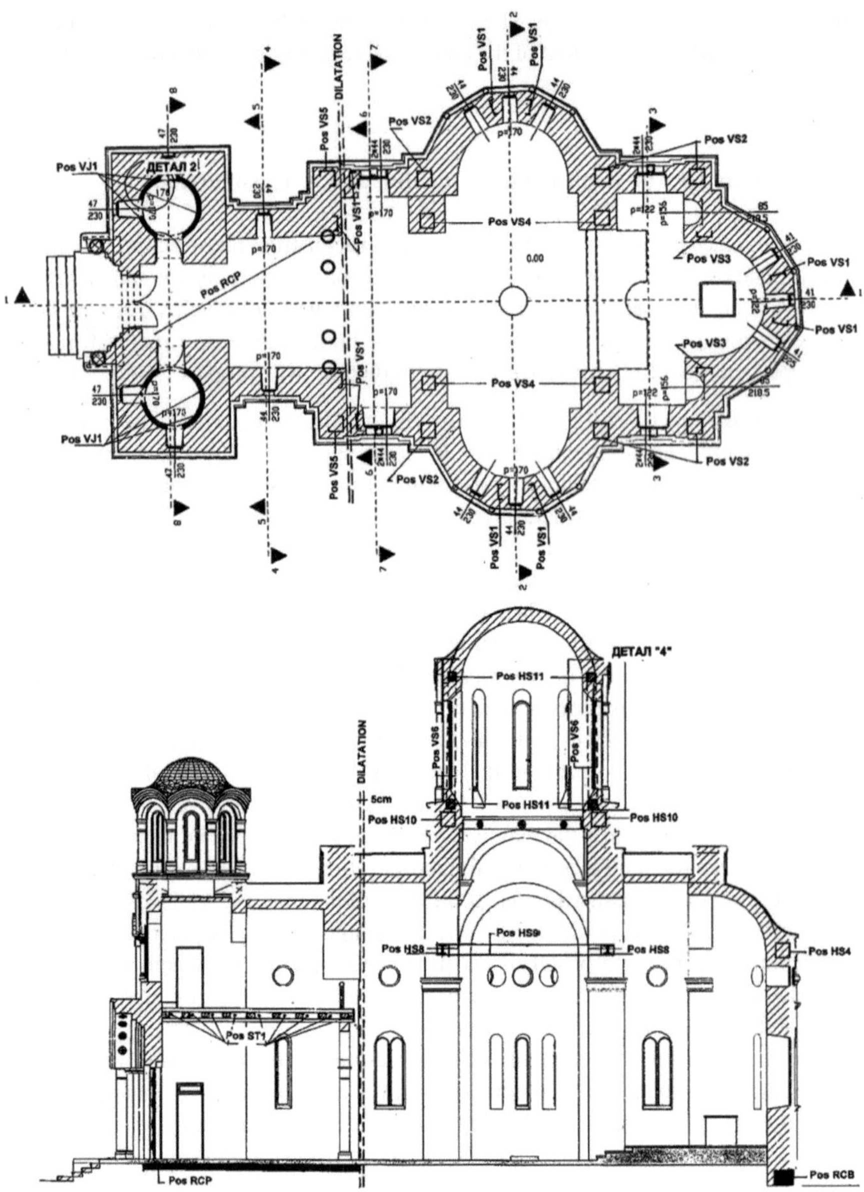

Fig. 8.15 Plan and cross section of the reconstructed church

Analysis of the load-bearing and deformation capacity: In accordance with the procedure and methodology developed at IZIIS (Sect. 8.3.1), the ratio of the ultimate storey shear force to the equivalent seismic force yields the factor of safety against failure. Both individual structural units of the church St. Athanasius have

Table 8.10 Load-bearing and deformation capacity

	Ki (kN/cm)		Q_u (kN)		$F = Q_u/S$	
	x-x	y-y	x-x	y-y	x-x	y-y
Results for repaired and strengthened part						
Storey 3	1,283	1,283	496	496	2.21	2.21
Storey 2	14,386	14,072	2,567	2,502	2.57	2.51
Storey 1	10,148	10,640	3,249	3,407	1.56	1.63
Results for reconstructed part						
Storey 2	1,158	1,158	1,323	1,058	2.86	2.29
Storey 1	16,498	5,448	4,439	4,283	1.36	1.31

been analyzed and the strengthening structural elements have been proportioned. The following quantities have been adopted as input parameters:

- Modulus of elasticity E = 1,200 MPa
- Shear modulus G = 460 MPa
- Ultimate compression strength f_c = 1 MPa
- Ultimate tensile strength f_t = 0.1 MPa.

The results show that the safety factor at first cracking and particularly at failure is greater than 1 for all the storeys. Table 8.10 summarizes results for both structural units (storey stiffness K_i, ultimate capacity Q_u, factor of safety against failure F_u) that represent the input parameters for dynamic analysis.

Dynamic analysis: With the masses lumped at two characteristic levels, a nonlinear dynamic analysis has been performed with storey hysteretic model obtained by summing up the elastoplastic characteristics of each of the bearing walls, whereas the load-bearing capacity of each of them has been limited to the lower value of bending and shear capacity (according to Sect. 8.3.3). To obtain the dynamic response, three different types of earthquake (Petrovac 1979, Ulcinj 1979 and El Centro 1940) with maximum input acceleration of 0.24g and return period of 1,000 years have been applied. Obtained as the results from the dynamic analysis are the storey displacements and ductility ratios required by the earthquake that have to comply with the design criteria defined in Sect. 8.3.4.

It has been concluded that, for both structural units, the absolute storey displacements are less than 1 cm. Hence, it can be asserted that such a designed repair and structural strengthening of the existing part of the structure, as well as the design of the part of the structure to be reconstructed, provides sufficient load-bearing and deformation capacity, i.e., that the dynamic behavior complies to the set out design criteria because a ductility of $\mu < 2$ has been obtained even for the maximum expected earthquake with a return period of 1,000 years.

Static and Equivalent Seismic Analysis by the finite element method: Based on the defined structural systems of the two individual structural units and the defined strengthening structural elements of the structure, a static and equivalent seismic

Fig. 8.16 3D model

analysis has been performed by using the finite element method and the SAP 2000 software. A moderately dense mesh of a total of 5,191 nodes and 1,901 elements has been adopted, reflecting the global geometrical characteristics of the model without paying attention to the inhomogeneity of the material (Fig. 8.16).

The bearing massive walls and the tambour walls have been modeled by a total of 938 SOLID 3D finite elements with eight nodes. The steel vertical and horizontal strengthening elements were modeled by 3D-FRAME or 3D-TRUSS elements, a total of 362 elements. The vaults and domes have been modeled with a total of 601 SHELL elements. A static analysis was then performed to compute the effect of dead loads and equivalent seismic forces according to the regulations.

The results from this analysis give an insight into the designed values of all the static quantities for the individual types of elements, justifying the use of the selected solution for repair and structural strengthening. They show that:

1. Both structural units constituting the integral structure have sufficient load-bearing and deformation capacity up to the target level of seismic protection;
2. Both structural units satisfy the design safety criteria. Under the expected design earthquake ($a_{max} = 0.20g$ with return period of 500 years), it is expected that the structure will behave completely in the elastic rage, while under the maximum expected earthquake ($a_{max} = 0.24g$ with return period of 1,000 years), it is possible that the structure suffers concentration of damage in the expansion joint as well as other nonstructural damage.

The church was reconstructed according to this methodology in 2003–2004 (Figs. 8.17 and 8.18).

Fig. 8.17 During reconstruction

Fig. 8.18 After reconstruction

8.4.3 Reconstruction and Seismic Strengthening of the Blown Up Cathedral Church of the Holy Trinity in Mostar

The Cathedral Church of the Holy Trinity in Mostar, built between 1863 and 1873, was shelled on June 7, 1992. On June 15 the belfry was torn down and the church set on fire and blown up. The remains of the church were cleared in 2005 (Fig. 8.19). Later, a decision on the renovation of the church involving full reconstruction and maximum possible use of the existing preserved material was made. Based on this decision, the main project on the Renovation of the Cathedral Church in Mostar (architecture and structure) was elaborated. The structure was designed to be constructed of massive stone masonry in cement lime mortar.

Renovation of the church started with construction of the newly designed, RC foundation over which reconstruction of the structure was planned (Fig. 8.20).

After categorization of the church as a structure of the first importance category, the Hydro-meteorological Institute of the Serbian Republic performed seismic microzonation and defined the seismic parameters of the considered location. For such defined seismic parameters, prior to the construction works, it was necessary to carry out analysis of the seismic stability of the structure.

Upon evidence on the necessity of strengthening, various strengthening options were proposed and analysed. Following the selection of the most adequate solution (as far as stability and economy are concerned), the stability of the strengthened structure under gravity and seismic effects was analysed. Three general states of the structure have been treated: (PS) – designed structure constructed of plain stone masonry; (HE) – strengthened structure by horizontal steel elements and (OS) – strengthened structure by horizontal and vertical steel strengthening elements (confined masonry). The applied methodology of analysis has been developed by IZIIS based on the most recent knowledge on the behavior of masonry structures, enriched with analytical and experimental experience and implementation of this knowledge in the reconstruction of important cultural-historic monuments (Shendova et al. 2011a, b, 2012). The results from the performed analyses have shown that the selected concept of strengthening of the structure enables optimization of the design structural system by adequate selection of strengthening elements and provides the necessary integrity and stability of the structure for the designed level of seismic protection.

Definition of Seismic Parameters: The elaboration of the Study for Seismic Microzoning of the Location of the Cathedral Church in Mostar according to the JUS and EC8 standards by the Republic Hydro-meteorological Institute of Banja Luka, enabled definition of the total seismic hazard at the church location and analysis of the response of the local soil to seismic effects for the purpose of definition of the parameters of the horizontal elastic spectrum of local soil response (Fig. 8.21), the corresponding time histories of acceleration and deformability properties of the soil.

Fig. 8.19 The original Holy Trinity church in Mostar and view of the torn down church

Fig. 8.20 Newly designed RC foundation in the course of construction

The seismic hazard at the location has been defined for return periods of 100, 475 and 1,000 years with PGA values of 0.07g, 0.16g and 0.22g, respectively. The local soil was categorised as type B, based on the data available from geological and geotechnical investigations. A set of six horizontal histories of acceleration, meeting the EC8 criteria, has been generated for the location. In accordance with the JUS standards, the numerical values of the seismicity factor Kc have been computed for the three return periods at the free surface of the terrain as 0.028,

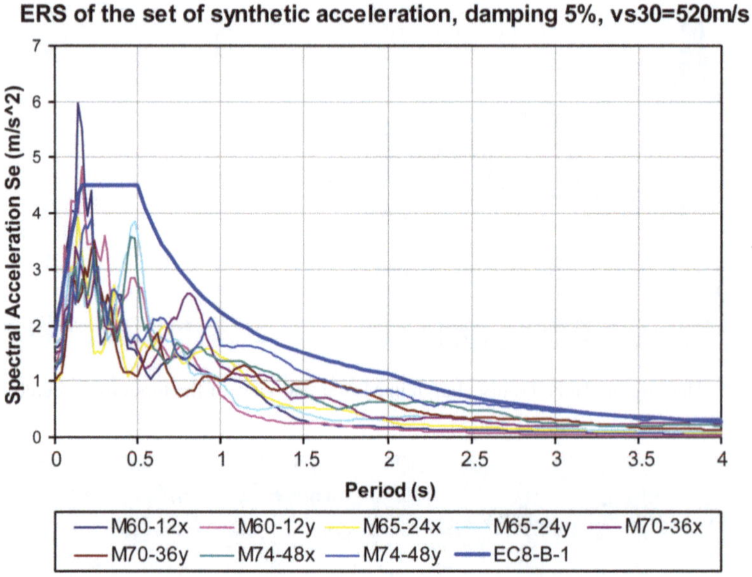

Fig. 8.21 Elastic response spectrum of the set of synthetic accelerations

0.062 and 0.088, respectively. With this, the input data for further computation of the stability of the structure according to the JUS and EC8 standards were defined for analysis of the load-bearing and deformation capacity and dynamic analysis taking into consideration actual earthquakes:

- JUS standards: Base shear coefficient $K = K_0 K_c K_d K_p$;

  K_0 (Importance factor) = 1.50 (for Importance Category I);
  K_c (seismicity factor) for 100, 475, 1,000 years: $K_c^{100} = 0.028$, $K_c^{475} = 0.062$, $K_c^{1,000} = 0.088$;
  K_d (for soil amplification) = 1.0;
  K_p (for ductility): $K_p^{\text{plain masonry}} = 2.0$, $K_p^{\text{confined masonry}} = 1.6$.
  Base shear coefficient, plain masonry: $K^{100} = 0.084$, $K^{475} = 0.186$; $K^{1,000} = 0.264$;
  Base shear coefficient, confined masonry $K^{100} = 0.067$, $K^{475} = 0.13$, $K^{1,000} = 0.211$

- Eurocode 8, (EC8): Base shear coefficient $K = \alpha S \beta_0 / q$;

  α (Peak ground acceleration, PGA, g's): $\alpha^{100} = 0.07$, $\alpha^{475} = 0.16$, $\alpha^{1,000} = 0.22$;
  S (soil factor for ground type B): $S_B = 1.20$;
  β_0 (spectral amplification factor for 5 % damping): $\beta_0 = 2.5$;
  q (behavior factor): $q^{\text{plain masonry}} = 1.5$, $q^{\text{confined masonry}} = 2.0$.
  Base shear coefficient, plain masonry: $K^{100} = 0.14$, $K^{475} = 0.32$, $K^{1,000} = 0.46$;
  Base shear coefficient, confined masonry: $K^{100} = 0.1$, $K^{475} = 0.24$, $K^{1,000} = 0.34$.

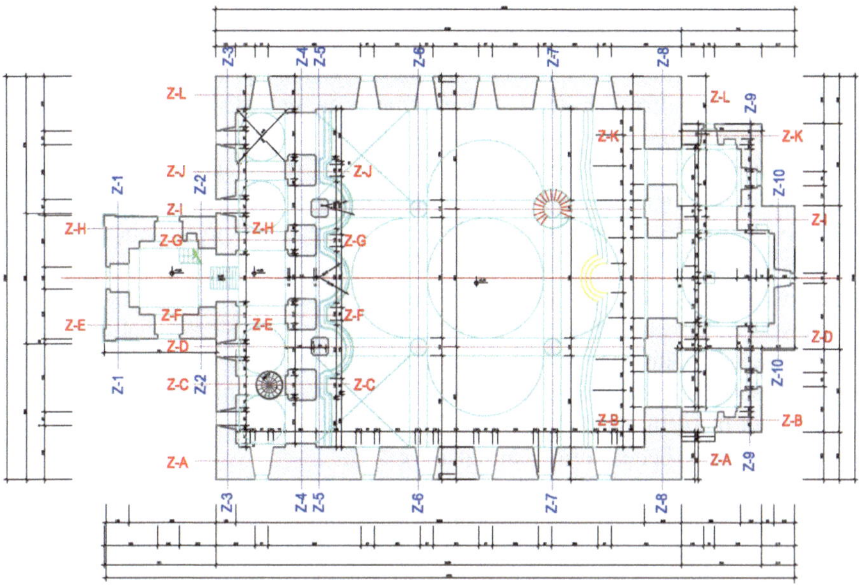

Fig. 8.22 Ground floor plan of the designed church structure of the Holy Trinity in Mostar

8.4.3.1 Concept for Three General Conditions of the Structure

With the elaboration of the main design for the renovation of the Cathedral Church by the Republic Institute for the Protection of Cultural-Historic and Natural Heritage, Banja Luka, the entire architecture and the principal structural system of the church have been defined (Fig. 8.22). The Cathedral Church has a cross-like plan with a high belfry in the western part and a triple apse in the eastern part. The total length of the church with the belfry and the altar is 45.32 m, while its width is 25.66 m. The cross-section of the arms of the inscribed cross is surmounted by the central dome with a dodecagonal plan with a span of about 9 m, on a tambour with a square plan. Over the altar, three domes rise with octagonal plans that are lower with respect to the central dome. Other two domes with octagonal plans are situated above the narthex. The width of the bearing walls varies from 130 to 206 cm, while the width of the tambour walls range from 40 to 60 cm. The inner width of the church naos is 21.53 m, while the height of the central vault is 17.96 m.

The church is designed as a massive masonry structure constructed of different types of stone in cement lime mortar (PS state). The main structural elements are the domes, the arches, the vaults, the wooden floor structures, the columns, the walls and the foundation. In order to decide on the mix design for the cement lime mortar based on the analysis of the church, three options have been adopted for it:

- Option 1: $f_c = 2$ MPa; $f_t = 0.2$ MPa, $E = 2{,}000$ MPa, $G = 0.25E = 500$ MPa;
- Option 2: $f_c = 4$ MPa; $f_t = 0.4$ MPa, $E = 3{,}500$ MPa, $G = 0.25E = 875$ MPa;
- Option 3: $f_c = 8$ MPa; $f_t = 0.8$ MPa, $E = 6{,}000$ MPa, $G = 0.25E = 1{,}500$ MPa.

After proving the necessity to improve the designed structure, the first option proposed (V4) is strengthening only with horizontal steel elements – ties made of rigid "L" and "I" profiles at five characteristic levels of the principal structure and at six levels of the belfry (HE state). These elements are placed along the length of the bearing walls in the transverse and longitudinal direction and are connected by welding where they touch each other. In this way, the integrity of the principal structure is considerably improved and contributes to the synchronized behavior of individual walls. The objective of placing these ties is to sustain the tensile stresses after exhaustion of the load-bearing capacity of the masonry and occurrence of the first cracks, and thus to prevent further damage to the walls.

However, the results from the performed analyses have shown that it is necessary to include additional vertical strengthening elements to provide the designed level of seismic protection (OS state). With this, the structure of the church constructed of plain masonry is turned into a confined masonry which has been proved to behave better during earthquake effects.

Based on the required strength and deformability characteristics of the elements and the system as a whole, a number of variant solutions have been considered and the most adequate (as far as stability and economy are concerned) has been selected (V5). The density and location of vertical elements have been defined according to the developed strengthening methodology (Sect. 8.2), the detailed analysis of the load-bearing system and the aim to place them in a way that does not disturb the architecture of the structure. Part of the vertical elements has been placed structurally to provide system lines in both orthogonal directions.

8.4.3.2 Comparison of Load-Bearing and Deformation Capacity of the Structure

Figure 8.23 shows the results on the characteristic level from the analysis of the load-bearing and deformation capacity described in Sect. 8.3.1 simultaneously for the three variant solutions of the structural system (V2, V4 and V5). The comparative presentation of the results obtained from this analysis clearly shows that the insertion of horizontal ties (HE-V4) along the length of the walls enables to increase of the load-bearing capacity and stiffness but reduces the deformability of the characteristic levels. However, including also vertical ties (OS-V5) at the ends of the walls and around the openings enables considerable increase of the strength and deformability of the structure at all levels in both orthogonal directions as a result of improved integrity and bending resistance. With this, it has been proved that the church structure constructed of confined masonry in both directions, possesses sufficient load-bearing capacity in accordance with the defined criteria since the strength of the most critical first level has been higher than the total seismic force according to both JUS and EC8 standards.

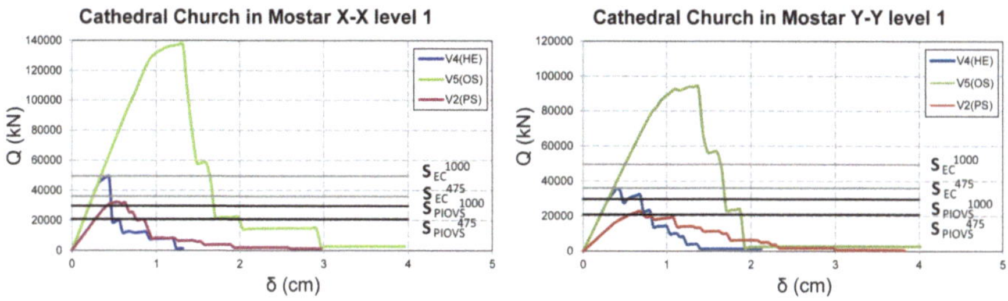

Fig. 8.23 Comparative presentation of load-bearing capacity, level 1

8.4.3.3 Comparison of Dynamic Response of the Structure for the Defined Seismic Parameters

To obtain the dynamic response of the structure, 9 different types of earthquakes have been used: 6 synthetic earthquake records defined by seismic microzonation as well as records of three other earthquakes, namely, Petrovac 1979, Ulcinj 1979, and El Centro 1940. The response has been investigated for the maximum input ground acceleration of $a_{max} = 0.16g$ and $a_{max} = 0.22g$ in accordance with the defined seismic hazard for return periods of 475 and 1,000 years. As a result of the dynamic analysis, displacements and ductility demanded by the earthquake ($\mu^{earthquake} = \delta^{max}/\delta^{y}$) are obtained and should comply with the design safety criteria, i.e., they should be less than $\mu^{100} < 1.0$, $\mu^{475} < 1.5$ and $\mu^{1,000} < 2.0$.

By comparative analysis of the dynamic responses, it can be concluded that the behavior of the structure constructed of confined masonry is considerably more favourable in respect to the other two variants (Fig. 8.24). Despite the strict design criteria, the demanded ductility of the structure strengthened by horizontal and vertical elements for all the analysed earthquakes is within the limits of the allowed ductility. It is only that the response of the fourth level in the longitudinal direction is more intensive than that allowed ($\mu > 1.5$ for $a_{max} = 0.16g$, $\mu > 2$ for $a_{max} = 0.22g$); despite this, the structure possesses the demanded ductility capacity. A drastic improvement of response is characteristic for the transverse direction, particularly for the first, the most critical level. While the first level in the case of the designed structure is deep in the nonlinear range under the maximum expected earthquakes ($\mu = 3$–6 for different earthquakes), in the conditions of a strengthened structure, it is in the elastic range of behavior.

8.4.3.4 Comparison of Stress–Strain State of the Structure

The results from the FEM analysis are presented below. The most important one concerning the efficiency of implementing horizontal and vertical elements are the main tensile stress for the applied seismic forces for a return period of 1,000 years according to the EC standards (Fig. 8.25).

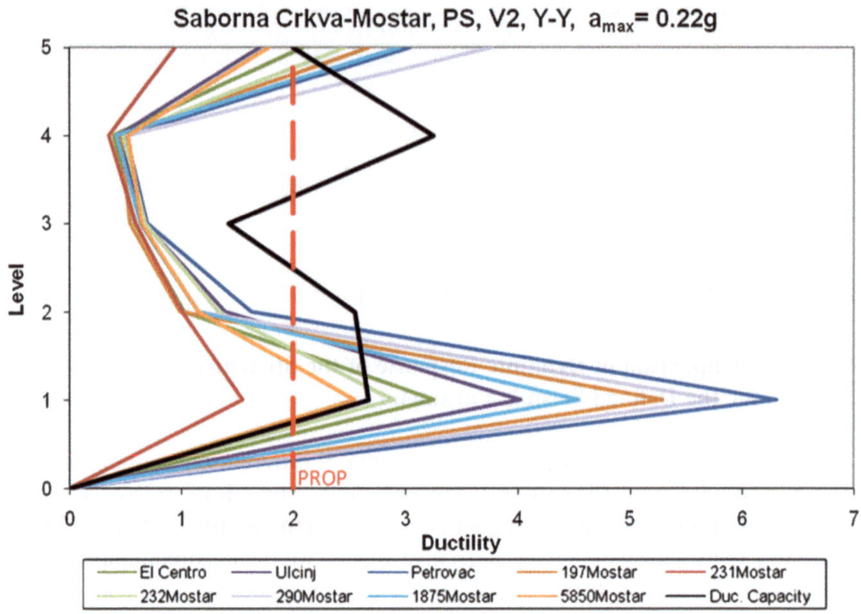

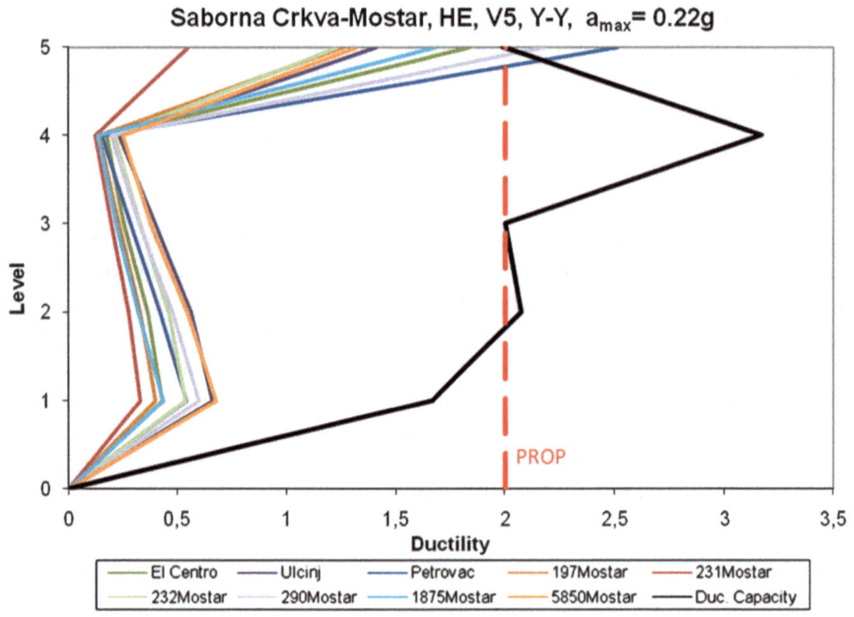

Fig. 8.24 Dynamic response to the maximum earthquake, (V2 and V5, transverse direction)

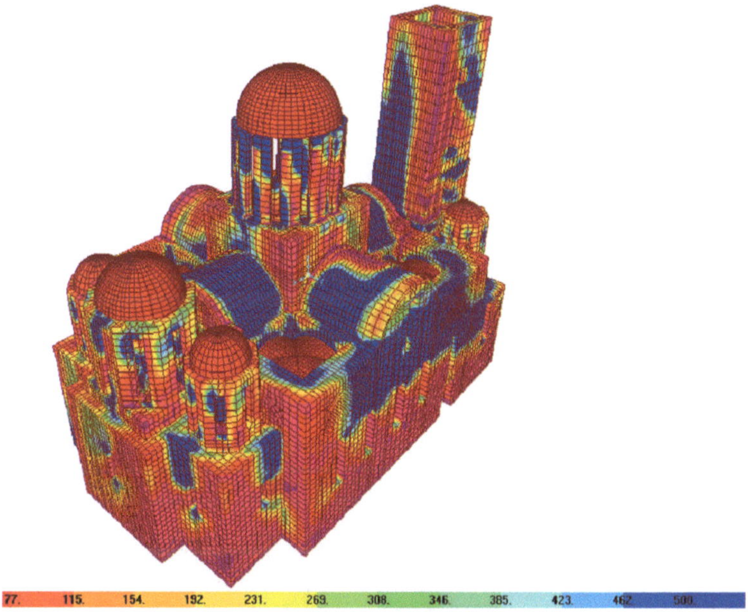

designed-V2

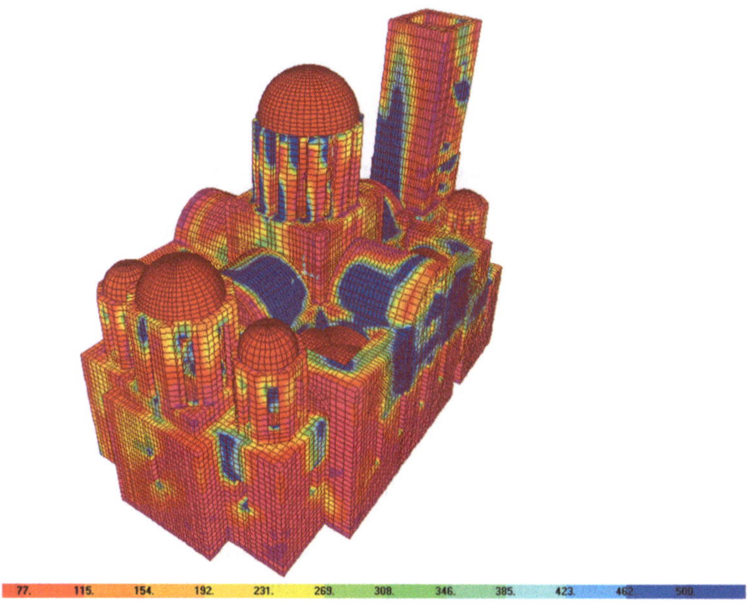

strengthened-V5

Fig. 8.25 Main tensile stresses (dead weight + $S_y^{EuroCode}$)

Fig. 8.26 Beginning of the construction of the church in May 2011

Fig. 8.27 Construction of the church up to the level of first horizontal ties

With this analysis of the stress state, one can identify the potential local instabilities where exhaustion of compressive or tensile strength or excessive deformations takes place. The dark blue zones shown in Fig. 8.25 indicate zones in which the tensile stress is higher than the tensile strength of the designed mortar, i.e., zones where occurrence of cracks could be expected under the maximum expected earthquake. The comparative presentation shows that these zones are reduced in the case of the strengthened structure in respect to the design structure considering both the JUS and the EC standards. This does not mean that the cracks will occur in reality, because they will be prevented by the presence of ductile elements (horizontal and vertical ties); but this cannot be modeled by the computer code, because it does not allow modeling nonlinearity of material for SOLID elements, i.e., excluding masonry with stresses greater than the ultimate ones. In reality, upon occurrence of the first crack in masonry, the placed ties will be activated, and thus extension of the cracks will be prevented, while the failure mechanism will be transferred in the lower zones instead, taking place by separation of individual walls.

Reconstruction of the cathedral church with the system constructed of confined masonry started in May 2011 (Figs. 8.26 and 8.27).

The analysis has shown that, with its strength, stiffness and deformability characteristics, the designed structure and the structure strengthened only with horizontal elements do not satisfy the seismic safety criteria and do not comply with the most recent knowledge on the behavior of masonry structures under gravity and seismic effects. Strengthening with horizontal and vertical steel elements turns the structural system from plain masonry into confined masonry that exhibits a much more favourable behavior under dynamic effects. All the performed analyses have shown that the designed strengthening system which complies with the conservation demands enables the necessary integrity of the structure at the characteristic levels and increases its strength, load-bearing and deformation capacity to the designed level of seismic protection.

8.5 Conclusion

Conducting shaking table tests on models in a realistic geometric scale is an excellent method to investigate the dynamic properties and structural characteristics, to verify the technology and method used for repair and strengthening, as well as to explore new technologies related to historical monuments. Experimental evidence helps to assess the suitability of the design choices and the effectiveness of the applied consolidation system. The original methodology using "ties and injection" for the repair and strengthening of Byzantine churches was developed and verified through ample experimental and analytical investigation.

The proposed and experimentally verified methodology was successfully implemented in the process of reconstruction and seismic strengthening of real historic monuments. The design process described in this paper is a useful tool when facing the complex problem of protection and conservation of constructions that form part of our cultural heritage.

References

Carydis PG, Mouzakis HP, Papantonopoulos C, Papastamatiou D, Psycharis N, Vougioukas EA, Zambas C (1996) Experimental and numerical investigations of earthquake response of classical monuments. In: 11WCEE, Paper No 1388, Acapulco

Danieli M, Bloch J, Ribakov Y (2008) Retrofitting heritage buildings by strengthening or using seismic isolation. In: Seismic engineering conference, MERCEA08, Reggio Calabria

De Canio G, Muscolino G, Palmeri A, Poggi M, Clemente P (2008) Shaking table tests validating two strengthening interventions on masonry buildings. In: Seismic engineering conference, MERCEA08, Reggio Calabria

Feilden BM (1982) Conservation of historic buildings. Butterworth Scientific, London

Feilden BM (1987) Between two earthquakes. ICCROM & GCI joint publication, Rome/Marina del Rey

Gavrilovic P, Ginell W, Shendova V et al (1991–1995) Seismic strengthening, conservation and restoration of churches dating from Byzantine Period in Macedonia. Joint research project, IZIIS – Skopje, RZZSK – Skopje, GCI –LA. Reports IZIIS 500-76-91, vol 1–12

Gavrilovic P, Shendova V, Ginell W (1999) Seismic strengthening and repair of Byzantine churches. J Earthq Eng 3(2):199–235
Gavrilovic P, Necevska-Cvetanvska G, Apostolska R (2001a) Consolidation and reconstruction of St Panteleymon Church in Ohrid. IZIIS report 2001
Gavrilovic P, Shendova V, Kelley S (2001b) Earthquake protection of Byzantine churches using seismic isolation. Macedonian–US joint research. Report IZIIS 2001
Gavrilovic P, Kelley S, Shendova V (2003) A study of seismic protection techniques for the Byzantine churches in Macedonia. J Assoc Preserv Technol APT Bull XXXIV(2–3):63–71
Gavrilovic P, Ginell W, Shendova V, Sumanov L (2004) Conservation and seismic strengthening of Byzantine churches in Macedonia. The Getty Conservation Institute, GCI Scientific Program Reports, ISBN 0-89236-777-6, J Paul Getty Trust, LA
Kelley SJ, Crowe TM (1995) The role of the conservation engineer. In: Kelley SJ (ed) ASTM STP 1258: standards for preservation and rehabilitation. ASTM, Philadelphia
Martelli A (2009) Development and application of innovative anti-seismic systems for seismic protection of cultural heritage. In: Protection of historical buildings, PROHITECH09, Italy
Necevska-Cvetanovska G, Apostolska R (2008) Consolidation, rebuilding and strengthening of St Pantelymon Church, Plaoshnik, Ohrid. Eng Struct 30:2185–2193
Oliveira C, Costa A (2010) Reflections on the rehabilitation and the retrofit of historical constructions. In: Earthquake engineering in Europe, 14ECEE, Skopje
Shendova V (1998) Seismic strengthening and repair of Byzantine churches in Macedonia. Doctoral dissertation, Skopje (in Macedonian)
Shendova V, Gavrilovic P (2004) Implementation of a methodology using "ties and injection" developed for repair and strengthening of historic monuments. In: IV international seminar on structural analysis of historical constructions, SACH 2004, Padova
Shendova V, Stojanoski B (2004) Main project on repair, strengthening and reconstruction of St Ahtanasius church in Leshok. IZIIS report 2004
Shendova V, Stojanoski B (2008) Quasi-static tests on wall elements constructed during the reconstruction of St Athanasius Church in Leshok. In: 14WCEE, Beijing
Shendova V, Rakicevic Z, Gavrilovic P, Jurukovski D (2006) Retrofitting of Byzantine church using passive base control system. In: 4th world conference on structural control and monitoring, San Diego
Shendova V, Bozinovski Z, Stojanoski B, Gjorgievska E (2011a) Analysis of seismic stability with technical solution for repair and seismic strengthening of the Orthodox church in Mostar. Report IZIIS 2011–22
Shendova V, Rakicevic Z, Krstevska L, Tashkov L, Gavrilovc P (2011b) Shaking table testing of models of historic buildings and monuments – IZIIS' experience. In: Fardis MN, Rakicevic ZT (eds) Role of seismic testing facilities in performance-based earthquake engineering – SERIES workshop. Springer, New York, pp 221–245. ISBN 978-94-00701976-7
Shendova V, Bozinovski Z, Garevski M, Stojanoski B, Gjorgievska (2012) Reconstruction and seismic strengthening of the blown up Cathedral church of the Holy Trinity in Mostar. In: 15WCEE, Lisbon
Tassios TP (2010) Seismic engineering of monuments, earthquake engineering in Europe. In: 14ECEE, Skopje
Tassios TP, Miltiadou Fezans A, Vintzileou E (2007) Proceedings of international symposium studies on historical heritage, Antalya

Chapter 9
Shaking Table Tests of a Full-Scale Two-Storey Pre-Damaged Natural Stone Building Retrofitted with the Multi-Axial Hybrid Textile System "Eq-Grid"

Lothar Stempniewski and Moritz Urban

Abstract This chapter reports about seismic testing on a full-scale two-storey building made of broken natural stones. The dynamic behavior for two stages (a) undamaged and (b) pre-damaged, repaired and strengthened was analysed. All tests took place on the 7 × 5.6 m^2 shaking table of Foundation EUCENTRE in Pavia, Italy. The used strengthening technique "eq-grid" was developed over 10 years at the Karlsruhe Institute of Technology (KIT). It consists of a multi-axial hybrid fibre textile embedded in a mortar matrix. In the first years 1:2 scale single masonry walls were quasi statically and/or pseudo-dynamically tested at the KIT lab to identify the adaptability of a mortar-textile-composite for strengthening different kinds of masonry structures including soft lime stone bricks and high strength natural stones. In the end a complete new strengthening system "eq-grid" was developed. To check the results of the conducted wall tests a full-scale two storey masonry building made of high strength natural stones was constructed at Foundation EUCENTRE. The building was uniaxially shaken at different levels of the same earthquake record. In the first stage the unstrengthened building was damaged in such a way that collapse was nearly reached. In the following stage the pre-damaged test sample was repaired and retrofitted using "eq-grid". As done in practice, the system was applied only on the outer surface of the building. The comparison of the measured and observed dynamic behavior showed a significant increase of the maximum acceleration and ductility. In the end a much better overall dynamic performance of the pre-damaged retrofitted building was achieved. These results fully confirmed the research work based on single wall tests.

L. Stempniewski (✉) • M. Urban
Department of Reinforced Concrete and Building Materials, Karlsruhe Institute of Technology, Gotthard-Franz-Strasse 3, Building 50.31, 76131 Karlsruhe, Germany
e-mail: Lothar.Stempniewski@kit.edu; Moritz.Urban@kit.edu

A. Ilki and M.N. Fardis (eds.), *Seismic Evaluation and Rehabilitation of Structures*, Geotechnical, Geological and Earthquake Engineering 26,
DOI 10.1007/978-3-319-00458-7_9,

9.1 Introduction

Masonry is the most used and most ancient building material worldwide. Due to the brittle behavior with regard to earthquake loads, most collapses after seismic events are observed with buildings made of this material. Therefore for more than 60 years many scientists have tried to study masonry structures and find practical and economic solutions for strengthening and rehabilitation. In addition to Ferro cement layers, air-placed-concrete, pre-stressing, needling and injections, a new kind of solution is given by using fibre reinforced plastics (FRPs). Schwegler (1994), Ehsani and Saadatmanesh (1996) and Ehsani et al. (1997, 1998) tested in their experimental works light weight fibre products in combination with high strength epoxy resins. They were able to increase the maximum loads but the behavior still remained brittle. In earthquake engineering the ductility behavior is considered more and more important. Therefore materials for textiles with higher ductility and cement-based mortar matrices are more suitable. Another advantage in comparison to epoxy resins is the better water vapour permeability. Kolsch (1998a, b) was one of the first authors to mention these facts. At KIT different two-stone and three-stone tests and cyclic in-plane shear wall tests were first conducted by Wallner (2008). More detailed in-plane shear wall tests with a new class of multi-axial hybrid fabrics were conducted by Münich (2010). Out-of-plane tests on small wall segments with cement based fibre bond can be also found in Papanicolaou et al. (2008). Detailed work can be read in El Gawady (2004) and Turek et al. (2004) for dynamic in-plane wall testing on a shaking table.

While many contributions about experimental work on single strengthened masonry test walls can be found, very few complete building tests have been carried out. Gülkan et al. (1990) conducted five single storey masonry building tests with three different time histories and loaded the last building also uniaxially in an angle of 30° to the orthogonal walls. Tomazevic (2000, 2009), Tomazevic et al. (2004) and Tomazevic and Gams (2010) reported about several 1:4 and 1:5 scale building tests with clay bricks, natural stones and aerated concrete stones. Benedetti and Pezzoli (1996) tested several strengthening techniques like Ferro cement layers and horizontal tendons on 13 buildings. The geometry of their two-storey natural stone building was similar to the one of this contribution and served as a good reference model.

9.2 Material Properties

The natural stones were taken from the L'Aquila region, where in 2009 an earthquake destroyed many buildings made of this type of material. The German code DIN EN 1926 was applied for the determination of the compression strength of these stones. Material tests were carried out on three specimens from 50 mm diameter boreholes with 50 mm length, from different stones. The average compression strength was 164.5 MPa. Results for all core samples are listed in Table 9.1.

Table 9.1 Test results of natural stone core samples

Specimen	Compression strength [MPa]
1	195.8
2	189.2
3	108.4

Table 9.2 Test results for the mortar properties

Specimen	Bending tensile strength [MPa]	Compression strength [MPa]
1	0.89	3.74
2	–	3.67
3	0.89	3.62
4	–	3.57
5	0.86	3.75
3	–	3.65

A ready mixed mortar Tassullo T30V on hydraulic lime base in bagged cargo was applied with 14 % additional water. Bending tensile strength and compression strength values were determined according to the German code DIN EN 1015-11 on 40 mm × 40 mm × 160 mm mortar samples. The test program foresees a three point bending test over a total length of 100 mm. The compression strength was determined after testing two pieces from uncracked locations with a size of 40 mm × 40 mm. The average compression strength was 3.67 MPa and the average bending tensile strength was 0.88 MPa. All results are summarised in Table 9.2.

Only for strengthening purposes Sikagard 720 EpoCem®, a cement-based mortar with epoxy additives, was used as matrix for the fabric. The assembling of the reinforcement consists of a 4 mm thick ground layer, the hybrid multi-axial fibre grid and a final 4 mm cover layer.

The multi-axial hybrid fabric made of glass and polypropylene fibres in four different fibre orientations (Münich 2010) was applied for practical reasons mainly from outside the building, in order to make up for the missing tensile strength of the masonry. With 5 cm wide stripes the tension forces were determined according to German code DIN EN ISO 13934-1. The corresponding tensile force is related to a fabric length of 1 m. The load-strain behavior of different orientations is given in Fig. 9.1.

As a result of the machine-broken stone material the surface had an unevenness of about ±2 cm. Therefore one additional layer of Sikagard 720 EpoCem® was applied as an equalising layer in order to smooth down the rough surface before applying the final system.

9.3 Cyclic Single Shear Wall Tests

Several single shear wall tests with different kinds of mortar and textile combinations were conducted to find the best strengthening solution for masonry under horizontal cyclic loading. In this chapter only two test results are discussed. Further information can be found in Münich (2010), Urban and Stempniewski (2012), and Urban (2013).

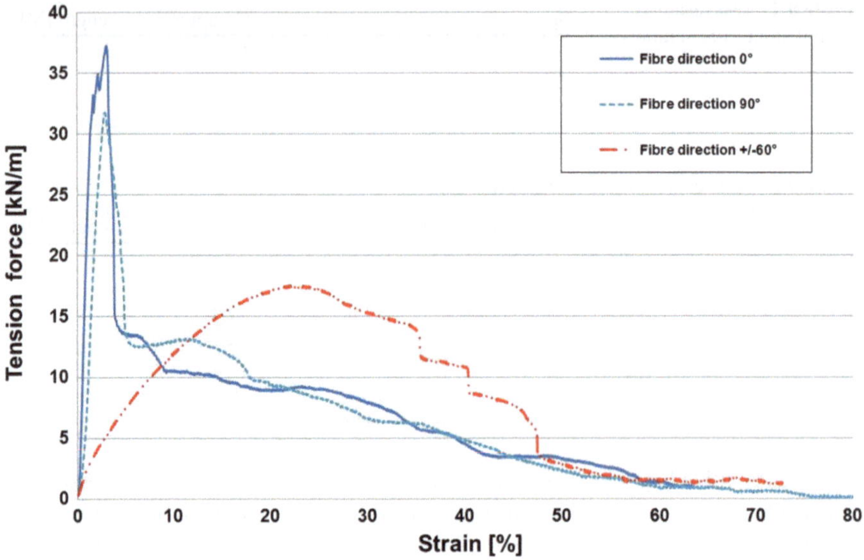

Fig. 9.1 Fabric properties for different fibre orientations (0° = vertical direction on the wall)

The test setup was chosen representing a typical storey-high wall scaled 1:2 in length but not in thickness. The specimens were 1.25 m × 1.25 m × 0.24 m. The shear loading in plane was in the strong inertia direction in combination with a vertical stress of 0.4 MPa, a value representing a usual wall at ground level in a two storey family house. A horizontal hydraulic piston loaded the walls in a displacement controlled manner in different steps. Each cycle was repeated three times over a stiff connected head beam. The resisting force and the displacement were measured at the head beam with a sampling rate of 2 Hz. Each sinus cycle had a time length of 120 s, which allowed visual observations of the failure mechanisms. As a consequence, a strain-dependent and velocity-dependent cracking mechanism was not considered in these tests; in the final tests, with dynamic earthquake loading, this effect was taken into account.

The first specimen (W1) was an unstrengthened wall (URM) composed of two individual wythes separated by a 3–6 cm mortar gap. The second wall (W2) was additionally reinforced with the aforementioned quadaxial hybrid fabric (eq-grid), only on one side. No specific connection of the fibres to the head and foot beam were chosen. Considering the roughness of the surface due to the fact that the stones used were machine-broken, the fibres were not straight in plane and had therefore from the beginning several imperfections. All measured and calculated parameters are summarised in Table 9.3. Only the envelope of the resulting hysteresis curves for both experiments are given in Fig. 9.2.

First cracks of the URM wall (W1) were observed two load steps before the maximum horizontal measured force of 129.6 kN was reached. After 16 mm horizontal displacement the wall failed due to the development of large diagonal

Table 9.3 Test results of masonry shear walls W1, W2

Specimen	Max. force	Δ	Max. displacement	Max. drift	Δ	Failure mode
(W1)	129.6 kN	–	16 mm	1.28 %	–	Shear (cross cracking)
(W2)	147.6 kN	+14 %	28 mm	2.24 %	+75 %	Shear (cross cracking)

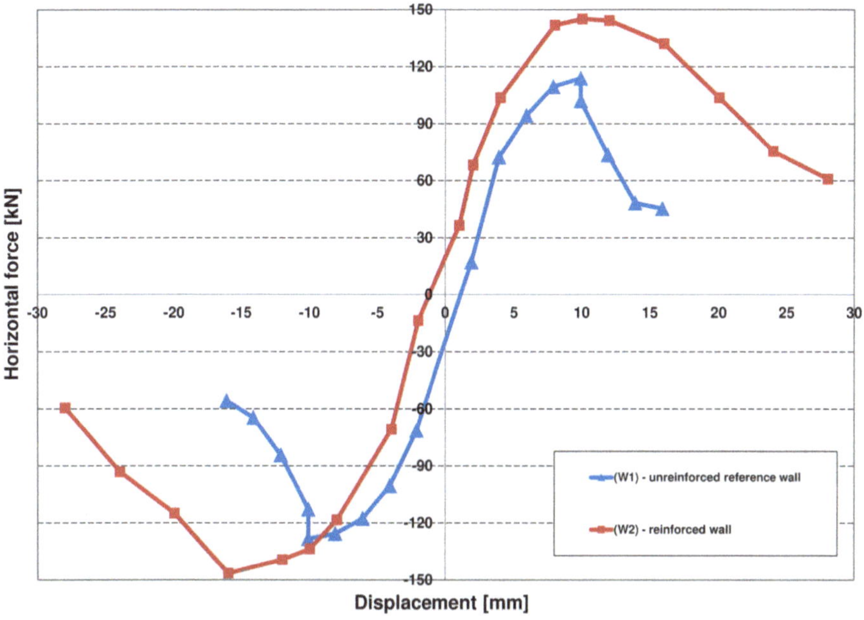

Fig. 9.2 Force-displacement envelopes for wall (W1) and (W2)

shear cracks over the complete cross section. One crack even continued through the middle of a single stone. In contrast to wall (W1) the first crack in the strengthened specimen was observed when the maximum force of 147.6 kN was detected in the middle of the wall. After 16 mm displacement – the maximum of the URM test specimen – several fine hairline cracks developed one after the other in straight X-shape form. The test was stopped at 28 mm displacement when some stones were dislodged at the right bottom corner.

This "brick-torsion-behavior" resulting from the eccentricity of the strengthening system was never observed by Schwegler (1994) and Wallner (2008), although they also used an unsymmetrical strengthening setup in their experimental works. One reason for this effect was the lower shear capacity between irregular applied mortar and stone material. Schwegler (1994) and Wallner (2008) tested regular mortar joints with a much better bonding than technically produced stones. The ductility and energy dissipation capacity was much higher by using the strengthening system. Damage pictures are given in Fig. 9.3.

Fig. 9.3 (a) Damaged URM wall; (b) damaged RM wall with fine cracks

9.4 Shaking Table Tests on a Full-Scale Two-Storey Building

9.4.1 Building Design and Test Setup

Pre-tests on 1:2 scale reinforced masonry walls showed clearly the possibility to increase significantly the ductility of the used natural stone. In the next step a more complex structure was tested under uniaxial dynamic shaking at Foundation EUCENTRE in Pavia (Peloso et al. 2012), in the framework of the FP7 project SERIES. A full-scale two-storey building was constructed on a classical reinforced concrete foundation. The floor over the ground level was a traditional timber construction. The overall design was similar to the buildings of the L'Aquila region, where the 2009 earthquake took place. The detailed plans, with dimensions, arrangement of openings and installed acceleration sensors are given in Fig. 9.4. The total building weight was calculated including the reinforced concrete foundation at 903.7 kN, below the limit of the shaking table of 1,400 kN.

A time history record from the Montenegro earthquake at 15.04.1979 with a magnitude of 6.9 Mw observed in an epicentral distance of 20 km was chosen as a representative strong motion earthquake for southern Europe. The time-history from the EW-direction recorded at Ulcinj station with peak ground acceleration (PGA) of 0.22g is depicted in Fig. 9.5.

The vertical component with a PGA of 0.17g could not be considered with the uniaxial shaking table. This means that the expected influence of the friction based shear behavior could not be taken into account. For more information with regard to this aspect see Urban (2013).

The test programme started at a low scaled acceleration level of the time-history and increased in each consecutive step. The tests were stopped just before the

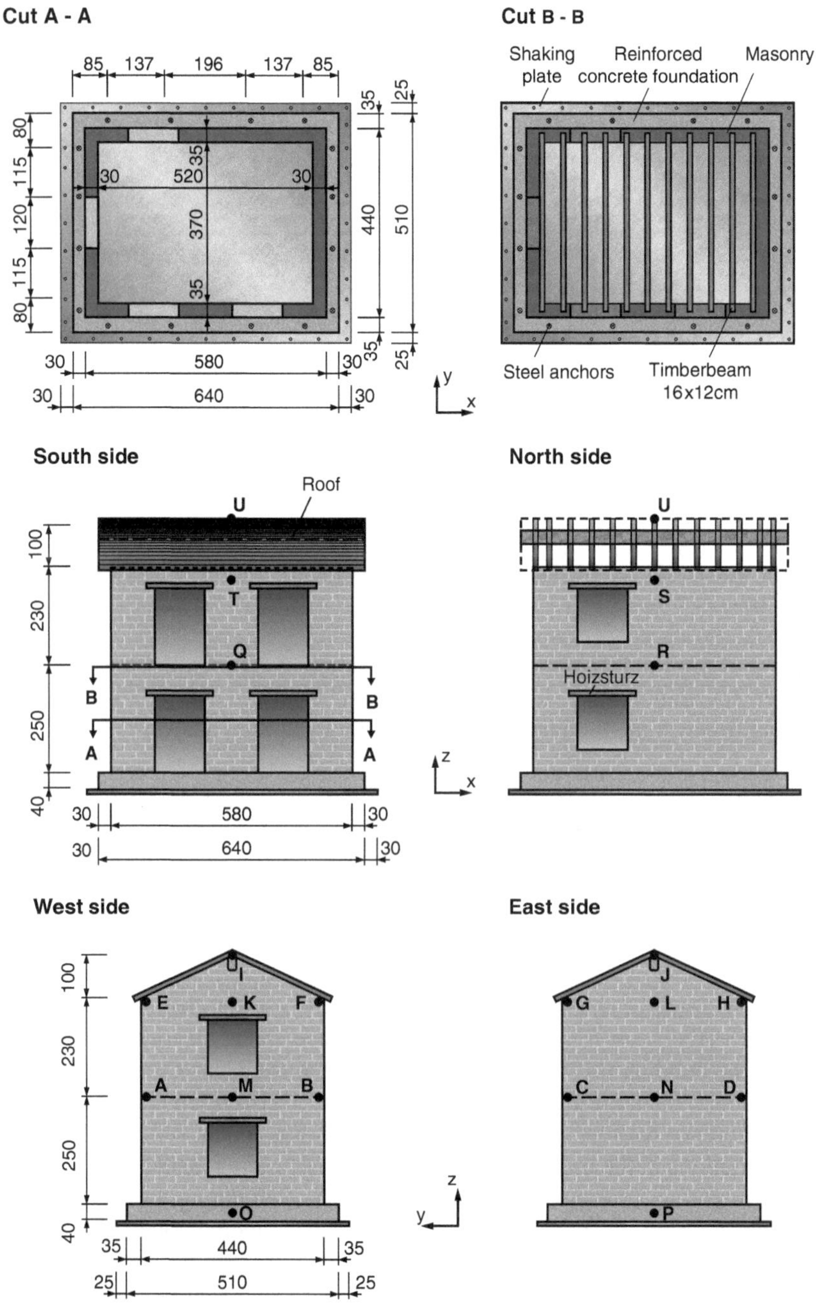

Fig. 9.4 Geometry and sensor location of the two-storey masonry building

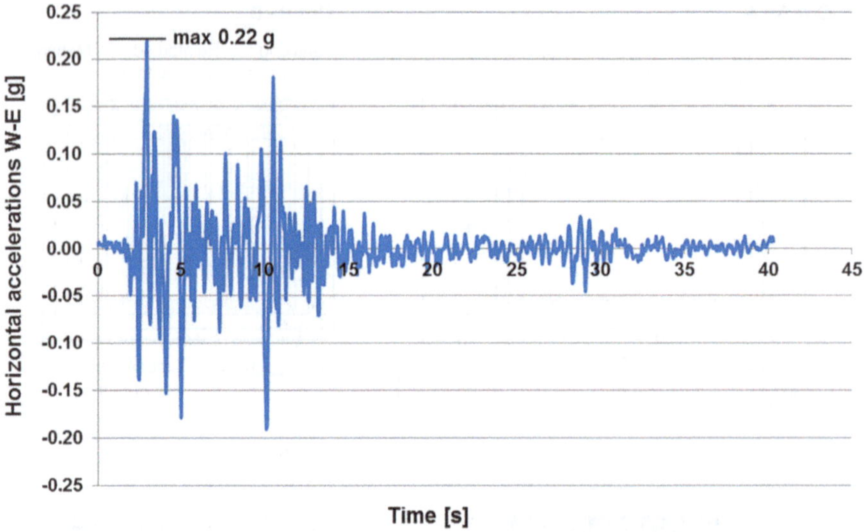

Fig. 9.5 Earthquake acceleration time-history "Montenegro" (1979) EW-direction (Ambraseys et al. 2001)

collapse of the building started. Only the acceleration data directly measured at the concrete foundation was considered for clear and unaltered data without any kind of slipping between the foundation and steel plate for further analysis.

9.4.2 Unstrengthened Building (G1) Results

After the construction stage of the 5.8 m tall unreinforced building several ambient vibration tests for system identification were conducted in the uncracked condition. The results from 12 channels were imported in the system identification software ARTeMIS [Extractor 5.3] for frequency and mode shape identification. The 3-D shell of the building was modelled with 25 triangle elements. Comparisons with a finite element model using ETABS software [v. 9.1.1] verified the results. Only a 12 % difference between the first two frequencies was found, which represents the usual variations in the material parameters and construction details.

One important problem was the analysis of the data during testing. Additional cracking, after strong nonlinear behavior started including rotation effects and frequency shifting, leads to approximate rough analysis with the linear algorithms. With the calculation method "stochastic subspace identification" (SSI) in ARTeMIS it was possible to identify and separate the first two modes in the x-direction (9.6 Hz) and y-direction (8.8 Hz) (definition after Fig. 9.6).

The tests were conducted in different load steps. The corresponding EW-PGA values measured directly at the foundation were 0.074g, 0.150g and 0.524g,

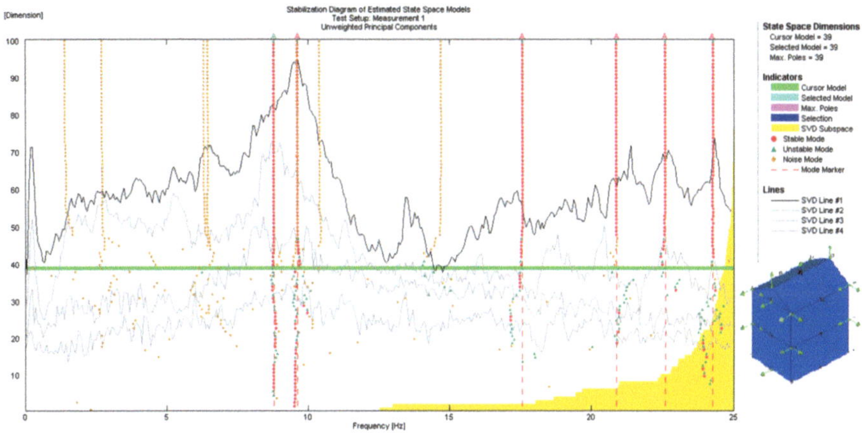

Fig. 9.6 Natural frequencies in ARTeMIS before testing

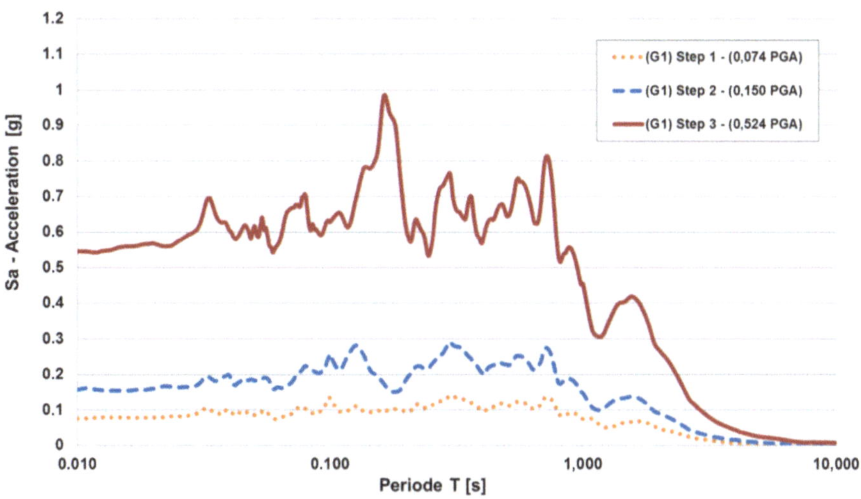

Fig. 9.7 Calculated response spectra for each load step with 5 % damping at building (G1)

respectively. The benefit of using this data instead of table values is the elimination of errors introduced by the whole tolerance at the foundation to table connection. In Fig. 9.7 the response spectra for the measured time history are depicted with 5 % damping. The first important period with T = 0.1038 s (9.63 Hz) in the x-direction at the plateau location confirms the method of increasing load based on PGA values.

After testing the unstrengthened reinforced masonry (URM) building – designed per (Magenes et al. 2010) and constructed by EUCENTRE staff – severe damage at more than seven locations were visible (Figs. 9.8 and 9.9). A characteristic attribute was the wooden floor between the two levels with its very low shear stiffness. This

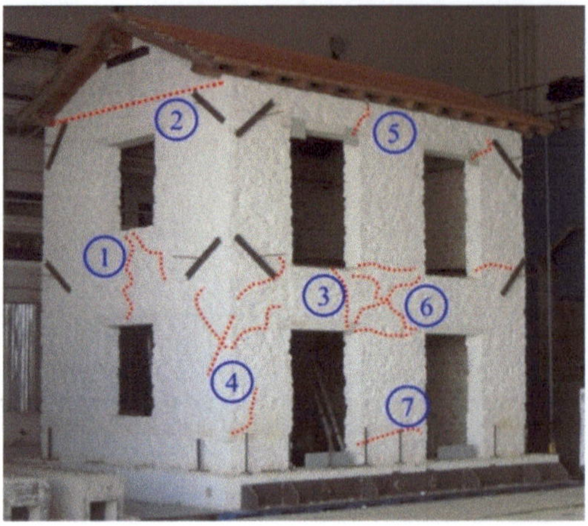

Fig. 9.8 Positions of cracks at the unreinforced building (G1)

Fig. 9.9 Out-of-plane bending cracks at the gable while testing at 0.524g PGA at position 2

was the reason why not all walls were rigidly coupled. After the development of significant cracks at position 3 and 5 (Fig. 9.8) the southern front side and the northern backside of the building showed very different behavior. Only a timber-beam directly over position 5 held the section together using nonlinear sliding effects. With observed crack width of more than 5 cm the building was just before collapse especially at the SW-corner. At position 1 and 2 out-of-plane cracking at the gable and between the windows at the west side was observed. Diagonal and horizontal shear cracking are the failure reasons at position 6 and 7, and developed after large deformations of the southern front part.

Large differences in the accelerations at the SW-corner and NW-corner over the height were measured. Figure 9.10 shows the interpolated accelerations between the sensor locations. The first sensor was placed directly on the foundation at a corresponding height of 0 m, the second at 2.5 m and the last sensor at 4.8 m. The straight interpolation lines illustrate a force reduction at the NW-corner and an increase at the front side especially in the second storey of the SW-corner. Notwithstanding the theory of stiff coupled walls after cracking of one side, a force

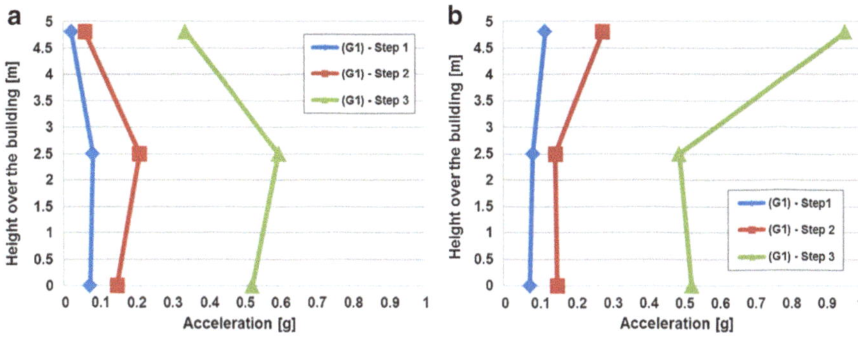

Fig. 9.10 Acceleration over the building height: (**a**) at the NW corner; (**b**) at the SW corner

increase at the stiffer wall side is the result. The authors argue that the cracked front side with its higher period is excited by higher accelerations like the response spectrum indicate. The other uncracked north side is less excited through a tuned damping effect of the south side. The wooden floor and the soft roof connecting them can be idealised as the spring between the two independent mass systems.

9.4.3 Strengthened Building (G2) Results

The nearly complete damaged building (G1) was repaired and strengthened after the tests using the fibre mortar system "eq-grid". This retrofitted structure (G2) was equipped again with different sensors and was tested a second time on the already described shaking table. First of all the broken timber slab was repaired with a new layer of 18 mm thick formwork panels. Most of the original nail connections were degraded through plastic deformations. Additional injections with steel bars linked the displaced roof with the masonry walls (Fig. 9.11a).

The main retrofitting work was done with the application of the "eq-grid" system. Therefore the total outer surface was covered with the 1.27 m wide fabric sheets, overlapping each other by at least 100 mm. A single additional horizontal sheet was applied at an angle of 90° to tie up the total building in one piece at the height of the wooden slab and to increase the reinforcement ratio between the openings. Like in the case of the wall tests no smooth plane of the fibres could be reached. Before fibre application it was necessary to smooth down the surface with a third putty layer of the same mortar type. The additional mass of the mortar surface was calculated with an area of 81.53 m^2, an average thickness of 10 mm and a density of 24 kN/m^3. Altogether an additional weight of 19.6 kN was applied and increased the total building weight by 2 % to at least 923.3 kN. A shift of the first two frequencies was measured after the repairing. After strengthening the first mode in the x-direction had a frequency of 9.95 Hz instead of 9.6 Hz and in the y-direction 11.75 Hz instead of 8.8 Hz.

Fig. 9.11 (**a**) Borehole drilling for steel anchor connection between masonry and roof; (**b**) application of textile layers in vertical orientation

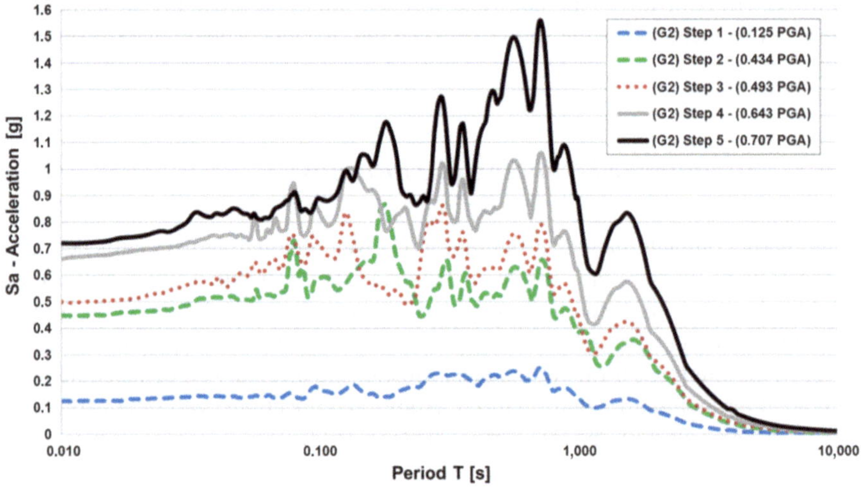

Fig. 9.12 Calculated response spectra for each load step for building (G2) (5 % damping)

Similar to the loading program applied to building (G1) five different load steps with 0.125g, 0.434g, 0.493g, 0.643g and 0.707g PGA were applied. The response spectra are depicted in Fig. 9.12.

No damage was observed after the first three load steps up to 0.493g PGA, which was about the maximum acceleration in the unreinforced condition. The retrofitting established more than the original performance of the building although the building was pre-tested until damage. First cracks occurred at position 3 at load stage 4 between the left openings (see Fig. 9.8). This damaged condition lead to a drop of the first natural frequency from 9.95 to 8.22 Hz followed by a further drop after the last load step of 0.707g PGA to 7.19 Hz. This frequency shift correlates with the crack development between the openings on the south side. While all

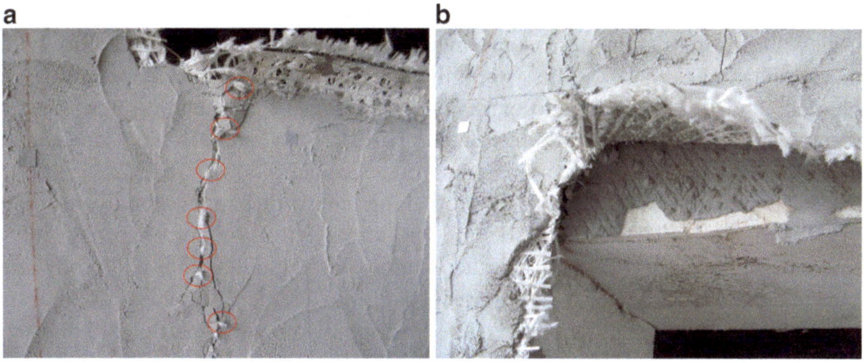

Fig. 9.13 (**a**) Horizontal crack in the glass fibres between the left openings at the south side; (**b**) delamination at the wooden beam over the door on the south side

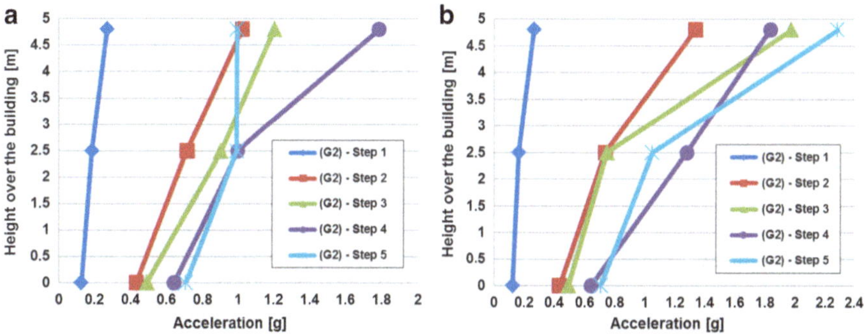

Fig. 9.14 Acceleration over the building height: (**a**) at the NW corner; (**b**) at the SW corner

polypropylene fibres with a higher strain rate were undamaged, only local horizontal glass fibres at position 3 failed; at the left corner near the left door at the ground floor delamination between the first mortar layer and the fibres was observed (Fig. 9.13). The very ductile behavior of the softer south side after cracking lead to a different deformation behavior. Like for the cracked unreinforced building (G1), a drop of the measured acceleration on the top of the west-north corner was observed shortly after the cracking in the final load step (Fig. 9.14). With ARTeMIS and after video analysis verification a new mode of the separated southern side with approximately 1.87 Hz was detected.

9.5 Conclusions

Traditional natural stone walls and buildings – similar to the construction technology used in the L'Aquila region in Italy – under original and strengthened condition were tested. New multi-axial hybrid fibre reinforcement "eq-grid" – developed for normal industrial bricks – was used for structural strengthening. Single shear wall tests with this high strength masonry material lead to a much higher ductility with more and finer cracks. This strengthening system was then applied to a pre-damaged full-scale two storey masonry building in a benchmark test for the reconstruction of existing buildings. In comparison to the unreinforced condition a minimum of 35 % increase in maximum ground acceleration was reached. The tests had to be aborted due to the limit of the shaking table with a maximum overturning moment of approximately 4,000 kNm. The horizontal glass fibres as main reinforcement cracked locally between the window parts, but diagonal polypropylene fibres held the cracked masonry without separation together. At the end of the tests there was less significant damage visible in comparison with the original building. In this case significant damage occurred at a PGA of about 0.524g. Different observations were made:

- the wooden floor lead to complex building behavior in the cracked condition;
- different accelerations were measured over the height at the corners after cracking;
- a full covering of the surface with multi-axial fibres significantly increased the load carrying capacity against horizontal earthquake;
- high elastic polypropylene fibres as additional fibre material lead to a much more ductile masonry behavior after cracking;
- out-of-plane damage was only observed at the unreinforced building;
- main cracks occurred at the building (G2) around the openings.

Acknowledgments The research leading to these results has received funding from the European Community's Seventh Framework Programme [FP7/2007–2013] for access to Foundation EUCENTRE under grant agreement n° 227887 [SERIES]. The support of Foundation EUCENTRE staff, headed by Prof. A. Pavese, and collaboration with D'Appolonia S.p.A. are also gratefully acknowledged.

References

Ambraseys N, Smit P, Sigbjörnsson R, Suhadolc P and Margaris B (2001) Internet-site for European strong-motion data. http://www.isesd.cv.ic.ac.uk, EVR1-CT-1999-40008. European Commission, Directorate-General XII, Environmental and Climate Programme. Bruxelles

Benedetti D, Pezzoli P (1996) Shaking table test on masonry buildings – results and comments. ISMES, Seriate Bergamo

Ehsani M, Saadatmanesh H (1996) Seismic retrofit of URM walls with fibre composites. TMS J 14(2):63–72

Ehsani MR, Saadatmanesh H, Al-Saidy A (1997) Shear behaviour of URM retrofitted with FRP overlays. J Compos Constr 1(1):17–25

Ehsani MR, Saadatmanesh H, Velazquez-Dimas JI (1998) Retrofit of clay brick walls with fibre composites. In: 11th European conference on earthquake engineering, Paris

El Gawady MA (2004) Seismic in-plane behaviour of URM walls upgraded with composites. Dissertation, EPFL, Lausanne

Gülkan P, Clough RW, Mayes RL, Manos GC (1990) Seismic testing of single-storey masonry houses: Part 1. ASCE J Struct Eng 116(1):235–274

Kolsch H (1998a) Carbon Fibre Cement Matrix (CFCM) overlay system for masonry strengthening. J Compos Constr 2(2):105–109

Kolsch H (1998b) Eine Laminatbeschichtung aus Kohlenstoffgeweben und Zementmatrix zur Mauerwerksverstärkung. Bauen mit Textilien Heft 2:13–18

Magenes G, Penna A, Galasco A (2010) A full-scale shaking table test on a two-storey stone masonry building. In: Proceedings of the 14th European conference on earthquake engineering, paper no 1432, Ohrid, Macedonia

Münich JC (2010) Hybride Multidirektionaltextilien zur Erbebenverstärkung- Experimente und Numerische Untersuchungen mittels eines erweiterten Makromodells. Dissertation, Department of Reinforced Concrete and Building Materials, Karlsruhe Institute of Technology, Karlsruhe

Papanicolaou CG, Triantafillou TC, Papathanasiou M, Karlos K (2008) Textile reinforced mortar (TRM) versus FRP as strengthening material of URM walls: out-of-plane cyclic loading. Mater Struct 41:143–157

Peloso S, Pavese A, Casarotti C (2012) Eucentre TREES lab: laboratory for training and research in earthquake engineering and seismology. In: Fardis MN, Rakicevic ZT (eds) Role of seismic testing facilities in performance-based earthquake engineering: SERIES workshop proceedings, vol 22, Geotechnical, geological and earthquake engineering. Springer, New York. doi:10.1007/978-94-007-1977-4_4

Schwegler G (1994) Verstärken von Mauerwerk mit Faserverbundwerkstoffen in seismisch gefährdeten Zonen. EMPA Bericht Nr. 229, EMPA, Dübendorf

Tomazevic M (2000) Shaking table tests of small-scale models of masonry buildings: advantages and disadvantages. Massivbau 2000: Forschung, Entwicklungen, Anwendungen, 67–83

Tomazevic M (2009) Shear resistance of masonry walls and Eurocode 6: shear versus tensile strength of masonry. Mater Struct 42:889–907

Tomazevic M, Gams M (2010) Seismic behaviour of confined autoclaved aerated concrete masonry buildings: a shaking table study, Mauerwerk 14 Heft 3, 153–160

Tomazevic M, Bosiljkov V, Weiss P (2004) Structural behaviour factor for masonry structures. In: 13th world conference on earthquake engineering, Paper no 2642, Vancouver

Turek M, Ventura CE, Kuan S (2004) In-plane shake table testing of FRP strengthened URM walls. In: 13th world conference on earthquake engineering, Paper no 3358, Vancouver

Urban CM (2013) Experimentelle Untersuchungen und Bemessungsansätze für Faserverstärktes Mauerwerk unter Erdbebenbeanspruchungen. Dissertation, Department of Reinforced Concrete and Building Materials, Karlsruhe Institute of Technology, Karlsruhe

Urban CM, Stempniewski L (2012) Comparison of different earthquake strengthening methods for masonry buildings. In: 15th world conference on earthquake engineering, Paper no 697, Lisboa

Wallner C (2008) Erdbebengerechtes Verstärken von Mauerwerk durch Faserverbundwerkstoffe – Experimentelle und numerische Untersuchungen. Dissertation, Department of Reinforced Concrete and Building Materials, Karlsruhe Institute of Technology, Karlsruhe

Chapter 10
Application of Mesh Reinforced Mortar for Performance Enhancement of Hollow Clay Tile Infill Walls

Pourang Ezzatfar, Barış Binici, Özgür Kurç, Erdem Canbay, Haluk Sucuoğlu, and Güney Özcebe

Abstract The use of mesh reinforcement with mortar on existing brick infill walls of reinforced concrete (RC) frames is a recommended seismic strengthening procedure in the Turkish Seismic Code (2007). The premise of the method lies in its ease of application and success in eliminating the out-of-plane failure of existing infill walls. The performance of the mesh reinforced mortar (MRM) application was investigated by pseudo-dynamic (PsD) and cyclic testing. A three-story-three-bay 1:2 scale RC frame with hollow clay tile (HCT) infills in the middle bay was first tested using a continuous pseudo-dynamic test method for three synthetic ground motions compatible with the Düzce city center response spectrum. The test specimen was code complaint. No significant structural damage besides some cracking in the boundary columns was observed but the infill walls almost collapsed. After removing the infill walls of the central bay, a new HCT wall strengthened with MRM was built and the rehabilitated frame was retested under a second series of PsD and reversed cyclic loading schemes. This Chapter reports the findings of the experimental study by placing special emphasis on the seismic response of the code compliant test frame.

10.1 Introduction

Brick infill walls constitute a large portion of building components. Taking the benefits of lateral load bearing capacity of brick infill walls can be considered as a cheap and effective solution for strengthening of damaged or undamaged RC structures. Past experimental studies have shown that infill walls do contribute to

P. Ezzatfar (✉) • B. Binici • Ö. Kurç • E. Canbay • H. Sucuoğlu • G. Özcebe
Department of Civil Engineering, Middle East Technical University,
Dumlupinar Bulvari 1, 06800 Cankaya, Ankara, Turkey
e-mail: e174831@metu.edu.tr; binici@metu.edu.tr; kurc@metu.edu.tr; ecanbay@metu.edu.tr; sucuoglu@metu.edu.tr; ozcebe@metu.edu.tr

A. Ilki and M.N. Fardis (eds.), *Seismic Evaluation and Rehabilitation of Structures*, Geotechnical, Geological and Earthquake Engineering 26,
DOI 10.1007/978-3-319-00458-7_10,

the lateral load carrying capacity of the structures and lead to higher overall lateral stiffness of the RC structures (Paulay and Priestly 2009). On the other hand, it is also a known fact that these elements are very brittle and during severe earthquakes they may break apart at low interstory drifts, which further causes dramatic reduction in stiffness and lateral load bearing capacity. Reinforcing the brick infill walls is an easy and feasible method to increase the capacity, in terms of strength and ductility. Application of mesh reinforced mortar (MRM) on existing brick infill walls of reinforced concrete structures has been considered as a strengthening method in Turkish Seismic Code 2007 (Ministry of Public Works and Settlement 2007). The purposes of applying this method are to prevent premature failure of infill wall, to keep the integrity of masonry infill, and to reduce the risk of out-of-plane collapse of masonry infill. A few experimental studies have been done to assess the seismic effectiveness of the MRM application. The effectiveness of the method has been proven by previous experimental researches by conducting cyclic loadings schemes on one-bay one-story and one-bay two-story RC frames. The study conducted by Alcocer et al. (1996) was one of the first experimental studies to assess reinforcing of brick infill walls as a repair/strengthening method. They tested damaged and undamaged test specimens with low transverse reinforcement and infilled by MRM using reversed cyclic loadings. The variables of this experimental study were the amount of reinforcing mesh, type and spacing of anchorage dowels. In-plane and out-of-plane behavior of weakly reinforced brick infill panels were investigated by Calvi and Bolognini (2001) using cyclic tests on one-bay one-story RC frames. They applied different reinforcing methods (horizontal reinforcement in mortar layer and external mesh in plaster) on brick infill panels. According to their study, the response is improved to a great extent when using external mesh in comparison with other reinforcing methods. In other independent studies, the most important parameters that affect the response of the MRM are found to be the type and compressive strength of covering mortar, mesh reinforcement ratio and type of anchorage of steel mesh (Acun and Sucuoğlu 2006; Altın et al. 2010). The increase in displacement ductility and lateral load bearing capacity, and significant reduction in strength deterioration were identified as the common outcomes of all studies mentioned above.

The current study focuses on the effectiveness of MRM as a repair method. In this context, a three story three bay RC frame with HCT infill walls in the middle bay was chosen as test specimen and tested under pseudo-dynamic loading protocols applied in two phases. Phase 1 loading lead to minor damage in frame elements and intensive damage in the brick infill wall of the first story. In order to assess the effectiveness of the MRM technique in increasing the lateral load bearing capacity and ductility of damaged RC frames, the damaged brick infill walls were replaced with new ones and enhanced by MRM on both sides in accordance with the Turkish Seismic Code (2007). As a second phase of loadings, the repaired frame was retested pseudo-dynamically. The details of the test specimens, test setup and instrumentation, the loading protocols and also the observed behavior of the specimens during the tests are given in the sections below.

10.2 Test Specimen

The test specimen is a half scale three-story three-bay RC frame (Fig. 10.1). The frame was designed and constructed according to the Turkish Seismic Code (2007). The dimensions of the frame were 4,500 mm in height and 6,950 mm in length in the direction of loading. The middle bay of the specimen was filled with hollow clay tile (HTC) infill walls in all stories.

Scaled HCT units with dimensions 190 mm × 100 mm × 95 mm were used in the fabrication of infill walls, Fig. 10.1e. The uniaxial compressive strength of a single HCT unit was 25 MPa. The mortar used in the construction of the infill wall had an average uniaxial compressive strength of 4.7 and 5.5 MPa for the phase 1 and phase 2 specimens, respectively.

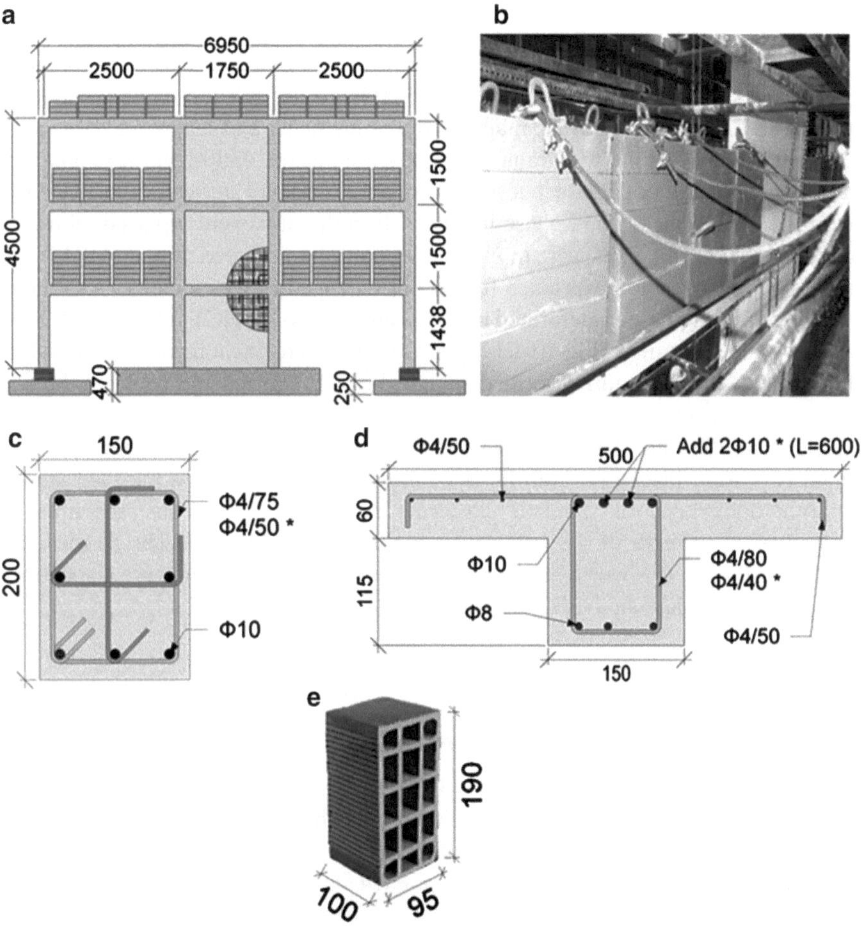

Fig. 10.1 (**a**) Test frame, (**b**) steel blocks, (**c**) column, (**d**) beam section, (**e**) hollow clay tile brick (* denotes additional reinforcement and change in spacing in end region; dimensions in mm)

Table 10.1 Mechanical properties of reinforcement

Bar diameter (mm)	Bar type	Tensile yield strength (MPa)	Tensile strength (MPa)
4	Plain (smooth)	240	340
5	Deformed (ribbed)	680	730
8	Deformed (ribbed)	430	560
10	Deformed (ribbed)	430	680

The amount of reinforcement and the tie spacing used were in compliance with the Turkish Seismic Code (2007) provisions. 8-mm and 10-mm deformed bars were used as longitudinal reinforcement and 4-mm plain bars were used as transverse reinforcement in the beams and the columns. Mechanical properties of reinforcing bars are given in Table 10.1. Average compressive strength of cylindrical concrete specimens was about 20.5 and 21.5 MPa for the phase 1 and phase 2 specimens, respectively. Figure 10.1 shows the dimensions of the test specimen, together with the section details. As shown in Fig. 10.1b, steel blocks were used to apply gravity loads.

The specimen was tested in two phases. During the first phase, three consecutive earthquakes were applied on the specimen by using pseudo-dynamic loading protocols. Low to moderate damage in frame elements and extensive damage in HCT infill walls of the test frame were observed at the end of the first phase of testing. The details of Phase 1 test are discussed in detail in Sect. 10.4.1. Hence, before applying the second phase loading, all infill walls were replaced with new ones and these infill walls were reinforced with mesh reinforced mortar (MRM). No additional repair action was taken for the pre-damaged frame elements. A 200 mm by 200 mm grid of 5-mm deformed bars was attached to the HCT infill on both sides as mesh reinforcement (Fig. 10.2). These meshes were anchored to the boundary frame by means of 8-mm deformed anchor dowels. The spacing between the two neighboring anchor dowels along the boundary beams and columns was kept constant at 300 mm. To fix the steel meshes to the HCT walls, 6-mm L-shaped plain bars were placed on both faces of the HCT infill walls. In the final step, a 25 mm thick layer of standard plastering mortar, having an average uniaxial compressive strength of 4.3 MPa, was applied on the steel meshes to form the MRM infill wall. It is worth noting that the plaster mortar used in phase 1 specimen was 10 mm and had an average compressive strength of 3.6 MPa.

10.3 Test Setup and Instrumentation

Lateral loads were applied to the specimen with three servo-controlled hydraulic actuators, mounted on the laboratory rigid wall at each story level (Fig. 10.3c). The actuators directly pushed the test frame at each story level and pulled it with rods tied to both ends of the frame.

During the tests, story displacements were measured by 6 linear variable differential transformers (LVDTs) installed at story levels (2 LVDTs at each floor level).

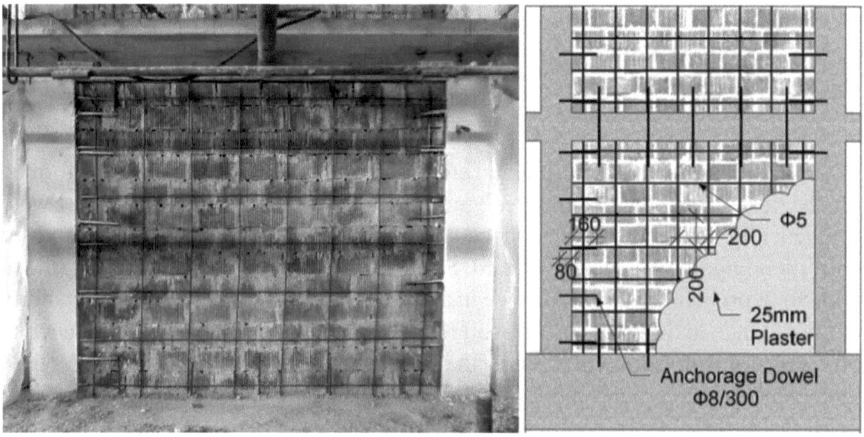

Fig. 10.2 MRM infill details

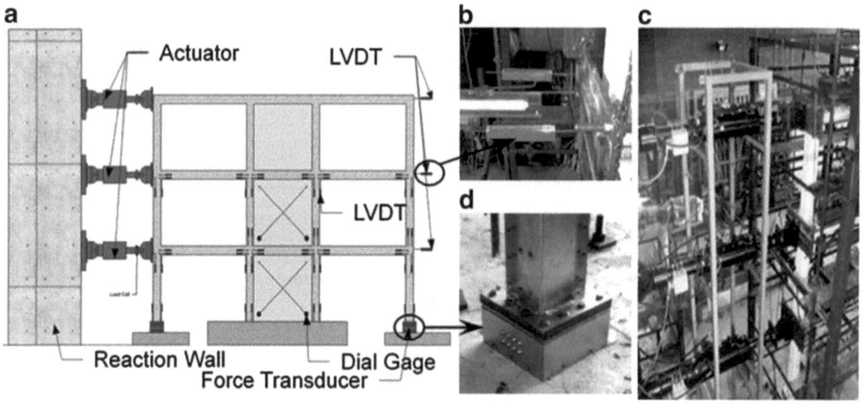

Fig. 10.3 Test setup and instrumentation: (**a**) test setup, (**b**) LVDTs, (**c**) actuators, (**d**) load cell

Column and beam end rotations were measured by LVDTs installed at the member ends of the first and second stories. Load cells were located between actuators and test frame to measure story forces. Shear deformation of infill walls were monitored by diagonally positioned LVDTs on infill walls of the first and second stories. Reactions (bending moment, axial force and shear force) at the base of external columns were measured using two special force transducers (Canbay et al. 2004). These transducers were manufactured, calibrated, and placed between the base of external columns and the foundation. Longitudinal reinforcements of external columns were welded to base plates that were connected to transducers. Transducers were fixed to the foundation block by using bolts.

10.4 Pseudo-Dynamic and Cyclic Tests

10.4.1 Phase 1

In the first phase of the experimental study, pseudo-dynamic tests were performed on the test specimen using three ground motions. The ground motions were synthetically derived from the Düzce city center site specific acceleration spectra corresponding to two levels of hazard. Spectral acceleration values were associated with 50 % probability of exceedance in 50 years for stiff soil (D1), 10 % probability of exceedance in 50 years for stiff soil (D2) and 10 % probability of exceedance in 50 years for soft soil (D3). Since the specimen has a scale factor of ½, time increments of the ground motion history during PsD tests was scaled by a factor of $1/\sqrt{2}$ according to the similitude law. The scaled ground motions are presented in Fig. 10.4a. The target pseudo-acceleration spectra of the above mentioned ground motions are presented in Fig. 10.4b.

The inter-story drift ratio (IDR) of each floor and base shear versus time diagrams are presented for ground motions D1, D2 and D3 in Fig. 10.5. The observed damage during these experiments is also shown in Fig. 10.5. Toe crushing and interface cracks between infill walls and boundary structural elements took place at the first story during the D1 ground motion. The maximum top displacement was 3.4 mm. The maximum IDR demands were 0.08, 0.1 and 0.08 % at the

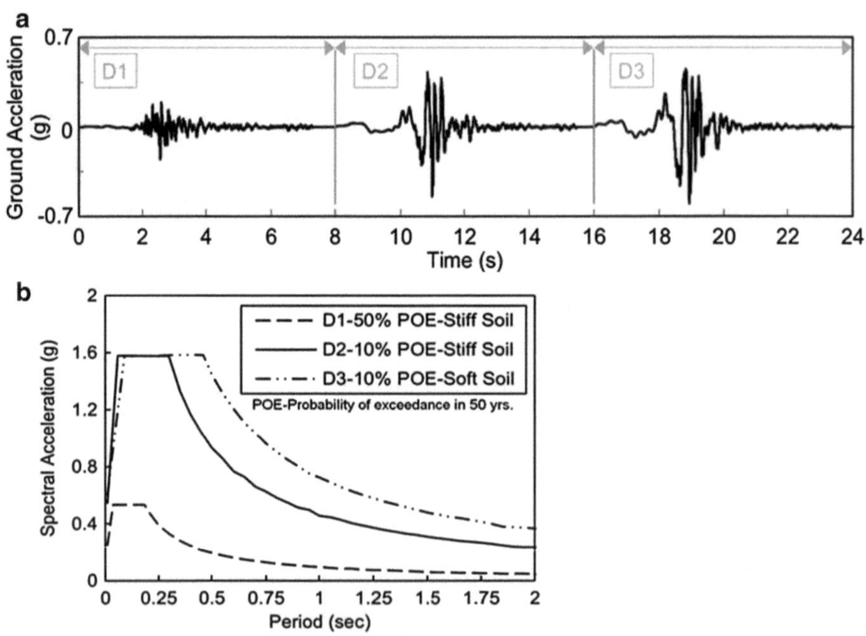

Fig. 10.4 Ground motions: (**a**) acceleration history, (**b**) target spectra of unscaled ground motions

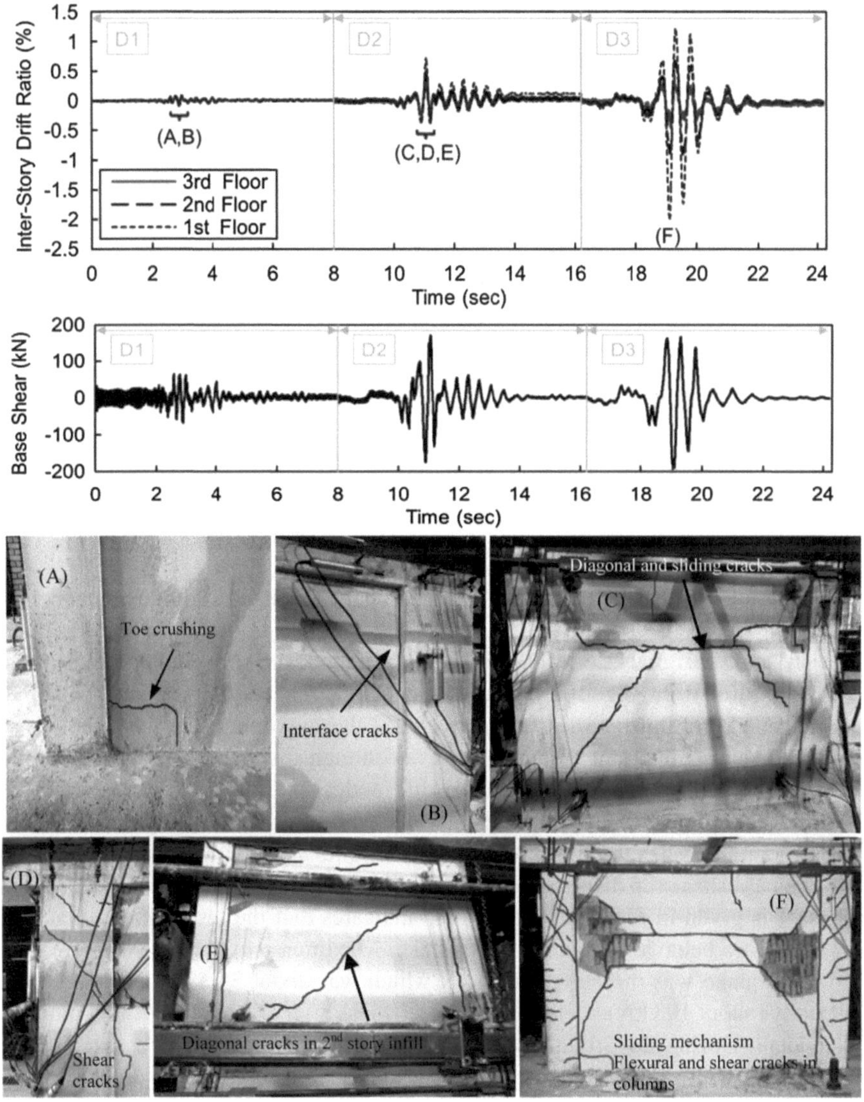

Fig. 10.5 Inter-story drift ratio and base shear demand histories, and main damage observed

first, second and third stories, respectively. At the end of the D1 ground motion the specimen undergoes minor damage, thus satisfying the immediate occupancy performance level. After the D2 ground motion, diagonal and sliding cracks in the 1st story infill wall were observed. Due to formation of compressive struts in the infill wall, diagonal cracks penetrated into the end regions of the boundary columns of the 1st story (Fig. 10.5). The IDR demands were 0.71, 0.5 and 0.37 % for first, second and third stories, respectively. The measured maximum top displacement

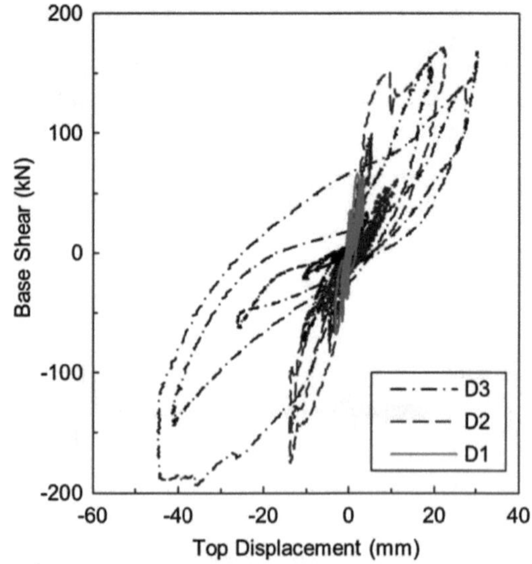

Fig. 10.6 Base shear-top displacement relationship

was 22.7 mm. Extensive damage due to diagonal cracks and sliding mechanism in the infill wall at the 1st story was observed during D3 ground motion. Moreover, flexural and shear cracks expanded along the height of the boundary columns of the 1st story infill wall (Fig. 10.5). The IDR demands reached 2, 0.89 and 0.46 % at first, second and third stories, respectively, at the maximum top displacement of 44.9 mm. It is important to note that, even though the frame was designed and detailed according to the Turkish seismic code, interior columns experienced significant damage because of the infill walls. At the end of the D3 ground motion, the structure satisfies the life safety performance level.

The base shear-top displacement relationship of the specimen for all loadings in phase 1 is given in Fig. 10.6. This figure indicates that the test frame had almost linear elastic behavior under D1 earthquake. Maximum base shear demand under D1 earthquake was measured as 67 kN which was about 35 % of the base shear capacity (about 193 kN). Deterioration in stiffness after D2 and D3 ground motions is evident. During D1 ground motion, only a minor damage was observed in the HCT infill wall, indicating that the wall significantly contributed to the lateral stiffness of the frame. As the damage accumulates during D2 and D3 ground motions, the contribution of the infill wall to lateral stiffness gradually decreases. Such stiffness degradation is mainly due to the diagonal and sliding cracks formed during these tests as shown in Fig. 10.5.

10.4.2 Phase 2

The objective of the second phase of this study was to investigate the seismic performance of MRM infill walls under low and moderate hazard levels. Therefore,

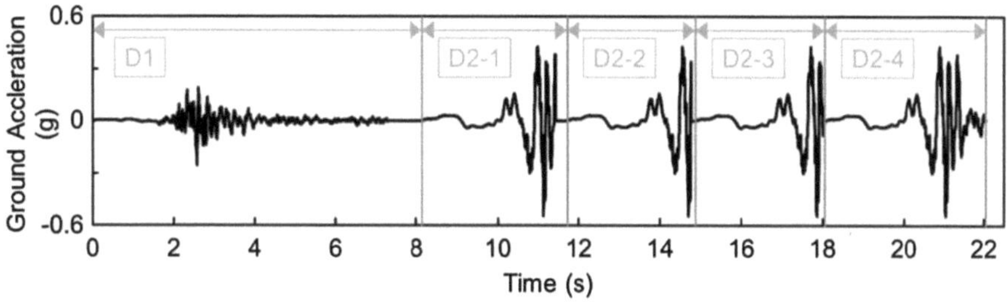

Fig. 10.7 Ground motions

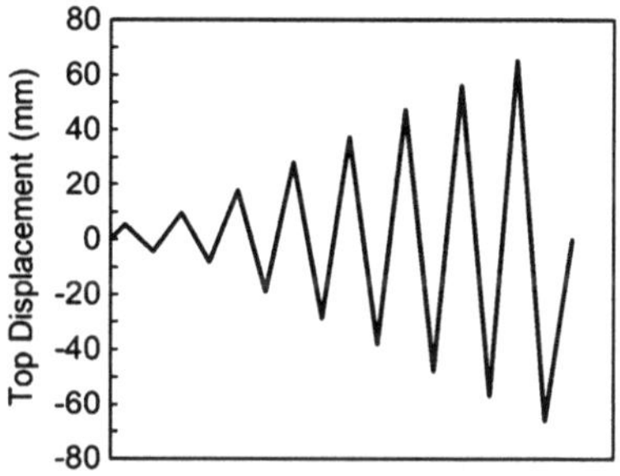

Fig. 10.8 Reverse cyclic loading protocol

after completion of the first phase of this study the damaged infill walls were replaced with MRM infill walls and the enhanced frame was tested pseudo-dynamically under D1 and D2 ground motions respectively. The application of the D1 motion was flawless. On the other hand, technical difficulties were encountered during the application of D2 motion. These difficulties were defeated in the fourth attempt. These successive D2 loadings were named as D2-1, D2-2, D2-3 and D2-4 respectively (Fig. 10.7).

In order to determine the lateral displacement and load bearing capacity of the strengthened system, a displacement control reversed cyclic loading up to 66 mm of top displacement was applied to the specimen after the pseudo-dynamic tests. Figure 10.8 shows the reverse cyclic displacement protocol for the top story of the test frame. The cyclic displacement histories applied to the first and second stories were consistent with the first modal shape of the test frame. They were determined using the method proposed by Molina et al. (1999) and Maia and Silva (1997) by taking the top story displacement protocol into consideration.

Figure 10.9 presents the inter-story drift ratio (IDR) of each floor and base-shear versus time for D1 and successive D2 ground motions, respectively. Main damage observed during these tests is shown in the same figure. At the end of D1 loading, tension cracks developing between infill wall and boundary elements at each story were the only visible damage. The maximum monitored top displacement was about 4 mm for this ground motion. The maximum IDR demands were measured 0.06, 0.11 and 0.11 % at first, second and third stories, respectively.

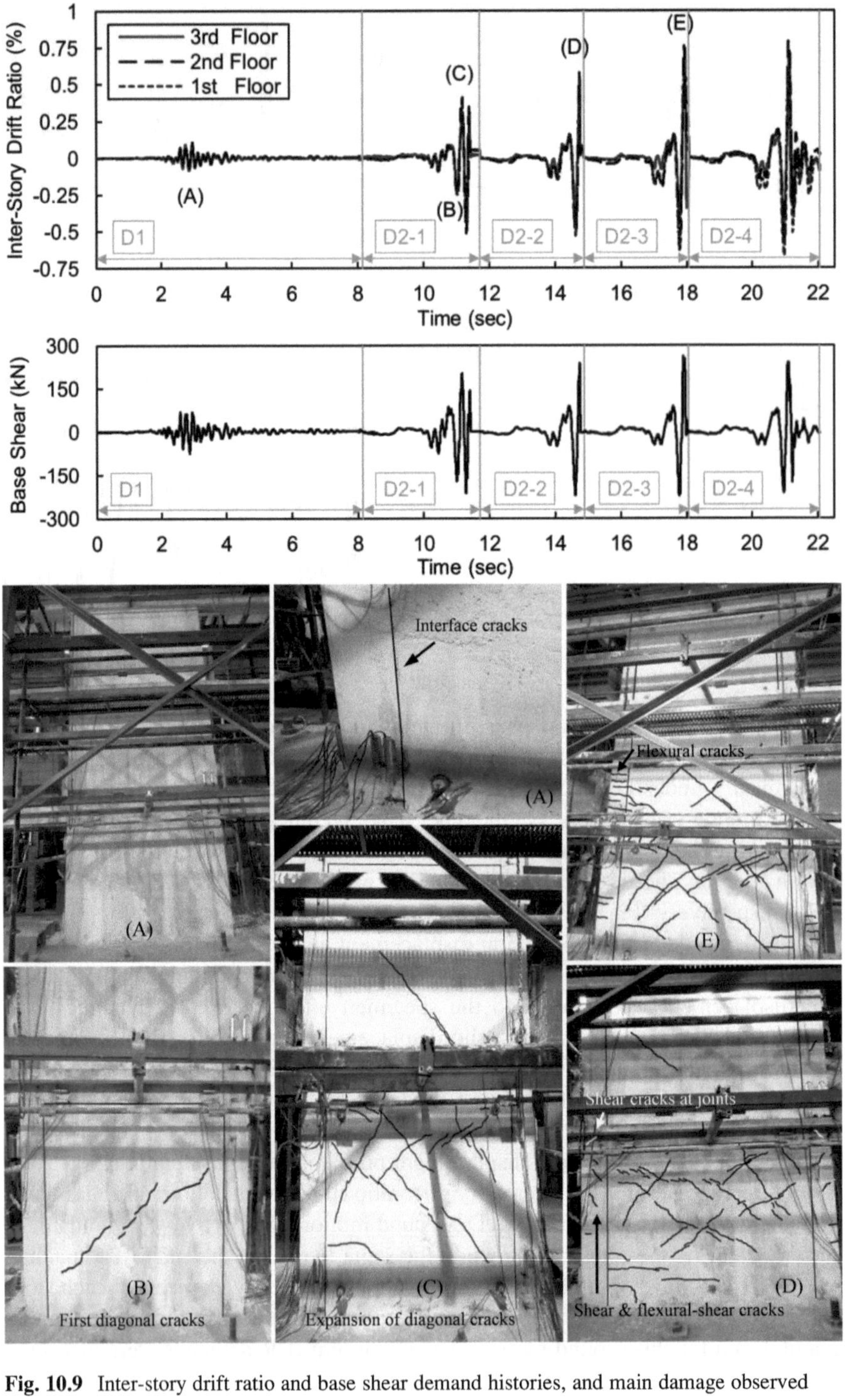

Fig. 10.9 Inter-story drift ratio and base shear demand histories, and main damage observed

Table 10.2 Summary of test results

Ground motion	Max IDR (%)			Max top displacement (mm)	Max base shear demand (kN)	Cycle stiffness (kN/mm)	
	1st story	2nd story	3rd story			A-Cycle	B-Cycle
D1	0.06	0.11	0.11	4	73.7	–	–
D2-1	0.32[a]	0.41[a]	0.32[a]	14.5[a]	202.5[a]	23.5	12
D2-2	0.51	0.58	0.57	23.4	237.7	11	10
D2-3	0.61	0.76	0.75	28.8	261.8	10	9
D2-4	0.62	0.79	0.70	29.9	240	8	8

[a]The first positive peak was selected as the maximum point of drift ratio, top displacement and base shear under the D1 ground motion

Expansion of diagonal cracks in MRM infill walls of 1st and 2nd floors was the main damage observed during D2-1 ground motion (Fig. 10.9). The IDR demands reached 0.32, 0.41, and 0.32 % at first, second and third stories, respectively, at the maximum top displacement of 14.5 mm. As the time span of D2 ground motion applications were all different (Fig. 10.7), the first 2.9 s of all D2 motions common in all tests were considered, in order to compare the maximum IDR demands of the test subjected to successive earthquakes, Fig. 10.9.

During the D2-2 ground motion, new diagonal cracks opened up in infill walls in addition to existing cracks. Moreover, new shear cracks were observed at the joints of the middle bay of the first floor. Shear and flexural-shear cracks formed in the boundary columns of the first story's infill wall. Flexural-shear cracks were also observed at the ends of some beams (Fig. 10.9). The IDR demands were measured as equal to 0.51, 0.58 and 0.57 % for first, second and third stories, respectively. The measured maximum top displacement was 23.4 mm.

The D2-3 ground motion caused additional diagonal cracks in the infill walls of the first and second stories. Furthermore, flexural cracks at the boundary columns of the second story's infill wall were extended (Fig. 10.9). The maximum top displacement measured during the test was 28.8 mm and the corresponding IDRs were 0.61, 0.76 and 0.75 % for the first, second and third stories respectively.

The D2-4 ground motion caused slight widening of the previously formed cracks. Maximum top displacement was 29.9 mm and resulted in 0.62, 0.79, and 0.70 % IDRs for the first, second and third stories, respectively. Comparing the maximum IDRs captured under the same ground motions, it is clear that these values gradually increased, Table 10.2. As can be expected, the stiffness of the structure progressively reduced due to the gradual increase in the observed damage under repeated earthquakes.

In order to investigate the stiffness degradation under successive strong ground motion excitations, two matching cycles were selected in each D2 ground motion excitations. The peak points of these cycles are marked as "A" and "B", as shown in Fig. 10.10. The slope of the line connecting the peaks of an individual A or B cycle was taken as the representative measure to study the stiffness variations after each earthquake. Figure 10.11 illustrates the stiffness variations in A and B cycles in all

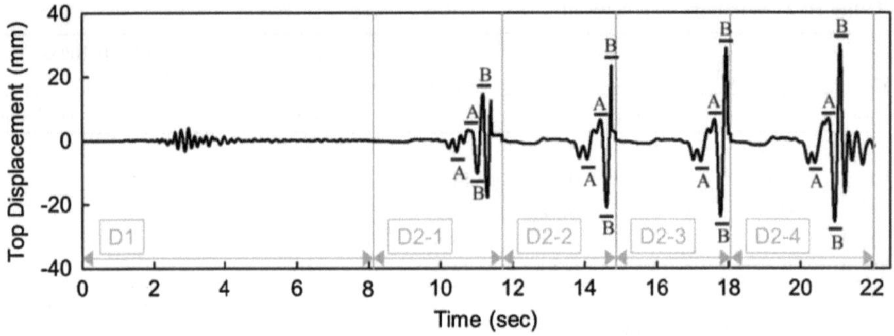

Fig. 10.10 Top displacement history during pseudo-dynamic loadings in phase 2

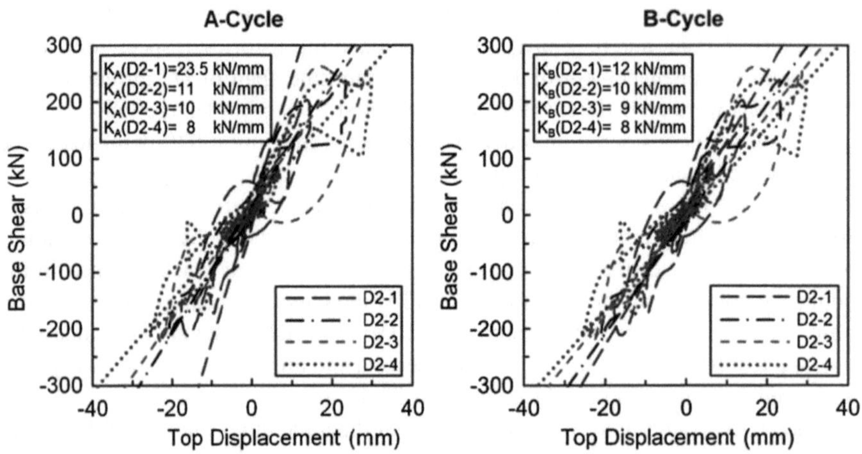

Fig. 10.11 Hysteresis behavior of the test frame and cycle stiffness of the frame in *A* and *B* cycles

D2 ground motions. In Fig. 10.11, the measured individual cycle stiffnesses are also provided for comparison. The peak-to-peak cycle stiffness in A-cycles of D2-1 and D2-4 loadings were 23.5 and 8 kN/mm (34 % of its initial value), respectively. Although the order of magnitude in stiffness degradation was not the same, similar observations were made on B-cycle stiffness values. The B-cycle stiffness values recorded after repeated D2 ground motion applications were 12, 10, 9 and 8 kN/mm, respectively. A nearly 50 % drop in peak-to-peak stiffnesses between A and B cycles of D2-1 loading was also found. While the peak-to-peak stiffness of A-cycle was 23.5 kN/mm, this value was calculated as 12 kN/mm for B-cycle. The expansion of diagonal cracks in infill walls, a clear indication of formation of compression strut, was the main reason for this stiffness drop. Although the specimen showed further softening in D2-2, D2-3 and D2-4 loadings, the order of this drop was not as severe as in the case of D2-1 loading. While the existing cracks were widened and developed to their full lengths in D2-2 loading, only a few new cracks

Fig. 10.12 Observed damage of enhanced frame at the end of the cyclic test

were observed in the structural members and infill walls of the test specimen during D2-3 and D2-4 loadings.

At the end of the pseudo-dynamic tests, a reversed cyclic loading was applied on the test specimen in order to investigate the residual lateral load carrying capacity and the displacement capacity of the enhanced frame.

The physical appearance of the test specimen at the end of the test is shown in Fig. 10.12. New flexure cracks and flexure-shear cracks along the height of the external columns developed during the cyclic loading. The specimen was able to mobilize the peak resistance of 252 kN that it displayed in D2-4 loading. Specimen reached its peak response at a top displacement of 37.5 mm. At this displacement level the IDRs of the first, second and third floors were 0.95, 1 and 0.65 %, respectively. Beyond this cycle the specimen started to display excessive strength and stiffness degradations. At the end of the test, shear failure of one of the internal columns at its upper end region was observed. This was due to formation of a

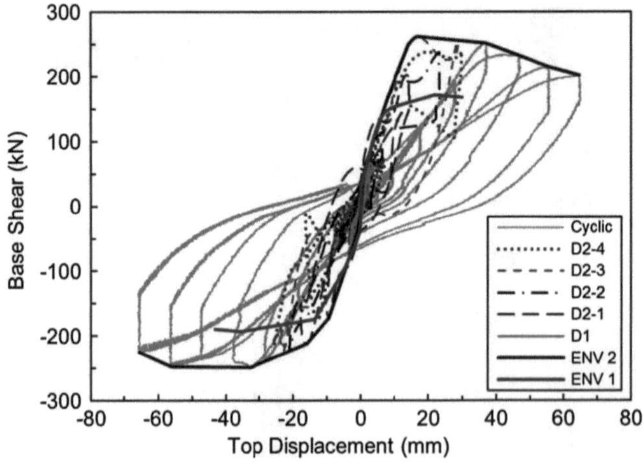

Fig. 10.13 Base shear-top displacement relationship of the specimen enhanced with MRM for all loadings in phase 2 and envelope of response in phase 1 and phase 2 of loadings

captive column upon crushing of the MRM infill wall in this region. Figure 10.13 presents the base shear-top displacement relationship of the specimen enhanced with MRM for all loadings. The envelopes of base shear-top displacement relationship of the specimen in phase 1 (ENV 1) and phase 2 (ENV 2) of the loadings are also shown in Fig. 10.13.

It must be noted that the maximum base shear demands measured in the first and second phases were 193.3 and 261.8 kN, respectively. The corresponding top displacements monitored during these tests, however, were significantly different. In phase 1 loading maximum base shear demand of 193.3 was reached at a top displacement of 44.9 mm during D3 loading. At this load stage the first story IDR was 2 %. In phase 2, however, at the instant when the maximum base shear demand was reached (during D2-3 loading) the top story displacement was 28.8 mm, which corresponds to a first story IDR of 0.61 %. This indicates that the MRM infill increased the base shear demand and also the capacity of the structure significantly.

Figure 10.14 compares the damage observed on the test specimen in two different phases of the experimental study. Figure 10.14a shows the specimen at the maximum first story IDR of 2 % that was measured during D3 ground motion in the first phase of this study. It should be remembered that the test specimen in this phase had plain HTC infill wall with no MRM application. At the end of this phase the HCT infill wall was heavily damaged and the frame members had suffered minor damage. In the second phase of this study the damaged HCT infill was replaced with a new one and MRM was applied on the infill wall on both sides. No repair was made on frame members. The damage observed on the test specimen at exactly the first story IDR of 2 % is shown in Fig. 10.14b. The two figures clearly indicate that the damage state of the test specimen at the identical first story IDR in both phases of the study were entirely different. In the first phase the specimen with plain HCT infill wall developed sliding shear mechanism, whereas the specimen

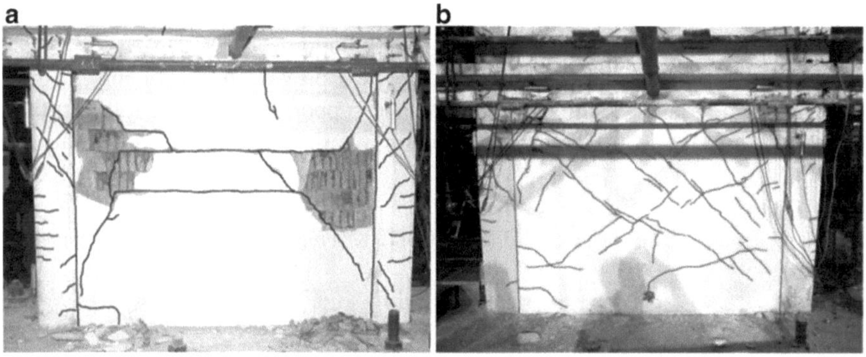

Fig. 10.14 Damage corresponds to 2 % IDR of first story level: (**a**) damage in frame with HCT, (**b**) damage in frame with HCT enhanced by using MRM

with HCT infill wall with MRM displayed well distributed diagonal cracks indicating the formation of compression strut within the wall. Moreover, less shear and flexural cracks opened up in boundary columns of enhanced infill, and the infill wall kept its integrity at this peak response level.

The observations reported in this work favor the use of MRM infill walls as an effective rehabilitation method of damaged and/or substandard RC frames located in seismic zones to mitigate the seismic risk of the built environments.

10.5 Conclusion

In this study, the performance of the Mesh Reinforced Mortar (MRM) application, which is one of the proposed methods for strengthening of RC structures in the Turkish Seismic Code (2007), was verified using pseudo-dynamic (PsD) tests. In this context, a three-bay, three-story frame in ½ scale was prepared and tested. The earthquake load was applied at each story level by means of servo-controlled actuators. The ground motions were synthetically derived from the Düzce city center site specific acceleration spectra. The hollow clay tile (HTC) masonry wall in the mid bay was enhanced with MRM.

The repaired frame performed satisfactorily under identical repeated earthquakes. It displayed a stable hysteretic response. The diagonal cracks on the infills were evenly distributed on the surface of the HCT infill walls with MRM. After formation of the primary diagonal cracks in infills, cycle stiffness was considerably decreased by 49 %. Consecutive earthquakes did not cause any significant decrease in cycle stiffness. A significant increase (61 %) in maximum roof displacement was observed in D2-2 ground motion with respect to the one observed in D2-1 ground excitation. The following two earthquake excitations caused slight increase in maximum roof displacement. By comparing base shear demands, it was concluded that there was no significant strength deterioration during successive application of strong ground motions.

The test results demonstrate that use of MRM provided strength enhancement and more importantly significant out of plane support to the brick wall. Furthermore, the damage to the boundary columns was delayed until large lateral drift demands.

Acknowledgments Funding provided by TUBITAK Project No: 108G034 is gratefully acknowledged.

References

Acun B, Sucuoğlu H (2006) Strengthening of masonry infill walls in reinforced concrete frames with wire mesh reinforcement. In: 8th national conference on earthquake engineering, San Francisco, paper 1852

Alcocer SM, Ruiz J, Pineda JA, Zepeda JA (1996) Retrofitting of confined masonry walls with welded wire mesh. In: 11th world conference on earthquake engineering, Acapulco, paper 1471

Altın S, Anıl Ö, Kopraman Y, Belgin Ç (2010) Strengthening masonry infill walls with reinforced plaster. Struct Build 163:331–342

Calvi GM, Bolognini D (2001) Seismic response of reinforced concrete frames infilled with masonry panels weakly reinforced. J Earthq Eng 5(2):153–185

Canbay E, Ersoy U, Tankut T (2004) A three component force transducer for reinforced concrete structural testing. Eng Struct 26:257–265

Maia NMM, Silva JMM (1997) Theoretical and experimental modal analysis. Research Studies Press/Wiley, New York

Ministry of Public Works and Settlement (2007) Specification for buildings to be constructed in seismic zones. Ankara

Molina FJ, Pegon P, Verzeletti G (1999) Time-domain identification from seismic pseudo-dynamic test results on civil engineering specimens. In: 2nd international conference on identification in engineering systems, Swansea

Paulay T, Priestly MJN (2009) Seismic design of reinforced concrete and masonry buildings. Wiley, Hoboken

Chapter 11
Shake Table Tests on Deficient RC Buildings Strengthened Using Post-Tensioned Metal Straps

Reyes Garcia, Iman Hajirasouliha, Kypros Pilakoutas, Yasser Helal, Yaser Jemaa, Maurizio Guadagnini, Mihail Petkovski, Philippe Mongabure, Mihaela Anca Ciupala, Nicholas Kyriakides, Christis Z. Chrysostomou, Alper Ilki, M. Saiid Saiidi, Lluis Torres, Nicolae Taranu, and Mihai Budescu

Abstract The European research project BANDIT investigated the effectiveness of a novel Post-Tensioned Metal Strapping (PTMS) strengthening technique at improving the seismic performance of deficient RC buildings using shake table tests. A full-scale two-story structure was designed with inadequate reinforcement detailing of columns and beam-column joints so as to simulate typical deficient buildings in Mediterranean and developing countries. Initial shaking table tests were carried out until significant damage was observed in the beam-column joints

R. Garcia (✉) • I. Hajirasouliha • K. Pilakoutas • Y. Helal • Y. Jemaa
M. Guadagnini • M. Petkovski
Department of Civil and Structural Engineering, The University of Sheffield,
Sir Frederic Mappin Building, Mappin Street, Sheffield S1 3JD, UK
e-mail: r.garcia@sheffield.ac.uk; i.hajirasouliha@sheffield.ac.uk; k.pilakoutas@sheffield.ac.uk; y.helal@sheffield.ac.uk; y.jemaa@sheffield.ac.uk; m.guadagnini@sheffield.ac.uk; m.petkovski@sheffield.ac.uk

P. Mongabure
CEA DEN/DANS/DM2S/SEMT/EMSI, 603/103 – PC n°112,
91191 Gif sur Yvette Cedex, Saclay, France
e-mail: philippe.mongabure@cea.fr

M.A. Ciupala
School of Computing, Information Technology and Engineering, University of East London,
4-6 University Way, Beckton, London E16 2RD, UK
e-mail: m.a.ciupala@uel.ac.uk

N. Kyriakides • C.Z. Chrysostomou
Department of Civil Engineering and Geomatics, Cyprus University of Technology,
P.O. BOX 50329, 3603 Limassol, Cyprus
e-mail: nicholas.kyriakides@cut.ac.cy; c.chrysostomou@cut.ac.cy

A. Ilki
Department of Civil Engineering, Civil Engineering Faculty,
Istanbul Technical University, Maslak, Istanbul 34469, Turkey
e-mail: ailki@itu.edu.tr

A. Ilki and M.N. Fardis (eds.), *Seismic Evaluation and Rehabilitation of Structures*, Geotechnical, Geological and Earthquake Engineering 26,
DOI 10.1007/978-3-319-00458-7_11,

of the bare frame. Subsequently, the damaged building was repaired and strengthened using PTMS and additional tests were performed. The results of this study show that the adopted strengthening strategy improved significantly the seismic performance of the substandard RC building under strong earthquake excitations.

11.1 Introduction

Extensive human and economic losses during past strong earthquakes in Mediterranean and developing countries (Turkey, 2003; Pakistan, 2005; China, 2008; Indonesia and Italy, 2009; Haiti, 2010) have highlighted the seismic vulnerability of substandard reinforced concrete (RC) buildings. Many of these buildings were designed to old standards using limited seismic provisions and often employing poor materials and construction practices. As a result, these deficient structures usually have inadequate lateral load resistance, insufficient energy dissipation capacity and high strength degradation that can lead to extensive damage and collapse during severe earthquakes. Thus, the strengthening of seismically deficient structures is a priority in many developing countries as it provides a feasible and cost-effective approach to reducing their vulnerability.

Previous research at The University of Sheffield (Frangou et al. 1995) has led to the development of a novel strengthening technique using external Post-Tensioned Metal Straps (PTMS). The PTMS technique involves the post-tensioning of high-strength steel straps (bands) around concrete elements using hydraulically-powered strapping tools as those used in the packaging industry. The straps are mechanically fastened with metal clips to maintain the desired post-tensioning force. As a result, the element is actively confined and its capacity and ductility are enhanced. Compared to other strengthening methods, the PTMS technique possesses advantages such as low material cost, ease and speed of application, ease of removing/replacing damaged straps, and flexibility to strengthen different types of structural elements. Whilst the PTMS technique has proven extremely effective in the strengthening of deficient RC beams (Frangou 1996), columns (Moghaddam et al. 2010a, b, see Fig. 11.1) and joints (Helal 2012), before it can be widely used in practice, it is necessary to verify its effectiveness on full-scale RC structures subjected to earthquake excitations.

M.S. Saiidi
Department of Civil and Environmental Engineering, University of Nevada, Reno, NV 89557, USA
e-mail: saiidi@unr.edu

L. Torres
Analysis and Advanced Materials for Structural Design, Polytechnic School, University of Girona, Campus Montilivi s/n 17071, Girona, Spain
e-mail: lluis.torres@udg.edu

N. Taranu • M. Budescu
Faculty of Civil Engineering and Building Services, "Gheorghe Asachi" Technical University of Iasi, 43 Bdul. Dimitrie Mangeron, 700050 Iasi, Romania
e-mail: taranu@tuiasi.ro; mbudescu@ce.tuiasi.ro

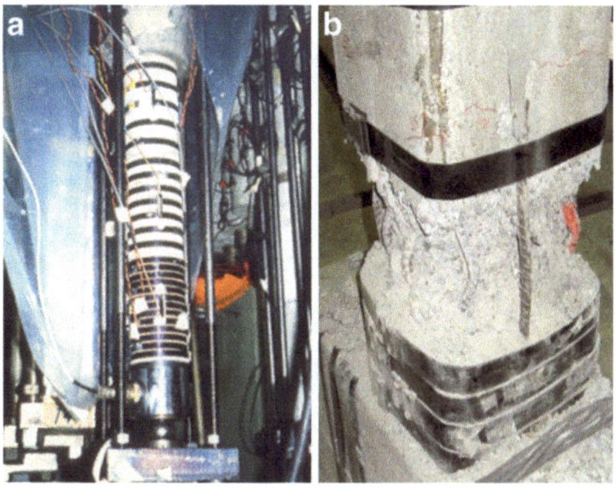

Fig. 11.1 PTMS strengthening of RC columns

The BANDIT project investigated experimentally the effectiveness of PTMS strengthening at improving the seismic performance of a seismically deficient full-scale RC building. The building was tested on the AZALEE shake table of the EMSI laboratory at CEA Saclay, France, as part of the EU-funded SERIES Program (Seismic Engineering Research Infrastructures for European Synergies). Inadequate detailing of reinforcement was provided in beam-column joints and columns so as to simulate typical deficiencies of buildings in Mediterranean and developing countries. The results of the first two phases of the BANDIT project are summarized in this paper. In the first phase, the bare building was tested under uniaxial shaking to produce significant (but still repairable) damage and evaluate its basic seismic performance. After the initial tests, the damaged building was repaired and subsequently strengthened using PTMS. In the second phase, the building was re-tested to examine the efficiency of the PTMS as a post-earthquake strengthening technique for substandard RC structures. An overview of the BANDIT experimental program is introduced in Sect. 11.2 of this chapter. The results of the shaking table tests performed in Phases 1 and 2 are presented in Sects. 11.3 and 11.4, respectively. The results are discussed in terms of the dynamic properties of the building and observed damage after the tests. Section 11.5 compares the seismic performance of the structure under different levels of earthquake excitations before and after the strengthening intervention. Concluding remarks are given in Sect. 11.6.

11.2 Experimental Program

11.2.1 Building Geometry

Figure 11.2 shows a general view of the BANDIT building. The building was a one-span two-story moment-resisting frame regular in plan and elevation similar to the structure tested as part of the European project ECOLEADER (Garcia et al. 2010). The span length was 4.0 m in both orthogonal X and Y directions, and the

Fig. 11.2 General view of BANDIT building in the X direction

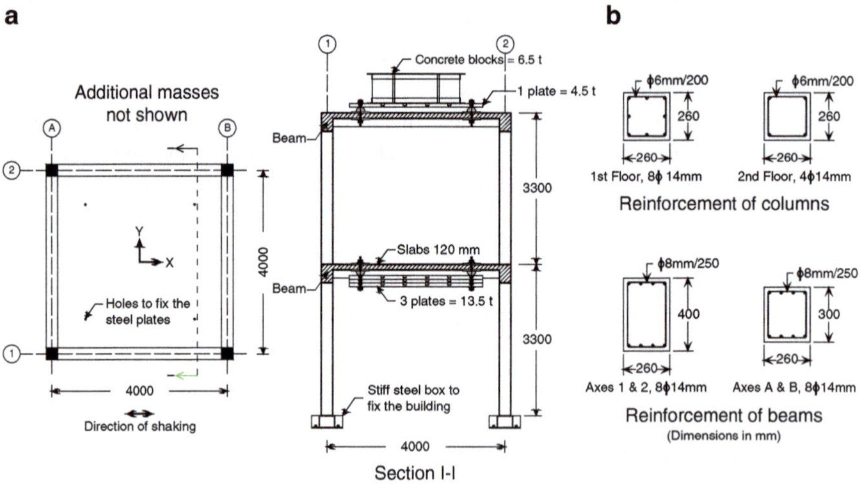

Fig. 11.3 BANDIT building (**a**) plan view and elevation; (**b**) reinforcing steel of structural elements

floor height was 3.3 m (see Fig. 11.3a). The columns had a cross section of 260 × 260 mm, as shown in Fig. 11.3. The longitudinal reinforcement of the 1st floor columns consisted of eight 14 mm bars placed along the column perimeter, and four 14 mm bars located at the column corners for the 2nd floor columns. This resulted in longitudinal reinforcement ratios (ρ_l) of 1.82 and 0.91 % for the 1st and 2nd floor columns, respectively. The columns were reinforced in shear using transverse stirrups of 6 mm diameter spaced at 200 mm centers. The stirrups were closed with 90° hooks instead of 135° hooks typically required by current seismic codes.

The geometry of the beams was different in the two orthogonal directions of the building, as shown in Fig. 11.3b. The cross section of the beams in the X direction was 260 × 400 mm, whilst the beams in the Y direction had a cross section of 260 × 300 mm. For all beams, the main flexural reinforcement consisted of four 14 mm bars at the bottom and four 14 mm bars at the top, resulting in a ρ_l of 0.65 and 0.90 % for the beams in the X and Y directions, respectively. This enabled the building to be tested and strengthened separately in each orthogonal direction. The beams had transversal reinforcement consisting of 8 mm stirrups spaced at 300 mm centers. The 120 mm thick concrete slabs were reinforced using two meshes (top and bottom) of 10 mm bars spaced at 100 mm centers. Additional masses were fixed to the slabs to simulate dead and live loads. Three steel plates with a total mass of 13.5 ton were fixed to the underside of the 1st floor slab. For the 2nd floor, 1 steel plate and 12 concrete blocks were clamped to the top of the slab adding a total mass of 11.0 ton. The masses were fixed using semispherical steel bearings to enable the free deflection of the slabs. The estimated self-weight of the building was 20.4 ton. This produced relatively low normalized axial load ratios of 0.05 and 0.03 for the 1st and 2nd floor columns, respectively.

The reinforcement detailing of columns and beam-column joints at 1st and 2nd floors are shown in Fig. 11.4. The beam reinforcement was anchored into the joint with a 230 mm anchorage length (approximately $17d_b$, where d_b is the bar diameter) and no hooks or bends were provided. This short anchorage length was insufficient to develop the yielding capacity of the 14 mm bars. The top beam reinforcement of the 1st floor joints was bent 90° into the joint. In order to replicate old construction practices, no transverse reinforcement was used in the joints. Moreover, the longitudinal bars of the columns were lapped just above the joint core as shown in Fig. 11.4a (lap length $l_b = 25d_b = 350$ mm).

Columns and beam-column joints are identified using an alphanumeric ID code that contains the location of the structural element in plan and elevation. The first number and letter of the ID stand for the axes' intersection at which the structural element is located, whilst the number after the hyphen denotes the floor number. For instance, the ID 1A-2 corresponds to the 2nd floor joint located at the intersection of axes 1 and A (refer to Fig. 11.3a).

Table 11.1 summarizes the axial load on columns (P), and the yield (M_y) and ultimate (M_u) flexural strengths of beams and columns obtained using moment-curvature analyses. The beams flexural strengths were computed considering the contribution of the slab (flange) and the top slab reinforcement. The table also includes the shear V_u corresponding to the development of M_u at the beams ends (considering the gravity load acting on the beams), and the shear resistance

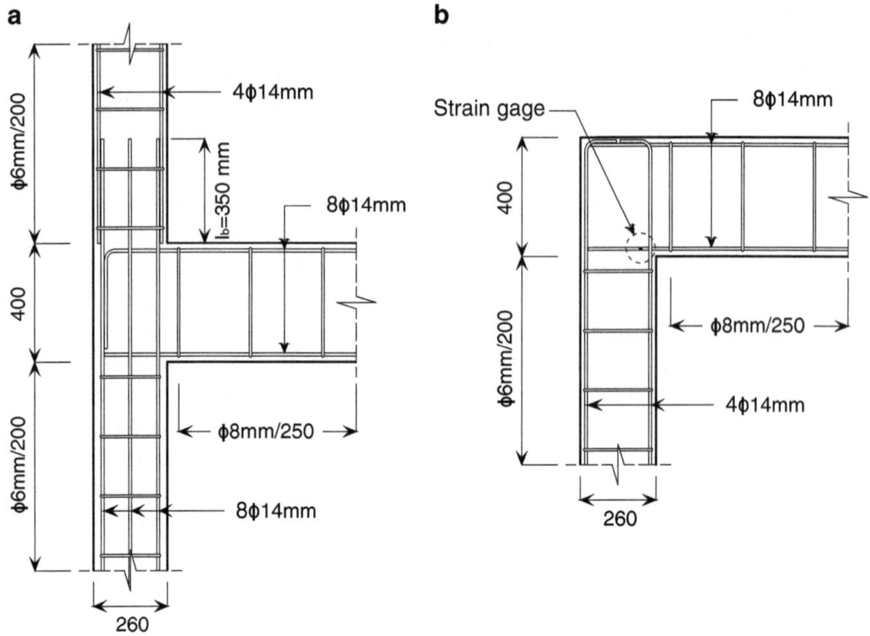

Fig. 11.4 Detailing of columns and beam-column joints at (**a**) 1st floor, and (**b**) 2nd floor, X direction

Table 11.1 Beams and columns strengths in the X direction

	P (kN)	M_y (kNm)	M_u (kNm)	V_u (kN)	$V_{R,s}$ (kN)
Column 2nd floor	51.9	37.8	40.7	24.7	77.2
Column 1st floor	110	63.3	77.2	46.8	77.2
Beams (sagging)	–	111	127	174	179
Beams (hogging)	–	156	167	174	179

$V_{R,s}$ provided by the steel stirrups computed using Eurocode 2 (CEN 2004a). The values in Table 11.1 show that the ratio of the ultimate flexural strength of column to beam elements ($\Sigma M_{u,cols}/\Sigma M_{u,beam}$) was approximately 0.93 for the 1st floor and 0.35 for the 2nd floor. Therefore, the behavior of the BANDIT building was expected to be dominated by a strong beam-weak column mechanism. Moreover, the columns and beam-column joints at the 2nd floor were expected to experience more damage during the initial shaking tests due to the short anchorage length used for the top and bottom beam reinforcement (see Fig. 11.4b). This is in agreement with the damage observed in the structure during the tests, as it will be discussed in the next sections. It should be noted that the provided transverse reinforcement was sufficient to resist the expected shear demand on the beams and columns ($V_{R,s} > V_u$).

Table 11.2 Mechanical properties of reinforcing steel

Bar size (mm)	f_y (MPa)	f_u (MPa)	ε_u (%)
6	574	604	3
8	544	572	3
10	513	587	7
14	526	616	8

11.2.2 Material Properties

Deformed bars were used for the longitudinal and transverse reinforcement. Table 11.2 reports the average yield strength (f_y), ultimate tensile strength (f_u), and ultimate bar elongation (ε_u) of the reinforcement. The mechanical properties of the longitudinal reinforcement were obtained from tensile tests on three bar samples.

Two batches of concrete were used to cast the building, one batch for each floor. The average compressive strength (f_c) and elastic modulus (E_c) for each batch were obtained from compressive tests on six standard cylinders (150 × 300 mm) and were: $f_c = 30.8$ MPa and $E_c = 23.9$ GPa for the 1st floor; and $f_c = 25.5$ MPa and $E_c = 21.7$ GPa for the 2nd floor. High-tensile, high-ductility steel straps with a cross section of 0.8 × 25 mm were utilized for the strengthening (PTMS) of the structure in this project. The average mechanical properties of the straps were obtained from coupons tested at The University of Sheffield, which gave a yield of $f_y = 1{,}000$ MPa and an elastic modulus $E_s = 230$ GPa. Ultimate strength and ultimate elongation of the straps were $f_u = 1{,}100$ MPa and 4.6 %, respectively. The ultimate strength compares reasonably well with the minimum tensile strength $f_u = 950$ MPa provided by the manufacturer (Megadyn® France).

11.2.3 Test Set-Up and Input Ground Motion

The structure was instrumented with horizontal and vertical displacement and acceleration transducers at each floor to monitor the displacement and acceleration histories during the tests. Foil-type strain gages were fixed onto the reinforcing bars at the locations where high strains were expected. Horizontal input shaking table tests were performed on the building using increasing levels of peak ground acceleration (PGA). A single artificial ground motion record was used based on the Eurocode 8 (EC8) ground type C spectrum (CEN 2004b). The total duration of the record was 30.0 s. The natural periods of the structure were obtained using white noise before and after each shaking test. The accelerations recorded at each floor were subsequently post-processed to identify the natural periods and damping ratios of the first two modes of vibration.

11.3 Test Results Phase 1: Bare Building

11.3.1 Dynamic Characteristics

Phase 1 of the BANDIT experiments consisted of uniaxial shake table excitations in the X direction (i.e. parallel to axes 1 and 2 in Fig. 11.3a). The main goal of these tests was to produce significant (but still repairable) damage. The PGA of the increasing uniaxial excitations ranged from 0.025g to 0.15g. Table 11.3 presents the 1st and 2nd mode natural periods (T_1 and T_2, respectively) at each PGA level, along with their increase (ΔT) with reference to the original period of the undamaged building. The table also shows the 1st and 2nd mode damping ratios (ξ_1 and ξ_2, respectively) computed from the white noise input signal after each test.

Table 11.3 shows that two tests were performed at a PGA level of 0.15g. The first test was halted after approximately 20.0 s due to resonance issues of the shake table. As some higher amplitude cycles of the ground motion record were not applied in the first test, the test was subsequently repeated to apply the full duration of the record (shown as 0.15g-b in the table).

The results in Table 11.3 clearly show the lengthening of the 1st and 2nd natural periods of the structure in the X direction as a consequence of damage accumulation. After the test PGA = 0.15g-b, the 1st and 2nd natural periods of the building increased by 83 and 64 % compared to the undamaged frame, respectively. Based on the 1st mode period, the residual stiffness after this test was calculated as 30 % of the original undamaged stiffness of the building, which indicates significant damage in the structure. It is also shown that, as expected, the 1st and 2nd mode damping ratios increased as damage progressed with the increasing levels of PGA.

11.3.2 Observed Damage in the Bare Building

No evident damage occurred during the shaking table tests up to PGA = 0.05g. As expected, the first cracks were detected at the beam-column joints of the 2nd floor after the tests at PGA = 0.10g. The subsequent tests at PGA levels of 0.15g

Table 11.3 Structural periods and damping ratios from tests Phase 1: bare building

Test	1st mode T_1 (s)	ΔT_1	2nd mode T_2 (s)	ΔT_2	1st mode ξ_1 (%)	2nd mode ξ_2 (%)
Undamaged	0.48	–	0.18	–	3.74	3.36
After PGA = 0.025g	0.53	+11 %	0.20	+12 %	2.64	1.86
After PGA = 0.05g	0.60	+24 %	0.22	+20 %	5.30	4.18
After PGA = 0.10g	0.68	+43 %	0.25	+38 %	5.04	3.95
After PGA = 0.15g	0.78	+62 %	0.27	+51 %	7.57	4.95
After PGA = 0.15g-b	0.88	+83 %	0.29	+64 %	6.29	5.55

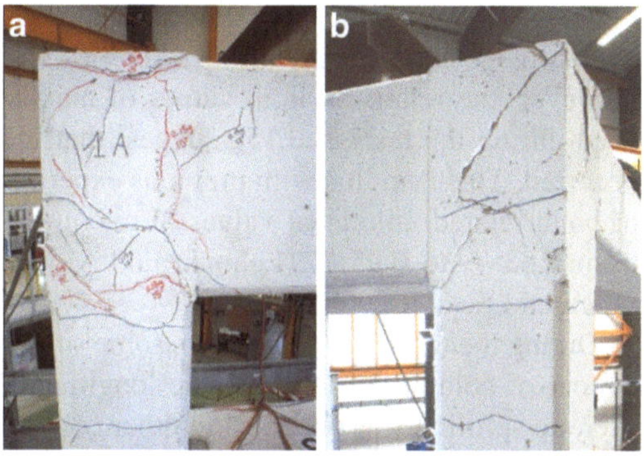

Fig. 11.5 Damage at the 2nd floor joints (**a**) joint 1A-2 and (**b**) joint 2A-2

Table 11.4 Bond stress results for bottom beam reinforcement of joint 2A-2

Test	Microstrains	Bar stress f_s (MPa)	Bond stress τ (MPa)	Norm. bond stress $\tau/\sqrt{f_c}$ ($\sqrt{\text{MPa}}$)
PGA = 0.025g	89	18	0.35	0.07
PGA = 0.05g	214	43	0.83	0.16
PGA = 0.10g	645	129	2.51	0.50
PGA = 0.15g	1,034	207	4.02	0.80
PGA = 0.15g-b	1,150	230	4.47	0.89

produced extensive damage at the joints of the 2nd floor, as shown in Fig. 11.5. In particular, joint 2A-2 (see Fig. 11.5b) experienced severe cracking and concrete spalling during the test PGA = 0.15g-b (crack width >10 mm). The damage observed suggested that a local failure occurred within the joint during the last shaking table tests. As a consequence of the excessive cracking, the beam reinforcement may have also pulled out from the joint. Possible debonding initiation was examined by computing the maximum bond stress developed along the interested bars using readings from a strain gage located at the beam reinforcement (see location in Fig. 11.4b).

Table 11.4 summarizes the maximum strain readings and corresponding bar stress (f_s) of the strain gage located on the bottom beam reinforcement at joint 2A-2. The value f_s was calculated assuming an elastic modulus of steel $E_s = 200$ GPa. The table also presents the uniform bond stress (τ) of the bar at each PGA level computed using the following equation:

$$\tau = \frac{f_s d_b}{4 l_b} \tag{11.1}$$

where d_b is the bar diameter (14 mm) and l_b is the distance from the tip of the bar to the location of the strain gauge (180 mm). The last column in Table 11.4 lists the bond stress normalized to the square root of the concrete compressive strength.

The results in Table 11.4 show that the bottom beam reinforcement remained within the elastic range during the series of shaking table tests in Phase 1. However, the relatively high values of maximum normalized bond stress developed during the tests at 0.15g suggest that localized debonding might have also occurred. This is in line with previous experimental studies that reported the onset of bond-related failures at values of maximum normalized bond stresses ranging from $0.83\sqrt{f_c}$ to $0.95\sqrt{f_c}$ (Lehman and Moehle 2000; Sezen and Moehle 2006; Sezen and Setzler 2008).

During the test PGA = 0.15g-b, some splitting cracks were clearly visible at the bottom of column 2A-2 where the longitudinal reinforcement was lapped. However, no splitting was observed in other columns. A detailed visual inspection also revealed some diagonal cracking at the 1st floor joints and horizontal cracking at the 2nd floor columns, mainly at the location of the transverse reinforcement. No significant damage occurred at the 1st floor columns. As the bare building experienced significant local damage within the 2nd floor joints, the experiments were halted after the test PGA = 0.15g-b to avoid a possible collapse.

11.4 Strengthening of the Building

After the tests in Phase 1, the damaged building was repaired and strengthened to perform further tests. The intervention included the following steps:

- Injection of cracks with epoxy resin
- Replacement of damaged and spalled concrete with high-strength repair mortar
- Welding of some bars of the bottom beam reinforcement to the column reinforcement to prevent further bar pullout
- Grinding of sharp joint corners (radius = 20 mm), and
- Local strengthening of beam-column joints and columns of both floors using PTMS

The local intervention with PTMS mainly aimed at increasing the strength capacity of columns and beam-column joints without modifying significantly the original stiffness characteristics of the building. A brief description of the PTMS strengthening is presented here. More details of the adopted PTMS strengthening strategy can be found in Garcia et al. (2012). In this project, the post-tensioned metal straps were installed according to the sequence described below (see also Fig. 11.6):

1. Steel plates were fixed into the columns and beams to anchor the straps.
2. Column ends were confined using horizontal straps at 50 mm centers (1 layer).
3. Beam ends were confined with vertical straps (1 layer).
4. Joints were strengthened with horizontal straps anchored to the steel plates (2 layers).

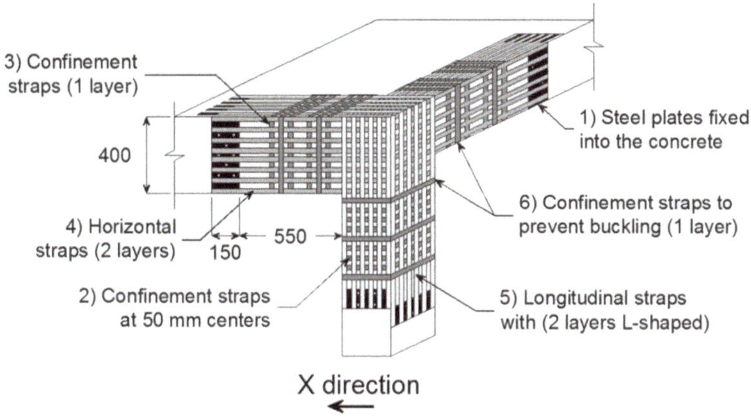

Fig. 11.6 PTMS strengthening at 2nd floor beam-column joints

Fig. 11.7 PTMS strengthening of joint 1A-2

5. Straps were installed along the outer face of the columns parallel to the columns axes (2 layers) to increase their flexural capacity.
6. Finally, beams and columns were confined with straps (1 layer) to prevent excessive buckling of the 2-layer horizontal and longitudinal straps during the tests.

Figure 11.7 shows a close-up view of a 2nd floor joint after the PTMS strengthening. It is shown in this figure that the steel straps provided an orthogonal confinement mesh around the beam-column joints and columns ends. The total strapping time for each joint varied from 2 to 3 h, which demonstrates the ease and speed of application of the proposed strengthening method. In addition, the added weight of the straps and steel plates did not increase significantly the total weight of the structure.

11.5 Test Results Phase 2: PTMS-Strengthened Building

11.5.1 Dynamic Characteristics

In the second phase of the project, the shake table tests were repeated on the PTMS-strengthened building starting at PGA = 0.05g and up to 0.35g (Table 11.5). The results show that the initial period of Phase 2 (T_1 = 0.61 s) was 30 % lower than the natural period after test PGA = 0.15g-b in Phase 1 (T_1 = 0.88 s, see Table 11.3). This implies that the structural stiffness of the building was substantially recovered as a result of the repairs and subsequent PTMS strengthening. Based on the results of Phase 1, a natural period of 0.61 s corresponds to the post-cracked behavior of the bare specimen after the test PGA = 0.05g (i.e. minor cracks).

After the test at PGA = 0.30g, the fundamental period of the strengthened building increased by 45 % as a consequence of damage accumulation. This increment in fundamental period is comparable to that obtained at the much lower PGA level of 0.15g in Phase 1. Hence, the local strengthening of joints and columns effectively controlled the stiffness degradation of the structure, which became critical only at higher PGA levels. Similar to the bare building, the 1st and 2nd mode damping ratios (ξ_1 and ξ_2, respectively) increased at higher levels of PGA.

11.5.2 Observed Damage in the Strengthened Building

Although only limited local damage was visible in the columns during the tests on the strengthened building, the lengthening of the natural periods indicated that significant softening had taken place in the structural elements. After the tests were halted, a thorough inspection of the building was carried out. No apparent damage of the PTMS strengthening or steel plates was visible. However, a few longitudinal straps at the 1st floor beam-column joints (parallel to the columns axes) showed a slight loss of post-tensioned force, possibly due to shearing-off of the

Table 11.5 Structural periods and damping ratios from tests Phase 2: strengthened building

Test	1st mode T_1 (s)	ΔT_1	2nd mode T_2 (s)	ΔT_2	1st mode ξ_1 (%)	2nd mode ξ_2 (%)
Initial	0.61	–	0.20	–	8.56	3.46
After PGA = 0.05g	0.64	+6 %	0.21	+8 %	6.71	5.65
After PGA = 0.10g	0.67	+10 %	0.22	+9 %	3.51	5.08
After PGA = 0.15g	0.68	+12 %	0.23	+16 %	6.79	5.20
After PGA = 0.20g	0.75	+23 %	0.23	+17 %	5.21	5.55
After PGA = 0.25g	0.78	+29 %	0.26	+28 %	7.47	9.40
After PGA = 0.30g	0.88	+45 %	0.26	+31 %	11.64	6.14
After PGA = 0.35g	1.01	+67 %	0.27	+35 %	13.36	3.76

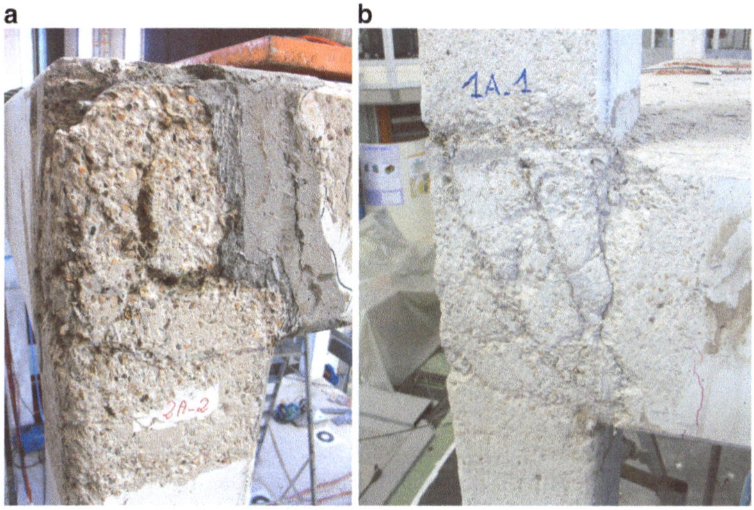

Fig. 11.8 View of joints (**a**) 2A-2, and (**b**) 1A-1 after removing the PTMS

metal clips used to secure the straps. The removal of the straps revealed more damage in the beam-column joints, particularly at joints 1A-1, 1A-2, 2A-1 and 2A-2 (see Fig. 11.8). The test results show that whilst the level of damage produced in Phase 1 was critical for the bare building, damage was effectively controlled during Phase 2, and the global stability of the strengthened building was never compromised.

The experimental results also show that during Phase 1, the inadequate reinforcement detailing led to significant local damage of the 2nd floor joints before the capacity of the 1st floor joints was fully mobilized. This was confirmed by the limited diagonal cracking observed at the 1st floor joints after the Phase 1 tests. Conversely, in Phase 2 the PTMS intervention increased significantly the shear strength of the 2nd floor joints and controlled better their structural deterioration. Accordingly, the seismic forces were redistributed among the building's elements and the 1st floor joints were subjected to higher seismic force demands, which in turn produced significant cracking (see Fig. 11.8b). This suggests that the proposed strengthening intervention allowed a better exploitation of the available member capacity and led to a more uniform damage distribution over the building height.

11.6 Seismic Performance of Bare and Strengthened Buildings

In spite of differences in opinion, maximum interstory drift ratio (IDR) has been widely used as a reliable performance criterion to evaluate damage in structural and non-structural elements and provide insight into the global behavior of a building.

Table 11.6 Maximum IDRs from BANDIT tests: Phases 1 and 2

	Floor no.	Bare building Phase 1 (%)	PTMS-strengthened Phase 2 (%)
PGA = 0.05g	2	0.30	0.29
	1	0.26	0.30
PGA = 0.10g	2	0.61	0.81
	1	0.55	0.69
PGA = 0.15g	2	1.69[a]	1.14
	1	0.82[a]	1.01
PGA = 0.25g	2	–	2.14
	1	–	1.59
PGA = 0.35g	2	–	2.80
	1	–	2.18

[a]Results from test PGA = 0.15g-b

Current seismic assessment guidelines such as FEMA 356 (2000) and ASCE/SEI 41–06 (2007) place limits on acceptable values of IDR implying that exceeding these limits is a violation of a performance level. Accordingly, ASCE/SEI 41–06 establishes maximum IDRs of 1, 2 and 4 % to the Immediate Occupancy (IO), Life Safety (LS), and Collapse Prevention (CP) performance levels, respectively.

Table 11.6 reports the maximum IDR values from the tests performed on the bare and PTMS-strengthened buildings. The results from the bare building show that the two floors had similar IDR at PGA levels of 0.05 and 0.1g. However, at PGA = 0.15g, the IDR of the 2nd floor was 106 % larger than that of the 1st floor. This increase of IDR can be attributed to the damage of the joints at the 2nd floor, as observed during the experiments.

The results in Table 11.6 show that at a PGA = 0.05g, the IDRs of both bare and PTMS-strengthened buildings were very similar. However, as a result of damage accumulation, IDRs of the 1st and 2nd floor of the PTMS-strengthened building at a PGA level of 0.10g were 33 and 25 % larger than those of the bare building. At PGA levels less than 0.15g, both bare and PTMS-strengthened buildings remained within the IO deformation performance level defined by ASCE/SEI 41–06. However, the local failure of the 2nd floor joints (and subsequent pulling of the beam bars) shows that the building was near collapse and needed serious intervention.

The effectiveness of the PTMS strengthening strategy is even more apparent when comparing the maximum IDRs of the bare and strengthened buildings at a PGA level of 0.15g. Whilst the maximum IDR of the bare building was 1.69 % at the 2nd floor, that of the strengthened building was reduced to 1.14 % (i.e. 33 % reduction of the critical IDR). This indicates that the strengthening intervention was very successful at controlling the excessive IDR of the 2nd floor and, consequently, preventing extensive damage in the joints at this floor. In contrast, the IDR of the 1st floor of the strengthened building was 23 % larger than that of the bare building (1.01 % vs. 0.82 %). This is consistent with higher level of damage observed in the joints at the 1st floor after removing the straps (see Fig. 11.8b).

Despite the relatively high intensity applied during the last test at PGA = 0.35g, the maximum IDRs of both floors of the strengthened building were near to the LS limit (IDR = 2 %). Therefore, it can be concluded that the adopted repair and PTMS strengthening strategy was very effective at improving the seismic performance of the building by increasing the strength and ductility of the structural elements and beam-column joints.

It should be mentioned that although the PTMS-strengthened building was capable of resisting shaking at higher PGA intensities and the straps were in general intact, the tests were halted after a PGA level of 0.35g to evaluate the structural damage. Moreover, as further shake table tests were planned on the other direction of the building, the degree of structural damage had to be maintained within repairable limits. The results of the latter tests will be presented in future publications.

11.7 Conclusions

This paper presented results from the two initial test phases of the BANDIT project. The effectiveness of a novel strengthening technique using Post-Tensioned Metal Straps (PTMS) was investigated experimentally through shake table tests on a seismically deficient full-scale RC building. In Phase 1, the building was tested in the bare condition. After the initial tests, the damaged building was repaired and strengthened using PTMS to perform additional tests in Phase 2. From the tests results, the following conclusions are drawn:

- The initial shake table tests on the deficient bare building produced significant damage at the beam-column joints of the 2nd floor. Conversely, the joints and columns at the 1st floor experienced limited damage. This confirms that the design of the bare structure was inadequate and would lead to significant damage at the 2nd floor before the capacity of the joints at the 1st floor could be fully exploited.
- The repair and adopted strengthening strategy using PTMS restored the initial post-cracked dynamic characteristics of the RC building. Moreover, the seismic capacity of the building was significantly increased. Whilst the bare building resisted a maximum PGA = 0.15g before a critical level of damage was reached, the PTMS-strengthened building sustained a maximum PGA of 0.35g.
- The repair and PTMS strengthening intervention increased the shear strength of the 2nd floor joints and controlled better the structural deterioration. Accordingly, seismic forces redistributed among the building's elements and the 1st floor joints were subjected to higher seismic force demands which produced some damage at these locations. Consequently, the PTMS allowed a more uniform damage distribution and a better exploitation of the available element strengths over the building height.
- The results show that the PTMS strengthening improved significantly the seismic performance of the deficient damaged building. For the same intensity at PGA = 0.15g, the repair and PTMS strengthening reduced the interstory drift

ratio (IDR) of the 2nd floor by 33 % compared to the bare building. Despite the relatively high intensity applied during the last test at PGA = 0.35g, the building remained very near the Life Safety performance limit (IDR = 2 %). Moreover, the global stability of the building was never compromised. Therefore, the proposed method is feasible and very attractive for quick post-earthquake strengthening of RC structures, especially in developing countries.

Acknowledgments The research leading to these results has received funding from the European Community's Seventh Framework Programme [FP7/2007-2013] for access to CEA (Commissariat à l'Energie Atomique et aux Energies Alternatives) under grant agreement n° 227887 [SERIES]. The first author gratefully acknowledges the financial support provided by CONACYT and partial support by DGRI-SEP. The authors gratefully acknowledge Thierry Chaudat from EMSI laboratory for his contribution at the early stages of the project. The strengthening was performed by Nuvia Travaux Spéciaux.

References

ASCE/SEI 41–06 (2007) Seismic rehabilitation of existing buildings. American Society of Civil Engineers, Reston

CEN Eurocode 2 (2004a) Design of concrete structures Part 1–1: general rules and rules for buildings. Comité Européen de Normalisation, Brussels

CEN Eurocode 8 (2004b) Design of structures for earthquake resistance Part 1: general rules, seismic actions and rules for buildings. Comité Européen de Normalisation, Brussels

FEMA 356 (2000) Prestandard and commentary for the seismic rehabilitation of buildings. Federal Emergency Management Agency, Washington, DC

Frangou M (1996) Strengthening of concrete by lateral confinement. PhD thesis, Department of Civil and Structural Engineering, The University of Sheffield, Sheffield

Frangou M, Pilakoutas K, Dritsos S (1995) Structural repair/strengthening of RC columns. Constr Build Mater 9(5):259–266

Garcia R, Hajirasouliha I, Pilakoutas K (2010) Seismic behaviour of deficient RC frames strengthened with CFRP composites. Eng Struct 32(10):3075–3085

Garcia R, Pilakoutas K, Guadagnini M, Helal Y, Jemaa Y, Hajirasouliha I, Mongabure P (2012) Seismic strengthening of deficient RC buildings using post-tensioned metal straps: an experimental investigation. In: Proceedings of the 15WCEE, Lisbon, Portugal

Helal Y (2012) Seismic strengthening of deficient RC elements using PTMS. PhD thesis, Department of Civil and Structural Engineering, The University of Sheffield, Sheffield

Lehman DE, Moehle JP (2000) Seismic performance of well-confined concrete bridge columns. Report No PEER-1998/01, University of California-Berkeley, 316 pp

Moghaddam H, Samadi M, Pilakoutas K (2010a) Compressive behaviour of concrete actively confined by metal strips, part B: analysis. Mater Struct 43(10):1383–1396

Moghaddam H, Samadi M, Pilakoutas K, Mohebbi S (2010b) Axial compressive behaviour of concrete actively confined by metal strips; part A: experimental study. Mater Struct 43(10):1369–1381

Sezen H, Moehle JP (2006) Seismic tests of concrete columns with light transverse reinforcement. ACI Struct J 103(6):842–849

Sezen H, Setzler EJ (2008) Reinforcement slip in reinforced concrete columns. ACI Struct J 105(3):280–289

Chapter 12
Bond Strength of Lap Splices in FRP and TRM Confined Concrete: Behavior and Design

Dionysios Bournas and Thanasis Triantafillou

Abstract The effectiveness of Fibre-Reinforced Polymer (FRP) and Textile-Reinforced Mortar (TRM) jackets was investigated experimentally and analytically in this study as a means of confining old-type reinforced concrete (RC) columns with limited capacity due to bond failure at lap splice regions. The local bond strength between lap spliced bars and concrete was measured experimentally along the lap splice region of six full-scale RC columns subjected to cyclic uniaxial flexure under constant axial load. The bond strength of two column specimens tested without retrofitting was found to be in good agreement with the predictions given by two existing bond models. These models were modified to account for the contribution of composite material jacketing to the bond resistance between lap spliced bars and concrete. The effectiveness of FRP and TRM jackets against splitting at lap splices was quantified as a function of jacket properties and geometry as well as in terms of the jacket effective strain, which was found to depend on the ratio of lap splice length to bar diameter. Consequently, simple equations for calculating the bond strength of lap splices in members confined with composite materials (FRP or TRM) are proposed.

D. Bournas (✉)
Department of Civil Engineering, The University of Nottingham,
Coates Building, University Park, Nottingham NG7 2RD, UK
e-mail: Dionysios.Bournas@nottingham.ac.uk

T. Triantafillou
Department of Civil Engineering, University of Patras, 26504 Patras, Greece
e-mail: ttriant@upatras.gr

A. Ilki and M.N. Fardis (eds.), *Seismic Evaluation and Rehabilitation of Structures*, Geotechnical, Geological and Earthquake Engineering 26,
DOI 10.1007/978-3-319-00458-7_12, © Springer International Publishing Switzerland 2014

12.1 Introduction

The majority of concrete structures in the seismic regions of the world has been constructed before enforcement of modern seismic design codes, and is, thus, inherently vulnerable to earthquakes. A significant inadequacy of existing concrete members lies in the mechanism of bond between longitudinal reinforcement and concrete. Bond-critical regions in concrete structures are for example at the base of bridge piers, where the reinforcement is lap spliced with starter bars projecting above the foundation, and the junction between the floor slabs or beams and the columns in reinforced concrete (RC) buildings, where the column reinforcement of two consecutive stories is lap spliced for ease of construction. However, according to modern seismic design philosophy, these locations are where the formation of plastic hinges is expected. To make matters worse, typical old-type RC columns include both short lap splices and low amounts of transverse reinforcement. As a consequence, the bond capacity of lap spliced bars is often the weak link; remedying it is essential for seismic upgrading of old concrete buildings.

The upgrading of existing RC columns through jacketing has become a popular technique in an increasingly large number of rehabilitation projects. The use of fibre-reinforced polymers (FRP) has gained considerable popularity among all jacketing techniques due to the favorable properties offered by these materials, namely high strength to weight ratio, corrosion resistance, ease and speed of application and minimal change of geometry.

The mechanism by which confinement with FRP jackets contributes to the enhancement of the bond strength between lap spliced bars in tension and concrete has been investigated in a limited number of studies. Harajli (2005) modified the Orangun equation (Orangun et al. 1977) and proposed a relationship for evaluating the thickness of the FRP jacket required for developing a desired steel stress at the lap splice bond failure. In another study, Hamad and Rteil (2006), based on experimental data of tension lap splices confined with FRP sheets in normal and high strength concrete specimens, proposed a new FRP confinement parameter, which accounts for the increase in bond strength due to the presence of FRP sheets. However in both expressions proposed by Harajli (2005) and Hamad and Rteil (2006), the calculated lap splice bond strength of FRP confined members depends exclusively on the mechanical characteristics of the jacket, irrespective of the lap splice geometric characteristics (lap length, bar diameter).

In the present study the authors investigate experimentally and analytically the use of jackets made of composite materials as a means of enhancing the (poor) bond resistance of lap splices, with a focus on understanding the role of the jacket as a confining element. Those jackets comprise either FRP or the new class of inorganic matrix composites, termed Textile-Reinforced Mortars (TRM) (Triantafillou et al. 2006; Bournas et al. 2007, 2009), as a means of enhancing the poor bond resistance of lap splices. The local bond strength between lap spliced bars and concrete is measured experimentally at the plastic hinge region (floor level) of large-scale RC columns subjected to simulated seismic loading. In this way the contribution of

FRP or TRM confinement as a means of improving the bond resistance of straight lap splices with deformed bars is assessed. Moreover, two existing analytical bond models are modified to account for the contribution of composite jackets to confinement. In addition, the effectiveness of FRP or TRM jackets against splitting at lap splices is quantified as a function of jacket properties and geometry as well as in terms of the jacket effective strain, which depends on the ratio of lap splice length to bar diameter. Finally, this chapter proposes simple design models for calculating the bond strength of lap splices in members confined with FRP or TRM jackets.

12.2 Experimental Investigation

12.2.1 Test Specimens and Experimental Parameters

Seven full-scale reinforced concrete column specimens with the same geometry, six with lap splicing of longitudinal bars at the floor level and one with continuous longitudinal reinforcement, were constructed and tested under lateral load (Fig. 12.1). The specimens were flexure-dominated cantilevers with a height to the point of application of the load (shear span) of 1.6 m (half a typical story height) and a cross section of 250 × 250 mm. The columns were fixed into a heavily reinforced 0.5 m-deep base block, 1.2 × 0.5 m in plan, within which the longitudinal bars were anchored with 90-degree hooks at the bottom. To represent old-type non-seismically designed and detailed columns, all specimens were reinforced longitudinally with four 14 mm-diameter deformed bars with an effective depth of 225 and 8 mm diameter smooth stirrups at a spacing of 200 mm, closed with 90-degree hooks at both ends. Together with the deficient reinforcing details the concrete cover of the spliced bars in the column section was chosen to a constant low value of 10 mm (hence the ratio of cover to bar diameter $c/d_b = 0.7$), which, especially for short lap splices, would induce splitting bond failures prior to yielding of longitudinal steel bars (e.g. *fib* 2000). The geometry of a typical cross section is shown in Fig. 12.1b.

The effectiveness of FRP or TRM jackets, applied at the ends of old-type RC columns was evaluated for two different lap lengths, which were selected equal to 20 and 40 bar diameters, as shown in Fig. 12.1b. Columns with the shorter lap lengths (Series L20d_...) are more representative of RC construction up to the late 1970s. These columns were designed as follows: One specimen was tested without retrofitting as control (L20d_C), the second one was retrofitted with a two-layered CFRP jacket (specimen L20d_R2) and the third one was retrofitted with an equal (to its FRP counterpart) stiffness and strength carbon fibre TRM jacket comprising four layers (specimen L20d_M4). Columns with longer lap lengths (Series L40d_...) are more representative of RC construction up to the late 1990s. These columns were given the notation L40d_C, L40d_R2 and L40d_M4 that is identical to Series L20d_..., except for the lap length. Note that the layers in the TRM-jacketed columns were twice as many compared with their FRP counterparts, resulting in

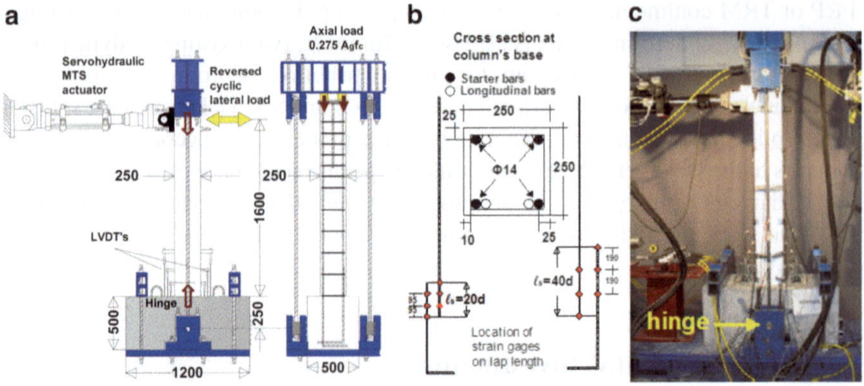

Fig. 12.1 (**a**) Schematic of test set up. (**b**) Cross section of columns and distribution of strain gages along the lap length. (**c**) Photograph of test setup

two "equivalent" confining systems, that is with the same stiffness and strength in the circumferential direction (as explained below, the fibres of the two jacketing systems in the circumferential direction were of the same type and nearly twice as many – per layer – in the FRP system compared with the TRM system).

The jackets extended from the base of each column (a gap of about 10 mm was left) to a height of 430 mm except for the two columns with longer lap splices (L40d_R2 and L40d_M4) where the jackets were extended to a height of 600 mm. The height of 430 mm was selected as it fulfills the requirements for exceeding the theoretical plastic hinge length (as per various models). In the case of columns with longer lap lengths, a height of 600 mm, which exceeds the plastic hinge length and the lap splice length of $40d_b = 560$ mm, was used. The overlapping length of the jacket was equal to 150 mm. Prior to jacketing, the four corners of the columns which received jacketing were rounded at a radius equal to 25 mm.

12.2.2 Materials, Strengthening Procedures and Test Setup

The longitudinal bars had a yield stress of 523 MPa, a tensile strength of 624 MPa and an ultimate strain equal to 12 %. The corresponding values for the steel used for stirrups were 351 MPa, 444 MPa and 19.5 %, respectively. For the specimens receiving TRM jacketing a commercial textile with equal quantity of carbon rovings in two orthogonal directions was used. Each roving was 3 mm wide and the clear spacing between rovings was 7 mm. The weight of carbon fibres in the textiles was 348 g/m^2, while the nominal thickness of each layer (based on the equivalent smeared distribution of fibres) was 0.095 mm. The mean tensile strength and elastic modulus of the carbon fibres (as well as of the textiles, when the nominal thickness is used) was taken from two datasheets and were equal to 3,800 MPa and 225 GPa, respectively.

For specimens receiving resin adhesive bonding, a commercial structural adhesive (two-part epoxy resin with a mixing ratio 3:1 by weight) was used with a tensile strength of 70 MPa and an elastic modulus of 3.2 GPa (cured for 7 days at 23 °C); those properties were provided by the manufacturer. For the specimens receiving mortar as a binding material, a commercial inorganic dry binder was used, consisting of cement and polymers at a ratio of about 8:1 by weight. The water: binder ratio in the mortar was 0.23:1 by weight, resulting in plastic consistency and good workability. The strength of mortar used in this study was obtained through flexural and compression testing according to EN 1015–11 (1993), using a servo-hydraulic MTS testing machine. The average flexural and compressive strength values were 6.51 and 20.8 MPa, respectively.

To simulate seismic excitation, the columns were subjected to lateral cyclic loading which consisted of successive cycles progressively increasing by 5 mm of displacement amplitudes in each direction. At the same time a constant axial compressive load was applied to the columns, corresponding to 27.5 % of the members' compressive strength. The lateral load was applied using a horizontally positioned 250 kN MTS actuator and the axial load was exerted by a set of four hydraulic cylinders with automated pressure self-adjustment, acting against two vertical rods connected to the strong floor of the testing frame through a hinge (Fig. 12.1a). Displacements, rotations and curvatures at the plastic hinge region were monitored using six rectilinear displacement transducers (three on each side, perpendicular to the piston axis) fixed at cross sections 1, 2 and 3, with a distance equal to = 130 mm, = 260 mm and = 450 mm, respectively, from the column base, as shown in Fig. 12.1a. The instrumentation also comprised a total of 12 strain gages for each column with lap splices, which were mounted on one pair of lapped bars (starter-longitudinal) per column side as follows (Fig. 12.1b). Measurements from the strain gages on each pair of starter-longitudinal bars were used to determine the average bond strength along the splice length.

12.3 Results and Discussion

12.3.1 Global Column Response

Detailed results in terms of load–displacement hysteresis loops, curvature and energy dissipation are given in Bournas et al. (2009). In this chapter the authors present only those (additional) test results related to bond strength at lap splices. The response of all columns tested is given in Fig. 12.2 in the form of load-drift ratio envelope curves. Key results are also presented in Table 12.1, which includes: (a) The peak resistance in the two directions of loading. (b) The average bond strength (corresponding to peak resistance) τ_{av}, developed over the splice length in the two directions of loading. (e) The observed failure mode.

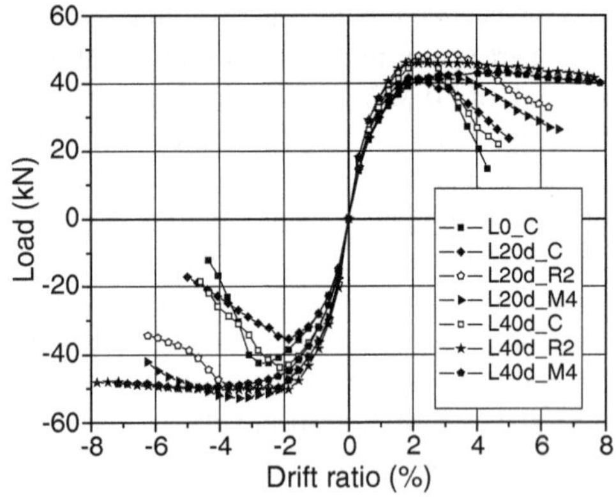

Fig. 12.2 Load versus drift ratio envelope curves

Table 12.1 Summary of test results of columns with lap-splices

Specimen notation	Peak force (kN) Push	Peak force (kN) Pull	Bond strength, τ_{av} (MPa) Push	Bond strength, τ_{av} (MPa) Pull	Failure Mode
L0_C	41.63	−42.48	–	–	Buckling of longitudinal bars
L20d_C	41.50	−36.62	4.42	–	Splitting bond failure followed by spalling of the concrete cover
L20d_R2	41.26	−52.86	6.78	6.86	Splitting longitudinal cracking followed by pull out bond failure of lapped bars
L20d_M4	48.46	−49.80	6.03	6.82	Splitting longitudinal cracking followed by pull out bond failure of lapped bars
L40d_C	46.26	−43.87	2.95	2.03	Splitting bond failure followed by spalling of the concrete cover
L40d_R2	42.97	−49.80	3.06	3.05	Conventional failure was not reached
L40d_M4	45.90	−50.48	–	2.93	Conventional failure was not reached

The performance and failure mode of all tested specimens with lap splices was controlled by flexure. The failure mode of the unretrofitted specimen with continuous longitudinal bars L0_C was controlled by buckling of longitudinal rebars above the column base, which led to direct lateral strength degradation. Both unretrofitted specimens with lap splices (L20d_C and L40d_C) experienced splitting bond failure. More specifically, significant longitudinal and horizontal splitting cracks developed along the splice length of lapped bars for both unretrofitted specimens L20d_C and L40d_C at drift ratios of 1.56 and 2.5 %, respectively, corresponding to peak lateral load. The length and width of the longitudinal cracks along the splice length was increasing at higher drift levels as the bond between reinforcing bars and concrete was deteriorating. As a consequence of this, the concrete under compression spalled along the base of specimens L20d_C and L40d_C, respectively, leading to substantial lateral strength degradation after peak lateral load. The drift ratio at failure sustained by unretrofitted columns L20d_C and L40d_C was 3.59 and 3.28 %, respectively.

FRP and TRM jacketed columns, with either short or long lap length, responded far better than their unretrofitted counterparts both in terms of strength and deformation capacity at failure. Confinement provided sufficient resistance against splitting cracks and lateral expansion of concrete. Specimens L20d_R2 and L20d_M4 (with short lap lengths) sustained reversed deformation cycles up to 6.3 % drift before failing due to pull-out bond failure of the spliced bars at an average bond strength between lap spliced bars and concrete of 6.8 and 6.4 MPa, respectively. Finally, in specimens L40d_R2 and L40d_M4 where the calculated bond stresses were much lower, namely 3.1 and 2.9 MPa, respectively, bond failures and spalling of concrete were suppressed until the end of the test at a drift ratio of 7.81 %.

For columns L20d_R2 and L20d_M4 the mean strength increase for both confining systems was 20.3 and 25.6 %, respectively, in comparison with the control specimen (L20d_C), while the corresponding increase in deformation capacity was 64.7 and 38.8 %, respectively. Columns with longer lap splices (L40d_R2 and L40d_M4) behaved in an identical manner until the end of the test at a drift ratio of 7.81 % (maximum stroke of piston was reached), resulting in an increase of the members' deformation capacity by a factor of more than 2.5. Peak resistance was practically the same as in the unretrofitted column, indicating that a lap splice length of 40 diameters is adequate for the development of the columns' full strength.

12.3.2 *Bond Strength of Lap-Spliced Bars*

The bond stress (τ) distribution between spliced bars and the surrounding concrete was calculated using the discrete strain readings along the splice length, as follows:

$$\tau_i = \frac{d_b E_s}{4} \frac{\Delta\varepsilon}{\Delta x} = \frac{d_b E_s}{4} \left(\frac{\varepsilon_i - \varepsilon_{l-1}}{x_i - x_{i-1}} \right) \tag{12.1}$$

where d_b = diameter of lapped bars; E_s = modulus of elasticity of steel; ε_i = axial strain of starter and longitudinal bars at discrete locations of strain gages; and x = coordinate along the splice length. The x axis is starting from the free end of the starter and longitudinal bars, respectively, as shown in Fig. 12.3a.

Based on Eq. (12.1) and by assuming zero strain at the free ends of spliced bars, the average bond strength along the splice length, corresponding to peak lateral force, was computed. By comparing the average bond strength of confined specimens (Table 12.1), it is clear that confinement was much more effective in terms of bond strength enhancement in the case of the shorter lap length. Additionally, FRP jackets

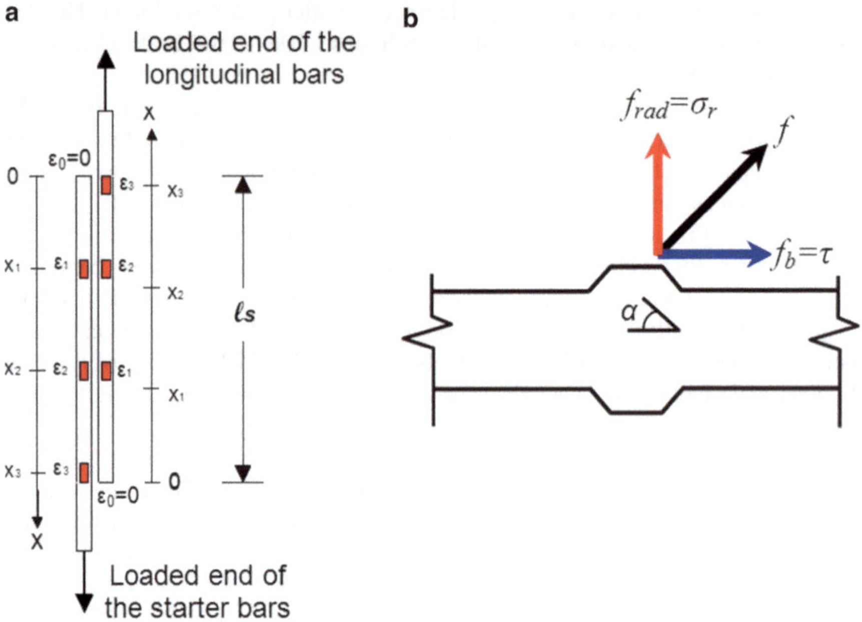

Fig. 12.3 (**a**) Position of the strain gages and symbols used in bond analysis. (**b**) Bond and radial stresses developed at lugs

were found to be slightly more effective, in terms of bond strength enhancement, in comparison with equal stiffness and strength TRM jackets.

The oblique bearing force exerted by the lugs of a bar (Fig. 12.3b) against bar pull out is higher in the case of the shorter lap splices (Series L20d_. . .) as the number of lugs participated over the splice length is smaller in comparison with the longer lap splices (Series L40d_. . .). Thus, the longitudinal component of this bearing force, which results in what is defined as bond stress (τ), is much higher for the columns with the lap length of 20 bar diameters. Correspondingly, the transverse component of this bearing force creates a radial stress (σ_r) which is responsible for splitting the surrounding concrete; this radial stress is also higher for the shorter lap length, especially near the lap splice ends. It is worth noting that steel plants worldwide construct deformed reinforcing bars with a rib angle α (Fig. 12.3b) of about 45°; consequently the two components of bond forces are approximately equal. For this reason the bond stress distribution presented in Fig. 12.6 is almost the same with the distribution of radial stresses (causing concrete splitting) along the lap splice length.

Overall, for columns with short lap lengths (L20d_R2 and L20d_M4) the increase in bond strength (and, approximately, in the radial stress, σ_r) for the two confining systems with FRP and TRM jackets was equal to 54 and 45 %, respectively. The corresponding increases for the specimens with long lap splices (L40d_R2 and L40d_M4) were 23 and 18 %.

12.3.3 ACI and fib Model Code Bond Models

Zuo and Darwin (2000), based on a database with 171 specimens containing bars not confined by transverse reinforcement and 196 specimens containing bars confined by transverse reinforcement, proposed a statistically based expression for calculating the lap splice bond strength τ_{max}. This formulation incorporates the effects of coarse aggregate quantity and type, and reinforcing bar geometry. This expression forms the basis for the bond recommendations of ACI Committee 408 (2003). The Zuo and Darwin (2000) expression is given as follows:

$$\tau_{max} = \frac{1}{\pi d_b \ell_s}\left\{[1.44\ell_s(c_{min}+0.5d_b)+56.3A_b]\left(0.1\frac{c_{max}}{c_{min}}+0.9\right)f_c{}^{1/4}+K_{tr,s}\cdot f_c{}^{3/4}\right\} \quad (12.2)$$

where f_c = compressive strength of concrete (MPa); c_{min} and c_{max} are the minimum and maximum values of concrete cover, where $c_{min} = \min(c_{so}, c_b, c_{si} + d_b/2)$, $c_{max} = \max(c_{so}, c_b)$ and the variables c_{so}, c_b, c_{si} are as defined in Fig. 12.4a; ℓ_s = splice length; A_b = area of longitudinal bars; and $K_{tr,s}$ = term representing the effect of confinement by steel stirrups, as follows:

$$K_{tr,s} = k_s\left(Ns_h\frac{A_{sw}}{s_h}\right)+743.6 = \frac{0.354t_r t_d}{n_s}\left(Ns_h\frac{A_{sw}}{s_h}\right)+743.6 \quad (12.3)$$

where $t_r = 9.6R_r + 0.28$ = term representing the effect of relative rib area; R_r = ratio of projected rib area normal to bar axis to product of nominal bar perimeter and center-to-center rib spacing; t_d (mm) = $0.78d_b + 5.6$ = term representing the effect of bar size; s_h = spacing of stirrups; A_{sw} = area of transverse steel reinforcement parallel to the direction of loading, namely the area of each stirrup or tie crossing potential plane of splitting adjacent to the reinforcement being spliced; f_{yw} = yield stress of stirrups; n_s = number of bars being spliced along plane of splitting; N = number of transverse reinforcing stirrups or ties crossing ℓ_s.

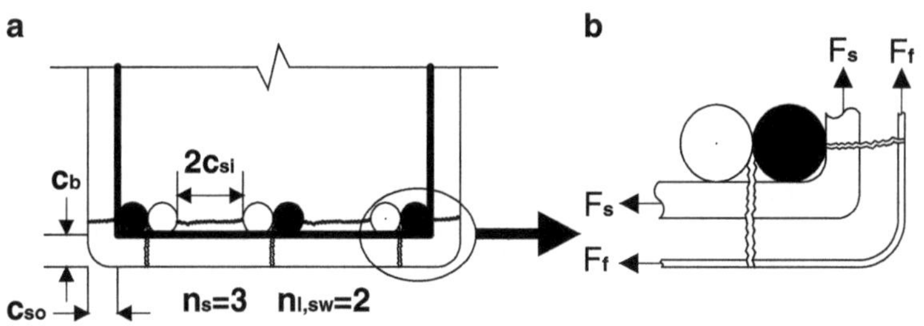

Fig. 12.4 (**a**) Definition of c_b, c_{si} and c_{so}. (**b**) Forces induced by stirrups and TRM or FRP jackets against side and face splitting cracks

More recently, Lettow and Eligehausen (2006) presented a semi-empirical expression, Eq. (12.4), for calculating the maximum stress f_{sm} that can be developed by the tensile bars in lap splice regions at splitting bond failure:

$$f_{sm} = 24.2\left(\frac{\ell_s}{d_b}\right)^{0.55}(f_c)^{1/4}\left(\frac{c_d}{d_b}\right)^{1/3}\left(\frac{c_{max}}{d_b}\right)^{0.1}\left(\frac{20}{d_b}\right)^{0.2}(1 + K_{tr,s}) \tag{12.4}$$

where $c_d = \min(c_{so}, c_b, c_{si})$ and $c_{max} = \max(c_{so}, c_b, c_{si})$. This equation has been based on an extended database of 793 test results including 402 specimens without transverse reinforcement and 391 specimens with transverse reinforcement, for which a quite good agreement with the experimentally measured steel stresses exists (the mean value of the experimental to analytical ratio is equal to 1.0 with a variation of 15 %). Equation (12.4), which was adopted (in a slightly revised form) in the new Model Code of *fib* (2012), can be applied for the calculation of the bond strength between lap spliced bars and concrete, as follows:

$$\tau_{max} = \frac{d_b}{4\ell_s}\left[24.2\left(\frac{\ell_s}{d_b}\right)^{0.55}(f_c)^{1/4}\left(\frac{c_d}{d_b}\right)^{1/3}\left(\frac{c_{max}}{d_b}\right)^{0.1}\left(\frac{20}{d_b}\right)^{0.2}(1 + K_{tr,s})\right] \tag{12.5}$$

$$K_{tr,s} = k_s\left(\frac{A_{sw}n_{l,sw}}{s_h}\right) = \frac{10}{d_b n_s}\left(\frac{A_{sw}n_{l,sw}}{s_h}\right) \tag{12.6}$$

where $n_{l,sw}$ = number of transverse reinforcing bar legs (stirrups or ties) crossing splitting cracks as defined in Fig. 12.4a. Use of the above equations is made with the following limitations: $20/d_b \leq 1.0$; $1.0 \leq c_d/d_b \leq 3.0$; $c_{max}/c_{min} \leq 5.0$ and $10K_{tr,s} \leq 0.4$.

The predictions of Eqs. (12.2) and (12.5) were compared with the experimentally measured bond strengths at the lap splice region of the two unretrofitted specimens L20d_C and L40d_C. In the case of the shorter lap length, the prediction – test ratios were 0.92 and 0.96 for Eqs. (12.2) and (12.5), respectively. The agreement between the analytical prediction of Zuo and Darwin's (2000) model and the experimental value was quite good, while the best match was obtained with Lettow and Eligehausen's (2006) model. For the longer lap splice of 40 bar diameters the prediction – test ratios were 1.08 and 1.22 for Eqs. (12.2) and (12.5), respectively. Here the best match was obtained for Zuo and Darwin's (2000) model.

Overall, it may be concluded that the predicted bond strengths according to Zuo and Darwin's (2000) and Lettow and Eligehausen's (2006) models are generally in very good agreement with the experimentally measured bond strengths for both lap lengths. For this reason and in view of the fact that these two approaches form the basis of code formulations (ACI 408 Committee and new *fib* Model Code), they are modified in the next section in order to account for the contribution of FRP or TRM confinement on the local bond strength between lap spliced reinforcing bars and surrounding concrete. However, the approach described in the next section is general and could be applied in the future to other bond strength models too.

12.3.4 Proposed Design Bond Models for TRM and FRP Confined Concrete

TRM and FRP jackets provide additional contribution to the transverse steel reinforcement tension resistance against splitting cracks, as illustrated in Fig. 12.4b. In order to account for this contribution of FRP or TRM confinement on the local bond strength of lap splices, it is reasonable to add a new parameter $K_{tr,j}$ to the transverse reinforcement parameter $K_{tr,s}$ provided by the stirrups. The proposed modified term $K_{tr,t}$, which accounts for the total confinement applied by both the contribution of stirrups and FRP or TRM jackets, is expressed as:

$$K_{tr,t} = K_{tr,s} + K_{tr,j} = (\text{Linear Function of } k_s(A_{sw}/s_h)) + k_f 2nt_f \tag{12.7}$$

where k_s = calibration factor for steel transverse reinforcement of each bond model as defined in Eqs. (12.3) and (12.6); k_f = calibration factor for the effectiveness of FRP or TRM jackets; n = number of layers of fibre sheet or textile layer; and t_f = thickness of one fibre sheet or textile layer. The proposed factor k_f is given by Eq. (12.8) for both models of Zuo and Darwin (2000) and Lettow and Eligehausen (2006), respectively.

Modified Zuo and Darwin (2000):

$$k_f = k_s h_f \frac{E_f}{E_s} \frac{\varepsilon_{f,ef}}{\varepsilon_{sw}} \tag{12.8a}$$

Modified Lettow and Eligehausen (2006):

$$k_f = k_s \frac{E_f}{E_s} \frac{\varepsilon_{f,ef}}{\varepsilon_{sw}} \tag{12.8b}$$

where h_f = height of FRP or TRM jacket; E_f elastic modulus of the jacket in the fibre (circumferential) direction; ε_{sw} = average effective strain of the stirrups in the circumferential direction and $\varepsilon_{f,ef}$ = average effective strain of the jacket in the fibre direction.

The proposed parameter $K_{tr,j}$ takes into account all the characteristics of the jacket, namely: the area of external FRP or TRM reinforcement ($2nt_f$) in the splice region; the modulus of elasticity of the jacket's material (E_f); and the average effective strain of the jacket in the circumferential direction ($\varepsilon_{f,ef}$). The jacket's height h_f is included only in proposed Eq. (12.8a), as it corresponds to the equivalent term Ns_h of stirrups used in Zuo and Darwin's (2000) model. Note here that an upper limit for h_f with respect to the

Table 12.2 Experimentally measured strains of stirrups placed in lap splice regions

Reference	Specimen type	ε_{sw}	ε_{yw}	$\varepsilon_{sw}/\varepsilon_{yw}$
Cairns and Arthur (1979)	Columns	0.0010	0.0017	0.60
Lukose et al. (1982)	Beams	0.0014	0.0023	0.63
Paulay (1982)	Columns	0.0015	0.0015	1.00
Panahshahi et al. (1992)	Beams	0.0011	0.0023	0.48
Valluvan et al. (1993)	Columns	0.0016	0.0024	0.66
Saadatmanesh et al. (1997)	Columns	0.0010	0.0018	0.55
Azizinamini et al. (1999)	Beams	0.0015	0.0021	0.71
Haroun et al. (1999)	Columns	0.0016	0.0022	0.73
Ma and Xiao (1999)	Columns	0.0015	0.0015	1.00
Melek and Wallace (2004)	Columns	0.0012	0.0024	0.50

lap length (h_f/ℓ_s), defined by pull out failure, has to be set, beyond which the increase of the jacket's height will have marginal effect, if any, on the bond capacity of the spliced bars. Based on the current experimental results, the determination of such a limit for h_f is not possible, as the interaction between τ_{max} and the ratio h_f/ℓ_s has not been investigated. On the other hand, application of Eq. (12.8b) – based on Lettow and Eligehausen's (2006) approach – for the evaluation of bond strength of RC members confined by FRP or TRM, is based on the assumption that the jacket's height is at least equal to the lap length.

The yield stress of the transverse reinforcement was removed from the two bond models presented above as it was found to have no effect on the lap splice bond strength. This effect was not measurable in most of the test results (included in the relevant databases), because splitting bond failures of the lap splice preceded yielding of stirrups. Nevertheless the percentage of activation of the transverse steel reinforcement in the circumferential direction against splitting cracks, which is quantified by the average effective strain of stirrups ε_{sw}, has to be addressed. In the present study, with the purpose of determining ε_{sw}, the authors made a careful interpretation of test results on RC members with lap splices. Of particular importance were those results corresponding to specimens in which strain gages were affixed on the stirrups crossing the lap length (Cairns and Arthur 1979; Lukose et al. 1982; Paulay 1982; Panahshahi et al. 1992; Valluvan et al. 1993; Saadatmanesh et al. 1997b; Azizinamini et al. 1999; Haroun et al. 1999; Ma and Xiao 1999; Melek and Wallace 2004). The determination of ε_{sw} was carried out for various lap lengths at the lap splice splitting bond failure. In general, the distribution of strains on stirrups over the splice region was non-uniform. Only the two outermost stirrups (loaded ends of starter and longitudinal bars) reached the yield strain at the lap splice bond failure, while the interior stirrups were strained below their yield strain. Thus, an average (over the lap length) experimental value for the effective strain ε_{sw} of stirrups, corresponding to members' bond failure, was considered for each experimental study; this value is given in Table 12.2. Note here that despite the variation of bond lengths and of splitting failure modes (side or face) among different researchers, the strains measured on stirrups were quite close, with a standard deviation of 0.0024 %. Hence an average value from all the experimental

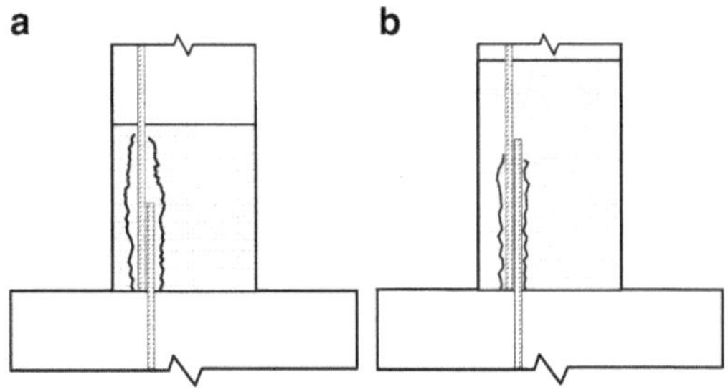

Fig. 12.5 Activation of FRP or TRM jacket against longitudinal splitting crack propagation in cases of (**a**) short and (**b**) long lap splice lengths

Table 12.3 Average lateral effective strain of FRP and TRM jackets

	Average lateral effective strain of the jacket, $\varepsilon_{f,ef}$			
	FRP		TRM	
Modified model	$\ell_s = 20d_b$	$\ell_s = 40d_b$	$\ell_s = 20d_b$	$\ell_s = 40d_b$
Zuo and Darwin (2000)	0.0044	0.0019	0.0034	0.0013
Lettow and Eligehausen (2006)	0.0031	0.0013	0.0023	0.0009

studies is approximately adopted for ε_{sw}, equal to 0.134 %, which is lower than the yielding strain of stirrups. This is consistent with the fact that the yield stress of stirrups was removed from the two aforementioned bond models.

The only term in the proposed modified bond models which is still to be addressed is the average effective strain of the jacket in the circumferential direction $\varepsilon_{f,ef}$. According to the authors' view, this value depends on the lap length to bar diameter ratio ℓ_s/d_b. For jacketed columns with short lap splices the radial stresses induced over the splice length (equal to bond stresses for $\alpha = 45^{\text{o}}$) are much higher in comparison with the tensile strength of concrete, especially at the lap splice end. Thus the splitting bond cracks propagate outside the lap splice region, until the radial stresses degrade below the tensile strength of concrete (3 MPa in the current study), in a length which increases as the initial lap length decreases. Hence for short lap splices the composite jacket's part which extends outside the lap splice region is activated more, as compared with longer lap splices, because it acts against the propagation of longitudinal splitting cracks in a more extended zone. This mechanism is illustrated in Fig. 12.5.

According to the experimental results of this study (Table 12.1), the average bond strength along the splice length was increased by 54 and 23 % for FRP confined specimens, and by 45 and 18 % for TRM confined specimens with respect to their unconfined counterparts, for the lap lengths of 20 and 40 bar diameters, respectively. Based on these increases of the bond strength and by using Eqs. (12.2), (12.3), (12.4), (12.5), (12.6), (12.7), and (12.8), the average strain in which the composite jackets were activated $\varepsilon_{f,ef}$ was determined indirectly. These average strains of FRP and TRM jackets are presented in Table 12.3 for the two lap lengths investigated here and for both bond models.

The average effective strain of the jacket in the circumferential direction $\varepsilon_{f,ef}$ for different values of lap length to bar diameter, ranging from 15 to 45, is provided by

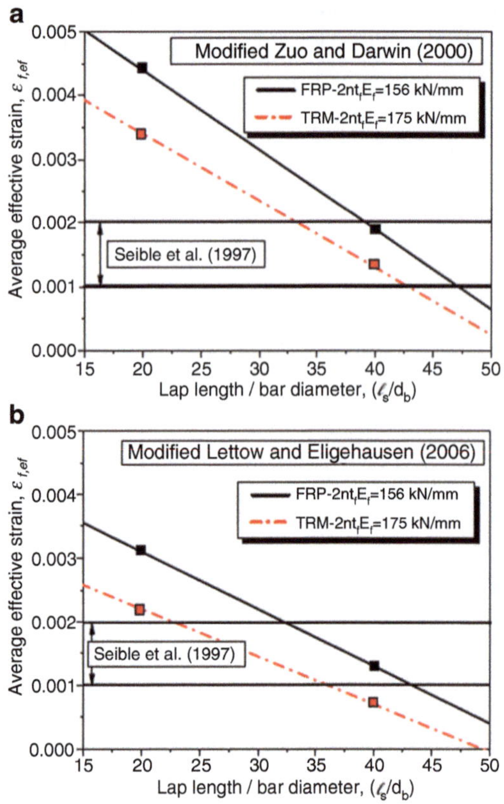

Fig. 12.6 Activation of FRP or TRM jacket against longitudinal splitting crack propagation in cases of (**a**) short and (**b**) long lap splice lengths

the diagrams of Fig. 12.6, assuming a linear fit. Of course these diagrams should be used with care, as they are based on limited test results and on specific values for the stiffness $2nt_f E_f$ of the composite jackets, equal to 156.4 and 175 kN/mm for FRP and TRM, respectively. More tests on FRP or TRM confined columns with different lap lengths are necessary in order to provide the best fit to the above diagrams. Moreover, other materials (e.g. glass, basalt) with different modulus of elasticity or jackets with a different number of layers may also result in different values for the effective strain.

Figure 12.6 illustrates a high activation (high $\varepsilon_{f,ef}$) of FRP or TRM jackets for short lap lengths and a lower one as the lap length increases. A key point here is that in the case of short splitting cracks (short lap splices) the activation of the jacket's part outside the lap length ℓ_s is enhanced, a fact quantified by the increased jacket effective strain $\varepsilon_{f,ef}$. In this way it is possible to estimate the effect of jacket confinement on RC members with lap splices as a function of lap splice length and jacket stiffness, contrary to current practice, based on the work of Seible et al. (1997), who proposed a fixed value for the circumferential jacket strain corresponding to the onset of splitting in the range 0.001–0.002 (insert in Fig. 12.6). It seems that using these strains (0.001–0.002) to a jacket design may be quite conservative for low values of ℓ_s/d_b, yielding an unrealistically large number of layers. The design equations proposed in this study for the effective strain in an FRP or TRM jacket at lap splice failures are summarized below for the two modified models:

Modified Zuo and Darwin (2000) model:

$$\varepsilon_{f,ef} = 0.0069 - 12.5 \cdot 10^{-5}(\ell_s/d_b) \quad \text{for FRP jackets} \tag{12.9a}$$

$$\varepsilon_{f,ef} = 0.0055 - 10.5 \cdot 10^{-5}(\ell_s/d_b) \quad \text{for TRM jackets} \tag{12.9b}$$

Modified Lettow and Eligehausen (2006) model:

$$\varepsilon_{f,ef} = 0.0049 - 9 \cdot 10^{-5}(\ell_s/d_b) \quad \text{for FRP jackets} \tag{12.9c}$$

$$\varepsilon_{f,ef} = 0.0037 - 7.5 \cdot 10^{-5}(\ell_s/d_b) \quad \text{for TRM jackets} \tag{12.9d}$$

It is worth noting that circumferential FRP strains at the base of rectangular columns with lap splices have actually been measured experimentally by a few researchers, namely Harajli and Dagher (2008) and ElGawady et al. (2010). The former reported a value of 0.00135 for a lap splice length equal to 30 bar diameters, whereas the latter reported values in the range 0.0013–0.0023 for a lap splice length equal to 36 bar diameters. Those independently measured strains are in reasonable agreement with the modified Lettow and Eligehausen (2006) equation, which predicts strains equal to 0.0022 and 0.0017 for the Harajli and Dagher (2008) and the ElGawady et al. (2010) test results, respectively. The comparison with the modified Zuo and Darwin (2000) is less favorable, a fact which adds more confidence to Eqs. (12.9c) and (12.9d) in comparison to Eqs. (12.9a) and (12.9b).

12.4 Conclusions

In the present study, the mechanism by which confinement with composite jackets (FRP and TRM) contributes to the enhancement of the bond resistance between lap spliced bars and concrete was investigated experimentally and analytically. The average bond strength at the critical regions of full scale columns subjected to seismic loading, based on direct strain gage measurements on the lap spliced bars, magnifies the validity of the relatively limited number of six tested specimens. Contrary to pull out bond tests, the test set up selected in this study reproduces realistically the bond conditions between lap-spliced bars and concrete at the plastic hinge region of columns subjected to seismic loading. More specific conclusions are summarized in a rather qualitative manner as follows:

- For both unconfined specimens with short and longer lap lengths splitting of the concrete cover at the lap splice region led to the sudden drop of the load resistance and to the decrease of columns' deformation capacity. The bond strengths measured experimentally for both unconfined specimens with short and long lap lengths were found in good agreement with the models of Zuo and Darwin (2000) and Lettow and Eligehausen (2006), respectively.

- Both FRP and TRM jackets resisted the propagation of longitudinal splitting cracks, resulting in the increase of the bond strength between lap spliced bars and concrete. In general, the external confinement with composite jackets and the enhancement of the local bond-slip relationship along the lap splice region resulted in the enhancement of the global response of the confined columns both in terms of strength and deformation capacity at failure.
- The contribution of FRP or TRM confinement on the local bond strength of lap splices is taken into account by proper modifications of the two aforementioned bond models. The proposed parameter for the modification takes into account all the characteristics of the jacket, namely: the area of external FRP or TRM reinforcement in the splice region, the effect of the modulus of elasticity of the jacket's composite material and the average effective strain of the jacket in the circumferential direction.
- The average effective strain in which the composite jacket is activated in the lateral direction decreases as the ratio ℓ_s/d_b increases. Hence, a rational design of the jacket thickness, to provide that steel yielding will precede bond failure of the rebar, is possible if one incorporates the dependence of the transverse jacket strains on the ratio ℓ_s/d_b. The most reliable approach for doing so relies on using the design Eqs. (12.9c) and (12.9d).

References

ACI Committee 408 (2003) Bond and development of straight reinforcing bars in tension (ACI 408R-03). American Concrete Institute, Farmington Hills, MI, 49 pp

Azizinamini A, Pavel R, Hatfield E, Ghosh SK (1999) Behaviour of lap-spliced reinforcing bars embedded in high-strength concrete. ACI Struct J 96(5):826–835

Bournas DA, Lontou PV, Papanicolaou CG, Triantafillou TC (2007) Textile-Reinforced Mortar (TRM) versus FRP confinement in reinforced concrete columns. ACI Struct J 104(6):740–748

Bournas DA, Triantafillou TC, Zygouris K, Stavropoulos F (2009) Textile-Reinforced Mortar (TRM) versus FRP jacketing in seismic retrofitting of RC columns with continuous or lap-spliced deformed bars. ASCE J Compos Constr 13(5):360–371

Cairns J, Arthur PD (1979) Strength of lapped splices in reinforced concrete columns. ACI Struct J 76(2):277–296

ElGawady M, Endeshaw M, McLean D, Sack R (2010) Retrofitting of rectangular columns with deficient lap splices. ASCE J Compos Constr 14(1):22–35

EN 1015–11 (1993) Methods of test for mortar for masonry – Part 11: Determination of flexural and compressive strength of hardened mortar. European Committee for Standardization. Brussels

fib (2000) Bond of reinforcement in concrete, Bulletin 10. International Federation for Structural Concrete, Lausanne

fib (2012) Model Code 2010 final draft, Bulletin 65. International Federation for Structural Concrete, Lausanne

Hamad BS, Rteil AA (2006) Comparison of roles of FRP sheets, stirrups, and steel fibres in confining bond critical regions. ASCE J Compos Constr 10(4):330–336
Harajli MH (2005) Behaviour of gravity load-designed rectangular concrete columns confined with fibre reinforced polymer sheets. ASCE J Compos Constr 9(1):4–14
Harajli MH, Dagher F (2008) Seismic strengthening of bond-critical regions in rectangular reinforced concrete columns using fibre-reinforced polymer wraps. ACI Struct J 105(1):68–77
Haroun MA, Feng MQ, Bhatia H, Baird K, Elsanadedy HM (1999) Structural qualification testing of composite-jacketed circular and rectangular bridge columns. Final report, California Department of Transportation/Department of Civil and Environmental Engineering, University of California at Irvine, Irvine, CA
Lettow S, Eligehausen R (2006) Formulation of application rules for lap splices in the new Model Code. Presentation, Task group 4.5 – Bond models, *fib* (federation internationale du Beton), Stuttgart, 13 Nov 2006
Lukose N, Gergely P, White RN (1982) Behaviour of reinforced concrete lapped splices for inelastic cyclic loading. ACI Struct J 79(5):355–365
Ma R, Xiao Y (1999) Seismic retrofit and repair of circular bridge columns with advanced composite materials. Earthq Spectra 15(4):747–764
Melek M, Wallace JW (2004) Cyclic behaviour of columns with short lap splices. ACI Struct J 101(6):802–811
Orangun CO, Jirsa JO, Breen JE (1977) Reevaluation of test data on development length and splices. ACI J 74(3):114–122
Panahshahi N, White RN, Gergely P (1992) Reinforced concrete compression lap splices under inelastic cyclic loading. ACI Struct J 89(2):164–175
Paulay T (1982) Lapped splices in earthquake-resisting columns. ACI Struct J 79(6):458–469
Saadatmanesh H, Ehsani MR, Jin L (1997) Seismic retrofitting of rectangular bridge columns with composite straps. Earthq Spectra 13(2):281–304
Seible F, Priestley MJN, Hegemier GA, Innamorato D (1997) Seismic retrofit of RC columns with continuous carbon fibre jackets. ASCE J Compos Constr 1(2):52–62
Triantafillou TC, Papanicolaou CG, Zissimopoulos P, Laourdekis T (2006) Concrete confinement with Textile-Reinforced Mortar jackets. ACI Struct J 103(1):28–37
Valluvan R, Kreger ME, Jirsa JO (1993) Strengthening of column splices for seismic retrofit of nonductile reinforced concrete frames. ACI Struct J 91(46):432–440
Zuo J, Darwin D (2000) Splice strength of conventional and high relative rib area bars in normal and high-strength concrete. ACI Struct J 97(4):630–641

Hamad BS, Rteil AA (2006) Comparison of roles of FRP sheets, stirrups and steel fibers in confining bond critical regions. ASCE J Compos Constr 10(4):330–

Chapter 13
Finite Element Modeling of Seismic Performance of Low Strength Concrete Exterior Beam-Column Joints

Danish Ahmed, Mohammed H. Baluch, Muhammad K. Rahman, and Alper Ilki

Abstract This paper presents a finite element simulation to capture the nonlinear response of a typical low strength beam-column joint tested at ITU and also results of the finite element modeling and experimental program conducted at KFUPM for beam-column joints in reinforced concrete construction. Finite element analysis is performed using the software DIANA, simulating the concrete response through Drucker-Prager plasticity with a tension cut-off as failure criterion and using for the reinforcing steel Von-Mises plasticity with multi linear isotropic hardening. The failure mode and deformation response of low strength concrete beam-column joints was predicted with a good correlation between the experimental and finite element results.

D. Ahmed
Department of Civil Engineering, King Fahd University of Petroleum and Minerals, Dhahran 31261, Saudi Arabia
e-mail: dahmed@kfupm.edu.sa

M.H. Baluch
Department of Civil and Environmental Engineering, King Fahd University of Petroleum and Minerals, Dhahran 31261, Saudi Arabia
e-mail: mhbaluch@kfupm.edu.sa

M.K. Rahman (✉)
Center for Engineering Research, Research Institute, King Fahd University of Petroleum and Minerals, Dhahran 31261, Saudi Arabia
e-mail: mkrahman@kfupm.edu.sa

A. Ilki
Department of Civil Engineering, Civil Engineering Faculty, Istanbul Technical University, Maslak, Istanbul 34469, Turkey
e-mail: ailki@itu.edu.tr

A. Ilki and M.N. Fardis (eds.), *Seismic Evaluation and Rehabilitation of Structures*, Geotechnical, Geological and Earthquake Engineering 26,
DOI 10.1007/978-3-319-00458-7_13,

13.1 Introduction

Most of the structures before the 1970s were designed for gravity load all over the world, and these structures performed well under gravity loads but their performance was questionable under earthquake loading. Several recent earthquakes such as in Taiwan (1999) and in Turkey (1999) caused extensive building damage and collapse of RC structures because of old design and poor reinforcement detailing. Shear failure of beam-column joints is defined as one of the causes of damage and collapse of these existing RC buildings due to poor reinforcement detailing within the beam column joint region. Several experimental and numerical investigations have been performed on the beam-column joints. Li and Kulkarni (2010) performed experimental and numerical investigation on RC wide beam-column joints when subjected to seismic loads. The experimental study was conducted by subjecting three full-scale wide exterior beam-column specimens to simulated seismic load. The experimental results were then used to validate a 3D-nonlinear finite-element model. Supaviriyakit et al. (2008) carried out a nonlinear finite element analysis of non-seismically detailed reinforced concrete beam-column connections under reversed cyclic load. Ibrahim and Mahmood (2009) presented an analysis model for reinforced concrete beams externally reinforced with fiber reinforced polymer (FRP) laminates using finite element modeling in an ANSYS environment. The finite element model was developed using a smeared cracking approach for concrete and three dimensional layered elements for the FRP composites. The results obtained from the ANSYS finite element analysis were compared with the experimental data for six beams. This paper has resulted from a collaborative research project between King Fahd University of Petroleum and Minerals (KFUPM) and Istanbul Technical University (ITU), in the area of seismic behavior of RC structures. The target objective has been to study the behavior of external beam-column joints with poor reinforcement detailing.

13.2 Test Program at ITU

Bedirhanoglu et al. (2010) tested eight exterior beam-column joints in two series of tests using low-strength concrete and plain reinforcing bars to represent the conditions of joints of existing deficient reinforced concrete building structures in Turkey. The configuration is common to all specimens; cross section of the columns and beams are 250 mm × 500 mm. Eight plain bars of 16 mm diameter were used in the column for longitudinal reinforcement and 8 mm diameter closed ties at a spacing of 150 mm were used as transverse reinforcement. For beams, four plain bars of 16 mm diameter were used for both the top and bottom longitudinal reinforcement and 8 mm diameter closed ties with spacing of 100 mm were used for transverse reinforcement in beam. The mechanical properties of the reinforcing bars and concrete are presented in Tables 13.1 and 13.2. Details of specimen geometry and reinforcement are given in Bedirhanoglu et al. (2010). The specimen referred to as JW2 was the major target for the numerical study.

Table 13.1 Mechanical properties of reinforcing bars

Reinforcement	Diameter (mm)	f_y (MPa)	$\varepsilon_y = f_y/E_s$	f_{smax} (MPa)	ε_{smax}	f_{su} (MPa)	ε_{su}
Φ16	16	333	0.0017	470	0.20	335	0.34
Φ8	8.4	315	0.0016	433	0.20	265	0.34

Table 13.2 Material properties of concrete

f'_c (MPa)	E_c (MPa)
8.3	13,000

13.3 Material Constitutive Models

13.3.1 Concrete Plasticity

The Drucker-Prager yield criterion is used to model the stress level at which yielding of concrete is initiated. The yield surface of the Drucker-Prager model is a circular cone in the deviatoric plane (Fig. 13.1) specified through the cohesion c and the friction angle φ by matching to Mohr-Column criteria. The Drucker-Prager yield criterion is given by

$$f(\sigma,\kappa) = \sqrt{\tfrac{1}{2}\sigma^T P\sigma} + \alpha_f \pi^T \sigma - \beta_C(\kappa) = 0 \tag{13.1}$$

with c(κ) the cohesion which is a function of the internal state variable κ. The projection matrix is equal to the matrix P defined in Eq. (13.2). σ is the 6 × 1 stress vector. The projection vector π is given by Eq. (13.3).

$$P = \begin{bmatrix} 2 & -1 & -1 & 0 & 0 & 0 \\ -1 & 2 & -1 & 0 & 0 & 0 \\ -1 & -1 & -2 & 0 & 0 & 0 \\ 0 & 0 & 0 & 6 & 0 & 0 \\ 0 & 0 & 0 & 0 & 6 & 0 \\ 0 & 0 & 0 & 0 & 0 & 6 \end{bmatrix} \tag{13.2}$$

$$\pi = \begin{Bmatrix} 1 \\ 1 \\ 1 \\ 0 \\ 0 \\ 0 \end{Bmatrix} \tag{13.3}$$

It may be noted that $\pi^T\sigma$ is the first stress invariant I_1 and $\sqrt{\tfrac{1}{2}\sigma^T P\sigma}$ is, $\sqrt{3J_2}$, where J_2 is the second invariant of the deviatoric stress tensor.

Fig. 13.1 3D yield surface of Drucker-Prager

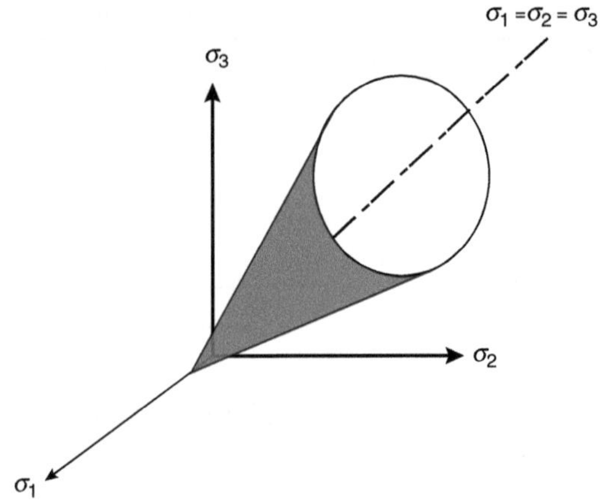

The scalar quantities α_f and β are given by (in order to make Drucker-Prager circle coincide with the outer apices of the Mohr-Coulomb criterion in the deviatoric plane (Owen and Hinton 1986))

$$\alpha_f = \frac{2\sin\phi(\kappa)}{3 - \sin\phi(\kappa)} \text{ and } \beta = \frac{6\cos\phi_o}{3 - \sin\phi_o} \tag{13.4a}$$

It may be shown that if one fits the Drucker-Prager model directly to concrete data for equal biaxial compression test $\left(\sigma_1 = 0, \sigma_2 = \sigma_3 = -f'_{bc}\right)$ and to the uniaxial compression test $\left(\sigma_1 = \sigma_2 = 0, \sigma_3 = -f'_c\right)$ that

$$\alpha_f = \frac{f'_{bc} - f'_b}{2f'_{bc} - f'_c} \tag{13.4b}$$

and using Eq. (13.4a) that

$$c = f'_c\left[\frac{1 - \sin\phi}{2\cos\phi}\right] \tag{13.4c}$$

The angle of internal friction φ can also be a function of the internal state variable for hardening hypothesis. The initial angle of internal friction is given by φ_o. The flow rule is given by a general non-associated flow rule $g \neq f$, with the plastic potential given by

$$g(\sigma, \kappa) = \sqrt{\frac{1}{2}\sigma^T P \sigma} + \alpha_g \pi^T \sigma \tag{13.5}$$

with the scalar α_g defined by the dilatancy angle ψ

$$\alpha_g = \frac{2\sin\psi(\kappa)}{3 - \sin\psi(\kappa)} \tag{13.6}$$

which results for the plastic strain rate vector in

$$\overset{\cdot}{\varepsilon}^{P} = \dot{\lambda}\left\{\frac{P\sigma}{2\psi} + \alpha_g \pi\right\} \tag{13.7}$$

with the scalar Ψ defined by

$$\psi = \sqrt{\tfrac{1}{2}\sigma^T P\sigma} \tag{13.8}$$

13.3.2 Hardening

The relation between the internal state variable κ and the plastic process is given by the hardening hypothesis. For the Drucker-Prager yield condition, usually the strain hardening hypothesis is considered. In the case of strain hardening, the relation between κ and incremental plastic strain is given in the principal space by

$$\dot{\kappa} = \sqrt{\tfrac{2}{3}(\dot{\varepsilon}_1^{P}\dot{\varepsilon}_1^{P} + \dot{\varepsilon}_2^{P}\dot{\varepsilon}_2^{P} + \dot{\varepsilon}_3^{P}\dot{\varepsilon}_3^{P})} \tag{13.9}$$

where

$$\begin{Bmatrix} \dot{\varepsilon}_1^{P} \\ \dot{\varepsilon}_2^{P} \\ \dot{\varepsilon}_3^{P} \end{Bmatrix} = \dot{\lambda}\left(\frac{1}{2\lambda}\begin{Bmatrix} 2\sigma_1 - \sigma_2 - \sigma_3 \\ -\sigma_1 + 2\sigma_2 - \sigma_3 \\ -\sigma_1 - \sigma_2 + 2\sigma_3 \end{Bmatrix} + \alpha_g \begin{Bmatrix} 1 \\ 1 \\ 1 \end{Bmatrix}\right) \tag{13.10}$$

Equation (13.9) can be elaborated to:

$$\dot{\kappa} = \dot{\lambda}\sqrt{1 + 2\alpha_g^2} \tag{13.11}$$

The flow rule determines the direction of plastic straining. The flow rule is termed "associative" if the plastic strain occurs in a direction normal to the yield surface. The plasticity model is termed "associated plasticity" in the case of associative flow rule. In the Drucker-Prager plasticity model, the associated plasticity is established by setting $\varphi = \psi$, in which case there will be a volumetric expansion of the material

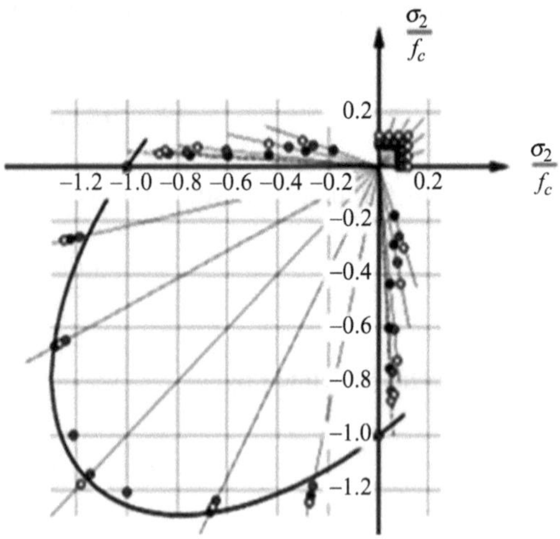

Fig. 13.2 Plane stress concrete failure plot in $\sigma_1 - \sigma_2$ space

with plastic strains. The non-associated plasticity is established by setting $\varphi > \psi$, resulting in less volumetric expansion. Clearly, if ψ is zero, there will be no volumetric expansion (De Witte and Kikstra 2010). The hardening rule describes the changing of the yield surface with progressive yielding so that the stress states for subsequent yielding can be established. Hardening can either be isotropic or kinematic. In isotropic hardening, the yield surface remains centered about its initial centerline and expands in size as the plastic strains develop. In kinematic hardening, the yield surface remains constant in size and the surface translates in the stress space with progressive yielding. The concrete behavior under monotonic loading has been modeled by many researchers using isotropic hardening. In this work, concrete was modeled as elastic, perfectly plastic with tension stiffening. For a normal strength quality concrete, the ratio between the uniaxial compressive strength and the biaxial compressive strength is approximately 1:1.16 as shown in Fig. 13.2, which results in a friction angle $\varphi \approx 10°$ (using Eq. (13.4b) and the first of Eq. (13.4a)) and a cohesion $c = 0.42f'c$ (using Eq. (13.4c)). It is recommended that this φ of 10° should be used to model concrete regions of biaxial and triaxial state of compressive stresses.

13.3.3 Concrete Cracking and Reinforcement

The cracking of the concrete can be specified as a combination of tension cut-off, tension softening, and shear retention (Fig. 13.3). The tension cut-off criterion is a linear stress cut-off. The tension softening criterion was set to be nonlinear tension softening with ultimate strain. The shear retention criterion was set to be a constant shear retention factor defined by a constant β value.

For reinforcement, Von Mises-Plasticity with isotropic hardening is used to match the behavior determined from testing.

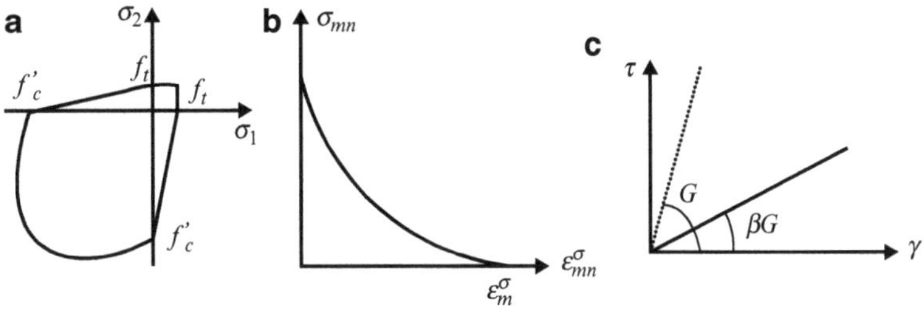

Fig. 13.3 (**a**) Linear tension cut-off, (**b**) nonlinear tension softening, (**c**) constant shear retention factor

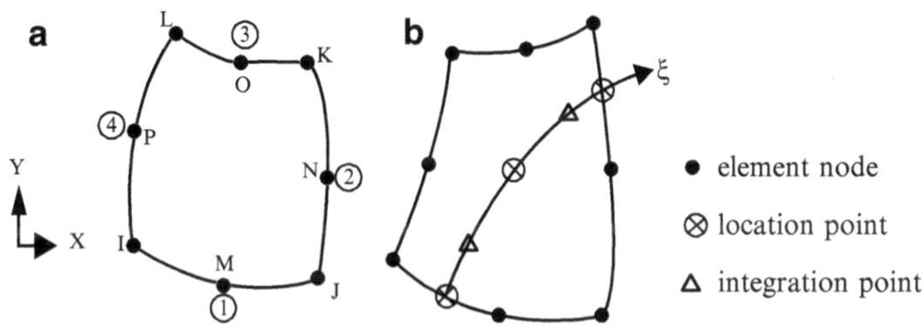

Fig. 13.4 (**a**) Eight-node quadrilateral isoparametric plane stress element; (**b**) embedded reinforcement element

13.4 Finite Element Meshing and Geometry

The concrete is modeled by eight-node quadrilateral isoparametric plane stress element CQ16M as shown in Fig. 13.4a. The rebar and stirrups are modeled as embedded reinforcement elements with perfect bond between rebar and concrete. The rebar embedded model is shown in Fig. 13.4b.

The creation of the geometry in DIANA is relatively straightforward; one way of doing it is by entering the coordinates to create points and later create lines by connecting the points. An alternative way is by creating the geometry in a CAD program and then importing it into DIANA, however this could create problems later on when assigning properties to the geometry. The geometry of the exterior beam column joint in 2D appears through its length and height. The thickness in the third direction is one of the inputs for the 2D plane stress element.

13.5 Finite Element Analysis of ITU Specimen

Material properties for concrete include $f_t = 0.73$ MPa at 28 days, $c = 2.15$ MPa (calibrated to match experimental data and lower than that determined from Eq. (13.4c)), $\Phi = \Psi = 10°$. Several researchers have used the angle of internal friction 30° when using Drucker-Prager yield criteria for modeling concrete. In this

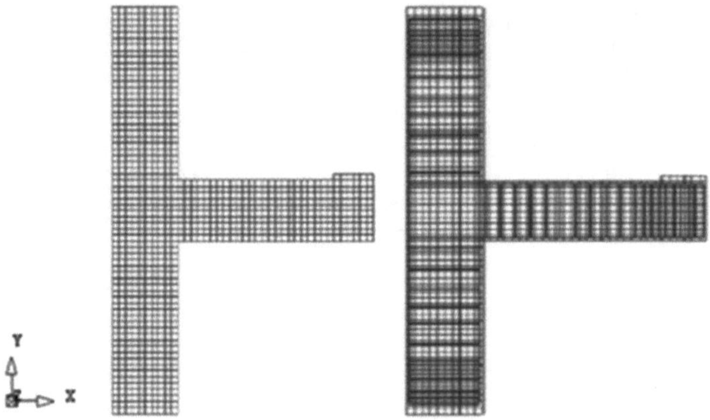

Fig. 13.5 Specimen with mesh and reinforcement

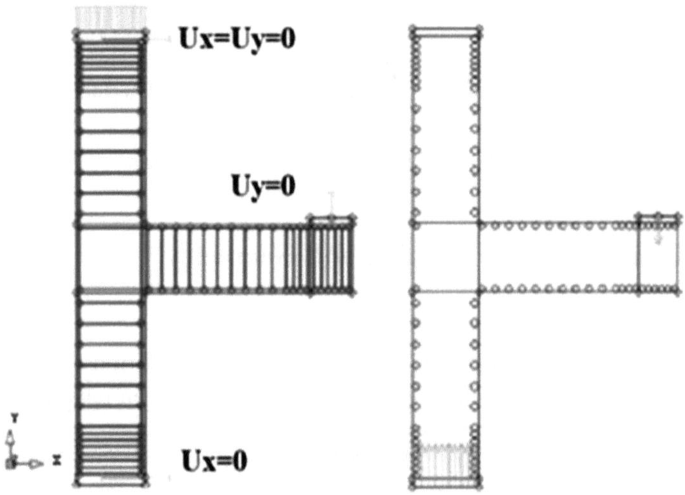

Fig. 13.6 Boundary conditions and loading details

problem of beam-column joint, a value of φ of 30° would result in much higher capacity than the experimentally determined failure load. The reason for this is that the joint region is in a biaxial state of stress, quite distinct from the beam region which is dominated by uniaxial compression and flexural related cracking in tension. The shear retention factor β is taken equal to 0.9 (smaller values lead to premature curtailment of the P-Δ curve), the modulus of elasticity of concrete E = 12 GPa, the steel hardening and the modulus of elasticity of steel taken from actual test results. The mesh scheme and density is shown in Fig. 13.5.

Figure 13.6 shows the top end of the column surface which is constrained in the x and y-direction. The bottom end of the column is constrained in the x-direction and free in the y-direction due to upward axial pressure (0.125f′c). The tip of the beam is constrained in the y-direction for application of displacement which is 50 mm.

13.6 Results from Experiment and Analysis

13.6.1 Beam Column Joint JW2

The results of the cyclic test of the joint carried out at ITU are shown Fig. 13.7a. For numerical simulation, the envelope of the load displacement curve is considered. The result of finite element simulation matches closely the test results, as shown in Fig. 13.7b.

Stresses in steel in the x and y directions at steel yielding (75 KN) and at ultimate load (86.91 KN) are shown in Fig. 13.8. Stresses in the top steel of the beam at steel yielding load and at ultimate load are 279 and 292 MPa, respectively. Steel stresses on the tension side of the column at steel yielding load and at ultimate load are 379 and 441 MPa, respectively. Stresses σ_x, σ_y and σ_{xy} in concrete at steel yielding load (75 KN) and at ultimate load (86.9 KN) are shown in Fig. 13.9. Stress σ_x at the bottom of the beam at steel yielding and at ultimate load is 5 and 7 MPa,

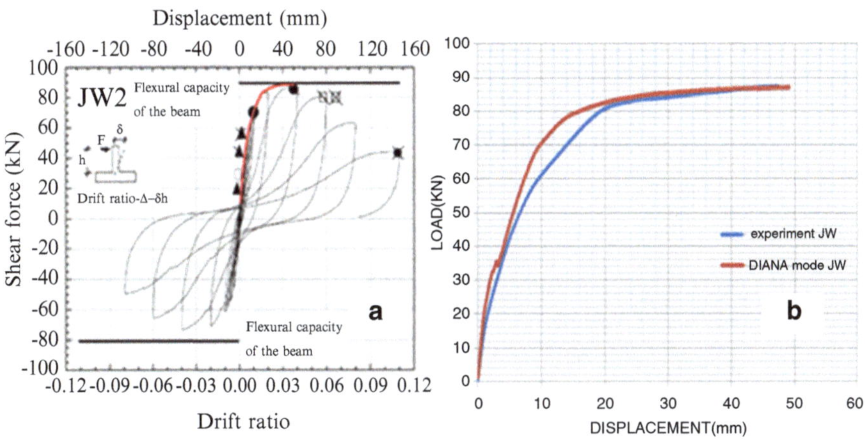

Fig. 13.7 (**a**) Envelope for hysteresis; (**b**) comparison of load–displacement graph from DIANA for specimen JW2

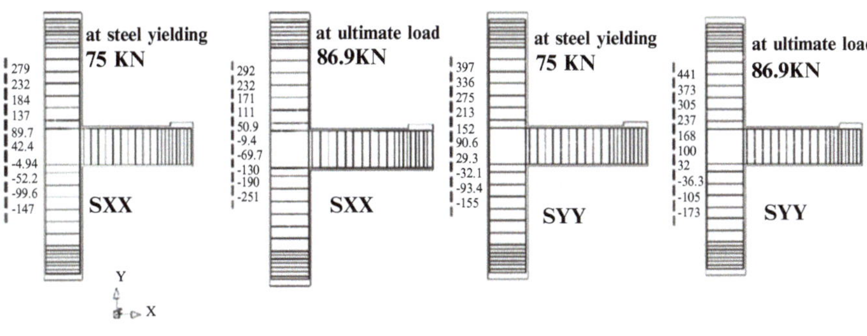

Fig. 13.8 Stresses Sxx & Syy in steel from DIANA

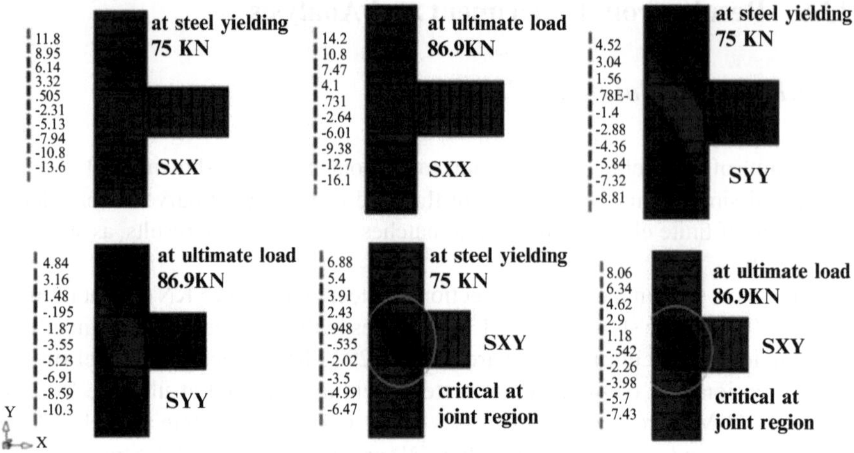

Fig. 13.9 Stresses Sxx, Syy & Sxy in concrete from DIANA

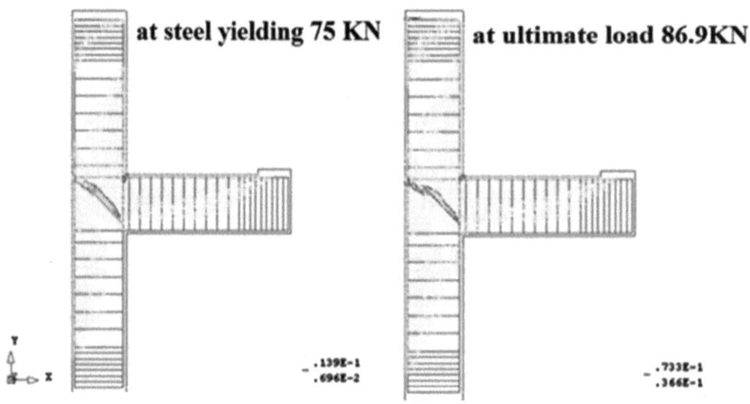

Fig. 13.10 Crack patterns from DIANA

respectively, which shows that concrete reaches its compressive strength at ultimate load at the bottom of the beam; the diagonal tension stresses in joint at steel yielding and ultimate are 0.95 and 1.18 MPa, respectively, which shows that concrete reaches its tensile strength in the joint region at ultimate load. The diagonal crack pattern at the joint closely resembles the crack pattern observed in the experimental program, as shown in Fig. 13.10.

13.7 Experimental Program at KFUPM

Beams and columns of the specimens were produced with the same dimensions, 250 mm × 300 mm. The cantilever length of the beam was 900 mm and the height of column 1,400 mm. The amount of longitudinal reinforcement was constant for both beam and column for the J-BI-18 specimen (see Fig. 13.11). Six 18 mm-diameter bars were used as longitudinal reinforcement in the column and 8 mm closed ties at a spacing of 75 mm as transverse reinforcement. For beams, three bars of 18 mm diameter were used for both the top and bottom longitudinal reinforcement and 8 mm closed stirrups with spacing of 75 mm as transverse reinforcement. The mechanical properties of the reinforcing bars and concrete were determined in the experimental program and are presented in Tables 13.3 and 13.4, respectively.

Specimens were tested in a self-reacting steel loading frame. An additional clamping system was added to the frame to hold the specimen at the top and bottom of the column and at the tip of the beam to apply the load of the hydraulic jack, as shown in Fig. 13.12. Two hydraulic jacks were used to apply the loads: jack (A) of 30 ton capacity at the top of the column for the axial load (maintained constant) and jack (B) of 10 ton capacity at the tip of the beam for cyclic loading on the beam, see Fig. 13.13.

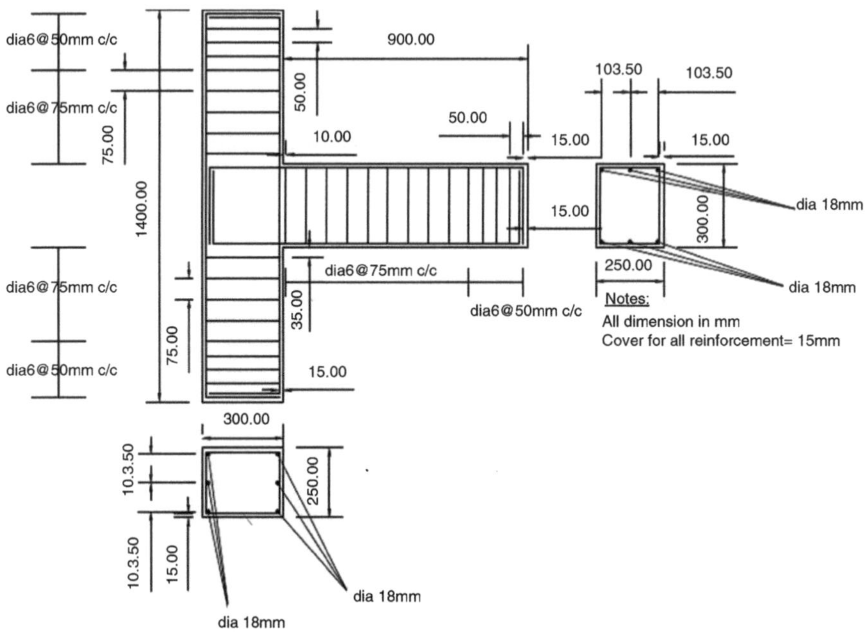

Fig. 13.11 Geometric and reinforcement details for J-BI-18

Table 13.3 Mechanical properties of reinforcing bars

Reinforcement	Dia (mm)	f_y (MPa)
Φ8	8	480
Φ18	18	660

Table 13.4 Material properties of concrete

f'_c (MPa)	E_c (MPa)
30	25,000

Fig. 13.12 Additional clamping system to hold the specimens during the test

Fig. 13.13 Hydraulic jacks used for testing beam column joints

13.8 Cyclic Tests for Type J-BI-18

13.8.1 Experimental Results

Type J-BI-18 specimen has the bent in joint bar anchorage details with 18 mm diameter flexural reinforcement for beam and column. Figure 13.14 shows the load and displacement relationships for J-BI-18 specimen for both push and pull directions. Specimen J-BI-18 was designed for the beam flexural capacity of 137 KN. However, as shown in Fig. 13.14, it collapsed at 99 KN in a joint failure mode in shear. Figure 13.15 represents the formation of crack in joint during the cyclic load test.

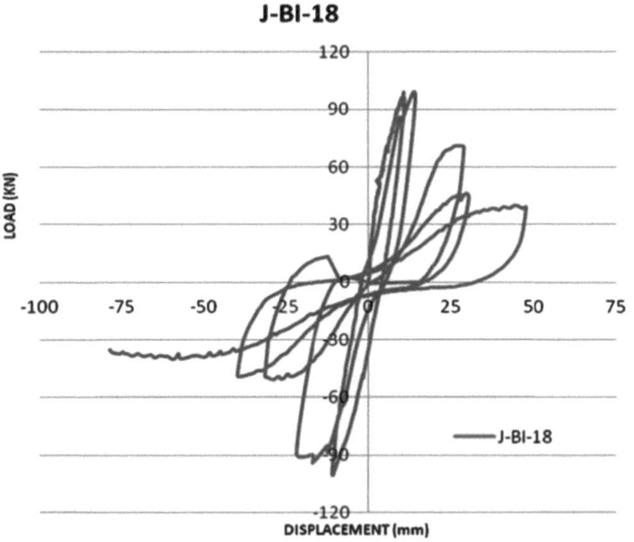

Fig. 13.14 Load–displacement graph for specimen J-BI-18

Fig. 13.15 Formation of crack in joint during cyclic load test of specimen J-BI-18

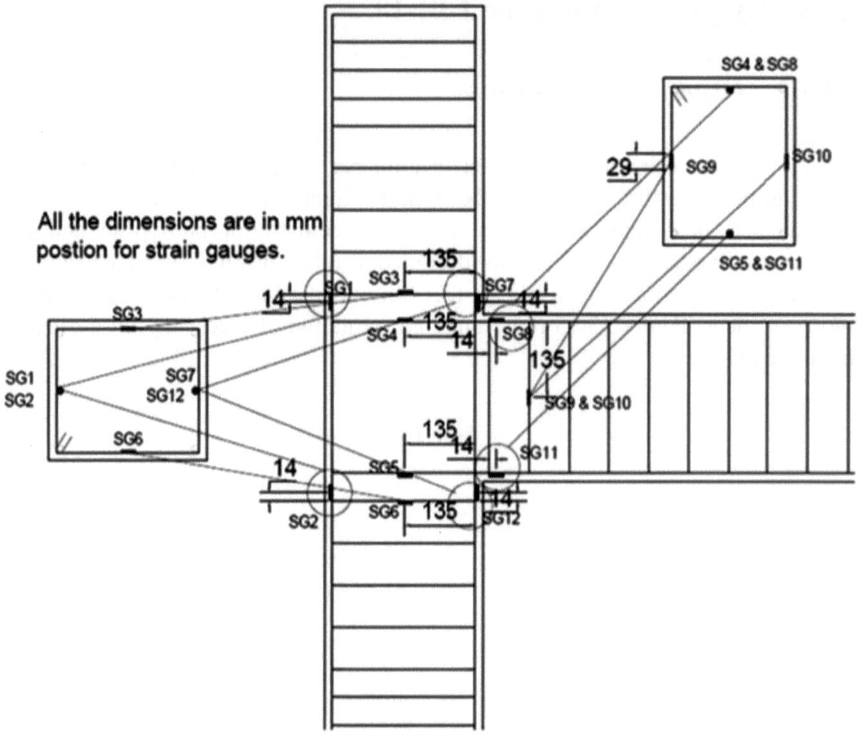

Fig. 13.16 Location of reinforcement strain gauges in specimen

Strain gauges SG8 and SG11 were installed to the top and bottom reinforcement of the beam, as shown in Fig. 13.16. The load vs. strain in steel during the test at the top and bottom reinforcement in the beam is shown in Fig. 13.17. These values of strains were used to calculate the stresses in steel and in concrete in the actual test and were compared to the corresponding quantities determined from the FEM analysis.

13.8.2 Mechanistic Model and Experimental Results of J-BI-18

At P = 99.2 KN, (collapse load of specimen): $\varepsilon_s = 0.002003 < \varepsilon_y, \varepsilon'_s = 0.000795 < \varepsilon_y$

where ε_s is the strain in the beam tension steel (SG8) and ε'_s the strain in the beam compression steel (SG11). Assuming linear strain variation across the beam (Fig. 13.18):

$T = \varepsilon_s \times E_s \times A_s = 0.002003 \times 200 \times 761.288 = 304.2$ KN (Force in beam tension steel)

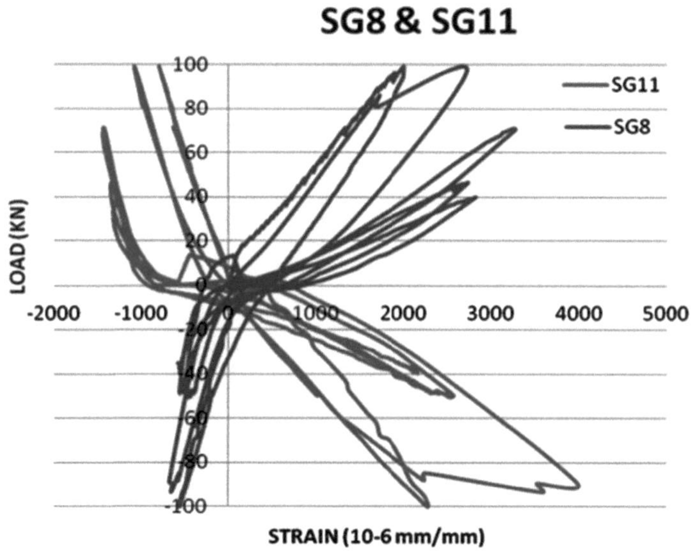

Fig. 13.17 Strain in beam top & bottom bar of specimen J-BI-18

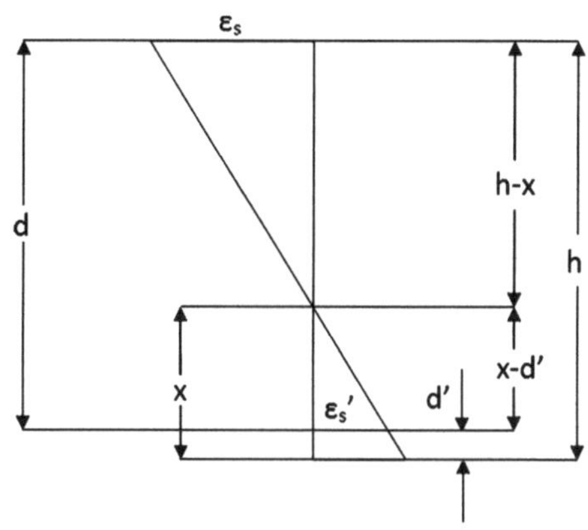

Fig. 13.18 Strain details in beam section

$C_s = \varepsilon_s' \times E_s \times A_s' = 0.000795 \times 200 \times 761.288 = 120.5$ KN (Force in beam compression steel).

$C_c = T - C_s = 304.2 - 120.5 = 183.7$ KN (Compressive force in beam concrete)

$C_c = \frac{1}{2} \times f_c' \times x \times b$ (assuming linear variation of compressive stress in beam), using $x = 94.5$ mm (Fig. 13.18), hence:

$f_c = \frac{2 \times 183.70}{250 \times 94.5} = 15.4 MPa$ (low f_c justifies the linear stress variation)

With reference to Fig. 13.19, the horizontal shear force in the joint is:

$$V_j = T - V_c. \text{ Using: } V_c = \frac{Plb}{l_c}, \text{ we obtain}$$

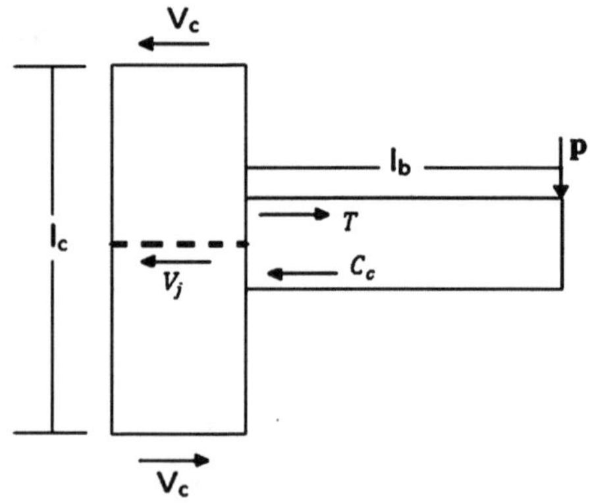

Fig. 13.19 Horizontal shear force in the joint

$$V_j = 304.2 - \frac{99.2 \times 900}{1,400} = 240.2kN > V_{cracking} = 115.6kN.$$

The joint shear capacity based on ACI recommendation with $\gamma = 15$ for exterior beam column joint is:

$$\phi V_{cap} = 0.75\gamma\sqrt{f_c'}A_j = 383.9KN$$

Based on computations, confirmed by reinforcement strain variations, it is concluded that the flexural capacity of the joint specimen was not reached. Further, the failure is identified as a joint shear failure, but at a failure load ($V_j = 240.2$ kN) less than the ACI recommended capacity ($V_{cap} = 383.9$ kN) for properly detailed joints.

13.8.3 New "γ" for Poorly Detailed Joints

We need to back calculate the value of new "γ" from the ACI empirical equation. So: $\gamma = \frac{240.2}{0.75\sqrt{f_c'A_j}}$, which yields: $\gamma = 9$.

Thus the poor detailing of the joint has resulted in a loss of 40 % of the shear capacity.

13.9 Finite Element Analysis of KFUPM Specimens

13.9.1 Input Parameters

The finite element analysis was performed using DIANA. The Drucker Prager model with associated plasticity was selected to model the concrete material. To satisfy the associative plasticity requirements, the values of Φ and ψ were taken to be the same and equal to 10°. However, by using the 10°-degree-based associated

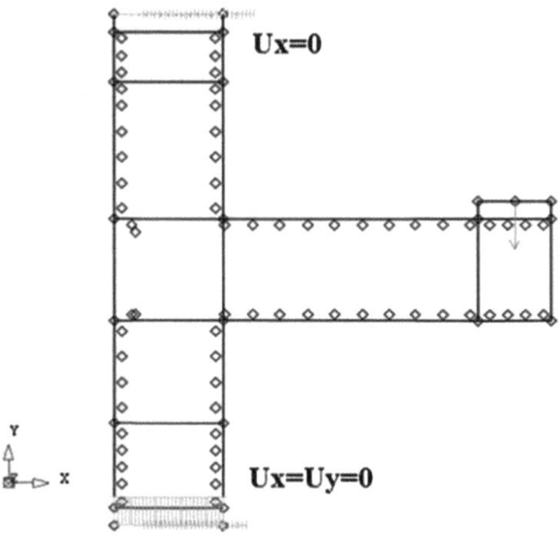

Fig. 13.20 Boundary conditions and loading details of specimens

theory, the DIANA model results in premature failure for specimen J-BI-18. In order to get better results, the non-associated Drucker Prager model was used with $\Phi = 30°$ and $\psi = 10°$ in the beam and column, and $\Phi = 10°$, $\psi = 4°$ in the joint zone. As discussed earlier, the joint zone stress field is dominated by compressive stresses in the biaxial range, in contrast to the beam region, which is closer to uniaxial compression combined with shear. Johansson (1996) noted that for highly under-reinforced sections, using $\Phi = 10°$ or $\Phi = 30°$ would give similar results; failure would be precipitated by considerable steel yielding and concrete compressive stress would be less than the compressive strength of concrete, whereas for other sections, where failure is initiated by concrete reaching concrete compressive strength and steel does not yield, $\Phi = 10°$ must be used. This is not surprising, as higher biaxiality or triaxial state of compressive stress requires the satisfaction of Eqs. (13.4b) and (13.4c).

In the DIANA model, the shear retention factor β must be assumed. For closed cracks, the coefficient is assumed to be 1.0; smaller β value resulted in premature failure. Also, a nonlinear tension softening was selected to describe the relation between the tensile stress and tensile strain at the cracked elements.

13.9.2 Boundary Conditions and Loading Details and Mesh

Boundary conditions were kept the same as in the actual test of beam-column joints, Fig. 13.20 shows the bottom end of the column surface, constrained in the X and Y direction. The top end of the column is constrained in the X-direction and free in the Y-direction due to downward axial force (150 KN). Near the tip of the free end of the beam, the point is constrained in the y direction because the loading method is displacement control. All the specimens were modeled with eight-node quadrilateral isoparametric plane stress element CQ16M with embedded steel (Fig. 13.21).

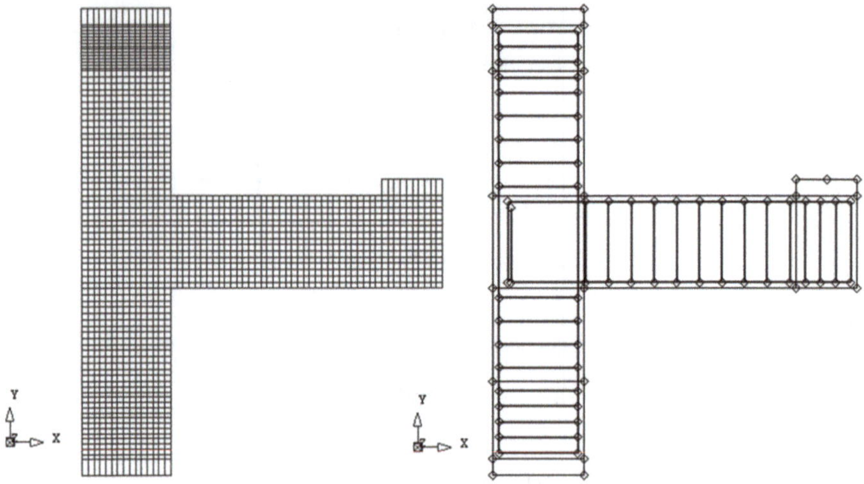

Fig. 13.21 Mesh and DIANA model for specimen J-BI-18

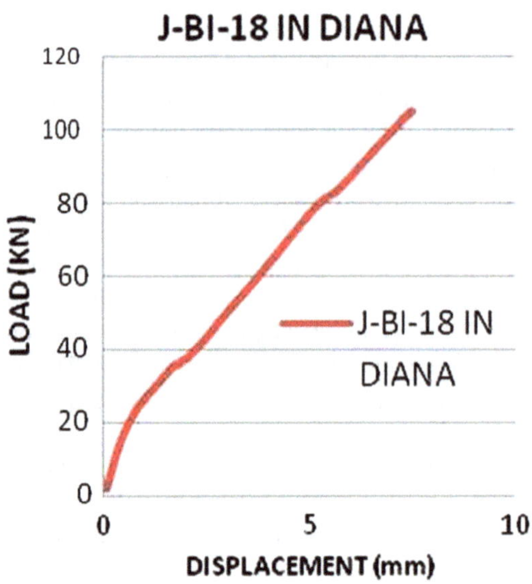

Fig. 13.22 Load–displacement graph for specimen J-BI-18 in DIANA

13.9.3 *Finite Element Analysis for J-BI-18*

Load–displacement graphs from DIANA were plotted to compare the results from the test and the finite element model. Results are shown in Figs. 13.22 and 13.23. For numerical simulation, the envelope of the load displacement curve is

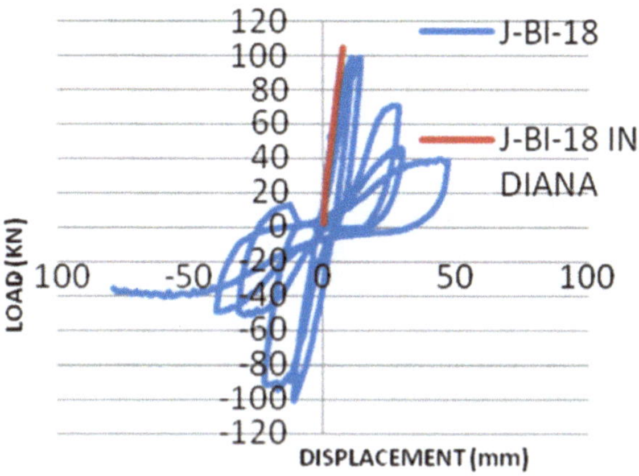

Fig. 13.23 Load–displacement graph comparisons

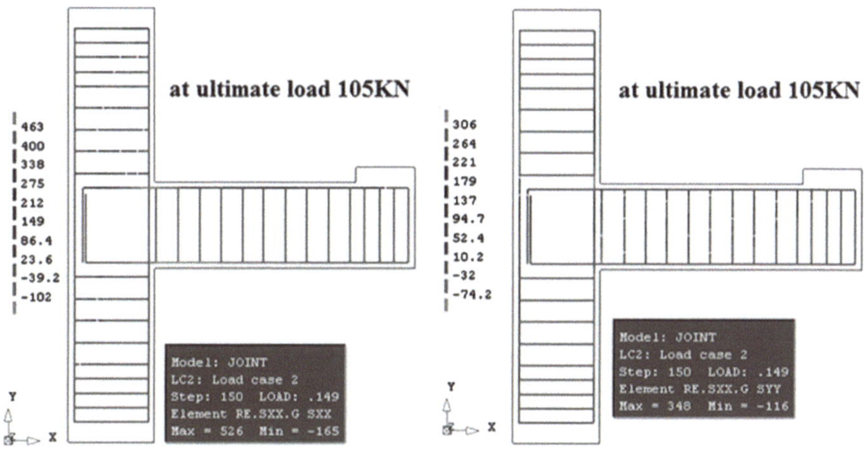

Fig. 13.24 Stress in steel Sxx & Syy for J-BI-18 in DIANA

considered. The result of the finite element simulation matches closely with experimental results, as shown in Fig. 13.23.

Stresses in steel at FEM model ultimate load (105 KN) are shown in Fig. 13.24. Stress in the top steel of the beam at ultimate load is around 460 MPa which is close to stress in the top steel of beam based on strain measurements (400 MPa). The FEM based stress in steel for column on the tension side at ultimate load is 348 MPa. Stresses σ_x and σ_{xy} in concrete at ultimate load (105 KN) are shown in Fig. 13.25. Stress σ_x at collapse load of the specimen in DIANA analysis at the bottom of the beam ranges between 15 and 20 MPa whereas in the actual experiment the stress in concrete at the bottom of the beam was found to be 15.3 MPa, as shown earlier. The diagonal crack pattern at the joint closely resembles the crack pattern observed in the experiment as shown in Fig. 13.26.

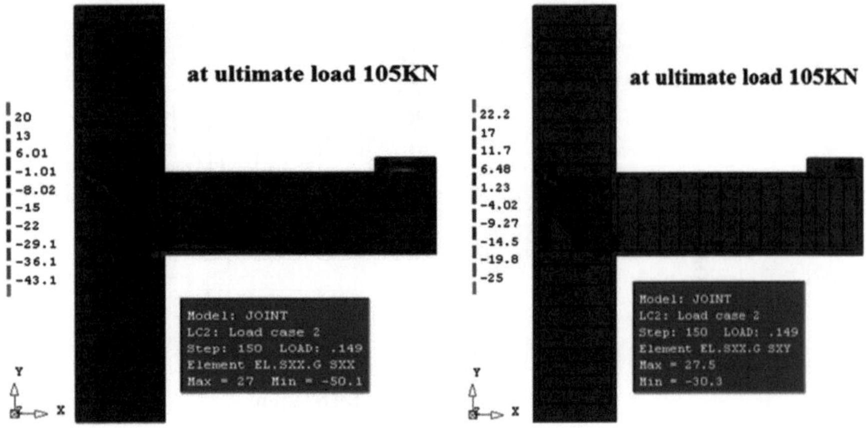

Fig. 13.25 Stress in concrete σ_{xx} & τ_{xy} for J-BI-18 in DIANA

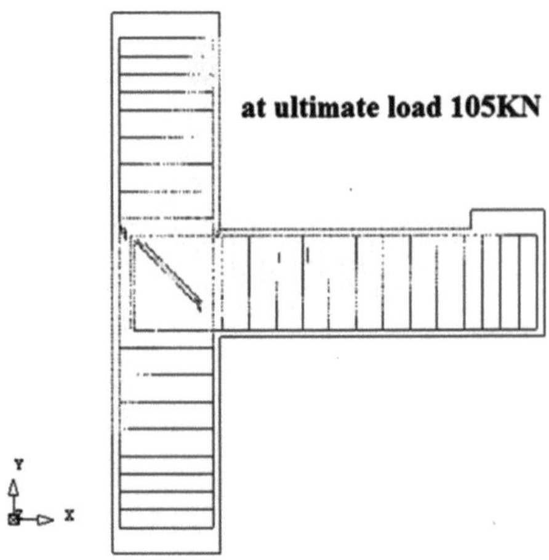

Fig. 13.26 Crack pattern for J-BI-18 in DIANA

13.10 Conclusions

Based on the experimental, mechanistic and computational modeling carried out for the ITU and KFUPM reinforced concrete beam-column joint specimens, the following conclusions may be drawn:

1. The actual mode of failure in the joint is dictated by the geometry of the beam, the column, and the joint, and the amount of reinforcement in the beam, together with reinforcement detailing of the joint.

2. Two cracking loads have been identified. One is associated with flexural cracking of the beam and the other with a diagonal crack in the joint which occurs when the maximum principal stress exceeds the tensile strength of concrete in the joint itself.
3. For the KFUPM specimen designed with relatively high reinforcement ratio $\rho = 0.01$ (J-BI-18), experimental results showed that the specimen collapsed due to failure of joint under shear, as the joint collapse load was lower than the flexural capacity of the beam (27.6 % lower). That was confirmed from the combined mechanistic/experimental computations and also further corroborated from DIANA results.
4. Based on experimental results of specimens with $\rho = 0.01$ (J-BI-18,) a value of "γ" of approximately 9 from the expression $\phi V_{nj} = 0.75\gamma\sqrt{f_c'}A_j$ was obtained, in contrast to the suggested values of "γ" from 12 to 15 in the ACI code for properly detailed joints subjected to seismic loading. In this study, the joint detailing did not meet ACI specifications as the work was simulating joints in old buildings not conforming to the ACI Code.
5. In order to study retrofitting of beam-column connections failing prematurely due to the collapse of joint, the ITU specimens with very low compressive strength concrete were modeled in the DIANA environment. The failure in this specimen occurred at load levels that corresponded closely to both the flexural strength of the beam and the shear capacity of the joint. This behavior was captured through the use of associated plasticity theory, with, a value for Ø that the authors recommend for stress states that have high biaxial or triaxial state of compressive stresses.
6. For the KFUPM beam-column joint specimen, the failure was initiated by joint shear failure. In order to model this in the DIANA environment using Drucker-Prager yield criteria, non-associated plasticity theory had to be used, with $\phi = 10°$ and $\varphi = 4°$ in the joint and $\phi = 30°$ and $\varphi = 10°$ in the beam and column zones. Such a partitioning of the structures into zones of biaxiality (Ø $= 10°$) in contrast to uniaxiality (Ø $= 30°$) appears to allow the use of the 2-parameter Drucker-Prager model in finite element simulation of complex structures.

Acknowledgements The authors would like to acknowledge the support provided by KFUPM through a research grant from the Office of the Deanship of Scientific Research, KFUPM, Dhahran 31261, S. Arabia, under Project No. RG1111-1&2.

References

Bedirhanoglu I, Ilki A, Pujol S, Kumbasar N (2010) Behavior of deficient joints with plainbars and low strength concrete. ACI Struct J 107(3):300–310

De Witte FC, Kikstra WP (2010) DIANA finite element analysis. User's manual. Analysis procedures. TNO DIANA bv, Delft

Ibrahim A, Mahmood M (2009) Finite element modeling of reinforced concrete beams strengthened with FRP laminates. Eur J Sci Res 30(4):526–541. ISSN 1450–216

Johansson M (1996) New reinforcement detailing in concrete frame corners of civil defense shelters non-linear finite element analyses and experiments. Licentiate thesis, Chalmers University of Technology, Division of Concrete Structures, Gothenburg, Sweden

Li B, Kulkarni SA (2010) Seismic behavior of reinforced concrete exterior wide beam-column joints. J Struct Eng ASCE 136(1):26–36

Owen DRJ, Hinton E (1986) Finite element in plasticity. Pineridge Press Limited, Swansea

Supaviriyakit T, Pimanmas A, Warnitchai P (2008) Nonlinear finite element analysis of non-seismically detailed interior RC beam-column connection under reversed cyclic load. ScienceAsia 34:049–058

Chapter 14
FRP Local Retrofit of Non-Conforming RC Beam-Column Joints

Andrea Prota, Marco Di Ludovico, Alberto Balsamo, Claudio Moroni, Mauro Dolce, and Gaetano Manfredi

Abstract Recent seismic events have clearly confirmed the vulnerability of existing reinforced concrete (RC) structures. In particular, field observation of structures damaged by L'Aquila earthquake strongly confirmed that premature failure of partially confined (i.e. exterior) beam-column joints was one of the main causes limiting the global structural seismic capacity. Poor attention to details and a lack of adequate transverse reinforcement typically lead to premature brittle shear failure of joints. To provide support to practitioners involved in the L'Aquila reconstruction process, a proper guideline which illustrates the design of local retrofit interventions on structural and non structural elements has been edited by the Italian Civil Protection Department (DPC) and the Laboratories University Network of Seismic Engineering (ReLUIS). In particular, a viable FRP strengthening strategy to increase the seismic performances of partially confined joints (design procedure and installation steps) is widely discussed and presented in the document. To validate the strengthening system recommended in this guideline, an experimental program has been carried out on as-built and FRP strengthened full scale corner RC joints (T shaped joints). After presenting the main guideline recommendations for local

A. Prota • M. Di Ludovico (✉) • A. Balsamo • G. Manfredi
Department of Structures for Engineering and Architecture,
University of Naples Federico II, 80125 Naples, Italy
e-mail: aprota@unina.it; diludovi@unina.it; albalsam@unina.it; gamanfre@unina.it

C. Moroni
Department of Civil Protection, Presidency of Council of Ministers, Via Vitorchiano 2, 00189 Rome, Italy
e-mail: claudio.moroni@protezionecivile.it

M. Dolce
Department of Structures for Engineering and Architecture,
University of Naples Federico II, 80125 Naples, Italy

Department of Civil Protection, Presidency of Council of Ministers, Via Vitorchiano 2, 00189 Rome, Italy
e-mail: mauro.dolce@unina.it

A. Ilki and M.N. Fardis (eds.), *Seismic Evaluation and Rehabilitation of Structures*, Geotechnical, Geological and Earthquake Engineering 26,
DOI 10.1007/978-3-319-00458-7_14,

strengthening of existing structures, the paper focuses on the experimental program activity; in particular, the specimen design strategy and test setup definition as well as the comparative analysis of the behavior of tested joints.

Abbreviations

b_c	column width
b_b	beam width
d	beam tip displacement
d_O	beam tip displacement for gravity load
f_c	concrete cylindrical compressive strength
f_{cm}	concrete average cylindrical compressive strength
h_c	column height
h_b	beam height
p_t	average principle tensile stress
$t_{f,eq}$	equivalent thickness of FRP dry fibres for each direction.
$A_g = b_c.h_c$	column gross concrete area
E_{eq}	energy dissipation associated to the equivalent elastoplastic system
L	beam free length
P	column axial load
V_c	column shear
$V_{c,MAX}$	experimental column shear peak value
V_{jh}	joint panel shear strength
ε_y	steel rebars yielding strain
ε_{FRP}	FRP strain
$\nu = P/(A_g.f_{cm})$	axial load ratio
μ	subassemblage ductility
Δ_u	ultimate drift
Δ_y	drift at yielding

14.1 Introduction

Devastating earthquakes such as the recent one in L'Aquila 2009 demonstrated the high vulnerability of reinforced concrete (RC) structures, especially of those designed with inadequate seismic rules. In most cases the high vulnerability was due to premature brittle failures in poorly detailed beam-column joints (Fig. 14.1a). Field observation of L'Aquila showed that these deficiencies strongly influenced the local and global structural performances. In particular, for partially confined beam-column joints, such as exterior ones, a brittle failure can be due to several factors such as lack of anchorage of beam rebars, absence or insufficiency of transverse reinforcement, casting sequence (i.e., beams and slabs cast at the same

a b

Fig. 14.1 (**a**) Joint failure; (**b**) FRP strengthening on an elementary school in Paganica, L'Aquila

time with columns below vs. beams and slabs cast after having cast the columns below).

Available literature studies (Paulay and Priestley 1992; Beres et al. 1996; Priestley 1997; Pampanin et al. 2002) have clearly shown that the joint panel behavior is affected by the presence of beams and/or slabs, longitudinal beam rebars anchorage length and type, beam-column dimensions, beams eccentricity and column axial load. Different capacity models have been proposed to evaluate beam-column joints first diagonal cracking or ultimate shear strength. In particular, several authors (Priestley 1997; Pampanin et al. 2002) suggested a simple expression to limit average principle tensile stress in the joint panel to values proportional to the square root of the concrete cylindrical compressive strength, $\sqrt{f_c}$, by a numerical coefficient depending on the beam internal reinforcement typology. This approach was adopted by several codes (i.e. DM 2008; EN 1998-1:2004); other authors and international codes (Fardis 2009; AIJ 1999; ASCE/SEI 41-06 2007; ACI 369R-11 2011) recommended semi-empirical limit values of joint panel shear strength, V_{jh}, taking into account joint typology, load direction, orthogonal beam confinement, aspect ratio and beam internal reinforcement. Several models are also proposed in order to reproduce shear stress or shear stress–strain behavior (Priestley 1997; Pampanin et al. 2002; ASCE/SEI 41-06 2007; ACI 369R-11 2011; Park and Mosalam 2012; Hakuto et al. 1995; LaFave and Kim 2011; Pantalides et al. 2002).

The effectiveness of different FRP systems as a strengthening solution for beam column joints has been widely investigated in the last decades (Gergely et al. 2000; Ghobarah and El-Amoury 2002, 2005; Antonopoulos and Triantafillou 2003; Prota et al. 2004; Tsonos 2008). It was observed that FRP effectiveness depends on several factors such as surface preparation, composite layout, geometrical and mechanical ratio, inclination of fibres, number of layers, mechanical anchorages, and presence of joint stirrups. In particular, Antonopoulos and Triantafillou (2003) showed that joint panel capacity may be significantly increased by using FRP sheets in two orthogonal directions provided that premature debonding is prevented through the use of mechanical anchors. The experimental tests carried out by Prota et al. (2004) confirmed that by changing the amount (numbers of plies and rebars) and location (column only or both column and joint) of FRP strengthening systems it is possible for a designer to modify the strength hierarchy by changing the beam-column joint failure mode. The experimental results on a full-scale RC structure (Di Ludovico et al. 2008) confirmed that a seismic retrofit approach using FRP systems could significantly increase the structural energy dissipation capacity. All this experimental and analytical work created a solid background towards a sound and safe use of composites in the seismic retrofit of existing RC structures. In particular, in the aftermath of the April 2009 L'Aquila earthquake, extensive local retrofit works based on FRP were executed to increase the seismic capacity of public and private buildings. FRP strengthening of partially confined joints and FRP wrapping of beams and columns ends to prevent brittle failure mechanisms and to increase the structural dissipation capacity were largely adopted on the school building stock (Fig. 14.1b) in order to quickly assure their use and to significantly reduce the seismic vulnerability of these strategic buildings. To provide support to engineers involved in the L'Aquila reconstruction process, a proper guideline which illustrates the design of local retrofit interventions on structural and non structural elements has been developed by the Civil Protection Department (DPC) and the Laboratories University Network of Seismic Engineering (ReLUIS): "Repair and strengthening of structural elements, infills and partitions" (DPC-ReLUIS Guidelines 2011). In particular, the typical FRP strengthening layouts of partially confined joints (design procedure and installation steps) are widely discussed and presented. The effectiveness of the strengthening system recommended in the guidelines has been investigated by means of experimental tests on as-built and FRP strengthened joints. The outcomes of these tests as well as a summary of local retrofit interventions presented in the guidelines are reported in the following sections.

14.2 Guidelines on Local Retrofit Interventions

The post-earthquake in situ inspections of buildings damaged by L'Aquila earthquake clearly showed that local interventions on structural members may significantly increase the global seismic capacity of existing structures. In particular, seismic actions dramatically pointed out the typical defects of structures designed according to non-seismic code provisions. These deficiencies should be carefully taken into account in the structural safety assessment. The "Repair and strengthening of structural elements, infills and partitions" guideline, specifically drawn up to support engineers involved in the L'Aquila reconstruction process, firstly outlines the typical damage found on both structural and non structural members of RC and masonry structures. Then, local retrofit interventions, both for structural and non structural members, are described by illustrating the installation and calculation procedures. In particular, local retrofit interventions on structural elements presented in the document are: local retrofit of RC beam-column joints; beams, columns and slabs strengthening by means of FRP systems, steel jacketing or prestressed steel ribbon (Active Confinement of Manufactured materials, CAM); strengthening interventions on masonry structures by means of traditional and innovative techniques; local strengthening of timber elements.

A wide section of the guideline focuses on the brittle failure mechanisms. In particular, according to experiences gained from examining the performances of RC structures after seismic events, a wide section focuses on the most common brittle collapse mechanisms resulting from shear failure of partially confined beam-column joints (i.e. exterior or corner joints on the perimeter of the structure; in some cases, where frames are only in one of the plan directions, they could be also interior). The typical joint failures observed in the L'Aquila post earthquake are described and a local strengthening procedure by means of FRP, steel jacketing or CAM technique is discussed.

The description of a strengthening scheme by means of FRP laminates for partially confined joints is reported in detail. The strengthening technique involves not only the shear strengthening of joint panels but also of each component of the beam-column joints in order to avoid premature failure mechanisms and, at the same time, to increase the structural local and global ductility. In particular, the shear failure prevention due to local effects of strong infills, the ductility increase of columns ends and the shear strengthening at ends of beams are also taken into account in the strengthening scheme. The strengthening scheme is outlined in the following points:

(i) Beam-column joint shear capacity increase against local effect of strong infills (see Fig. 14.2). Observation of seismic damage confirms that the shear loading due to the infill strut force at the column joint interface can cause significant damage to the joint panel (i.e. pseudo-horizontal crack at the concrete recast region or diagonal crack in the joint panel). In particular, in order to withstand the horizontal component of the infill strut force, Steel Reinforced Polymer (SRP) composites in the form of uniaxial systems can be installed around the beam-column joint both in the case of corner or exterior joints (see Fig. 14.3a). The strengthening phase can be completed by applying L-shaped quadri-axial FRP laminates at the beam-column connection, Fig. 14.3b;

(ii) Shear capacity increase of beam-column joint panel. It can be achieved through the application of composites with fibres placed along the principal tensile stresses (i.e. quadriaxial FRP laminates) as depicted in Fig. 14.3c;

(iii) Confinement of columns ends. It allows to significantly increase the deformation capacity in plastic hinge zones with a corresponding enhancement of the global structural ductility. Indeed, FRP wrapping increases the ultimate compressive strain of concrete, thus determining an increase of cross-section ultimate curvature corresponding to a member rotational capacity increase. The confinement is also effective to prevent longitudinal bars buckling and to sustain the shear action, at the top of the column, due to the infill strut force. FRP uniaxial laminates can be installed as reported in Fig. 14.3d;

(iv) shear capacity increase of beams: the use of U-wrap FRP laminates can increase the shear capacity of beams at the ends (in the zone of maximum

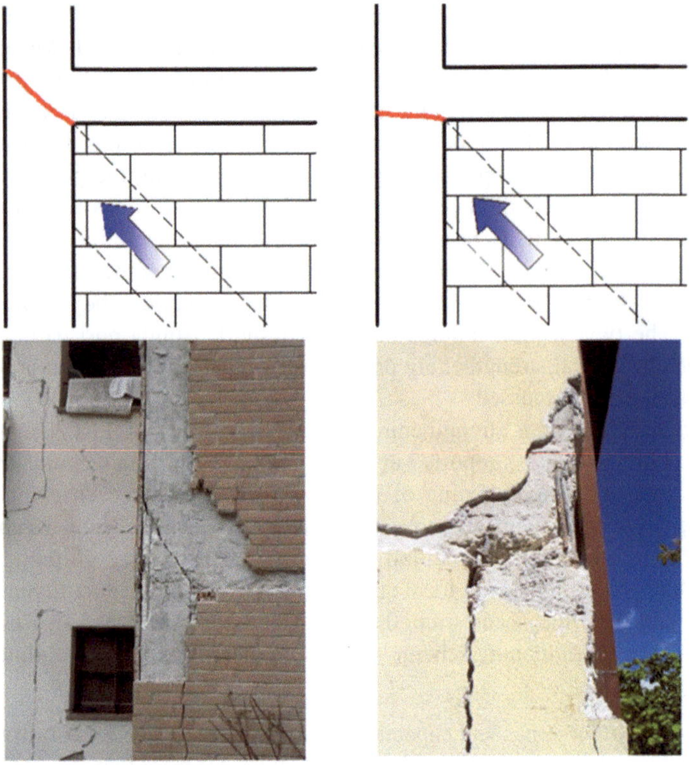

Fig. 14.2 Effects of strong infills on the joint panel

shear demand in case of seismic action) and, at the same time, can be very useful in order to provide mechanical anchorage to the quadriaxial FRP panel sheet applied on the joint, see Fig. 14.3e; they also allow to prevent premature debonding of the panel FRP external reinforcement and thus to increase the effectiveness of the whole strengthening scheme.

A field application of strengthening is depicted in Fig. 14.3f.

14.3 Experimental Program

The experimental program consists of two tests on RC full scale T-shaped joints: "as-built" specimen, T_C; and FRP strengthened specimen, T_FRP. The tests were specifically designed in order to validate the FRP strengthening technique recommended in the post-L'Aquila earthquake guideline for repair and strengthening of damaged structures.

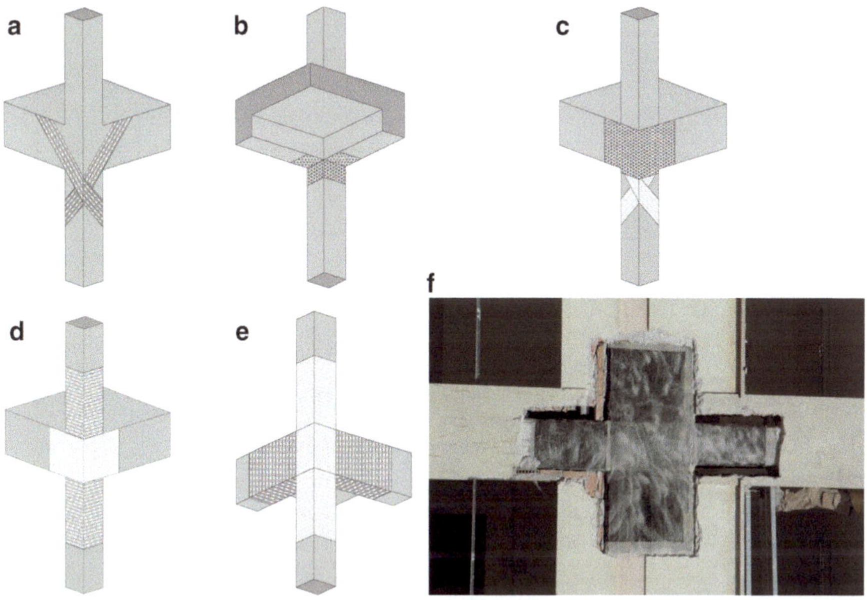

Fig. 14.3 FRP strengthening system details (DPC-ReLUIS Guidelines 2011): (**a**) SRP uniaxial system to sustain horizontal actions due to strong infills; (**b**) L-shaped FRP laminates at beam column connection; (**c**) Shear capacity increase of beam column joint panel; (**d**) Column end confinement; (**e**) Beam end FRP wrapping; (**f**) FRP strengthened joint

14.3.1 Specimen Design

In order to reproduce beam-column subassemblies representative of typical existing constructions, a three storey regular RC structure with a storey height of 3.4 m and three-bay frames spanning at 5 m was designed for gravity loads only. Thus, poorly detailed beam columns subassemblies without stirrups in the joint panel, with beam and column length of 1.80 and 3.40 m, respectively, were designed for the experimental program (Fig. 14.4a). The beam and column lengths were designed to be representative of typical inter-storey height and beam portion up to a zero point of bending moment diagram in frames designed for gravity loads (Fig. 14.4b). Members dimensions and reinforcement ratios were also aimed at reproducing subassemblies with weak column and strong beam condition; moreover they were designed to achieve shear failure in the joint panel before yielding of both beam and columns reinforcement under simulated seismic actions. Thus, a square column with a side length of 300 mm has been adopted with a longitudinal reinforcement ratio equal to 0.9 % corresponding to 4ϕ16 in the cross section corners. The beam cross section is 500 mm deep and 300 mm wide with 5ϕ16 on the top side and 3ϕ16 on the bottom side (reinforcement ratio of 0.7 and 0.4 %, respectively); the beam

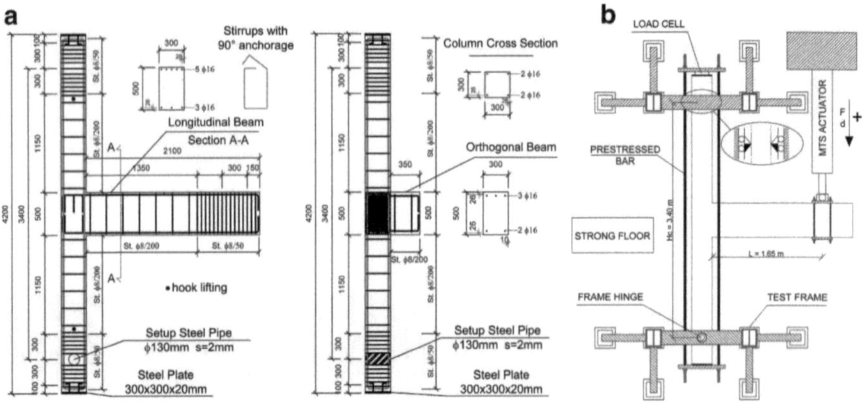

Fig. 14.4 (**a**) Specimens details; (**b**) test setup (*dimensions in mm*)

longitudinal reinforcement is anchored in the joint panel with 90° standard hook (effective straight length equal to 200 mm). In order to predict the behavior of subassemblies, linear and non linear analyses have been carried out taking into account the test setup boundary conditions. The analyses provided the following failure sequences for the subassemblies:

- Joint panel shear failure: joint shear capacity V_{jh} (corresponding to the average principle tensile stress $p_t = 0.42\sqrt{f_{cm}}$ (Priestley 1997), with f_{cm} average concrete cylindrical compressive strength) is reached for column shear $V_c = 42.8$ kN;
- Top column rebars yielding at $V_c = 55.9$ kN;
- Beam rebars yielding at $V_c = 58.2$ kN.

Furthermore, an orthogonal beam stub has been designed in order to represent a typical corner joint (Fig. 14.4a). The longitudinal reinforcement of this beam was designed considering the actions provided by the external infill only (3ϕ16 on the top side and 2ϕ16 on the bottom side). Both beam and column have ϕ8 stirrups spaced at 200 mm (a lower spacing has been adopted in the load application zones).

One specimen, namely T_C, was tested in the as-built configuration. Then the second specimen, namely T_FRP, was tested in the strengthened configuration in order to investigate the benefits of the recommended FRP layout, see Fig. 14.3b–e. Obviously, SRP composites were not installed since infills action was not simulated in the test. One layer of quadriaxial CFRP, extended for 200 mm on beams, was used for joint panel shear strengthening, Fig. 14.5a. Column ends confinement was provided by means of uniaxial CFRP wraps extended for 750 mm on column. U-shaped uniaxial CFRP wraps were extended for 750 mm on beam, Fig. 14.5b. An FRP Strengthening system was installed according to the DPC-ReLUIS Guidelines (2011) for field applications.

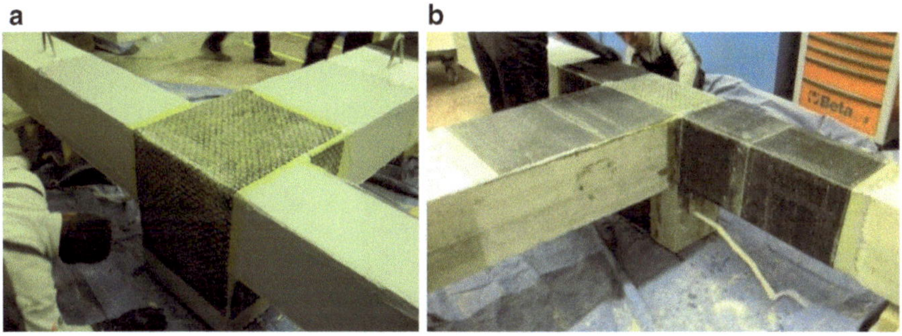

Fig. 14.5 (a) Joint panel strengthening; (b) column and beam strengthening

14.3.2 Material Properties

A low concrete strength was designed (i.e. 15–20 MPa) to simulate the application on existing RC buildings in which poor quality concrete is usually found. Concrete average compressive strength provided by means of three tests on cylindrical specimens at 28 days after casting were $f_c = 16.3$ MPa and $f_c = 17.7$ MPa, for T_C and T_FRP, respectively. Steel longitudinal rebars had an average yield stress equal to 470 MPa (yield strain $\varepsilon_y = 2.4$ ‰). FRP laminate properties were provided by the manufacturer. Quadri-axial CFRP sheets with a unit weight of 380 g/m^2, thickness of dry fibres, $t_{f,eq}$, of 0.053 mm, Young modulus of 230 GPa and ultimate strain of 1.5 % were used for joint panel strengthening. Uni-axial CFRP sheet with a unit weight of 300 g/m^2, $t_{f,eq}$, of 0.166 mm, Young modulus of 230 GPa and ultimate strain of 1.5 % were also used for column confinement and to mechanically anchor the joint panel reinforcement at the end of the beams (this reinforcement also works as beam shear strengthening).

14.3.3 Test Setup and Instrumentation

The concrete casting was executed with the beam-column subassemblage horizontally placed on a plane parallel to the strong floor (Fig. 14.6a). A constant axial load was exerted on the column by a set of four vertical prestressed rods connected to strong steel plates; in particular, a constant value of axial load ratio $\nu = 0.2$ (that is, axial load, P, divided by the product of the column gross concrete area A_g and f_c) was adopted. At the column end, a load cell was placed to monitor the axial load during the test. Each specimen was constrained to the strong floor by means of two rigid steel frames, with a steel roller placed inside the lower column end to simulate pin connection. The column top was constrained to a rigid frame by means of two steel rollers (Fig. 14.6a). As the specimen lied on the laboratory structural floor, steel spherical hinges were placed between the beam end and floor to limit friction and to allow tip beam free movement.

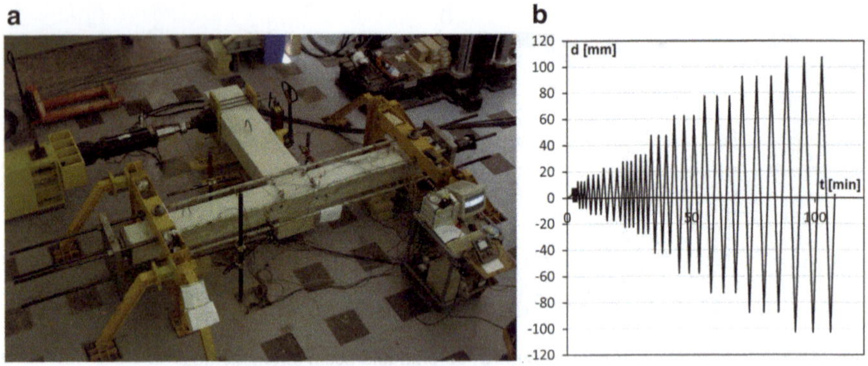

Fig. 14.6 (**a**) Test setup; and (**b**) load protocol

To simulate gravity loads, a preload of 19.2 kN was applied on the beam together with the axial load on the column, while to reproduce seismic excitation, the beam was subjected to transverse cyclic loading which consisted of three successive cycles progressively increasing by 5 mm (up to 30 mm) or 15 mm (up to failure) of displacement amplitudes in each direction (Fig. 14.6b). This loading arrangement, with column ends held in a vertical line, may effectively reproduce seismic actions on a beam-column subassemblage provided that P-Δ effects due to column axial load are negligible (Park 1992).

Real-time measurement of the structural response was achieved by using an electronic data acquisition system. Several instruments were used to monitor the specimen behavior: linear variable displacement transducers (LVDTs) to monitor the global deformed shape and the joint panel shear strain; a potentiometer on tip beam to monitor the actual displacement imposed by the actuator to the specimen; strain gauges on internal steel rebars to record their strains at ends of beam and columns; and strain gauges on CFRP quadriaxial sheet along fibre directions to monitor FRP strains at joint panel.

14.4 Experimental Results

Experimental results for each specimen are discussed with reference to:

- the specimen global behavior (crack patterns, failure modes, column shear-drift hysteresis loop, subassemblies energy dissipation, and stiffness); and
- strains attained on the internal steel rebars and FRP joint panel strengthening.

The main experimental results, for both positive and negative load action, are reported in Table 14.1. The maximum shear recorded on the column, V_{CMAX}, as well as the relevant drifts have been initially evaluated. Top column interstorey drift has been computed as the ratio between beam tip displacement (total displacement

Table 14.1 Experimental results

Spec.	f_{cm}	Load sign	$V_{c,MAX}$	ΔV_c[a]	Drift ($V_{c,MAX}$)	μ (Δ_u/Δ_y)	Δμ[a]	E_{eq}	ΔE[a]	ε_{FRP} ($V_{c,MAX}$)	Failure mode
[–]	[MPa]	[–]	[kN]	[%]	[%]	[–]	[%]	[kN mm]	[%]	[‰]	[–]
T_C	16.3	+	43.76	–	1.31	3.87	–	132.7	–	–	JS
		–	36.88	–	1.31	4.90	–	98.9	–		
T_FRP	17.7	+	56.13	29.9	2.38	4.28	10.6	199.7	50.5	10.2	CH/FD
		–	45.16	26.6	2.38	5.25	7.1	148.1	49.8	5.8	

Failure modes: *JS* joint shear, *CH* column flexural hinging, *FD* FRP debonding
[a]Means computed with respect to T_C

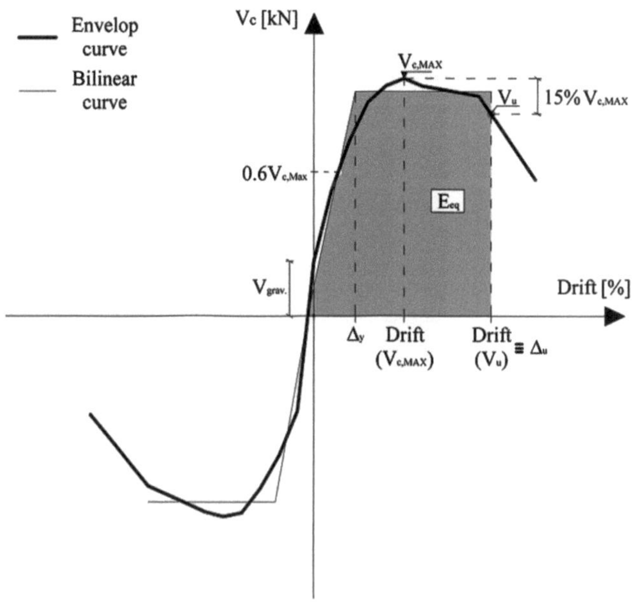

Fig. 14.7 Parameters derived from experimental V_c-Drift relationship

d, minus gravity load displacement, d_0) and the actuator distance from column axis, $L + h_c/2$ ($L = 1{,}650$ mm and $h_c = 300$ mm), see Fig. 14.4b:

$$Drift(\%) = \Delta = \frac{(d - d_0) * 100}{L + (h_c/2)} \tag{14.1}$$

Subassembly ductility, μ, is evaluated as Δ_u/Δ_y, where Δ_u is the imposed drift when the load carrying capacity undergoes a 15 % reduction in load, see Fig. 14.7. Drift at yielding, Δ_y, is defined as the drift corresponding to the beginning of the plastic behavior on the equivalent elastoplastic system (Park 1988). The stiffness of the equivalent elastoplastic system has been identified by the intersection with

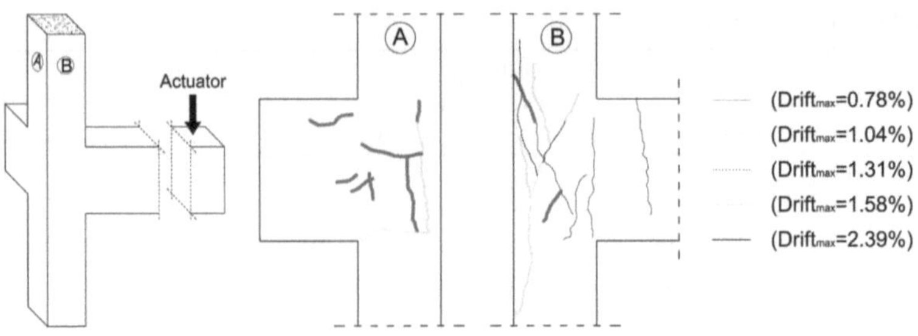

Fig. 14.8 T_C crack pattern

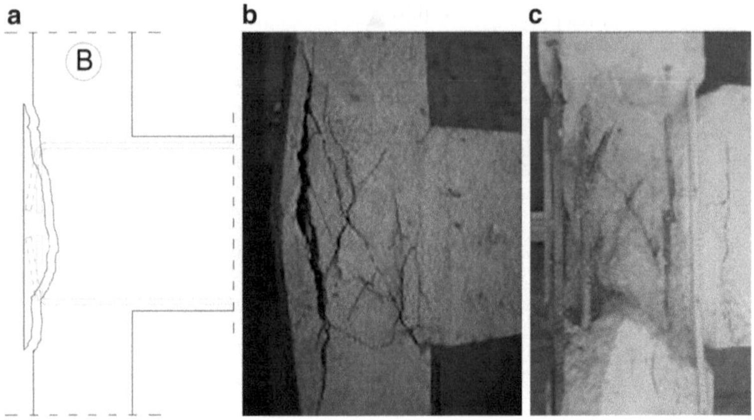

Fig. 14.9 T_C failure mode

the experimental hysteresis loop envelope at $0.6V_{c,MAX}$ (D.M. 2008). To further assist the specimens' performance analysis, the energy dissipation associated to the equivalent elastoplastic system, E_{eq}, has been also computed (see Fig. 14.7) and reported in Table 14.1. For each parameter reported in Table 14.1, the relevant percentage differences between the FRP strengthened specimen and the as-built one are computed. Joint panel FRP strengthening strain records at peak column shear, $\varepsilon_{\mathrm{FRP}}$, are also reported in Table 14.1.

14.4.1 "As-Built" Specimen (T_C)

The crack pattern observed on T_C is depicted in Fig. 14.8.

The subassemblage failure mode was characterized by large and deep diagonal cracks and concrete "wedge" spalling off, as typically found in post-earthquake in situ inspections, see Fig. 14.9, as well as in available experimental test reports (Priestley 1997; Pantelides et al. 2002).

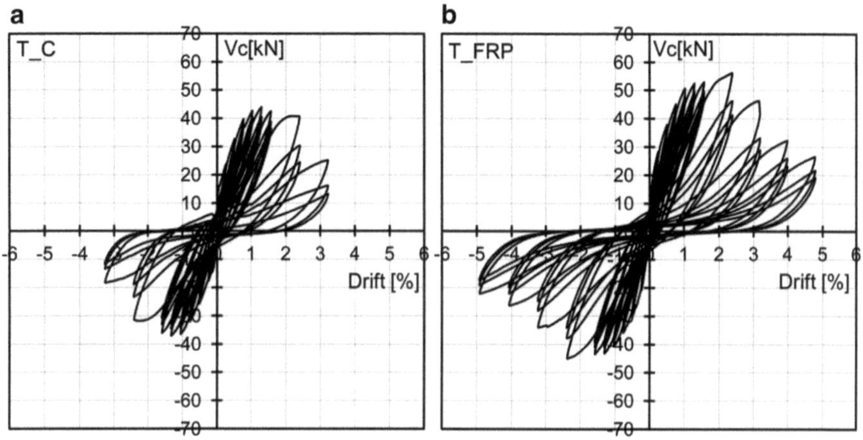

Fig. 14.10 T_C column shear vs. drift (**a**); T_FRP column shear vs. drift (**b**)

The V_c-*Drift(%)* relationship is reported in Fig. 14.10a. Note that the cyclic behavior was asymmetrical since the beam internal top and bottom reinforcement were different (i.e. 5ϕ16 on the top side and 3ϕ16 on the bottom side). Furthermore the higher peak shear value was attained in the positive load direction due to the initial positive preload applied on the beam to simulate existing gravity loads. Initial hairline cracking in joint panel started at the beginning of the third load cycle, for drifts of about 0.5 % and column shear of about 33 kN. Beyond this point, in the drift range of 0.78–1.38 %, the joint panel cracks progressively increased in terms of number and width as shown in Fig. 14.8 leading to a significant stiffness degradation. Subassemblage peak strength was attained for a drift of 1.31 % at which joint panel large diagonal cracks were observed.

The post-peak stage was characterized by strength degradation due to significant opening of beam longitudinal rebars bent into the joint region. This led to concrete "wedge" spalling off which became significant at a drift of 2.39 %. At this stage, a significant pinching effect due to residual shear strain in the joint panel was recorded (see the hysteresis loops in Fig. 14.10a). At a drift of 3.3 %, the beam rebars anchorage opening led to complete concrete spalling, Fig. 14.9b, and hairline cracking also involved the concrete stub which simulates orthogonal beams in existing buildings. To examine the crack dimensions in the joint core, damaged concrete was removed after the test; the core inspections showed two deep diagonal cracks leading to the joint panel splitting in four parts, Fig. 14.9c.

The strain gauge readings showed that no yielding was attained in the columns and beams internal rebars; thus, according to theoretical predictions, the subassemblage failure was due to shear.

Fig. 14.11 T_FRP failure mode

14.4.2 FRP Strengthened Specimen (T_FRP)

The main experimental outcomes related to the test on the FRP strengthened specimen are reported in Table 14.1; hysteresis loops are reported in Fig. 14.10b. With respect to the T_FRP specimen, the column peak shear was 56.13 kN at a drift of 2.38 %. The experimental hysteresis loops show that a significant strength drop, due to FRP premature debonding, was attained. Figure 14.11a shows that FRP debonding started from the free end of the uniaxial U-shaped CFRP wrap. This happened at a positive drift of 2.38 % as clearly indicated by the FRP tensile strain values recorded at each cycle peak drift (see Fig. 14.12a).

A contraflexure point in the joint panel FRP strains trend was observed at a drift of about 2.38 %. The maximum strain on the FRP reinforcement was about 6 ‰ in the horizontal fibres (strain gauge s.g.f.#1, Fig. 14.12) and about 10 ‰ in the diagonal ones (strain gauge s.g.f.#2, Fig. 14.12a). The experimental maximum FRP strain on the joint panel was higher than that suggested by CNR-DT 200 (2004) for design purposes (i.e. 4 ‰).

On the T_FRP specimen, column rebars yielding was attained at a drift of about 2 %. Since no internal rebar yielding was recorded on the column rebars of the T_C specimen, it can be concluded that the strengthening solution changed the subassemblage failure sequence by delaying the joint panel shear failure after the column end yielding. This led to a ductility increase of about 10 % and to an energy dissipation increase of about 50 % with respect to T_C (see Table 14.1).

Full FRP panel debonding was achieved at a drift of 3.3 %, see Fig. 14.11a. Debonding started at the end of the quadriaxial FRP panel (at 200 mm from the beam column connection). Then, quadriaxial fibres along the joint panel started to be cut. Indeed, when the first debonding happened, the outer fibres on joint panel still had an effective anchorage length. In this phase (between first FRP debonding and full FRP debonding) the specimen strength capacity gradually reduced. FRP joint panel strengthening was removed after the test; the concrete

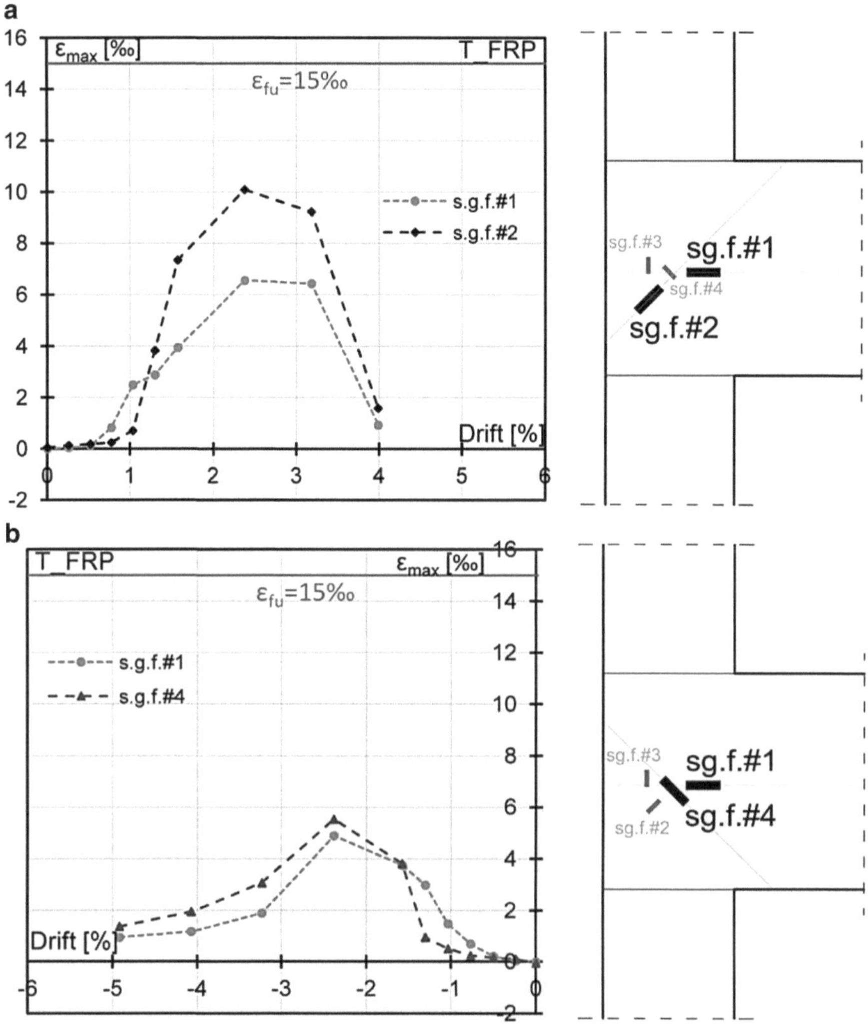

Fig. 14.12 Joint panel FRP strains: (**a**) strain gauges #1, 2 positive load direction; (**b**) strain gauges #1, 4 negative load direction

substrates state is shown in Fig. 14.11b. More internal cracks were observed with respect to the control specimen; however the crack width was lower in the case of T_FRP with respect to T_C.

14.5 Discussion and Conclusions

The comparison between the experimental envelope curve of the as-built specimen (T_C) and that of the FRP strengthened specimen (T_FRP) is shown in Fig. 14.13. The figure clearly shows that FRP laminates did not affect the subassemblage global

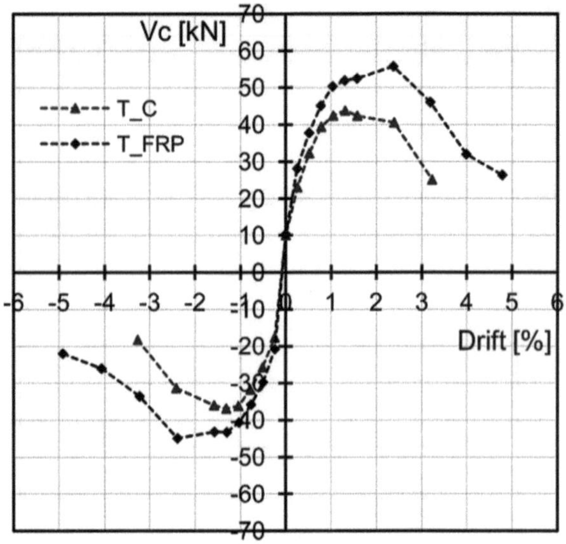

Fig. 14.13 Force drift envelopes for reference and strengthened specimens

stiffness, especially at initial cycles (i.e. up to about 0.5 % drift that is before concrete cracking on T_C joint panel). Then, a slight slope difference in the experimental curves is visible. This is mostly due to the confinement effect provided by FRP laminates to the joint panel region, which significantly reduced the concrete cracking development.

The experimental tests provided evidence that joint panel tensile failure was achieved on the as-built specimen; the absence of joint panel internal ties led to the anchorage of beam rebars opening with a corresponding concrete cover wedge spalling off. This effect strongly promoted a sudden post peak strength degradation. By contrast, FRP strengthening strongly delayed this phenomenon. It allowed a strength increase of about 30 % to be attained, see Table 14.1. Indeed, the 30 % increase in column shear corresponded to the attainment of column rebars yielding, beyond which no more shear forces can be supported by the column. Thus, FRP strengthening modified the subassemblage failure mode from brittle, due to joint panel shear failure, in a more favorable ductile hinge on top column. In particular, the subassemblage ductility on the strengthened specimen increased by about 10 % with a corresponding subassemblage energy dissipation increase of about 50 % (related to the equivalent elasto-plastic system, E_{eq}). The recommended strengthening layout, properly designed to facilitate field application phases, allowed to achieve a maximum strain in the FRP reinforcement of about 10 ‰ which is more than two times greater than what is allowed by current design provisions, 4 ‰.

The experimental tests clearly pointed out the potential of the recommended FRP solution to increase the seismic performances of beam-column joints designed according to non-seismic code provisions. The experimental validation of this local strengthening technique may strongly encourage the use of local interventions based on the use of FRP reinforcement to significantly increase both the local and global seismic capacity of RC structures.

Acknowledgments This study was performed in the framework of PE 2010–2013: joint DPC-ReLUIS project, Task 3.1. The materials used for joints strengthening were donated by Mapei Spa, Milano.

References

ACI 369R-11 (2011) American Concrete Institute, committee 369: guide for seismic rehabilitation of existing concrete frame buildings and commentary, February 2011, USA

AIJ (1999) Architectural Institute of Japan. Design guideline for earthquake resistant reinforced concrete buildings based on inelastic displacement concept, Tokyo

Antonopoulos CP, Triantafillou TC (2003) Experimental investigation of FRP-strengthened RC beam-column joints. ASCE J Compos Constr 7(1):39–49

ASCE/SEI 41–06 (2007) Seismic rehabilitation of existing buildings. American Society of Civil Engineers, Reston, pp 428

Beres A, Pessiki SP, White RN, Gergely P (1996) Implications of experiments on the seismic behaviour of gravity load designed RC beam-to-column connections. Earthq Spectra 12(2):185–198

CNR-DT 200 (2004) National Research Council: guide for design and construction of externally bonded FRP systems for strengthening existing structures. CNR, Rome, 13 July 2004

Di Ludovico M, Balsamo A, Prota A, Manfredi G (2008) Comparative assessment of seismic rehabilitation techniques on a full-scale 3-storey RC moment frame structure. Struct Eng Mech 28(6):727–747

DM 14 Gennaio 2008 (DM 2008) Technical codes for structures (in Italian). GU n°29, Feb 2 2008, Supplemento Ordinario n°30, Rome

DPC/ReLUIS (2011) Guidelines for repair and strengthening of structural elements, infill and partitions (in Italian) by Dolce M, Manfredi G, Doppiavoce Dec 2011. ISBN 978-88-89972-29-8, pp. 194

European Standard (EN 1998–1:2004) Eurocode 8: design of structures for earthquake resistance. Part 1: general rules, seismic actions and rules for buildings, Brussels, December 2004

Fardis MN (2009) Seismic design, assessment and retrofitting of concrete buildings based on EN-Eurocode 8, vol 8, Geotechnical, geological and earthquake engineering. Springer, Dordrecht. ISBN 978-1-4020-9841-3

Gergely I, Pantelides CP, Reaveley LD (2000) Shear strengthening of RC T-joints using CFRP composites. ASCE J Compos Constr 4(2):56–64

Ghobarah A, El-Amoury T (2002) Seismic rehabilitation of beam-column joint using GFRP sheets. Eng Struct 24:1397–1407

Ghobarah A, El-Amoury T (2005) Seismic rehabilitation of deficient exterior concrete frame joints. ASCE J Compos Constr 9(5):408–416

Hakuto S, Park R, Tanaka H (1995) Retrofitting of reinforced concrete moment resisting frames. Research report 95–4, Department of Civil Engineering, University of Canterbury, Christchurch

LaFave JM, Kim J (2011) Joint shear behaviour prediction for RC beam-column connections. Int J Concr Struct Mater 5(1):57–64

Pampanin S, Calvi GM, Moratti M (2002) Seismic behaviour of RC beam-column joints designed for gravity loads. In: Proceedings of the 12th European conference on earthquake engineering, London, September 2002. Paper reference 726

Pantelides CP, Clyde C, Reaveley LD (2002) Performance-based evaluation of reinforced concrete building exterior joints for seismic excitation. Earthq Spectra 18(3):449–480

Park R (1988) Ductility evaluation from laboratory and analytical testing. In: Proceedings of the 9th world conference on earthquake engineering, vol VIII. Tokyo-Kyoto, pp 605–616

Park R (1992) Simulated seismic load tests on reinforced concrete elements and structures. In: Proceedings of the 10th world conference on earthquake engineering, Balkema, Rotterdam, pp 6681–6693. ISBN 9054100605

Park S, Mosalam KM (2012) Parameters for shear strength prediction of exterior beam-column joints without transverse reinforcement. Eng Struct 36:198–209

Paulay T, Priestley MJN (1992) Seismic design of reinforced concrete and masonry buildings. Wiley, New York, pp 250–303

Priestley MJN (1997) Displacement-based seismic assessment of reinforced concrete buildings. J Earthq Eng 1:157–192

Prota A, Nanni A, Manfredi G, Cosenza E (2004) Selective upgrade of under-designed reinforced concrete beam-column joints using carbon fibre- reinforced polymers. ACI Struct J Title n 101 (S69) (2):699–707

Tsonos G (2008) Effectiveness of CFRP-jackets and RC-jackets in post-earthquake and pre-earthquake retrofitting of beam-column sub-assemblages. Eng Struct 30(3):777–793

Chapter 15
Seismic Rehabilitation of Concrete Buildings by Converting Frame Bays into RC Walls

Michael N. Fardis, Antonis Schetakis, and Elias Strepelias

Abstract Infilling certain frame bays with RC is popular as a seismic rehabilitation technique. Unless the connection between the old concrete to the new ensures monolithic behavior, this technique is not covered by codes. To avoid penalizing the foundation of the new wall with a very high moment demand, the new concrete should not be thicker than the old frame members. A cost-effective connection of these members to a thin new web is proposed, with design and detailing conforming to current codes for new structures. For practical reasons footings of added walls are often small and weakly connected to the others; so, they uplift and rock during earthquakes. A model is proposed for uplifting footings in 3D, comprising two pairs of nonlinear-elastic springs in a cross layout. It is applied for nonlinear static or dynamic analyses of three buildings with added walls. Analyses of a prototype, regular 4-storey building show the benefits from uplift to the added walls and certain adverse effects on some columns, as well as the lack of a clear positive effect of adding tie-beams. Applications to a 7- and a 2-storey real building with extreme but typical irregularities exemplify the real-life restrictions in the use of added walls and show their limits for the improvement of seismic performance; certain deficiencies in flexure or shear remain in both buildings and are corrected at very low cost with local FRP jackets without new analysis of the building.

15.1 Introduction

Added concrete walls are very popular for seismic retrofitting of concrete buildings. A simple, cost-effective way of adding them is by infilling with reinforced concrete (RC) some bays of the existing frame, especially at the perimeter. The technique is

---

M.N. Fardis (✉) • A. Schetakis • E. Strepelias
Department of Civil Engineering, Structures Laboratory, University of Patras, P.O. BOX 1424, 26504 Patras, Greece
e-mail: fardis@upatras.gr; schetakisantonis@yahoo.gr; ilstrepelias@upatras.gr

A. Ilki and M.N. Fardis (eds.), *Seismic Evaluation and Rehabilitation of Structures*, Geotechnical, Geological and Earthquake Engineering 26,
DOI 10.1007/978-3-319-00458-7_15,

widely applied, but there are still open issues about the retrofit design and some aspects of the seismic response of the retrofitted building.

A prime question concerns the design and detailing of the connection between the added web and the surrounding frame members. Codes for seismic retrofitting (e.g., Part 3 of Eurocode 8) do not provide guidance, neither specific rules for the modeling and the verification of frame bays converted into RC walls depending on the connection type and its details. The only feasible way for code-compliant design of the retrofitting is to aim at a composite wall that acts as fully monolithic and may be designed as such. For monolithic behavior, the new web is often chosen thicker than the beams and columns around the frame panel being infilled; the new reinforcement is placed outside the old members and detailed as in a new wall and the slabs are pierced for across-floor continuity of the new concrete and vertical bars. This normally gives a new wall much stronger than strictly necessary for the retrofitting of the building. The overstrength aggravates the second open issue for the practical application of added walls, notably their foundation.

Old RC buildings most often have isolated footings, sometimes with tie-beams. To accommodate the large moment resistance of the new wall, its footing is normally large, encapsulating the footings of any columns incorporated in the wall and possibly connected to other footings via tie-beams, depending on accessibility and other practical difficulties. To reduce the intervention to the foundation by limiting the size of new footings and their connection to the rest of the foundation, the moment resistance at the base of the new wall should be kept as small as possible, by having just a new web added between the existing columns without encapsulating them into jackets. However, if, to reduce cost and facilitate access, the footing of such a "thin" added wall is kept small and weakly connected to the other footings, it will uplift and rock during the seismic response, possibly preventing the new wall from reaching its moment resistance at the base. A question is then how uplift of the new footings affects the overall seismic response and performance of the building and the effectiveness of the added walls, and how it may be taken into account in practical retrofit design. Note that the problem is less acute for new buildings, as the designer normally fixes the base of large walls at a deep foundation beam or in a box-type foundation comprising perimeter walls connected at top and bottom by rigid diaphragms throughout the plan; if the superstructure consists of frames and some walls, the designer avoids foundation systems consisting of isolated footings, because their very different rocking casts doubts about the analysis results. So, the question about the effects and modeling of footing uplift pertains mainly to buildings needing rehabilitation.

This Chapter contains a contribution to the cost-effective application of added walls in an engineered, technically sound way, fully compliant to current codes for seismic retrofitting and requiring the normal professional effort for retrofitting projects. Pilot applications to real buildings exemplify real-life restrictions in the application of added walls and show its limits and the supplementary measures needed. A monotonic or cyclic nonlinear spring model is developed and applied to uplifting footings; it is complete in the sense of accounting for uplift of the centroids of the footing and the wall which accompany rotation, yet simple enough to computationally implement for nonlinear static or dynamic seismic analysis in 3D. In view of the reality of footing uplift, to ease the burden on the foundation structurally light added walls are promoted, nowhere thicker than the frame members they are connected to, yet behaving as monolithic with them and amenable to the use of simple, established expressions for the effective stiffness, the moment or shear resistance and the cyclic deformation capacity of monolithic walls (e.g., those in Part 3 of Eurocode 8). This helps the designer avoid the uncertainties associated with attempts to account in an approximate, yet complex, way for the composite nature of RC-infilled frames, including the separation or slippage at the interfaces between the old and the added concrete.

15.2 Frame-Web Connection for Monolithic Behavior

For a composite wall which conforms to codes for concrete structures and may conveniently be taken in the analysis and the verifications as monolithic, each bar of the two-way reinforcement in the new web that terminates at the interface with an existing frame member should be fully anchored into the member past that interface. Figure 15.1a shows an "orthodox" way to achieve this: the web bars (in dashed line) are lap-spliced one-to-one to starter bars with the same diameter (in thin continuous line). Each starter bar is epoxy-grouted into the frame member for the length needed to anchor the full yield force of the web bar lap-spliced to the starter bar. Shorter, larger diameter dowels (depicted in Fig. 15.1a with their thickness) cross the interface to transfer the shear; a single row of dowels, epoxy-grouted at (about) mid-width of the interface and all-along the perimeter of the new web, is normally sufficient. This solution can be designed and detailed to conform to codes for new structures as described below, but the large number of bars to be epoxy-grouted in the surrounding frame members imply high cost and considerable labour.

An "indirect" code-compliant connection, shown in Fig. 15.1b, is promoted in this Chapter as an alternative; the small-diameter starter bars are omitted and the dowels play that role instead. To this end, each dowel extends into the web by at least the lap length of the web bar it anchors into the frame member and is dimensioned for simultaneous tension and (dowel) shear. To satisfy the codes to the letter, the clear distance between the dowel and the web bar it splices may not exceed the maximum the code allows for lap-spliced bars (four bar diameters, or 50 mm, whichever is smallest, according to Eurocode 2). If it exceeds this value, codes often require to increase the lapping by the excess distance. This increased lap length is denoted by $l_{o,d}$ and is the embedment length of the dowel into the new web. With the dowels in a single row at mid-width of the interface, the clear distance to the nearest web bar is minimized if each dowel is placed in-between two web bars on opposite faces of the wall and lap-splices both. This gives as many dowels along each side of the frame panel as the web bars in each curtain, which may be insufficient for the transfer of the web shear force, unless the dowel diameter increases considerably. It may be more practical to have twice as many

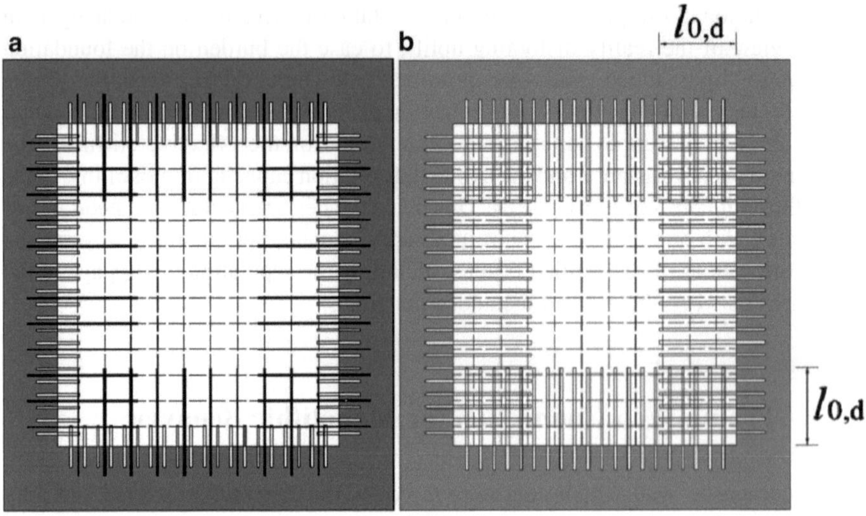

Fig. 15.1 Connection of web to surrounding frame members: (**a**) dowels all-along the interface, plus starter bars anchored into the frame members and lap-spliced to same-diameter horizontal and vertical bars of the new web; (**b**) only dowels all-along the interface, each one doubling as starter bar, lap-spliced to smaller diameter web bars over the lap length of these bars

dowels as the web bars in each curtain, placed at half the on-centre distance of the web bars and at quarter-points along the wall surface between two such bars, as in Fig. 15.1b. Then the number ratio of dowels to spliced web bars is 1:1. Note, though, that the clear distance of spliced bars increases considerably. The dowels should be epoxy-grouted in the existing frame member with at least the minimum embedment length of a dowel (typically 8-times the dowel diameter d_d), or that needed to anchor the full yield force of the web bar(s) which the dowel lap-splices – whichever is largest. Their shear force resistance may be found as follows:

According to (*fib* 2012), the design value of shear resistance of a bar with diameter d_d and design yield stress f_{yd}, working as a dowel per Fig. 15.1a in a concrete with design strength f_{cd}, increases with the slippage at the interface, s, as:

$$F_d(s) \approx F_{0,\max}\sqrt{\frac{s}{s_{\max}}} = 1.6\frac{\pi d_d^2}{4}\sqrt{f_{cd}f_{yd}}\sqrt{\frac{s}{s_{\max}}} \tag{15.1}$$

Its maximum value, $F_{0,\max}$, is attained at a slippage $s_{\max}$ between $0.1d_d$ and $0.2d_d$. If the dowel doubles as a starter bar for n web bars of diameter d_{bw} (the cases $n = 1$ or 2 were discussed above), it develops a tensile stress equal to $\sigma_s = nf_{yd}(d_{bw}/d_d)^2$ in order to anchor the design yield force of these web bars into the frame member. The dowel's maximum shear resistance in Eq. (15.1) is then reduced to (*fib* 2012):

$$F_{d,\max} = F_{0,\max}\sqrt{1-\left(\frac{\sigma_s}{f_{yd}}\right)^2} = F_{0,\max}\sqrt{1-n^2\left(\frac{d_{bw}}{d_d}\right)^4} \tag{15.2}$$

The slippage at the interface associated with almost full development of the dowel shear resistance is sufficiently large to mobilize the shear resistance of each one of the two existing columns at the edge of the new wall, with design value $V_{Rd,c}$, at the time the total design shear capacity of n_d dowels along the top or bottom side of the new web panel is activated. The interface slip, s, needed to resist the shear force demand, $V_{d,i}$, on a horizontal section through the top or bottom of the new web, may be estimated from:

$$V_{d,i} = 2V_{Rd,c} + n_d\left(1.6\frac{\pi d_d^2}{4}\sqrt{f_{cd}f_{yd}}\right)\sqrt{\left(1-n^2\left(\frac{d_{bw}}{d_d}\right)^4\right)\frac{s}{s_{\max}}} \tag{15.3}$$

Dowels should be placed along the interfaces of the new web and the two columns at the same density as along its top and bottom sides.

To determine the design shear force demand on the dowels around the new web, $V_{d,i}$, the designer should establish what is attained first:

1. a plastic hinge at – or near – the connection of the wall to the footing, or
2. the uplift resistance of the footing from the ground, along with plastic hinges at the ends of tie-beams framing into the footing in the wall's strong direction.

For potential no. 1, the cross-section that controls the moment resistance of the wall is first identified:

- For the connection in Fig. 15.1a the base section may be taken as controlling (provided that the embedment of the dowels into the new web is at least what is needed for their dowel action). Its moment resistance, M_{Rwo}, is computed from section analysis, accounting fully for the contribution of all vertical web bars anchored into the footing via starter bars they are lap-spliced to, as well as of the vertical bars of the two old columns at the edges of the new wall.
- For the connection in Fig. 15.1b, the wall section at the tip of the long dowels controls; as it is crossed by the same fully anchored vertical bars as the wall base in the connection in Fig. 15.1a, it has the same moment resistance M_{Rwo}. The moment resistance of the controlling section is transferred to the wall base as a moment equal to $M_{Rwo}/(1 - l_{o,d}/L_{so})$, where $l_{o,d}$ is the embedment length of the long dowels into the new web and L_{so} the wall shear span (moment-to-shear-ratio) at its base under the seismic lateral forces. L_{so} is typically taken equal to $H_w/2$, where H_w is the distance of the base section from the top of the wall in the building.

For potential no. 2 above the maximum moment that can develop around the centre of the footing base before "overturning" starts is:

$$M_0 = 0.5 B_f N_{tot} + \left(1 + \frac{B_f}{L_{tb}}\right) \sum M_{Rtb} \qquad (15.4)$$

with:

- B_f: length of the footing in the strong direction of the new wall;
- N_{tot}: total vertical force transferred to the ground (the axial force at the base of the wall plus the weight of the footing and of any soil borne by it);
- ΣM_{Rtb}: sum of moment resistances of any tie-beams framing into the footing (in opposite faces, for two tie-beams);
- L_{tb}: clear length of the tie-beam between the footings it connects.

This "overturning" moment is transferred to the wall base as a moment equal to $M_0/(1 + h_f/L_{so})$, where h_f is the vertical distance from the base of the wall to the underside of the footing and $L_{so} \sim H_w/2$. Then the maximum moment that can develop at the base section of the wall is:

$$\text{If} \quad \frac{M_{Rwo}}{1 - \frac{l_{o,d}}{L_{so}}} \le \frac{M_0}{1 + \frac{h_f}{L_{so}}} \rightarrow \quad \max\, M_{wo} = \frac{M_{Rwo}}{1 - \frac{l_{o,d}}{L_{so}}} \qquad (15.5a)$$

$$\text{If} \quad \frac{M_{Rwo}}{1 - \frac{l_{o,d}}{L_{so}}} > \frac{M_0}{1 + \frac{h_f}{L_{so}}} \rightarrow \quad \max\, M_{wo} = \frac{M_0}{1 + \frac{h_f}{L_{so}}} \qquad (15.5b)$$

where $l_{o,d}$ is taken as zero for the connection in Fig. 15.1a, or as the length of the long dowels in the web for that in Fig. 15.1b. This value of max M_{wo} is translated into a design shear at storey *i* of the new wall as:

$$V_{d,i} = \frac{\max\, M_{wo}}{L_{s,i}} \qquad (15.6)$$

where $L_{s,i}$ is the wall shear span at the base of storey *i* under seismic loading, commonly taken as $L_{s,i} \sim H_{w,i}/2$, with $H_{w,i}$ the distance of the base of storey *i* from the wall top in the building. If Eq. (15.5a) controls over (15.5b), a plastic hinge is likely to form near the wall base, and the design shear force up the wall from Eq. (15.6) should normally be magnified for higher modes after yielding at the base – e.g., in the way specified in Part 1 of Eurocode 8 for the walls of DC H buildings.

This design approach is applied in (Strepelias et al. 2013) to three 4-storey single-bay 0.75:1 scale frames, Pseudo-dynamically (PsD) tested after infilling with a RC web flash with the lateral sides of the surrounding frame members. It has also been used in (Chrysostomou et al. 2013) for the two added walls of a full-scale 4-storey frame building PsD-tested at the ELSA lab within the SERIES project. The PsD tests of the composite walls in (Strepelias et al. 2013) and of the retrofitted 4-storey building in (Chrysostomou et al. 2013) have shown essentially monolithic, flexure-controlled behavior, with very little slippage or opening at the interfaces between new and old concrete for both connection types in Fig. 15.1.

15.3 Simple Numerical Modeling of Uplifting Footings

A practical way to model soil compliance in 3D under a rigid footing is by connecting the centre of its underside to the soil with three springs: a vertical spring and a rotational one per vertical plane of bending of the vertical element. So long as there is full contact to the soil, expressions for the elastic impedance of these springs can represent the response of the elastic soil continuum and account for the 3D geometry of the footing and its embedment much better than a bed of linear Winkler springs. However, once uplift starts, lumped springs do not capture neither the decrease in rotational stiffness due to reduced contact, nor the upward displacement of the footing centre due to rotation about a horizontal axis not passing through it anymore, nor the larger absolute vertical displacement of the ends of beams framing into the uplifting side of the footing or the wall it supports, compared to those connected to the down-going side, etc. To capture such effects, the three elastic lumped springs are replaced in this work by two pairs of nonlinear vertical ones, each pair near opposite ends of the footing in the two orthogonal vertical planes of bending. The stiffness of each spring in a pair is derived from the rotation, θ, and the vertical displacement at the centre of the footing, δ_o, given in (Crémer et al. 2002) in terms of a monotonically increasing moment M. These functions, fitted by Crémer et al. (2002) to results of nonlinear 2D Finite Element analyses of uplifting strip footings on inelastic soil, represent the physical phenomena better than a bed of no-tension Winkler springs under the footing. The so-derived secant relations between: (a) the force $F = M/B$ in each vertical spring near the edge of the footing and (b) the spring vertical displacement are:

- at the uplifting edge:

$$\delta_1 = \frac{B\theta_o}{2}\left(\frac{\frac{FB}{M_o}}{2-\frac{FB}{M_o}} + \ln\left(2-\frac{FB}{M_o}\right)\right) \tag{15.7a}$$

- at the down-going edge:

$$\delta_2 = -\frac{B\theta_o}{2}\left(1-\ln\left(2-\frac{FB}{M_o}\right)\right) \tag{15.7b}$$

where B is the footing width in the vertical plane of bending, M_o the moment causing uplift to start and $\theta_o = M_o/K_{\theta o}$ the elastic rotation from the elastic rotational impedance $K_{\theta o}$ of a footing in full contact with elastic soil. If N is the vertical load and N_u the bearing capacity of the concentric footing, then (Crémer et al. 2002):

$$M_o \approx 0.25BN\exp\left(-2.5\frac{N}{N_u}\right) \tag{15.8}$$

If the footing's safety factor for vertical loads is high, i.e., $N << N_u$, there is very little hysteresis in cyclic loading and the cyclic M-θ relation is nonlinear-elastic, returning to about zero displacement at zero moment or force and dissipating very little energy. Then Eqs. (15.7) may be applied also in nonlinear dynamic analysis, with the twin springs taken as nonlinear elastic.

In a seismic analysis in 3D, with each footing uplifting and rocking in both horizontal directions, a pair of nonlinear springs per Eqs. (15.7) is used in each direction, in a cross layout with respect to the centre of the footing in plan. Before uplifting, the composite stiffness of the two pairs should reproduce the vertical spring and the two rotational ones in horizontal directions X and Y, K_{zo}, $K_{\theta x,o}$ and $K_{\theta y,o}$, respectively. To this end, the two springs with vertical stiffness $K_{zo}/4$ in the plane through direction X may be placed at a distance from the centre $B_x/2 = \sqrt{(2K_{\theta x,o}/K_{zo})}$; those through direction Y at a distance from the centre $B_y/2 = \sqrt{(2K_{\theta y,o}/K_{zo})}$. These values of B_x and B_y should be used as B in Eqs. (15.7) for the pair. Note that the force F in Eqs. (15.7) pertains to the couple producing the moment as $F = M/B$; therefore, the component $N/4$ due to the vertical load should be added or subtracted from the total force of the vertical spring before entering Eqs. (15.7). The two pairs of vertical springs are uncoupled and employ a constant value of M_o – hence a fixed N-value per Eq. (15.8).

The nonlinear spring model described above has been computationally implemented in the computer code ANSRuop (www.ansruop.net) and used for the nonlinear static (pushover) or dynamic analyses in Sect. 15.4.

15.4 Applications to Three Buildings

15.4.1 Introduction

The first case concerns a very regular prototype building. Its aim is to carry out parametric analyses and draw conclusions of wider applicability. The other cases concern real buildings: one medium-rise and slender, the other low-rise and squat, both with seemingly extreme, yet very typical, irregularities. Their objective is to see and face all the restrictions and difficulties encountered in practice and investigate the limits and true effectiveness of adding walls for seismic rehabilitation. All walls are added at the perimeter, to facilitate intervention to the foundation and enhance the torsional rigidity about the vertical, in order to reduce undesirable twisting. Note that 3D case studies can capture effects and phenomena not reflected in the idealized planar systems with uplifting footings used in the past analytical studies: in the present cases the new walls are much fewer than the existing vertical members; footings are of various sizes and uplift differently; beams which frame into the added walls at right angles oppose their uplift – albeit less than in the case of interior walls with beams framing into all four sides, etc.

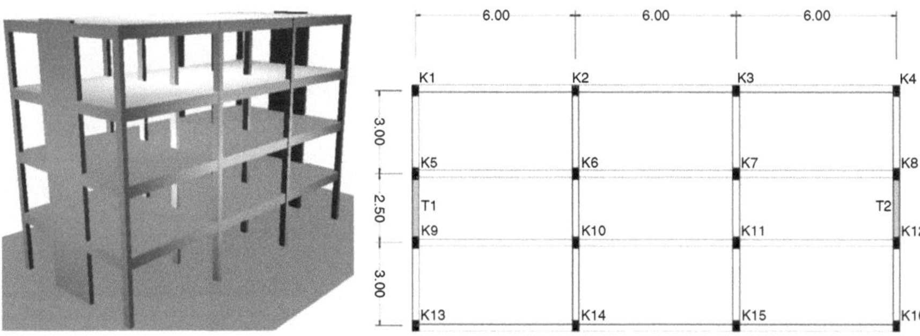

Fig. 15.2 Four-storey prototype frame building with added walls

The nonlinear modeling, analysis and evaluation of the building performance conform with the general guidance and any specific rules in Part 3 of Eurocode 8. Rayleigh damping equal to 5 % of critical at the two lowest modal periods in 3D accounts for sources of damping besides the post-yielding hysteretic energy dissipation in the nonlinear cyclic models of RC members, e.g., for damping in the soil and its interface with the foundation, where the nonlinear elastic springs of Eqs. (15.7) do not dissipate energy. The connection between the new concrete and the old members follows Sect. 15.2 and is presumed to achieve monolithic behavior. So, the expressions for the effective stiffness, the moment or shear resistance and the cyclic deformation capacity of monolithic walls in Part 3 of Eurocode 8 are used to model the composite walls and evaluate their performance.

Nonlinear springs per Sect. 15.3 are placed in a cross layout under each footing. Soil strength and the size of footings provide a safety factor of at least 5 against bearing capacity failure under vertical loads; so, permanent deformation of the soil under the toe of the uplifting footing are small and the soil is not critical.

15.4.2 Prototype 4-Storey Building

The building is shown in Fig. 15.2. The original building is a pure RC frame with poor detailing and little seismic resistance. The central bays of the outer frames in the short direction are infilled with RC flash with the surrounding frame members. In the framework of SERIES, this building, but with only the two outer frames where the walls were added, was PsD-tested at the ELSA lab (Chrysostomou et al. 2013), with fixed base and under a unidirectional ground motion along the new walls conforming to the Type 1 elastic spectrum of Eurocode 8 for Ground Category C. Because in the test building the reinforcement of the added walls at the two opposite sides is asymmetric, so is that in the prototype building.

A model of the building per the second paragraph of Sect. 15.4.1 is subjected here to unidirectional nonlinear analyses: static with heightwise triangular lateral loads (see Fig. 15.3a), or dynamic (Fig. 15.3b–d) under 14 historic records

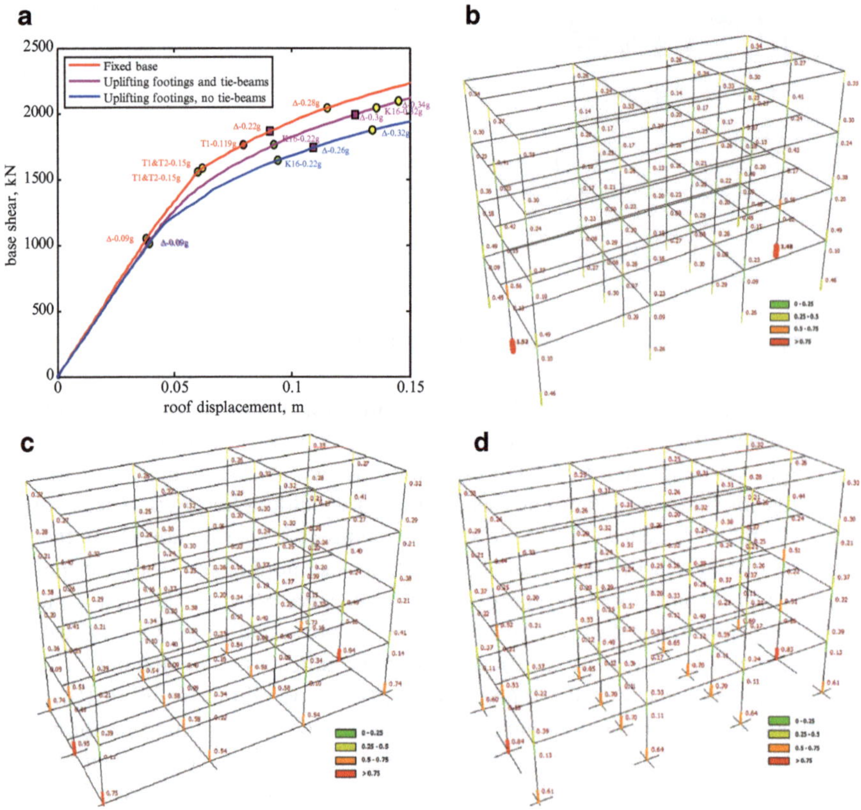

Fig. 15.3 Results of nonlinear analyses of 4-storey prototype building parallel to the walls: (**a**) base shear-roof displacement curves from pushover analysis for fixed base or uplifting footings with tie-beams or without; first attainment of yielding (*green circle*), significant damage (*yellow*), near collapse in flexure (*red*) or shear (*purple square*) per Eurocode 8-Part 3 in any beam (Δ), column (K) or wall (T) and corresponding causative PGA on rock; (**b–d**) average demand-to-limit capacity (DI) per Eurocode 8-Part 3 for the significant damage limit state in vertical members from dynamic analyses under 14 motions conforming to 0.25g-spectrum: (**b**) fixed footings, (**c**) uplifting footings and tie-beams, (**d**) uplifting footings, no tie-beams. Thickness of line at member end proportional to DI; *red*: DI > 0.75, *orange*: 0.75 > DI > 0.5, *green*: DI < 0.25

modulated to individually fit with a 10 % tolerance the Type 1 elastic spectrum of Eurocode 8 on Category C Ground for 0.25g Peak Ground Acceleration (PGA).

The base of all vertical elements was taken either fixed, or on 0.8 m-deep footings on nonlinear springs of the type described in Sect. 15.3. The footings are 1.5 m × 4 m in plan under the two walls, or 1.5 m-square under the columns. They are considered either with two-way 0.25 m × 0.6 m tie-beams, or without any. Figure 15.3a shows the base shear vs. roof displacement curves from pushover analysis for fixed base or for uplifting footings, with or without tie-beams. Points on each curve depict the first yielding in any beam (Δ), column (K) or wall (T) in the building (green circles), the first attainment of the "Significant Damage" limit state in Part 3 of Eurocode 8 per type of member (yellow circles) and the first attainment of the "Near collapse" limit state in flexure (red circles) or shear (purple squares) in any beam, column or wall. Next to each point is given the PGA on rock of a motion conforming to the Ground C Type 1 elastic spectrum in Eurocode 8 that induces that value of roof displacement.

The curves in Fig. 15.3a show that footing uplift does not significantly soften the response but protects the walls from yielding or any other damage, delays shear failure or other damage in the beams (especially if there are tie-beams), but hastens yielding at the base of columns. These conclusions are confirmed by the average nonlinear dynamic analysis results in Fig. 15.3b–d, which show in addition that tie-beams increase the demands on the walls and the outer frame columns, but decrease them on interior ones. Results for beams (not shown in Fig. 15.3) show that footing uplift does not have a marked effect on damage to them.

15.4.3 Real 7-Storey Building

The 7-storey building is shown in Fig. 15.4. The five columns of its façade in storeys 2–5 (at the bottom side in Fig. 15.4a) do not continue to the foundation but are supported at the tip of cantilevering beams. The top floor is set back along the façade and the left-hand side; the second from the top is set back along the façade. Owing to the eccentric location of the few shear walls and the elevator shaft, the eccentricity between the centres of mass and stiffness is large at all storeys. The first torsional mode has longer period than the first translational one in the X-direction; so, the building is torsionally flexible.

Because footings have smaller size in the X-direction than in Z (Fig. 15.4c), their uplift impacts less the capacity curves and seismic performance of the as-built building in Fig. 15.5 for pushover analysis in X (rows 1 and 2 of figures in Fig. 15.5) than in Z (rows 3 and 4). As in Fig. 15.3, uplift is much more beneficial for vertical members (rows 2 and 4 of figures in Fig. 15.5) than for the beams (rows 1 and 3). Notwithstanding any benefits from uplift, members grossly violating the Significant Damage limit state of Part 3 of Eurocode 8 in flexure or shear under the 475-year earthquake with a PGA of 0.25g, as shown by the average results of the top row in Fig. 15.6 from nonlinear dynamic analyses for 14 bidirectional ground motions compatible with the elastic spectrum at the site. So, three walls are added at the only bays on the perimeter where the architectural function of the building is not seriously compromised (Fig. 15.4a, c, d). Witness in Fig. 15.4d the termination of the added wall at the lower-right corner at the 5th floor due to the setbacks at the two upper storeys. The walls added in the weak X-direction enhance significantly the global stiffness (see top two rows of figures in Fig. 15.5), but increase further the static eccentricity between the centres of mass and stiffness. However, the three new walls reduce the first torsional period below both fundamental translational ones. The small and non-systematic impact of uplifting on beam performance (rows 1 and 3 of figures in Fig. 15.5) is retained in the retrofitted building, alongside its

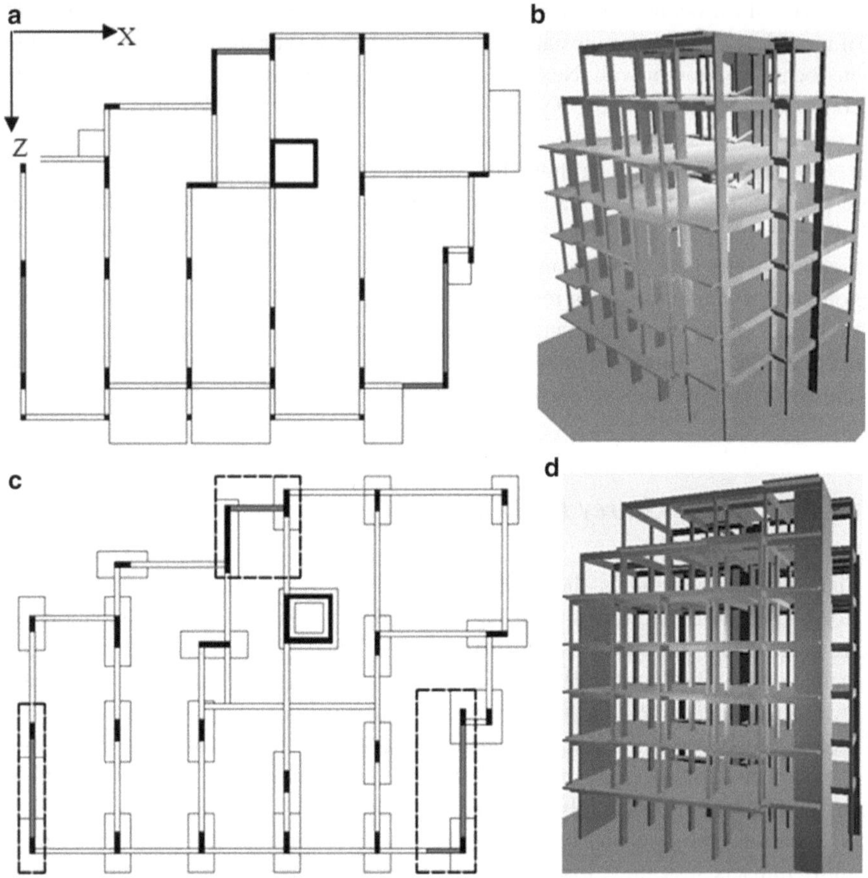

Fig. 15.4 Seven-storey building: (**a**) plan of storeys 2–5 with the three walls added at the perimeter shown hatched; (**b**) axonometric, as-built; (**c**) plan of foundation with walls added shown hatched and new footings with a *dashed outline*; (**d**) axonometric, with three added walls

marked benefit to vertical members (rows 2 and 4). For fixed footings, two of the new walls, the elevator shaft and few columns at the setbacks fail to meet the significant damage limits in flexure of Part 3 of Eurocode 8 under the 0.25g-PGA bidirectional motions; the elevator shaft is also inadequate in shear. If footings uplift, these requirements are met, except in the elevator shaft and few columns at the setbacks. As a matter of fact, the average results of nonlinear dynamic analyses for 14 bidirectional spectrum-compatible ground motions in the two lower rows of figures in Fig. 15.6 show that uplifting spreads moderate-to-significant flexural damage to a larger number of vertical elements in the retrofitted building, while reducing the extent of shear damage; it also decreases the peaks of both flexural and shear damage in the building. Similar effects are found for the beams (not shown in Fig. 15.6).

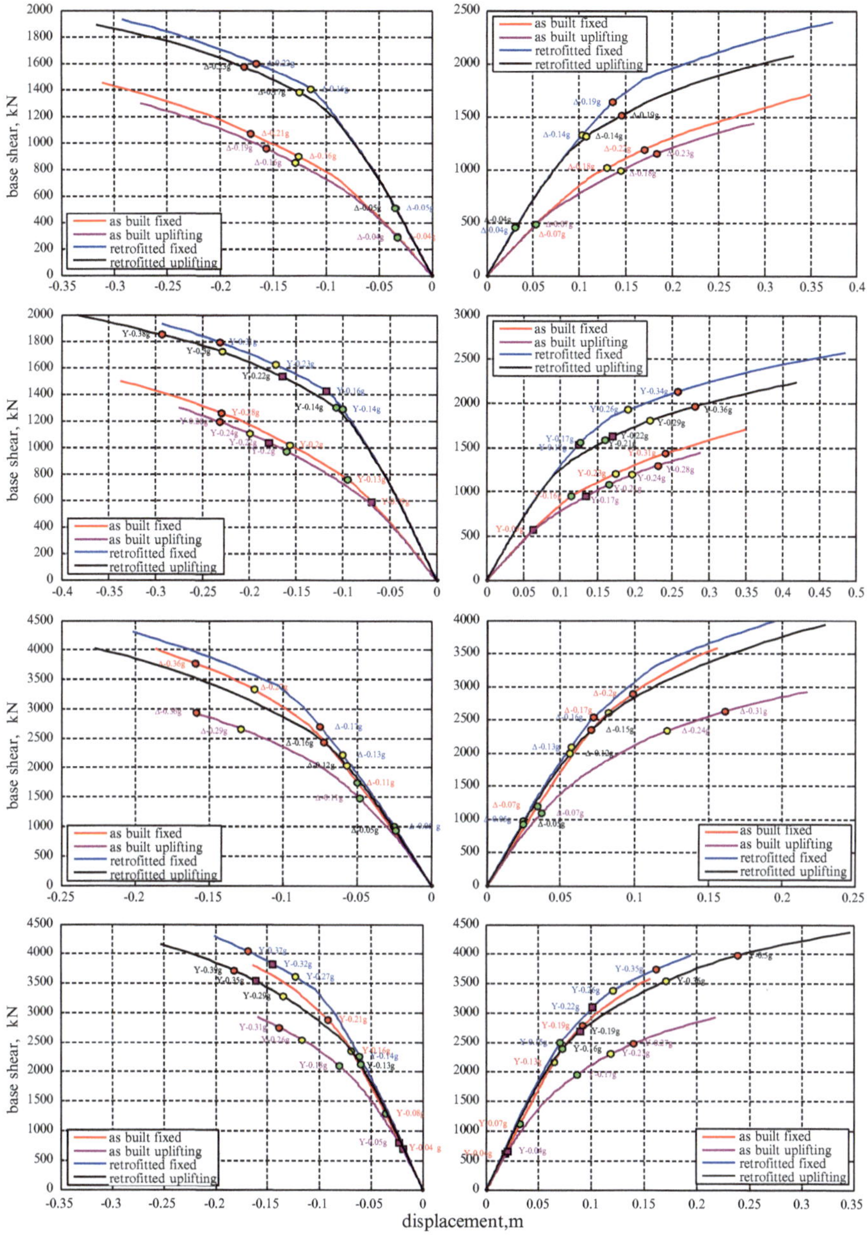

Fig. 15.5 Base shear-roof displacement curves from pushover analysis of 7-storey building in directions ± X (rows 1, 2) or ± Z (rows 3, 4); points show first attainment of yielding (*green circle*), significant damage (*yellow*) or near collapse in flexure (*red*) or shear (*purple square*) per Eurocode 8-Part 3 in beams Δ (rows 1, 3) or vertical elements Y (rows 2, 4) and causative PGA on rock

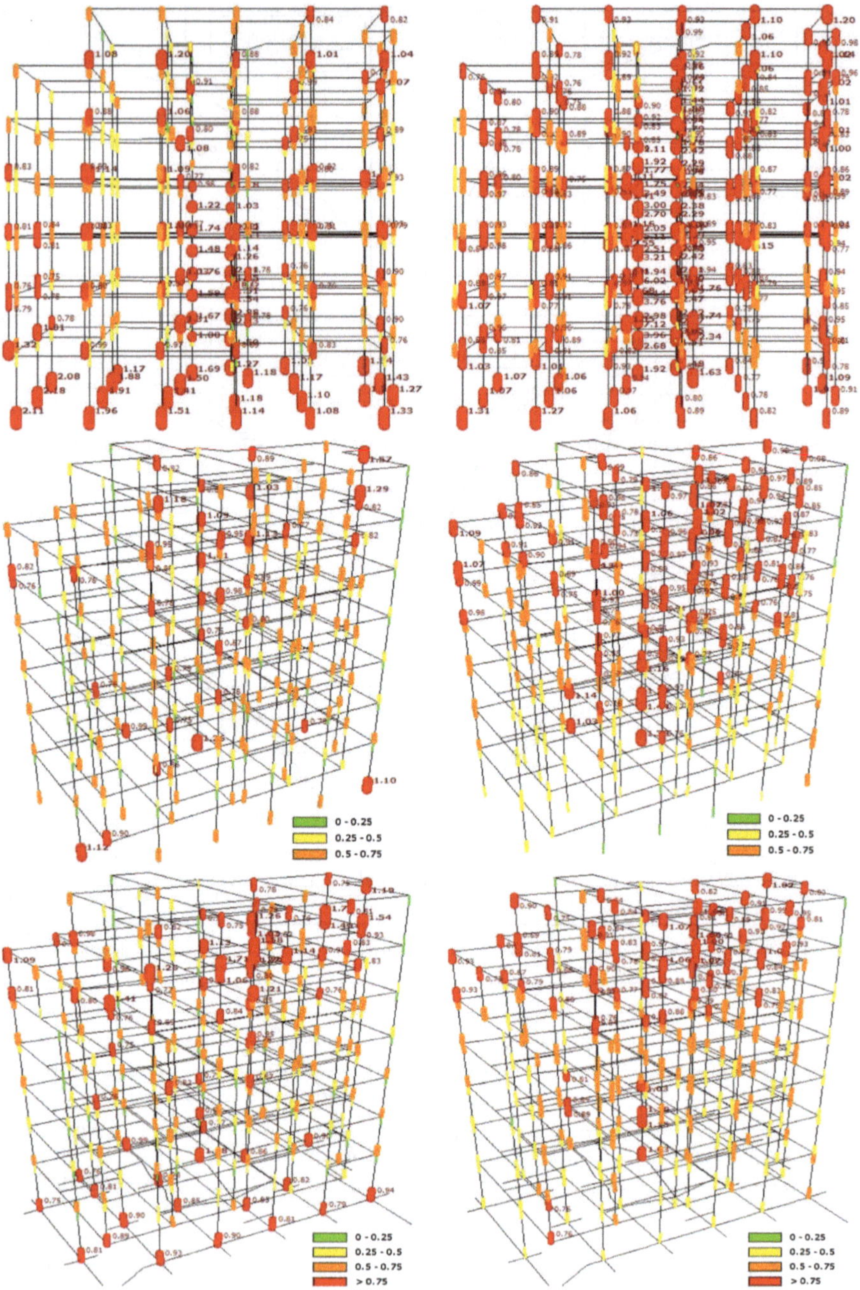

Fig. 15.6 Damage Index (DI): demand-to-limit capacity per EC8-3 for significant damage limit state at vertical member ends in 7-storey building: (*top*) as-built, fixed footings; (*middle*) with added walls, fixed footings; (*bottom*) with added walls, uplifting footings; (*left*) DI of flexural deformations; (*right*) DI in shear. Line thickness at member end proportional to DI; *red*: DI > 0.75, *orange*: 0.75 > DI > 0.5, *yellow*: 0.5 > DI > 0.25, *green*: 0.25 > DI

It is clear from Figs. 15.4, 15.5, and 15.6 that there is only so much that adding walls can do to improve the performance of the 7-storey building. Fortunately, deficiencies in flexure or shear that remain in vertical elements after such retrofitting are limited in spatial extent and severity – demands do not overly exceed the limit capacity per Part 3 of Eurocode 8 for the Significant Damage limit state. Flexural deficiencies can be fixed by wrapping a 0.6 m-length of the member end in question with unidirectional Carbon Fibre Reinforced Polymer (CFRP); those in shear are corrected by adding 100- or 150 mm-wide CFRP strips to the lateral faces of the element, all along its length where the shear resistance falls short. The number of CFRP plies is limited to the minimum required to increase the limit capacity per Part 3 of Eurocode 8 for the Significant Damage limit state to reach the mean demand from the nonlinear dynamic analyses under the 14 bidirectional 0.25g, spectrum-compatible ground motions (i.e., reduce the Damage Index values in the lower two rows of figures in Fig. 15.6 below 1.0). This phase of retrofit design is easy, as the limit capacity in flexure or shear in Part 3 of Eurocode 8 – i.e., the denominator of DI – is an explicit function of the thickness and strength of the CFRP layers; new analysis is not required for the seismic demands (the numerator), as the effective stiffness and the moment resistance of members are not materially affected by the addition of CFRP in the circumferential direction.

If uplifting is neglected, the supplementary intervention with CFRPs is limited to six members at a cost of €1,900; if it is taken into account, the additional fixing concerns 11 members (different from those fixed when uplifting is ignored, except for the elevator shaft) at a cost of about €1,600. The additional cost is very little compared to the three added walls (without their footings), which cost €52,500 for the connection detail in Fig. 15.1a, or about €37,000 for that in Fig. 15.1b. Some deficiencies remain in the beams, mainly in the upper storeys. However, in view of their lower significance for the integrity of the 7-storey building, the costly task of correcting them may not be worthy. In fact, practically their deficiencies disappear if these beams are taken as "secondary seismic" elements, which increases their flexural capacity limits per Part 3 of Eurocode 8 by 50 % and, in addition, they are checked at the Near Collapse limit state, implying a further relaxation by a factor of 4/3 to their ultimate flexural deformation in Eurocode 8.

15.4.4 Real 2-Storey Building

Figure 15.7 shows the 2-storey building (with attic under the sloping roof) of the third case study. Witness the long L-shaped shear wall supporting the staircase. The results of pushover analysis in Fig. 15.8 show that the lateral stiffness and resistance in the X-direction (along the wall, rows 1 and 2 in Fig. 15.8) is several times larger than in the Z-direction (rows 3 and 4); indeed, they well exceed those shown in Fig. 15.6 for the 7-storey building. Due to the very central location of that wall, the torsional period is much longer than the first translational ones in the two main directions in plan, X and Z, suggesting a strong torsional sensitivity.

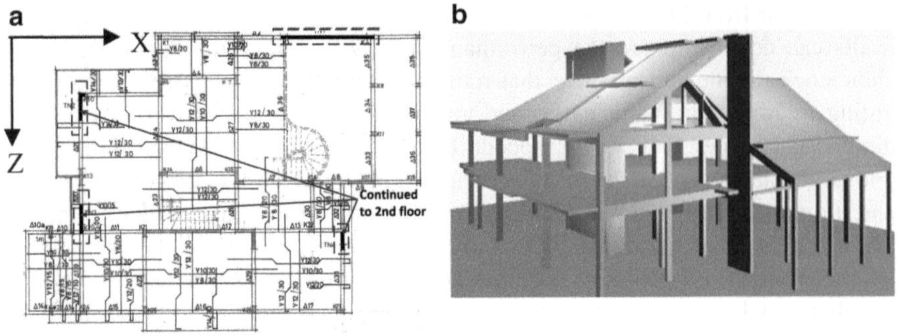

Fig. 15.7 Two-storey building: (**a**) ground storey plan with the four walls added at the only possible locations on the perimeter; (**b**) axonometric, as-built

Because of the small size of all footings in the weak Z-direction, uplifting affects very little the capacity curves and the seismic performance of the as-built structure for pushover analysis in Z (see rows 3 and 4 in Fig. 15.8). By contrast, as the footing of the squat wall is very long in the X-direction, its uplift has a major impact on the capacity curves and the seismic performance of the as-built structure for pushover analysis in that direction (rows 1 and 2 in Fig. 15.8). Figure 15.8 and the top row of figures in Fig. 15.9 suggest that, despite being low-rise and notwithstanding the very large stiffness and resistance imparted to it by the squat wall in the X-direction, the as-built structure fails to meet the requirements of Part 3 of Eurocode 8 for the Significant Damage limit state at several vertical members and few beams under a 475-year earthquake with a PGA of 0.25g.

To improve performance, four walls are added at the only positions on the perimeter where they do not seriously impair architectural function (Fig. 15.7a). The three walls which are parallel to the weak Z-direction have small size and are monolithically connected to a frame column only at one edge. The only one which is in the strong X-direction extends throughout a full frame bay but is limited vertically to the ground floor, as there is no second floor at that point of the perimeter; it aims to increase the torsional stiffness of the system about a vertical axis and balance the eccentricity of the existing squat wall. Both goals are met and the first torsional period drops well below the translational one in Z, but still exceeds that in X, albeit by much less than in the as-built system. In fact, this may be the main benefit from the added walls: as shown in Figs. 15.8 and 15.9, flexural damage is less wide-spread and much more limited in severity, but there is little improvement concerning the "Significant damage" limit state in shear per Part 3 of Eurocode 8 – the force-controlled squat wall near the centre still fails in shear. The performance of some beams (not shown in Fig. 15.9), which is marginal in the as-built structure, is acceptable after adding the walls. The problems remaining in vertical members are corrected by wrapping with CFRP the ends of six columns which are still deficient in flexure and by adding CFRP strips to the lateral faces of the squat central wall. The CFRPs cost about €2,900, much less than the new walls, which cost (without the footings) €12,200 for the connection in Fig. 15.1a or

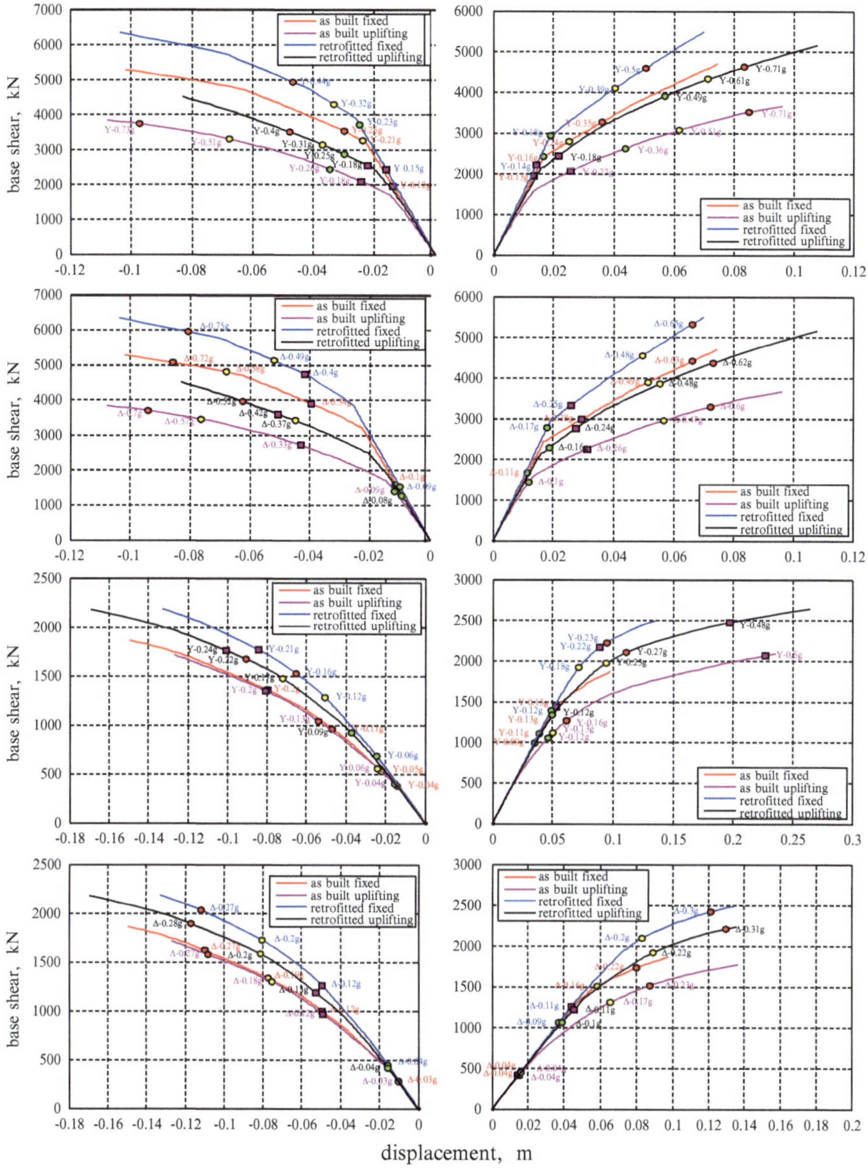

Fig. 15.8 Base shear-roof displacement curves from pushover analysis of 2-storey building along ± X (row 1, 2) or ± Z (row 3, 4); points show first attainment of yielding (*green circle*), significant damage (*yellow*) or near collapse in flexure (*red*) or shear (*purple square*) per Part 3 of Eurocode 8 in vertical elements Y (rows 1, 3) or beams Δ (rows 2, 4) and causative PGA on rock

€9,700 for that in Fig. 15.1b. Considering the difference in size and the limited improvement in performance under the 475-year earthquake, the retrofit cost of the 2-storey building is disproportionately high in comparison to the 7-storey one.

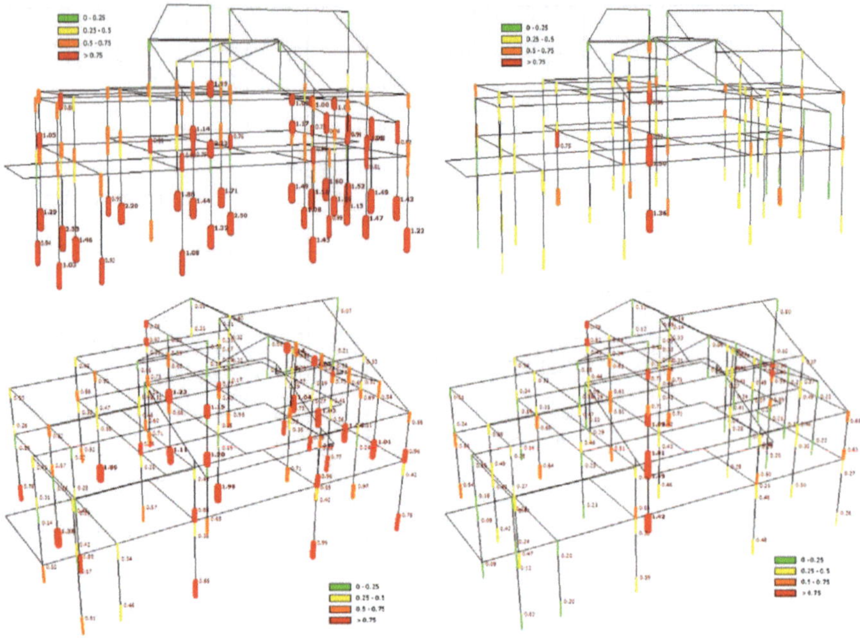

Fig. 15.9 Damage Index (DI): demand-to-limit capacity per EC8-3 for Significant Damage limit state at vertical member ends in 2-storey building on fixed footings from nonlinear dynamic analyses under 14 motions conforming to 0.25g-spectrum: (*top*) as-built; (*bottom*) with added walls; (*left*) DI for flexural deformations; (*right*) DI in shear. Line thickness at member ends proportional to DI; *red*: DI > 0.75, *orange*: 0.75 > DI > 0.5, *yellow*: 0.5 > DI > 0.25, *green*: 0.25 > DI

15.5 Conclusions

This Chapter deals with the addition of RC walls to deficient RC buildings in a way that conforms to codes and is cost-effective and feasible in the context of current retrofit practice. Through pilot applications to real buildings, it also exemplifies real-life restrictions in the use of added walls, shows their limits and suggests supplementary measures. As a main problem of these walls is the cost and difficulty of the construction of their foundation, the focus is on minimizing the size and moment resistance of a new wall, to reduce the demands it places on its footing. This is served by walls consisting of a new RC web which fills a bay between two columns and is not thicker than them or the beams between these columns. A connection of the new web to these members through dowels, which double as anchors of the new web bars into the old members as shown in Fig. 15.1b, may be designed to current codes using Eq. (15.3) and gives savings of 20–30 % of the total cost of the new wall above the footing, compared to the cleaner solution of separate dowels and anchors depicted in Fig. 15.1a. The resulting macroscopically monolithic behavior of the new web with the surrounding frame members greatly simplifies the design and modeling of the composite wall and the evaluation of its seismic performance, as it allows to use for it simple, well-established expressions for the properties (effective stiffness, moment or shear resistance and cyclic deformation capacity) of monolithic walls, such as those in Part 3 of Eurocode 8.

Existing buildings commonly have isolated footings; the foundation of the new wall is normally a footing encapsulating those of the columns incorporated in the wall and possibly connected to other footings via tie-beams. Owing to space or accessibility limitations and other practical difficulties, the footing of the added wall is often small and weakly connected to the others – especially as the new walls are normally at the perimeter. So, it will uplift during the response, possibly before the base of the wall reaches its moment resistance. The footings of a building with a few, often dissimilar, walls and many possibly different columns, have very different sizes, vertical loads and seismic moment demands and exhibit very different rocking response. So, the distribution of seismic demands and damage between the various elements of the superstructure may differ radically from that in a building with fixed footings. These important effects may easily be taken into account in nonlinear analysis, static or dynamic, by using under each footing two pairs of nonlinear springs in a cross-like layout, each spring following the nonlinear-elastic law of Eqs. (15.7) and (15.8), reflecting in good approximation not only partial uplift of the footing, but also moderate nonlinearities in the soil.

The application of the footing uplift model to a regular 4-storey building with four parallel 3-bay frames and two outer bays infilled with RC into walls, has quantified the clear benefit of uplift to the added walls at the detriment to some columns (but not to the beams) and the lack of a clear positive effect of tie-beams, or even a certain negative one. Applications to a slender, 7-storey real building and to a 2-storey squat one, both with seemingly extreme, yet very typical, irregularities, have demonstrated the difficulty to find places on the perimeter for effective new walls which do not severely compromise architectural function. They have shown that added walls are not a panacea: some deficiencies in flexure or shear do remain and require supplementary interventions. Local FRP jackets are ideal against remaining flexural deficiencies at member ends while sets of FRP straps may be wrapped around still shear-deficient members: besides being cost-effective, they may readily be dimensioned to upgrade member capacity to the level of seismic demand without new analysis of the retrofitted building, as circumferential FRPs do not change the member effective stiffness or moment resistance. Some aspects of the two real cases have shown the inefficacy of seismic retrofitting techniques if the layout of the as-built structure does not favour seismic resistance.

Acknowledgement This work has received funding from the Research Promotion Foundation (RPF) of Cyprus for Project ASTI/0308(BIE): "Seismic vulnerability and strengthening of existing privately owned buildings" of RPF's FP 2008.

References

Chrysostomou CZ, Kyriakides N, Poljanšek M, Taucer F, Molina FJ (2013) RC infilling of existing RC structures for seismic retrofitting, Chapter 17. In: Ilki A, Fardis MN (eds) Seismic evaluation and rehabilitation of structures. Springer, Dordrecht

Crémer C, Pecker A, Davenne L (2002) Modelling of nonlinear dynamic behaviour of a shallow strip foundation with macro-element. J Earthq Eng 6(2):175–212

fib (2012) Model Code 2010 – Final draft, vol 1: Bulletin 65, vol 2: Bulletin 66. fédération internationale du béton (fib), Lausanne

Strepelias E, Palios X, Bousias SN, Fardis MN (2013) Experimental investigation of concrete frames infilled with RC for seismic rehabilitation. ASCE J Struct Eng. doi:1061/(ASCE) ST.1943-541X.0000817

Chapter 16
Pseudo-Dynamic Tests of 4-Storey Non-Ductile Frames with RC Infilling of the Bay

Elias Strepelias, Xenophon Palios, Stathis N. Bousias, and Michael N. Fardis

Abstract Three 4-storey reinforced concrete (RC) frames at a scale of 3:4 were converted into slender RC walls by filling the bay between the columns with an RC web of the same thickness, which were pseudo-dynamically tested. The specimens were presumed to belong to a non-ductile frame building, the rest of the building been substructured in the test as elastic. Two code-conforming designs were applied to connect the web to the surrounding frame members. Behavior and failure were dominated by flexure, but U-shaped FRP jackets were necessary at the two edges of the composite wall to avoid premature failure near the base due to poor detailing of columns. Slippage or separation at interfaces between the web and the surrounding frame members were minor for both connection details. In one specimen the critical plastic hinge did not form at the base, but in a horizontal band coinciding with the 1st-storey beam of the frame, where the vertical beam stirrups are much weaker than the vertical wall reinforcement; however, this change in failure mode did not adversely affect the global deformation capacity.

16.1 Introduction

It is popular in seismic retrofitting of concrete buildings to add reinforced concrete (RC) walls, which reduce, thanks to their stiffness, the storey drifts and hence structural and non-structural seismic damage. A simple and cost-effective way of adding walls is by infilling with RC strategic bays of the frame, especially at the perimeter of the building. If the new wall takes up a full frame bay, it can incorporate the beams as well as both columns, at the location normally taken up by the wall boundary elements. The web is totally new and should be fastened to the

---

E. Strepelias (✉) • X. Palios • S.N. Bousias • M.N. Fardis
Department of Civil Engineering, Structures Laboratory, University of Patras,
P.O. BOX 1424, 26504 Patras, Greece
e-mail: ilstrepelias@upatras.gr; xenpal@upatras.gr; sbousias@upatras.gr; fardis@upatras.gr

A. Ilki and M.N. Fardis (eds.), *Seismic Evaluation and Rehabilitation of Structures*, Geotechnical, Geological and Earthquake Engineering 26,
DOI 10.1007/978-3-319-00458-7_16, © Springer International Publishing Switzerland 2014

existing beams and columns so as to transfer to them the full web shear and the tensile capacity of the two-way web reinforcement. Poor detailing of the connection between the old members and the new web may lead to premature failure and low ductility. Moreover, if the connection between the existing and the new parts is not integral, the behavior is uncertain; modelling and verification of the wall as a single, integral element is then of questionable reliability. For integral behavior the wall is often built thick enough to encapsulate the existing beams and columns; its reinforcement is placed outside these members and detailed as in a new wall; epoxy-grouted dowels are placed throughout the interface of the old and the new concrete. Even when it does not encapsulate the beams, the new wall may encapsulate the columns in order to confine them, especially if they have deficient lap splices of vertical bars. Full encapsulation and reliance on new reinforcement placed and detailed as in a new wall allows to model and verify the resulting composite wall as a monolithic new member. However, such a composite wall may end up much stronger than needed for upgrading the building as a whole, or may create difficulties at the foundation. To reduce this overstrength and to avoid piercing through the slabs for continuity of the vertical bars across floors, the new web may be chosen not thicker than the beams and columns of the frame. In such a scheme the shear connection of the web to the surrounding frame members may be the critical link. Even if this connection is very good, integral behavior of the old and the new concrete cannot be taken for granted and there is still some uncertainty about the magnitude of the force- and moment-resistance of the composite system and its deformation capacity.

Although the common field practice is to have the new wall encapsulate the frame members, past work on frames converted into walls by infilling with RC has been limited to specimens with a new web (much) thinner than the surrounding frame members – possibly owing to difficulties in testing walls of large resistance. Past testing reviewed below has been carried out in Japan, Turkey and the USA.

Highlighted in (JCI 2007) are cyclic tests of 27 single-storey, single-bay specimens at a scale of 1:3 to 1:4 and of monolithic companions. The thickness of the RC infill was from 25 to 60 % of the width of the frame members. Failure was always brittle and due to shear, sometimes with sliding at the interface. In 17 specimens the infill was connected to the surrounding frame via epoxy-grouted dowels; compared to a monolithic counterpart, their shear resistance was on average 13 % lower and their ultimate deformation 55 % higher. In six specimens the connection was through shear keys or ungrouted dowels, giving on average a shear resistance 20 % less than in an integral wall and an ultimate deformation 115 % higher. However, in absolute terms the ultimate deformation was very low, because failure was always in shear, before flexural yielding. Four other specimens already had a thin web, whose thickness was increased by 100 % or 150 % without direct connection to the frame members; on average their shear resistance was 8 % less and their ultimate deformation 13 % less than in a monolithic specimen.

Anil and Altin (2007) tested one-storey, one-bay specimens at 1:2 or 1:3 scale with an infill wall thickness equal to 25 % to one-third of that of the frame members. Altin et al. (1992, 2008), Kara and Altin (2006), Sonuvar et al. (2004) tested two-storey, one-bay specimens at 1:3 scale with infill wall thickness of one-third or 40 % that of the frame members. Epoxy-grouted dowels extending well into the RC infill were mostly used for the connection, but a design procedure was not presented; failure was in shear or a combination of shear and flexure before flexural yielding, at deformations normally less – sometimes much less – than expected or inferred from monolithic specimens. Specimens without positive connection of the web to the frame members gave a much lower force resistance than companions with a positive connection. Two-storey, 3-bay, 1:3 scale frames were tested by Canbay et al. (2003) and Erdem et al. (2006), with the middle bay infilled with an RC web of thickness equal to 63 % of the width of the frame members; long epoxy-grouted dowels were used for the connection. Despite the rather low shear span ratio of these walls, the failure mode was dominated by flexure. The ultimate deformation was less, but not much less, than expected from a monolithic wall specimen.

Experimental work in the USA has focused on infilling with precast or masonry units. Klingner and Bertero (1978) and Bertero and Brokken (1983) tested cyclically several 1:3 scale, one-bay, 3-storey frames with various infill materials, often different in the three storeys: solid clay bricks, hollow cement or clay blocks (with the hollows filled with mortar), full-thickness normal-weight or lightweight concrete, etc. The infills had two-way reinforcement lapped with dowels epoxy-grouted in the surrounding frame, or welded-wire fabric anchored into it. Despite this reinforcement, the specimens behaved like frames infilled with unreinforced masonry: the infills separated from the frame, a diagonal compression strut formed in most of them and the frame members failed in a combined bending and shear mode. Kahn and Hanson (1979) tested a one-storey, one-bay 1:2 scale specimen with a cast-in-place wall which was half as thick as the columns. Shear sliding occurred in a layer of drypack between the top of the web and the beam. Compared to a companion, not exactly similar, monolithic specimen, the wall had about 25 % less shear resistance and deformation capacity. Frosch et al. (1996) tested a 2:3 scale, two-storey, one-bay frame infilled with precast panels much thinner than the columns and the beams. Short steel pipes embedded in the surrounding frame members crossed the interface into the space between adjacent panels, to transfer the shear as dowels and allow the vertical bars of the infill (placed in the grouted strips between adjacent panels) to be continuous from storey-to-storey and the horizontal ones to be epoxy-grouted into the columns. The weak link were the 20-bar-diameter lap splices of the frame columns at floor levels; to provide, in spite of them, a reliable moment resistance of the composite wall and control its value so that flexure governed instead of shear, vertical external tendons on either side of the old columns were post-tensioned between the footing and the specimen top. Diagonal cracks formed in the webs, but the interfaces with the frame members or between adjacent precast panels appeared to work as monolithic. For low post-tensioning, the tendons yielded after the column splices failed and a flexural crack opened wide at the base; for high post-tensioning, diagonal compression failure of the infill took place. In either case, peak resistance did not drop by more than 20 % during the load cycling, so the low peak drifts reached cannot be taken as ultimate values. Finally, in the 1:4 scale one-storey, one-bay specimens tested by Pulido et al. (2004) on a shaking table, the wall had thickness one-half of the width of the frame members. It was connected to the surrounding frame through long dowels along three sides; a gap left to the beam soffit was crucial for the non-integral behavior and the failure: the leeward column separated from the wall and the column failed in shear, but the wall was almost inert.

For practical reasons most past specimens had low aspect ratio, not typical of multi-storey, slender walls used in practice. As the behavior and failure mode even of monolithic squat walls is normally governed by shear, most past RC-infilled frame specimens had inherently low ductility, ultimate deformation and energy dissipation capacity. The small thickness of the RC infill in past tests with respect to the width of frame members further penalized shear resistance of the composite wall and the web-frame connection.

The present paper studies experimentally the seismic behavior of multistorey, nonductile RC frames, infilled with an RC web which has the same thickness as the frame members and is connected to them in a code-conforming way. The aim is to achieve a ductile, macroscopically monolithic behavior, which would allow to simplify the design, the modeling and the verifications of the composite new member by considering it as a monolithic, flexure-controlled wall. The composite walls tested have aspect ratio (slenderness) over 3.0. So, they represent better the multistorey, slender, flexure-critical walls of current practice than the test specimens of the past and do not suffer from low shear resistance due to a thin web.

16.2 Test Specimens

Three single-bay RC-infilled frame specimens were tested. Each one was a 3:4 scale model of the central bay of a 3-bay, 4-storey frame infilled with RC into a wall (see Fig. 16.1a). The 3-bay frame represents in turn the outer frame in a 4-storey building with four similar parallel frames at a spacing of 6 m from each other (4.5 m in the scaled building). The central bay of each outer frame is infilled with RC, to resemble as closely as possible a monolithic RC wall.

The 4-storey frame specimens which represent the central bay were cast first in a horizontal position, with their foundation beam anchored to the strong wall of the laboratory. Member sections and reinforcement are shown in Fig. 16.1b, c. Stirrups were smooth with 90°-hooks and 10 mm cover. Column bars were lap-spliced at each floor over 34-bar-diameters. Each frame was then infilled with an RC web as thick as the width of the frame members. Measured values of material strengths are given in Table 16.1.

Table 16.2 presents the two-way reinforcement of the new webs (different in each specimen and, in general, in each floor) and the resulting wall cyclic shear resistance at the 1st storey per CEN (2005). The 8 mm bars at 260 mm centers used in all storeys of specimen no. 3, in the three upper ones of specimen no. 2 and the

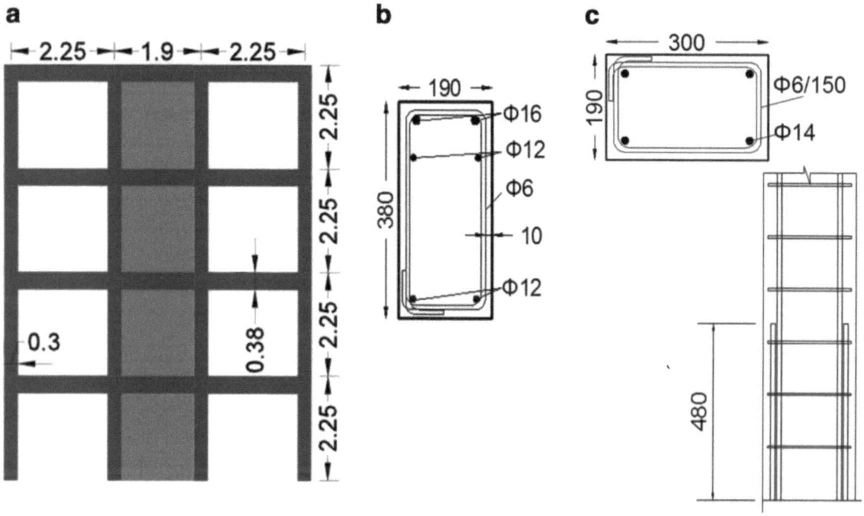

Fig. 16.1 Geometry and reinforcement of nonductile frame specimen: (**a**) overall dimensions (in m) of 3-bay frame whose central bay is infilled with RC; (**b**) beam section and reinforcement, with the contribution of the parallel slab bars in an effective slab width lumped in the top reinforcement; (**c**) column section and reinforcement, including lap-splices at floor levels (in mm)

Table 16.1 Material strengths (MPa)

	Wall specimen no. 1			Wall specimen no. 2			Wall specimen no. 3		
	Frame	New web	Dowels	Frame	New web	Dowels	Frame	New web	Dowels
f_c	27.9	27.7		24.4	26.2		25.6	23.6	
f_y	Ø6[a]: 283	Ø 8: 545	Ø14: 551	Ø6[a]: 283	Ø8: 552	Ø14: 551	Ø6[a]: 283	Ø8: 571	Ø12: 533
	Ø12: 542	Ø10: 504	Ø16: 540	Ø12: 542		Ø16: 567	Ø12: 542		Ø14: 504
	Ø14: 551			Ø14: 557			Ø14: 507		Ø16: 567
	Ø16: 523			Ø16: 523			Ø16: 523		

[a]Smooth bars

two upper storeys of specimen no. 1, give the minimum steel ratio of 0.2 % specified in present-day seismic design codes for the web of new walls, but violate the maximum bar spacing in these codes.

Two options were tried for a monolithic connection between the new web and the frame members (Fardis et al. 2014):

(a) tensile anchorage of the web reinforcement into the surrounding frame and separate dowels for shear transfer at the interface, or
(b) only dowels serving both ends.

Table 16.2 Web reinforcement and dowels for the connection to the frame (in mm) and nominal shear strength they provide at storey 1 (in kN)

	Wall specimen no. 1		Wall specimen no. 2		Wall specimen no. 3	
	Two-way web reinforcement (each face)	Dowels (single row)	Two-way web reinforcement (each face)	Dowels (single row)	Two-way web reinforcement (each face)	Dowels (single row)
Storey 4	Ø8/260	Ø14/130	Ø8/260	Ø16/580	Ø8/260	Ø16/580
Storey 3	Ø8/260	Ø16/130	Ø8/260	Ø16/580	Ø8/260	Ø16/580
Storey 2	Ø8/170	Ø16/85	Ø8/260	Ø16/130	Ø8/260	Ø12/130
Storey 1	Ø10/185	Ø16/92.5	Ø8/170	Ø16/85	Ø8/260	Ø14/130
Shear strength	926	828	724	860	511	405

Option (a) is an "orthodox" solution, strictly following the letter of design codes (CEN 2004), but is labour-intensive because many bars are epoxy-grouted into the frame members:

- Every bar in each curtain of two-way reinforcement of the new web which terminates at the interface with a frame member is fully anchored into that member. To this end, the bar is lap-spliced with a same-diameter starter bar, which is in turn epoxy-grouted into the frame member over the embedment length needed for the development of the bar's full yield force. Lap splices are dimensioned per (CEN 2004).
- A single row of shorter, larger diameter bars are epoxy-grouted at mid-width of the interface and all-along the perimeter of the new web, to transfer the shear across the interface as dowels – shear lugs. Note that two rows of dowels may be difficult to place in a narrow web and frame members and may violate the minimum lateral cover a dowel needs for its full shear resistance. These bars extend on each side of the interface over the minimum embedment length of a dowel: 8-times the diameter. As the crack width at peak response is expected (and confirmed) to be large over most of the critical horizontal interface, shear friction was not relied upon; the dowels were dimensioned per (*fib* 2012) for the full shear, without roughening the interface. After all, there is an incompatibility of the displacements which accompany the peak resistance of dowel action and shear friction: the resistance of dowels increases with interface slippage, reaching its peak value at a slippage of 10–20 % the dowel diameter; by contrast, shear friction disappears at large cyclic displacements.

Option (b) is an "indirect" connection: the small-diameter starter bars are omitted; the dowels play the role of starter bars by extending into the web by at least the lap length of the web bar they intend to anchor into the frame member. The dowel bar is dimensioned in this case per (*fib* 2012) for simultaneous tension (as an anchor) and shear (as a dowel); it is epoxy-grouted into the frame member for 8-times its diameter, or for the length needed to anchor the full yield force of the web bar(s)

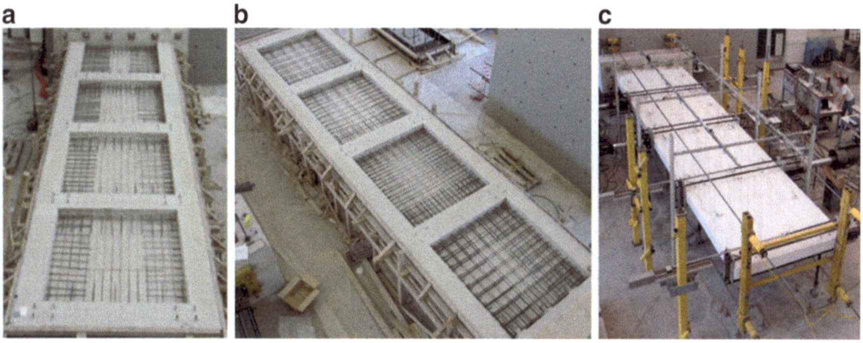

Fig. 16.2 Specimen no. 1: (**a**) frame with long starter bars and short dowels per option (a) along the top and one lateral side of each panel and the long dowels per option (**b**) along the other two; (b) two-way reinforcement placed in the new web; (**c**) specimen in testing position with actuators and sensors attached

which the dowel bar splices – whichever is largest. By placing one dowel in-between two web bars on opposite faces of the wall and lap-splicing it to both at the same time, the clear distance between lap-spliced bars (the web bar and the dowel) is minimized. However, the resulting number of dowels may not be enough to transfer the web shear force, unless their diameter increases considerably. So, twice as many dowels as the web bars per curtain were placed: at mid-width of the frame member and at quarter-points between two web bars. As an exception, the two upper storeys of specimens no. 2 and 3 had only two dowels per side, but of larger diameter.

Table 16.2 lists the dowels connecting the web to the frame in each specimen and the nominal shear strength they provide at the base section per (*fib* 2012), including the shear capacities of the columns per (CEN 2004). In specimen no. 1 the two connection options were compared by using option (a) (long starter bars and short dowels) at the top and one lateral side of each storey panel and option (b) (long dowels alone) at the bottom and the opposite lateral side (Fig. 16.2a). The dowels in the two options have the same diameter and density: their lap-splicing to the small diameter web bars in option (b) penalises very little their shear capacity; their length increases to the lap length of the web bars, as this is augmented per (CEN 2004) owing to the larger bar spacing in each pair of lapped bars.

As shown in Fig. 16.2c, the walls were tested in a horizontal position, supported at each storey by a pair of vertical steel studs between the strong floor and the bottom face of the specimen, with swivels at each end to permit unimpeded horizontal displacement of the wall.

16.3 Testing

The specimens were subjected to pseudo-dynamic testing under the 15 s-long Y component of the Herzegnovi record in the 1979 Montenegro earthquake, scaled to a peak ground acceleration (PGA) of 0.25g and modulated as in Fig. 16.3a to

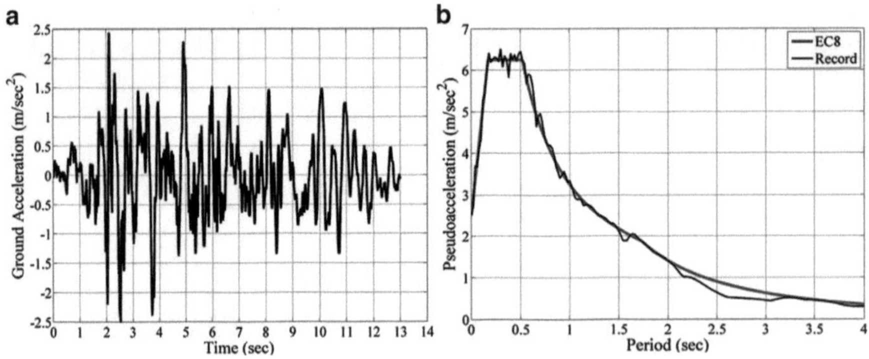

Fig. 16.3 Ground motion applied in pseudo-dynamic tests: (**a**) acceleration time-history after Y component of Herzegnovi record in 1979 Montenegro earthquake; (**b**) 5 %-damped elastic spectrum compared to target Type 1 spectrum in Eurocode 8 for ground type C (*factor* $\sqrt{0.75}$ *applied on the time- and period axes to account for the specimen scale*)

conform to the Eurocode 8, 5 %-damped elastic spectrum of Type 1 for ground C, as demonstrated in Fig. 16.3b. A factor of $\sqrt{0.75} = 0.866$ was applied on the time-axis to account for the specimen scale according to the Cauchy and the Froude similitude laws. Notwithstanding the good overall matching of the target spectrum, the frequency content varies during the record duration: between 8.5 and 9.5 s the record in Fig. 16.3a exhibits a clear acceleration pulse with a frequency of 1.1 Hz (0.9 s period). As we will see later on, this has interesting implications for the pseudodynamic response of the test structures.

The tests aim to simulate the response of the full retrofitted building, including the two unretrofitted inner 3-bay frames. In the scaled building each floor had a mass of 2 × 43,375 kg. As only one wall of the building was physically tested, the simulation included the non-tested part of half the building which is tributary to that wall – corresponding to 45 m^2 of floor area in the scaled building or 80 m^2 in the prototype. The non-tested part was numerically simulated as an elastic substructure (hybrid test).

The substructure includes one interior frame and the two outer columns and beams of the exterior frame to which the tested wall belongs. Its 4 × 4 elastic stiffness matrix relates the floor forces applied on the wall to the floor displacements. It was established from (one-half) the difference in the total column shears below and above a floor and the (difference in) floor displacements, from elastic analysis of the full scaled building under lateral floor forces with inverted triangular distribution. This analysis considered the floor diaphragms as rigid and used as elastic rigidity of a member the secant-to-yield-point stiffness of its shear span (: moment-to-shear ratio at the end section).

For technical reasons, only two out of four degrees of freedom (DoFs) were included in the pseudo-dynamic tests: those of floors 2 and 4. The other two DoFs were statically condensed. One-half of the tributary mass of 43,375 kg of each one of the 1st and 3rd floors was lumped to the nearest DoF.

An axial load of 465 kN was computed at the wall base from the analysis of the scaled building for gravity loads. As shown in Fig. 16.2c, it was applied concentrically to all four storeys, using four jacks near the four corners of the specimen, which shared the same pressure and acted against vertical post-tensioned rods connected to the strong wall next to the specimen footing. The fluctuation of the axial load during the test was monitored; it was not so large as to have an effect on the behavior.

A network of 96 sensors at selected locations monitored:

- the slippage and the separation at the interfaces of the web and the frame members, via sets of three potentiometers in triangular layout, each set at different position along the interface;
- the deformations of the diagonals of the web panels;
- the deformation along six consecutive zones on the outer faces of the 1st storey columns and at the base of the 2nd storey ones;
- strains in the dowels which connect the 1st storey web panel to the foundation beam; and
- the horizontal displacements of each floor.

16.4 Experimental Results

16.4.1 Wall Specimen No. 1

Specimen no. 1 was subjected first to two quasi-static cycles to a peak displacement at the top of 60 or 70 mm, under lateral forces at all four floor levels with inverted triangular distribution. Hysteresis loops of the global response are shown in Fig. 16.4.

Two major flexural cracks opened: one at the base section, the other across the ends of the long dowels connecting the web to the footing, Fig. 16.5a. In addition, there was heavy damage of the edge regions where the outermost bars of the frame columns were lap-spliced to starter bars from the footing over a length of 34 bar-diameters, as shown in Fig. 16.5a–c. This caused an instant drop of the base moment, evident in Fig. 16.4a at a top drift ratio of ± 0.4 %. All damage was repaired after the test. The two major flexural cracks were epoxy-grouted; any spalled or loose concrete at the lap-splice region at each edge was replaced with epoxy mortar; corners were chamfered to a 30 mm radius for a height of 500 mm from the base and two plies of Carbon Fibre Reinforced Polymer (CFRP) were applied over that height (Fig. 16.5d). Each end of these open CFRP jackets extended 100 mm past the inner corner of the corresponding frame column and was epoxy-bonded to the surface of the new web. In the pseudo-dynamic test of the same specimen that followed, the U-shaped CFRP jackets – including their anchorage – were very effective in preventing or minimizing the apparent damage of the two edge regions at the wall base. So, specimens no. 2 and 3 had similar jackets

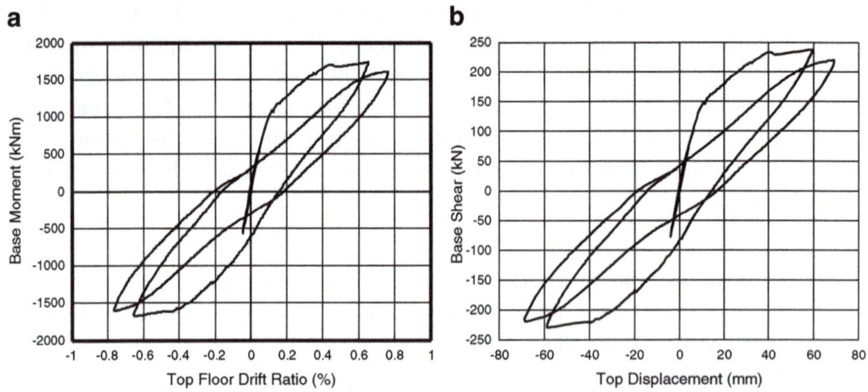

Fig. 16.4 Specimen no. 1 in cyclic loading: (**a**) base moment vs. top drift ratio; (**b**) base shear vs. top displacement

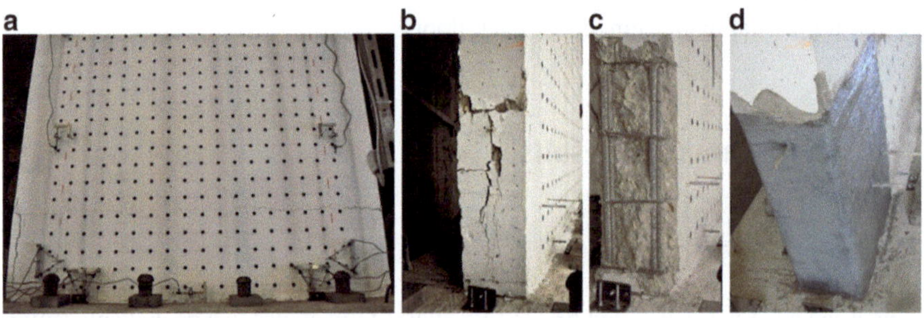

Fig. 16.5 Damage to specimen no. 1 after quasi-static cyclic test: (**a**) flexural crack across the ends of the long dowels and splitting crack along the lap-splice at the *left edge*; (**b**) damaged lap-splice region at one edge of the wall; (**c**) lap-spliced bars without the spalled or loose cover; (**d**) CFRP jacket around wall edge (*specimen shown rotated, as if it were vertical*)

applied to these regions from the outset. This type of jacket is strongly recommended for practice if the base of the existing frame columns has poor detailing (short lapping at floor level; sparse stirrups with 90-deg hooks, etc.) and is not incorporated into the thickness of the new wall. Unless the upper floors have enough moment resistance at the base to preclude plastic hinging there, CFRP jackets should be added to them as well.

In the pseudo-dynamic test all repaired cracks re-opened, but the one at the base section became dominant. At the time when the drift ratio at the top of the wall was 1.0 % (Fig. 16.6), the base section reached its ultimate deformation in flexure (conventionally identified with a 20 % drop in the moment resistance of that section compared to its peak prior value). At that point the ultimate chord rotation of the shear span of the base section (: moment-to-shear ratio, measured during the test to be 6.75 m on average) was 10.5 mrad, the crack at the base had opened to 2.5 mm and at the bottom of the 2nd storey to 1 mm.

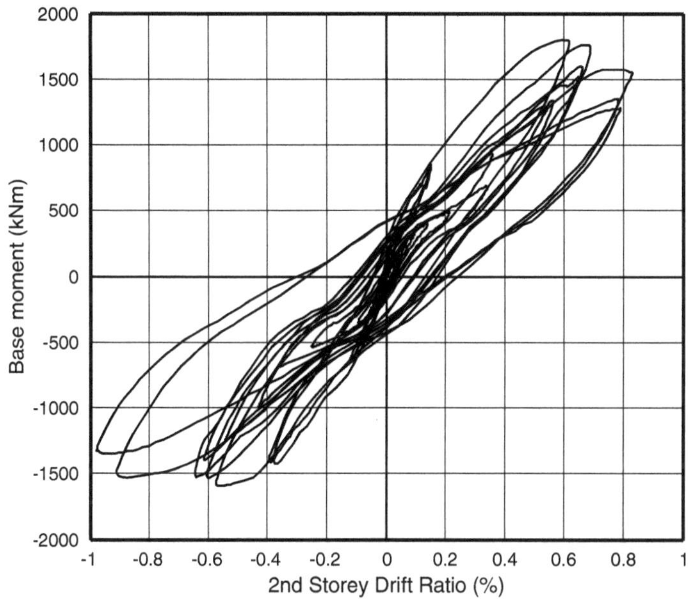

Fig. 16.6 Response of specimen no. 1 in pseudo-dynamic test: base moment vs. top drift ratio

Slippage reached 1.2 mm along the base section, but did not exceed 0.2–0.3 mm along any other horizontal or vertical interfaces at the 1st or 2nd storey, except at mid-length of a vertical interface at the 2nd storey where values as high as 0.9 mm were measured (Fig. 16.7). Similar or even higher slippage was locally measured at horizontal or vertical interfaces of the 3rd and 4th storeys, but not all-along them. In these upper storeys, measured slippage was larger along the horizontal interfaces where the connection followed option (b) than along the opposite side with the detail of option (a). By contrast, the difference between opposite vertical interfaces was not systematic. Large slippage values were sporadic and unsystematic, but also essentially one-sided, as they did not reverse to significant values. Taking all these observations into account, it was concluded that large interface slippage was not an important feature of the behavior, even wherever the connection option (b) was used. So, this simpler detail was exclusively used along all four sides of each storey panel in specimens no. 2 and 3.

Despite having reached ultimate conditions, specimen no. 1 sustained with little further strength degradation another quasi-static test of six cycles of lateral forces with inverted triangular distribution over all four floors (Fig. 16.8). Old cracks re-opened despite been epoxy-grouted, but the major one at the base monopolized the deformations: when the top drift reached 1.2 %, its width exceeded 7 mm at one edge and 4 mm at the opposite. Interface slippage reached 3 mm at the base, but was modest at all other interfaces.

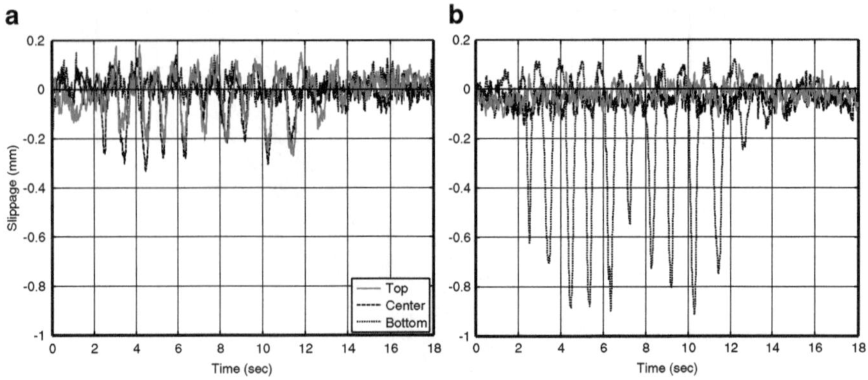

Fig. 16.7 Slippage measured along north side column-web interface of specimen no. 1 at: (**a**) 1st storey; and (**b**) 2nd storey

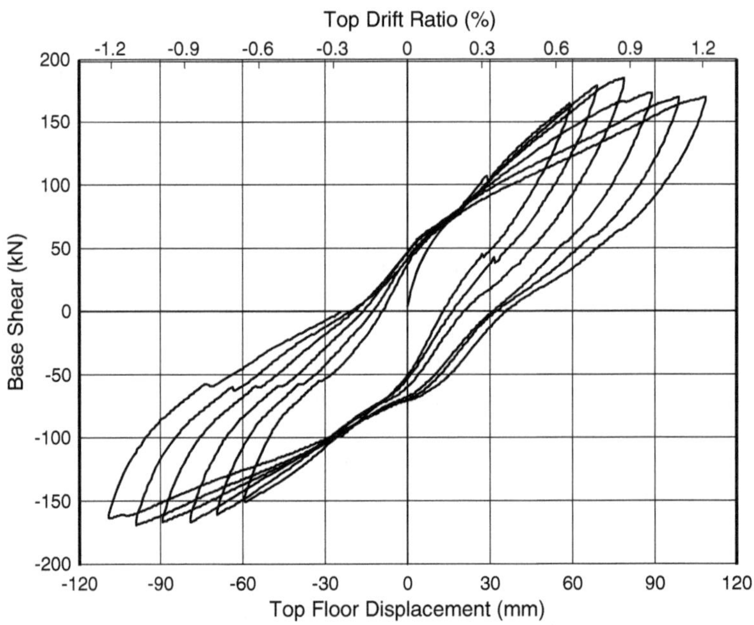

Fig. 16.8 Wall specimen no. 1: base shear-top displacement response in cyclic test to failure

16.4.2 Wall Specimen No. 2

Specimen no. 2 was tested pseudo-dynamically as virgin, with the U-shaped CFRP jacket applied from the outset around its two edges to a height of 500 mm from the base. Cracking of the 1st storey started at essentially the same locations as in specimen no. 1, but was followed by damage of the – unretrofitted – lap-splice

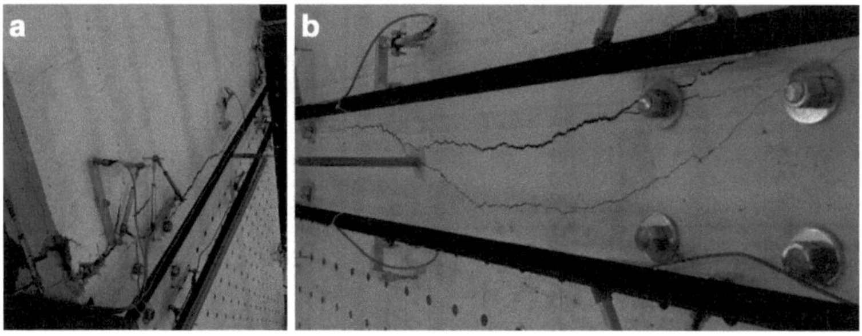

Fig. 16.9 Damage to specimen no. 2 during pseudo-dynamic test: (**a**) damaged lap-splice region at the base of the 2nd storey extending into horizontal flexural damage along the 1st storey beam; (**b**) close-up of flexural damage in the 1st storey beam

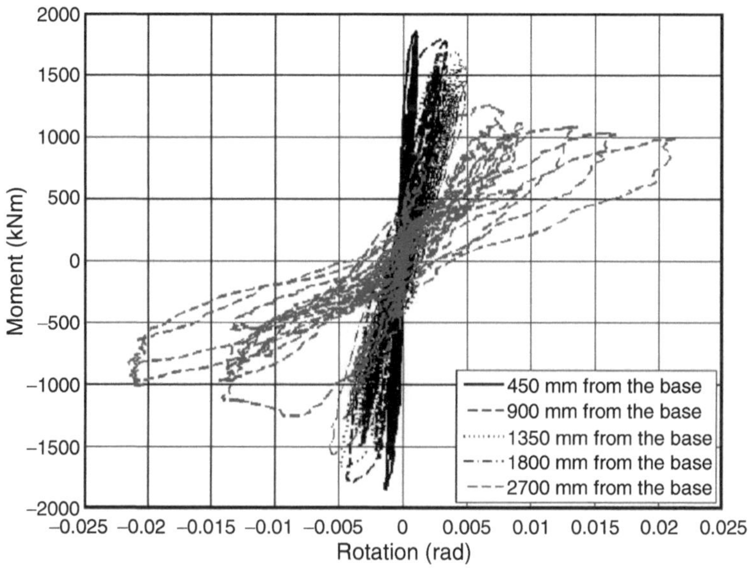

Fig. 16.10 Bending moment vs. rotation with respect to the base in pseudo-dynamic test of specimen no. 2

regions at the base of the 2nd storey (Fig. 16.9a), which instantaneously propagated downwards into the 1st storey beam and horizontally over almost the full length of the beam between the columns (Fig. 16.9b). From then on, all flexural deformations of the wall concentrated in a horizontal crack-band at about mid-depth of the 1st storey beam, as depicted in Fig. 16.10 showing the applied moment vs. the respective rotation at several sections along-height (ranging from 450 to 2,700 mm from the base section, 2,250 mm being the distance of the beam top from the wall base).

Peak crack width reached 7 mm at some point along the beam; slippage there exceeded 6 mm (Fig. 16.11a); at the interface with the base of a 2nd storey column

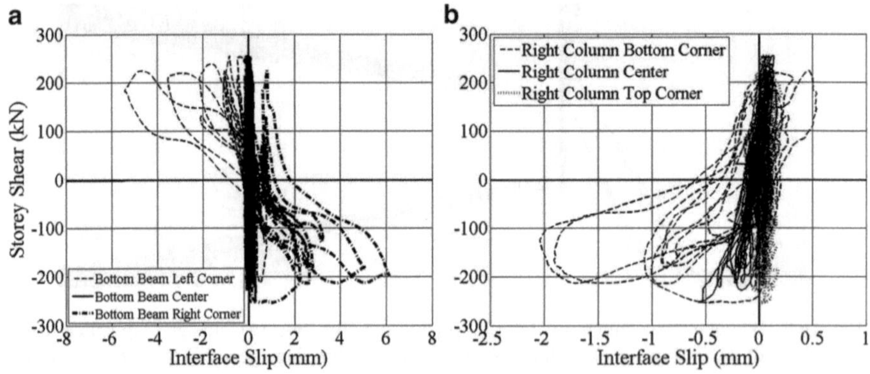

Fig. 16.11 Storey shear vs. interface slip in pseudo-dynamic test of specimen no. 2: (**a**) at the interface of the web of the 2nd storey to the beam between the 1st and 2nd stories; (**b**) at the interface between the web and a column in the 2nd storey

it reached 2 mm (Fig. 16.11b). By contrast, at the base of the wall the peak crack width was about 0.9 mm, peak slippage about 0.2 mm and dowel strains were well below yielding. When the top drift ratio reached 1.25 %, the chord rotation of the wall shear span at the horizontal level through the 1st storey beam mid-depth (: moment-to-shear ratio, measured during the test to be 5 m on average) reached an ultimate value of 8 mrad (identified by a 20 % drop in the moment resistance of that section compared to its peak value ever). Apart from the failure zone at the 1st storey beam (Fig. 16.11a) and the adjacent region at the vertical interface of the 2nd storey column (Fig. 16.11b), interface slippage did not exceed 0.2–0.3 mm anywhere in the specimen, despite the use of connection option (b) throughout.

16.4.3 Wall Specimen No. 3

The pseudodynamic test of this specimen was interrupted at about 5.5 s. The damaged specimen was subjected to a full-duration pseudo-dynamic test, exhibiting softer response than specimens no. 1 or 2.

Wall no. 3 had less web reinforcement than specimen no. 2 and much less than no. 1 (see Table 16.2). Nevertheless, its overall behavior in the pseudo-dynamic test was very similar to that of specimen no. 1: two major flexural cracks opened early on, one at the base section and the other across the ends of the long dowels connecting the web to the footing, with the one at the base soon becoming dominant (Fig. 16.12).

Like in specimen no. 1 and in contrast to no. 2, the flexural deformation of the wall was evenly spread, as demonstrated by the plots of moment vs. rotation-with-respect-to-the-base shown in Fig. 16.13 for several levels up to the bottom of the 2nd storey. When the drift ratio at the top reached 1.2 %, the wall base section

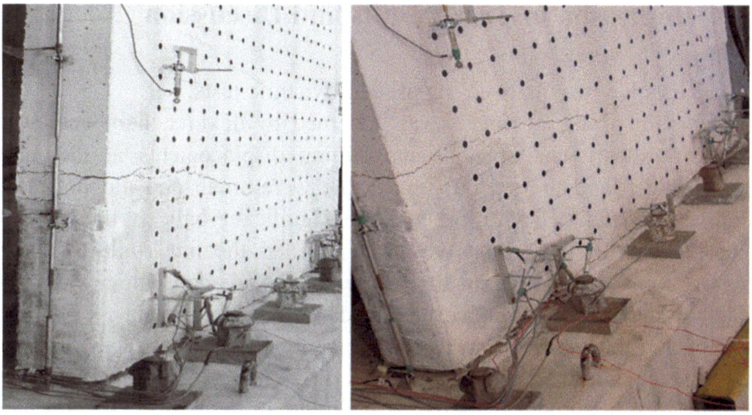

Fig. 16.12 Specimen no. 3: major flexural crack at the base and slightly inclined crack above the ends of the lap-splices in the edge columns and above the ends of the long dowels connecting the web to the base

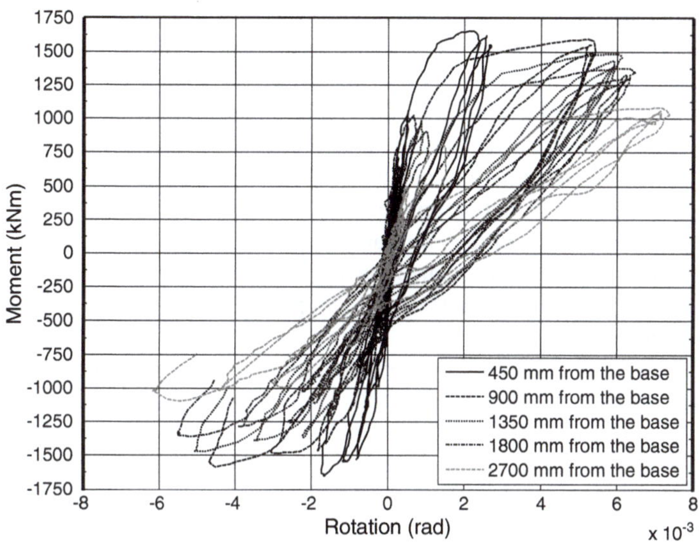

Fig. 16.13 Bending moment vs. rotation with respect to the base in first pseudo-dynamic test of specimen no. 3

attained its ultimate deformation (i.e., its moment resistance had dropped by 20 % with respect to its peak value ever); at that point the ultimate chord rotation of the shear span of the base section (: moment-to-shear ratio, measured during the test to be around 6.75 m on average) was 9.2 mrad. The peak crack width at the base was 12 mm at one end and 9 mm at the opposite; at the other interfaces it was normally below 0.1 mm and did not exceed 0.3 mm. Slippage reached 4 mm at the base, but was less than 0.1–0.2 mm over any other horizontal or vertical interface.

16.5 Comparison of Test Results and Discussion

Table 16.3 lists key results from the pseudo-dynamic tests.

The behavior of all specimens was governed by flexure; their webs were not distressed by shear. The opening and sliding displacements at the interfaces between the web and the surrounding frame were small and did not affect the global response and behavior. They were significant only at those horizontal interfaces where the governing crack formed and the final failure took place, which is a common feature in all flexure-controlled monolithic walls and not particular to the composite ones of this study. All specimens stayed uncracked in the two upper storeys, apart from hairline separation at the interfaces. Figure 16.14 compares the cracking and damage patterns of the three specimens.

The full time-histories of floor displacements in these tests are depicted and compared in Fig. 16.15. Witness the resurgence of the displacements of specimens no. 1 and 2 between 9 and 11 s of the response, due to the acceleration pulse in the record pointed out at the first paragraph of Sect. 16.3. The two specimens had at that time an effective period of around 0.9 s and experienced resonant response. Owing to its smaller amounts of web steel and dowels and to a certain damage it sustained during the discontinued prior test, specimen no. 3 had a longer effective period at that time and the resurgence of its response was more subdued.

Although wall specimen no. 2 had web reinforcement intermediate between the two other specimens, it had radically different behavior and failure mode. Practically all its flexural deformations and the final failure took place within a horizontal band coinciding with its 1st storey beam. That zone was damage-free in the two other specimens: their flexural deformations, damage and final failure took place in the vicinity of the base, at the connection to the footing and near the level where the long dowels connecting it to the web stopped. This difference in behavior is elaborated in the following.

Within the depth of the beam, the vertical reinforcement distributed between the compression and the tension chords of the wall comprises just the legs of the beam stirrups. By contrast, at and beyond the interface of a beam with the web above and below, that role is played by the (stronger and heavier) dowels. At a horizontal section of the web outside the length of the dowels, the vertical reinforcement

Table 16.3 Summary of pseudo-dynamic test results

Wall	Cracking moment, M_{cr}, kNm	Moment at yielding of the base, M_y, kNm	Peak base shear, kN	Peak moment at mid-depth of 1st storey beam, kNm	1st storey drift ratio at base yielding, $\delta_{y,i}$, %	Top drift ratio at base yielding, $\delta_{y,t}$, %	Peak drift ratio of critical storey, $\delta_{u,i}$, %	Peak top drift ratio, $\delta_{u,t}$, %
No 1	1,100	1,650	1,300	310	0.26	0.40	0.96[a]	1.05
No 2	1,050	1,700	1,412	344	0.26	0.39	1.52[b]	1.28
No 3	1,050	1,400	1,225	255	0.20	0.29	0.79[a]	0.90

[a]At 1st floor
[b]At 2nd floor

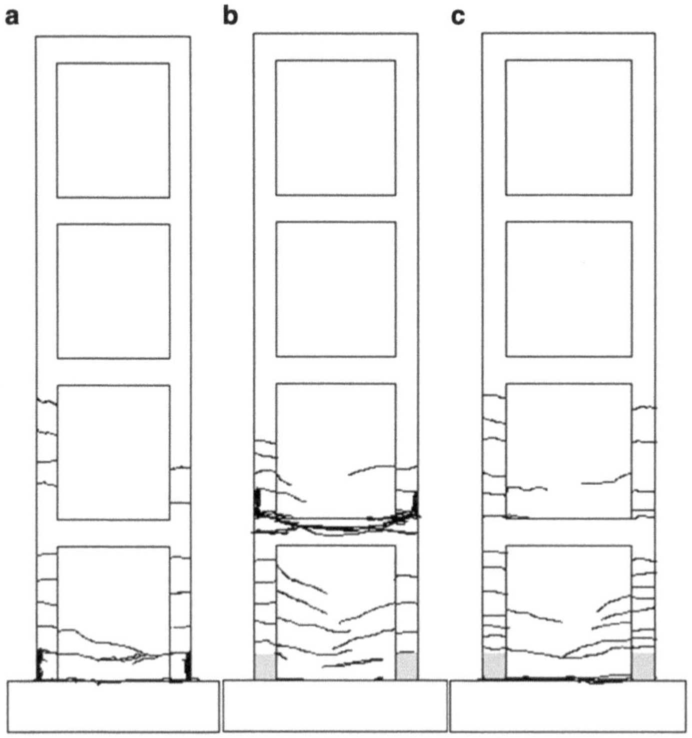

Fig. 16.14 Major cracks and damage in specimen: (**a**) no. 1; (**b**) no. 2; (**c**) no. 3

between the tension and compression chords consists of the web bars – which have larger yield stress and cross-sectional area than the beam stirrups. Owing to its light distributed reinforcement, a horizontal section of the wall within a horizontal band coinciding with its 1st storey beam yields in bending almost right after it cracks and soon reaches its ultimate moment resistance, which is not much higher than the yield moment, because the plain bars of the beam stirrups have little tensile resistance. By contrast, the heavier distributed vertical reinforcement of all horizontal wall sections above and below the beam depth have much larger margin between cracking and yielding and more pronounced strain-hardening afterwards. So, once the tension chord of a section through the 1st storey beam yields right after it cracks, the beam stirrups – being of lower grade – yield too, almost all-along the beam. This happens despite the fact that the base region of the wall may crack first under its larger response moment and its tension chord may yield earlier than in a section through the 1st storey beam. So, further flexural deformations take place at the level of the 1st storey beam and ultimate conditions are reached there, in lieu of the base region.

The deformations measured and the moments derived at various levels of the wall from the measured storey forces show that, in specimens no. 1 and 3, at the time the wall yielded near the base, the band of the wall within the depth of the 1st

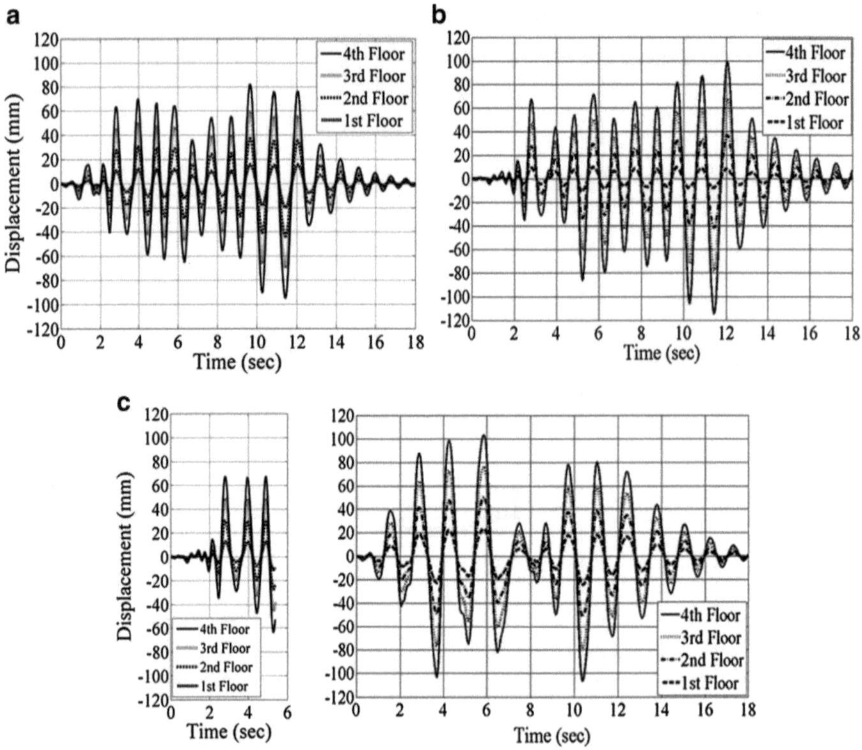

Fig. 16.15 Time-histories of floor displacements in pseudo-dynamic test of specimen: (**a**) no. 1; (**b**) no. 2; (**c**) no. 3 (first, interrupted test and full-duration test of damaged specimen)

storey beam was still uncracked and stayed so throughout the test. The horizontal interfaces of that beam and the web above and below, albeit much stronger than a horizontal section through the beam, apparently did not transfer to the zone between them a moment above the cracking moment (listed in Table 16.3); the tensile forces in the bars and dowels of the web which are anchored in the beam past these horizontal interfaces apparently did not make up for the lack of concrete tensile strength across the interface. So, inelasticity was concentrated in the base region of these two specimens. By contrast, at about the instant the base region of specimen no. 2 started to yield, the zone of the wall within the depth of its 1st storey beam cracked in flexure, yielded immediately and experienced a major excursion into strain-hardening. The deformation measurements show that this happened at 5 s into the response (witness the change in the proportion between the 2nd and 1st storey displacements at and after 5 s, compared to the previous history of the same specimen and the full response history of specimens no. 1 and 3). From that point on, the horizontal zone within the depth of the 1st storey beam kept accumulating inelastic deformations, acting as a fuse for the base region, shielding it from moments or rotations any higher than those at the instant the 1st storey beam band cracked. Note that the flexural failure of the 1st storey beam region in lieu of the base did not impair much the overall deformation capacity of specimen no. 2: a comparison with the test results of specimens no. 1 and 3 shows that, had failure of specimen no. 2 not occurred where it did, it would have taken place anyway at the base, maybe at a lower top drift but under higher lateral forces compared to specimen no. 1.

The failure mode of specimen no. 2 cannot be precluded in a real structure merely by the continuation of the frame beams into adjacent spans: that specimen shows that the horizontal crack band at about a beam's mid-depth may wind around the joints and into the end regions of the frame columns above or below, instead of continuing straight into the adjacent spans. There are various options against this failure mode in beams with stirrups of a lower tensile capacity per unit cross-sectional area of the wall than the vertical bars of the new web, or the tensile strength of the frame concrete:

1. the dowels are made continuous through the beam depth, from the web above to the one below, or
2. the individual dowels driven separately from the beam top and soffit (to avoid drilling continuous vertical holes that stay clear of top and bottom beam bars) overlap sufficiently across the beam, or
3. the lack of continuity of non-overlapping ends of individual dowels is bridged by applying to the lateral surfaces of the beam vertical FRP strips anchored well above and below those ends.

16.6 Conclusions

RC-infilling of selected frame bays throughout the height of RC buildings is very appealing for seismic rehabilitation, provided that the resulting composite walls have a flexure-controlled, ductile mode of behavior and failure and may be treated in analysis and design as monolithic. In this paper, a wall web with thickness not larger than the width of the frame beams or columns was used, without RC jackets added around the boundary columns, in order to limit the moment resistance of the composite wall and the demand it places on its foundation. Economic ways were sought to connect the web to the frame members, which can achieve nearly-monolithic behavior and are covered by common dimensioning and detailing rules for new RC construction.

Three 4-storey, one-bay frame specimens were infilled with RC; in one of them two options were employed to connect the RC web to the surrounding members. The walls were considered as part of a seismic retrofitting of a non-ductile frame building and were tested pseudo-dynamically for a 0.25g seismic excitation of the entire building, with the rest of it substructured as elastic. Some conclusions are:

- The response and failure of all specimens was flexure-controlled, without shear distress of the web nor marked opening or sliding displacements at the interface of the web and the beams and columns.
- A significant reduction in the web reinforcement from specimen no. 1 to no. 3 did not have a marked effect on behavior and performance.
- A simple connection, with fewer long and large-diameter bars transferring from the web to the surrounding frame members the shear (as dowels) on one hand and the tension of the web bars (as anchors) on the other, was found to be as effective as a connection comprising (a) many long starter bars transferring the tension of the web bars and (b) short dowels for the shear.
- Premature failure of the existing edge columns due to poor detailing and lap splicing at floor level can be prevented through open, U-shaped FRP jackets (with as few as two FRP plies) at the two edges of the composite wall, in lieu of RC jackets converting the existing columns to ductile boundary elements.
- If separate vertical dowels are driven into the beam top from the storey above and its soffit from the storey below without their ends overlapping vertically, a beam with sparse or low grade stirrups may be the weak link instead of the wall base: flexural deformations may concentrate in a horizontal band within the beam depth, and eventual failure may occur there. Although this problem did not have as dramatic an impact on the three wall specimens as might be expected, in practice it should be avoided via proper detailing or continuity of the vertical dowels through the beam depth.
- With the connection between the web and the surrounding frame members employed in the three specimens, the composite wall may be taken in modeling and verifications as monolithic.

Acknowledgments The research leading to these results has received funding from the European Community's 7th Framework Program [FP7/2007-13] under grant agreement n°204697 and from the Greek Earthquake Planning and Protection Organization (EPPO). The assistance of graduate student Nicos Stathas in part of the testing campaign is acknowledged.

References

Altin S, Ersoy U, Tankut T (1992) Hysteretic response of reinforced-concrete infilled frames. ASCE J Struct Eng 118(8):2133–2150

Altin S, Anil O, Kara E (2008) Strengthening of RC nonductile frames with RC infills: an experimental study. Cem Concr Compos 30:612–621

Anil O, Altin S (2007) An experimental study on reinforced concrete partially infilled frames. Eng Struct 29:449–460

Bertero VV, Brokken S (1983) Infills in seismic resistant building. ASCE J Struct Div 109(6): 1337–1360

Canbay E, Ersoy U, Ozcebe G (2003) Contribution of reinforced concrete infills to seismic behaviour of structural systems. ACI Struct J 100(5):637–643

CEN (2004) European Standard EN 1992-1-1: 2004 Eurocode 2: design of concrete structures, Part 1–1: general rules and rules for buildings. European Committee of Standardization, Brussels

CEN (2005) European Standard EN 1998–3:2005 Eurocode 8: design of structures for earthquake resistance, Part 3: assessment and retrofitting of buildings. European Committee of Standardization, Brussels

Erdem I, Akyuz U, Ersoy U, Ozcebe G (2006) An experimental study of two different strengthening techniques for RC frames. Eng Struct 28:1843–1851

Fardis MN, Schetakis A, Strepelias E (2014) Seismic rehabilitation of concrete buildings by converting frame bays into RC walls, Chapter 15. In: Ilki A, Fardis MN (eds) Seismic evaluation and rehabilitation of structures. Springer, Dordrecht, pp 261–280

fib (2012) Model Code 2010 – Final draft, *fib* Bull 65 and 66. Federation Internationale du Beton, Lausanne

Frosch R, Li W, Jirsa J, Kreger M (1996) Retrofit of non-ductile moment resisting frames using precast infill wall panels. Earthq Spectra 12(4):741–760

JCI (2007) Seismic rehabilitation of concrete structures. In: Sugano S (ed) International Publication Series IPS-2. American Concrete Institute, Farmington Hills

Kahn LF, Hanson RD (1979) Infilled walls for earthquake strengthening. ASCE J Struct Eng 105(ST2):283–296

Kara ME, Altin S (2006) Behaviour of reinforced concrete frames with reinforced concrete partial infills. ACI Struct J 103(5):701–709

Klingner RE, Bertero VV (1978) Earthquake resistance of infilled frames. ASCE J Struct Div 104(ST6):973–989

Pulido C, Saiidi MS, Sanders D, Itani A, El-Azazy S (2004) Seismic performance of two-column bents – Part II: retrofit with infill walls. ACI Struct J 101(5):642–649

Sonuvar M, Ozcebe G, Ersoy U (2004) Rehabilitation of reinforced concrete frames with reinforced concrete infills. ACI Struct J 101(4):494–500

Chapter 17
RC Infilling of Existing RC Structures for Seismic Retrofitting

Christis Z. Chrysostomou, Nicholas Kyriakides, Martin Poljanšek, Fabio Taucer, and Francisco Javier Molina

Abstract The effectiveness of seismic retrofitting of multi-storey multi-bay RC-frame buildings by converting selected bays into new walls through infilling with reinforced concrete (RC) was studied experimentally at the ELSA facility of the Joint Research Centre in Ispra (Italy). A full-scale structure was tested with the pseudo-dynamic method. It consisted of 2 four-storey (12 m tall) three-bay (8.5 m long) parallel frames linked through 0.15 m slabs. The central bay (2.5 m) of each frame is infilled with a RC wall. The frames were designed and detailed for gravity loads only and are typical of similar frames built in Cyprus in the 1970s. Different connection details and reinforcement percentages for the two infilled frames were used in order to study their effects in determining structural response. The results of the pseudo-dynamic and cyclic tests performed on the specimen are presented, and conclusions are drawn.

17.1 Introduction

The construction of new walls is the most effective and economic method for retrofitting multi-storey reinforced concrete (RC) buildings, especially those with pilotis (soft-storey). Their structural and economic effectiveness increases when selected bays of an existing RC frame are fully infilled. Most of the experimental research work performed in the last decades has focused on other frequently used

C.Z. Chrysostomou (✉) • N. Kyriakides
Department of Civil Engineering and Geomatics, Cyprus University of Technology, P.O. BOX 50329, 3603 Limassol, Cyprus
e-mail: c.chrysostomou@cut.ac.cy; nicholas.kyriakides@cut.ac.cy

M. Poljanšek • F. Taucer • F.J. Molina
European Laboratory for Structural Assessment, Institute for the Protection and Security of the Citizen, Joint Research Centre, European Commission, I-21027 Ispra, VA, Italy
e-mail: martin.poljansek@jrc.ec.europa.eu; fabio.taucer@jrc.ec.europa.eu; francisco.molina@jrc.ec.europa.eu

A. Ilki and M.N. Fardis (eds.), *Seismic Evaluation and Rehabilitation of Structures*, Geotechnical, Geological and Earthquake Engineering 26,
DOI 10.1007/978-3-319-00458-7_17,

types of retrofitting, in particular on fibre reinforced polymers (FRP) and concrete jackets. Research on the use of RC infill walls has mainly targeted on what is feasible: testing of one- to two-storey specimens. However, data is lacking for taller full-scale specimens that reflect real life applications, due to the practical difficulties associated with the high forces needed for the tests. Regarding code provisions, Eurocode 8 – Part 3 fully covers retrofitting with FRP or concrete jackets, while it does not address the retrofitting of RC frames with the addition of new walls created by infilling selected bays. The KANEPE (2012) guidelines in Greece refer to the design of such walls only in terms of forces, providing tools for calculating their deformations (at yield and failure) and stiffness only if they are integral with the bounding frame.

Experimental research on reinforced concrete frames converted into walls by infilling with RC has been carried out almost exclusively in Japan and Turkey. The experiments in Japan (Hayashi et al. 1980; Higashibata et al. 1978; Kato et al. 1984; Shiohara et al. 1984; Sugano (ed.) 2007; Masuo et al. 1998; Takeyama et al. 1998; Sugimoto et al. 1999) were performed on 27 1:3 to 1:4 scale single-storey one-bay RC-infilled frames with RC infill walls with a thickness of 26–60 % (on the average 43 %) of the width of the frame members. The test results were compared in most cases with monolithically cast specimens of the same geometric characteristics (in which the frame and the infill wall were cast at the same time and integrally connected). The connection of the RC infill to the bounding frame was done by means of epoxy-grouted dowels (17 specimens), or through mechanical devices, such as shear keys and dowels without epoxy (6 specimens). In four other test campaigns the thickness of a pre-existing thin wall was increased by 100–150 % without any direct connection of the new wall with the bounding frame. The failure mode of all the specimens was in shear (including sliding at the interface). It is interesting to note that for epoxy-grouted dowels the force resistance of the infilled frame was on average 87 % of the integral one, while for the mechanical connections it was 80 % on average. For the increased thickness of an existing thin infill wall, the force resistance was on average 92 % of the monolithic specimen, while the displacement at failure was on average 13 % smaller than for the integral specimen. For the epoxy grouted dowels and for the mechanical connection the ultimate deformation was on average 55 and 115 % larger than in the integral specimen, respectively. The results show that although a deformable connection gives a somewhat reduced strength with respect to the monolithic case, the ultimate deformation of the retrofitted structure is considerably increased.

Concerning the specimens tested in Turkey, those of Teymur et al. (2008), Anil and Altin (2007) were single storey one-bay 1:2 and 1:3 scale, with RC infill thickness 25 and 33 % of the width of the frame members. Those of Altin et al. (1992), Turk et al. (2003), Canbay et al. (2003), Sonuvar et al. (2004), Kara and Altin (2006) were two-storey one-bay scaled at 1:3, with infill wall thickness 33 and 40 % of the width of the members of the bounding frame. The RC infill was in most cases fully connected on the perimeter with dowels; in some cases (Teymur et al. 2008) there was a gap between the infill and the columns; in some other cases there was no connection other than simple bearing. Altin et al. (1992) proposed to weld the rebars of the infill to those of the members of the frame, instead of using dowels. Only Altin et al. (1992) included some monolithic specimens, but not exactly similar to the infilled ones. Finally, the specimen of Erdem et al. (2004, 2006) was two-storey three-bay scaled at 1:3, with the middle bay infilled with a wall with 63 % thickness of the width of the frame members. The connection was made with epoxy grouted dowels and the failure mode was predominantly flexural. In all other cases the single storey walls failed in shear, while the two storey walls failed by a combination of flexure and shear sliding at the base.

The test specimens used in the experiments above correspond to walls with failure modes dominated by shear, with low aspect ratios not representative of multi-storey slender walls. In fact, the failure mode of multi-storey slender walls is controlled by bending and the design is governed by the formation of a plastic hinge at the base. In such a case, shear will not have a detrimental effect on displacement and energy dissipation capacity. In addition, it has been shown numerically (Eibl and Keintzel 1988; Keintzel 1990) that higher modes may increase considerably the shear forces at the upper floors of a wall after the formation of a plastic hinge at the base. This aspect has never been studied experimentally even in integral walls, because their height and number of storeys has not been large enough to allow higher mode inelastic response. Another common element of past tests is the smaller thickness of the RC infill wall relative to the width of the frame members. As a result, the weak link of the structural system is either the infill wall in diagonal compression, or its connection with the surrounding frame.

In order to start filling the gap of knowledge regarding infilling of existing RC frames with RC walls, the effectiveness of seismic retrofitting of multi-storey multi-bay RC-frame buildings by converting selected bays into new walls through infilling with RC was studied experimentally at the European Laboratory for Structural Assessment of the Joint Research Centre in Ispra (Italy). The present research was carried out as Transnational Access project SERFIN within the framework of the project "Seismic Engineering Research Infrastructures for European Synergies" (SERIES), funded by the 7th Framework Programme of the European Commission. The consortium was integrated by the Cyprus University of Technology (co-ordinator), the Ecole Central de Nantes, DENCO and the University of Cyprus. In the first part of the paper the design of the bare-frame specimen is presented and in the second part the details of the design of the RC infills are given. The results of the testing campaign are presented and conclusions are drawn.

17.2 Description of the Specimen

The specimen was designed based on a four-storey prototype building structure consisting of four three-bay frames spaced at 6 m, with RC infilling of the exterior frames only. The specimen was designed at full-scale to represent the two exterior frames of the prototype structure, spaced at 6 m and linked to a 0.15 m thick RC slab (Fig. 17.1b).

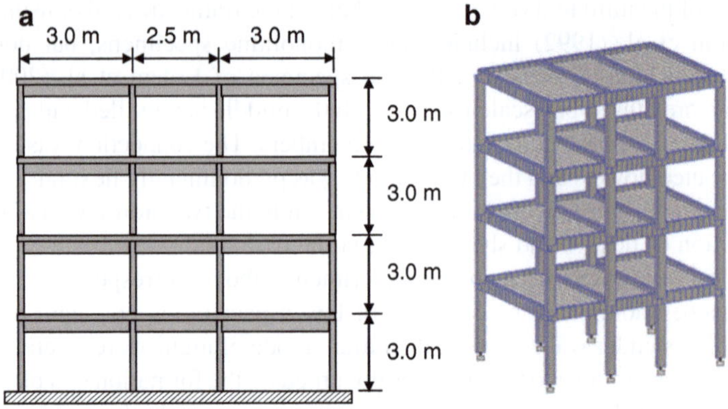

Fig. 17.1 Full scale specimen without the RC infill (**a**) Longitudinal section with dimensions, (**b**) 3D elevation of the FE model

The dimension of the specimen in the direction of testing was 8.5 m (two exterior bays of 3.0 m and a central bay of 2.5 m), with an inter-storey height of 3.0 m and a total height, excluding the foundation, of approximately 12.0 m (Fig. 17.1a). The cross section of the columns was 0.25 m by 0.40 m (long dimension along the plane of loading). The beams were of rectangular section with a depth of 0.50 m and a width of 0.25 m (along the plane of the frame and perpendicular to it).

17.2.1 Design of Structure for Gravity Loads Only

The proposed structure represents the construction practice of the late 1970s and early 1980s in Cyprus. Building structures at that time were designed for gravity loads only, since there were no provisions for earthquake loading. There was no specific design standard in Cyprus and the authorities accepted standards used in other countries such as CP110 and BS8110, DIN, Greek Code, US code, etc.

For the design of the mock-up it was decided to use the provisions of BS8110 (1983) which are very close to those of CP110 (BSI 1972), with very minor differences. In Cyprus, the transition from CP114 (BSI 1957), which was an allowable stress design, to BS8110 was made without going through the CP110 phase. The mock-up was designed such that all reinforcement details conformed to CP110:1972 and BS8110:1983.

The material properties used in the mock-up were constrained by the availability of materials in the Italian and European market. Concrete C20/25 was used for both the frame and the walls, with a unit weight of 25 kN/m^3 and a modulus of elasticity $E = 30{,}000$ MPa. Deformed steel reinforcement with characteristic yield strength f_{yk} equal to 400 and 450 MPa was used for all the members of the RC frame and the slab, and for the RC infill and dowels connecting the wall to the bounding frame members, respectively. The 400 MPa characteristic yield strength steel represents the one used in Cyprus construction practice in the 1970s and 1980s, while the

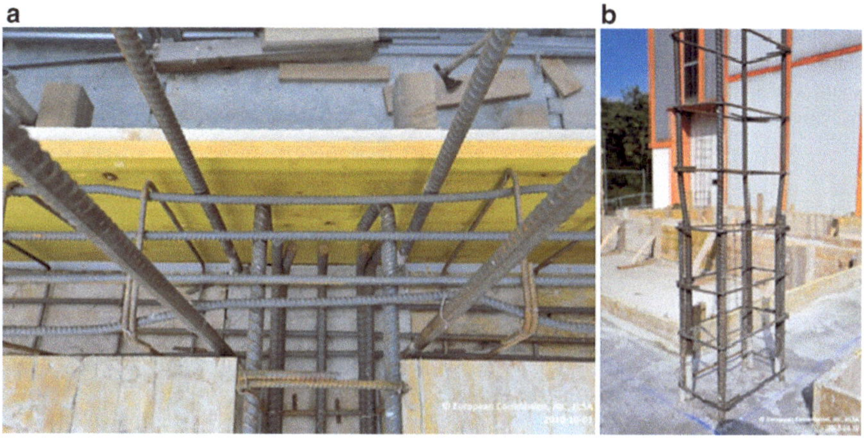

Fig. 17.2 (a) Reinforcement details for beams, (b) Reinforcement details for columns

450 MPa was the closest available in the Italian market to substitute for the 500 MPa steel that would be used today in the walls for retrofitting such a structure.

The self-weight was calculated using the unit weight of concrete specified above. The imposed dead load was 3 kN/m^2, including the load of masonry infill walls, and the live load was 1.5 kN/m^2. The above loads were combined using partial factors of safety of 1.4 for the permanent loads, and 1.6 for the live load. Material partial factors of 1.5 and 1.15 were used for concrete and steel, respectively.

The reinforcement details of the beams and columns are shown in Fig. 17.2a, b, respectively. For the beams of the frame 4Φ12 bars were used for the top and bottom longitudinal reinforcement, running along the total length of the beams. Shear links Φ8 were placed at 200 mm intervals starting at 50 mm from the face of the column. The longitudinal reinforcement of the transverse beams consisted of 2Φ20 and 5Φ20 at top and bottom, respectively. Shear links Φ8 were spaced at 100 mm in the transverse beam to ensure the transfer of forces from the actuators to the frames. The columns were reinforced with 4Φ20 bars lapped along a length of 0.55 m measured from the top face of the slab (Fig. 17.2b). This represents a compression lap, in line with a design for gravity loads only, and is expected to fail when subjected to tension forces. Shear links Φ8 were spaced at 200 mm in the column, starting at a distance of 50 mm from the top face of the slab.

Similar to the transverse beams, the slab was considered as an element facilitating the transfer of forces from the actuators to the two parallel frames, therefore the reinforcement was considerably increased. Although a nominal reinforcement of Φ10/200 was required by the standard, Φ10/100 was specified in order to ensure adequate transfer of forces. This was necessary to avoid damage to the slabs due to high concentration of forces from the lateral load application from the actuators during the PsD tests, which was achieved through a loading U interface as explained in the experimental setup section of the paper.

Fig. 17.3 Elevation of the specimen in the lab. The wall shown on the *right* is the south wall and the one on the background is the north wall

17.2.2 Reinforcement Details of the RC Infill Walls

In order to facilitate the study of the effect of as many parameters as possible, the walls in the two frames, which had a thickness of 0.25 m equal to the width of the beams and columns of the bounding frame, were reinforced with different amounts of reinforcement, with the north frame being the stronger of the two. Figure 17.3 depicts the south and north frames, with the south one being on the foreground and the north in the background. Hence, the east direction is the one towards the reaction wall and the west the one away from the wall.

Two parameters were examined: (a) the amount of web reinforcement in the walls and (b) the connection detail between the wall and the bounding frame. The web reinforcement in the two walls is shown in Table 17.1. Regarding the connection with the bounding frame, two distinct connection details were used. In the first detail, the web bars are connected to the surrounding frame through lap-splicing with same diameter starter bars epoxy grouted into the frame members. Short dowels are then used in order to transfer the shear at the interface between the wall and the frame members. This detail was used to connect the wall at the bottom beam and east column of the central bay at the 1st and 2nd floors of the north frame (Fig. 17.4a, starter bars shown here only for the bottom beam), while for the south frame it was used to connect the wall at the bottom beam of the 1st and 2nd floors, and the west and east columns of the 1st and 2nd floors, respectively.

Table 17.1 Reinforcement details for the RC infill walls

N wall

Story	Web bars	Embedment of web starter bars, mm		Dowel Φ mm	Dowel embedment, mm			
					Bottom and east		Top and west	
		Wall	Frame		Wall	Frame	Wall	Frame
1	Φ12@200	600	230	Φ20	160	160	600	190
2	Φ10@200	500	170	Φ20	160	160	500	160
3	Φ8@200			Φ18	400	145	400	145
4	Φ8@200			Φ16	400	130	400	130

S wall

Story	Web bars	Embedment of web starter bars, mm		Dowel Φ mm	Dowel embedment, mm							
					Bottom		East		Top		West	
		Wall	Frame		Wall	Frame	Wall	Frame	Wall	Frame	Wall	Frame
1	Φ10@200	500	170	Φ20	160	160	500	160	500	160	160	160
2	Φ8@200	400	120	Φ18	145	145	145	145	400	145	400	145
3	Φ8@200			Φ16	400	130	400	130	400	130	400	130
4	Φ8@200			Φ16	400	130	400	130	400	130	400	130

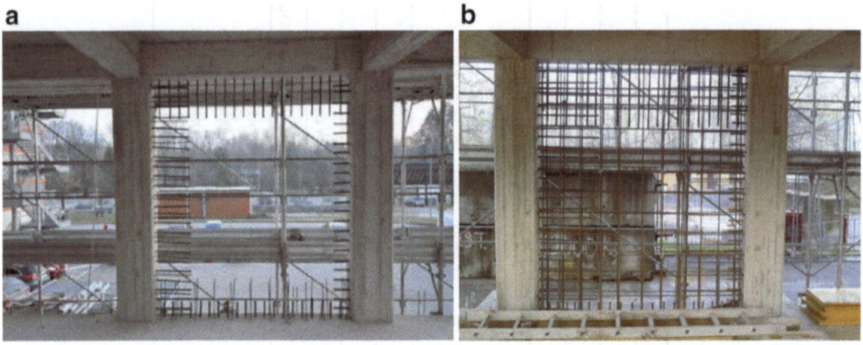

Fig. 17.4 (**a**) Dowels and starter bars, (**b**) Dowels, starter bars and web reinforcement

In the second detail, longer dowels were used to act both as dowels and as anchorage of the web panel to the surrounding frame; to this end, the dowels are considered as lap-spliced with the nearest – smaller diameter – web bars. However, in this case, the clear distance between the dowel and the nearest web bar violates the maximum clear distance of 50 mm or 4Φ between lapped bars, as specified in Eurocode 2 (CEN 2004). This detail was used to connect the wall at the top beam and west column at the 1st and 2nd floors of the north wall (Fig. 17.4a), while for the south frame it was used to connect the wall at the top beam of the 1st and 2nd floors, and the east and west columns of the 1st and 2nd floors, respectively. In the 3rd floor of both the north and south frames only the second detail was used, while for the 4th floor only two dowels per wall interface were used to provide safety against out of plane falling of the wall. The reinforcement details for the dowels and the starter bars are shown in Table 17.1. The completed wall reinforcement (including web, starter bars and dowels) for the 1st floor of the north wall is shown in Fig. 17.4b. In all cases the dowels were positioned along the centreline of the elements (i.e. at 0.125 m from the face of the wall).

Since the lapping of the column reinforcement can only take compression forces, a lap splice failure in tension would be highly detrimental to the whole experiment. In order to safeguard against this type of failure and allow the test to be performed without any premature failure, it was decided to reinforce the bounding columns of the wall at the 1st floor with three-sided CFRP (carbon fibre reinforced polymer) for a height of 0.60 m from the base of the column (Fig. 17.5).

17.3 Instrumentation

The test structure was instrumented with 108 potentiometric displacement transducers, 22 inclinometers, 8 Heidenhain linear encoders and 8 load cells. The pattern of instrumented zones was symmetrical for the two frames. The north ground floor wall and its bounding columns and beams was monitored with a pair of high resolution, water cooled Long Wording ForPco (PCO) Edge digital cameras.

Fig. 17.5 Reinforcing the column with CFRP against column lap-splice failure

Displacement transducers were installed to measure local displacements in critical areas. In particular, transducers were placed to monitor: slip and crack opening between all walls and their bounding beams and columns, the displacements between the ground floor walls and the foundation beams, and the shear deformations of the two ground floor walls. Displacement transducers were also installed to measure at all storeys the vertical elongation of the bounding columns.

Inclinometers were used to measure the rotation of beams and columns at the first floor. They were placed at the centre joints and on beams and columns 30 cm away from the joints. Inclinometers were also placed at selected columns 30 cm above the foundation beam. Heidenhain linear encoders were installed on two reference frames to measure the horizontal displacement of the two frames at each floor in the direction of testing.

The part of the structure in the field of view of the two digital cameras was painted with a random speckled pattern. During post processing of the images this allows to recreate a 3D displacement field of the area covered by the pattern and to identify the occurrence of cracks and follow their progression. This system allows an accurate measurement of displacement and crack openings, which will be an invaluable resource of data for the study of the local behavior of the specimen. At this point both the post processing of the images and the examination of the local data are under way and will be reported in a future publication.

17.4 Experimental Set-Up

The foundation slab of the specimen was clamped to the strong floor of ELSA by a series of prestressed steel bars. The short side of the structure was parallel to the reaction wall as seen in Fig. 17.6.

In the pseudo-dynamic (PsD) tests eight actuators (two 1,000 kN ones in the two top floors of the specimen and two 500 kN ones at the lower two floors, Fig. 17.6) applied the controlled displacement at each floor. At each floor there were two

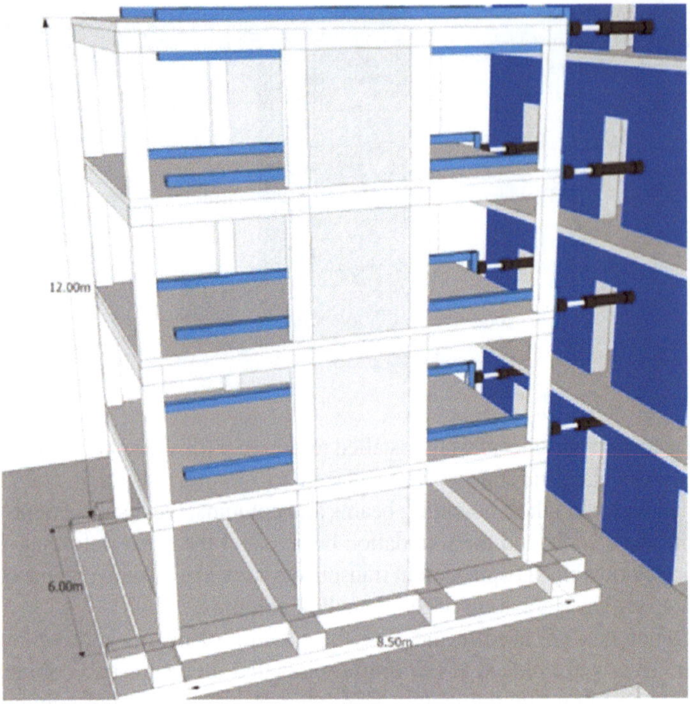

Fig. 17.6 SERFIN structure in ELSA

actuators symmetrically positioned at an axial distance of 4.05 m. The actuators were braced against the reaction wall and at the specimen end pinned to a loading U interface consisting of two 8.5 m HEB 200 longitudinal beams welded to a transverse element at the actuator connection. One of the beams rested on the top surface of the slab (Fig. 17.6), while the second was positioned beneath the transverse beams. The steel beams were clamped at three points along their length with prestressed bars passing through the slab. This configuration allowed a near to uniform distribution of load from the actuators to the floor slab.

The displacement control of each floor was done though linear encoders that provided floor displacement data by which the actuators were controlled.

The displacement control typically used for PsD tests encountered some difficulties of stability due to the high stiffness of the specimen (owing to the presence of the RC infill walls) that showed an eigenfrequency of approximately 30 Hz at the 4th mode of vibration of the prototype. The test was made stable by using low values of the proportional parameter of the control. A numerical model of the control system was also used to improve the tuning of the proportional integral derivative (PID) control loop (Molina et al. 2012). In order to guarantee stable behavior in the PsD response at the highest mode of the prototype and minimum error of the modes present in the response, the test was 800 times slower with respect to the real duration of the accelerogram used as input.

The testing campaign consisted of two pseudo-dynamic tests and a cyclic test. The gravity load on the structure, corresponding to the earthquake load combination of permanent and reduced live loads was equal to 192 kN per floor, and was applied by means of 15 plastic barrels per floor filled with water and distributed on the slabs of the test specimen (the amount of water was computed by taking into account the gravity load contribution of the loading system used for transferring the load from the actuators to the slab – as explained in the previous paragraph) (Fig. 17.3). Because of the slow nature of the PsD method, the water did not show any dynamic effects during the tests.

17.5 Test Campaign

In a PsD test on-line computer numerical models are combined with actual measurements of the properties of a structure. To simulate the response of a structure under seismic loading the computer running the PsD simulation takes an accelerogram as input. For the test campaign, the first 15 s of the transverse component of the Herzeg Novi accelerogram was adapted to the EC8 response spectrum of a Type B soil, digitized in 0.005 s steps and scaled to a target peak ground acceleration.

In a PsD test it is assumed that the response of a structure can be determined by a discrete model with a limited number of degrees of freedom (DoF). In this test campaign four DoFs were selected: the horizontal displacements of each storey with the assumption that all the mass is concentrated at the selected DoFs (i.e., the floor slabs). The equations of motion for such an idealized system are second order differential equations which can be expressed in matrix form:

$$M \cdot a(t) + C \cdot v(t) + r(t) = f(t) \tag{17.1}$$

where M is the mass matrix, C is the viscous damping matrix, $r(t)$ the internal (restoring) force vector and $f(t)$ the external force vector on the structure. Horizontal displacements of the controlled DoFs were solved for a prototype time step of $0.005/2{,}000 = 2.5 \times 10^{-6}$ s using the explicit Newmark time integration method. These displacements were then computed and applied on the structure by horizontal actuators at each storey with a laboratory time step of 0.002 s corresponding to the sampling of the controllers. The forces measured by the load cells in the actuators, following the application of the controlled displacements, represent the restoring forces that are fed back to the computer and that are used in the next time step of the calculation. Restoring forces are thus obtained from the specimen's response and reflect its state of damage. Since the inertial and viscous damping forces are modelled in the computer, the test does not have to run in real time scale. Hysteretic damping is automatically accounted for through inelastic deformation and damage progression of the test structure; consequently no viscous damping matrix was used (Molina et al. 2011). During the PsD test campaign the equation of motion was solved for restoring forces coming from the north frame only (calculated from static equilibrium of the load cell force measurements at each floor) and multiplied by a factor of two (the south frame is considered equal to the north frame in the numerical model). Equal displacements were applied to the two frames, in order to maintain zero rotation along the horizontal plane of the floor. The PsD test method used for the test campaign was continuous, which reduces problems of material relaxation and avoids load overshoot (Pegon et al. 2008).

The mass used in the equations of motion of the PsD test correspond to the total mass of the prototype, namely 156 tons per floor. In this way, the two tested frames were subjected to the same dynamic forces of a prototype building with negligible contribution of the two inner RC frames. Within the testing campaign two PsD tests and one cyclic test were run. The accelerogram was scaled to a maximum acceleration of 0.10g and 0.25g for the first and second PsD tests, respectively. For the final cyclic test, a history of displacements was imposed at the fourth floor, while maintaining a triangular distribution of loads along the height of the north frame and zero rotation at each of the four floors. The aim of this test was to explore the final capacity of the specimen up to a 20 % drop of peak strength of the structure.

17.6 Global Results

The 0.10g test was designed to induce minimum damage to the structure. After a visual inspection, no visible cracks on the columns or walls could be noted. Some hairline cracks that appeared on the surface of the wall at maximum displacement closed down at the end of the experiment. The recorded maximum top storey displacement was equal to 24 mm (towards the reaction wall) and −25 mm in the opposite direction (Fig. 17.7a). It should be noted that the variation of the displacements at the first and second stories is shown only up to about 11 s, since a problem was encountered with the data acquisition system. The variation of the base shear with the top displacement in each of the frames is shown in Fig. 17.7b. As it can be observed from the figure, there is very little difference between the two frames. The maximum positive shears were 645 and 574 kN and the maximum negative shears were −634 and −625 kN, for the south and north frames, respectively. Based on the results of the test and the observed damage, it can be considered that both walls reached their cracking moment.

The 0.25g test was designed to study the performance of the specimen at its ultimate capacity. The maximum top storey displacements were 109 and −93 mm (Fig. 17.8a). Some differences were observed in the base shear between the two frames. As it can be observed in Fig. 17.8b, the maximum base shear in the positive direction (towards the reaction wall) was 1,074 kN for the south frame and 1,036 kN for the north one; a larger difference was observed for the negative base shear: −843 kN for the south frame and −1,011 kN for the north frame (at the same displacement time step), providing an indication that the south frame had suffered larger levels of damage than the north frame. This was confirmed by a crack that

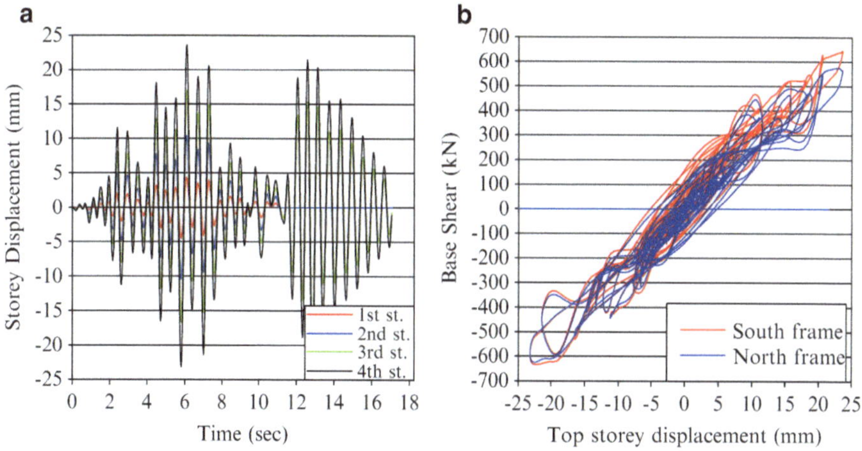

Fig. 17.7 Experiment at 0.10g acceleration (**a**) Variation of the storey displacements with time, (**b**) Base shear versus top storey displacement for the north and south frames

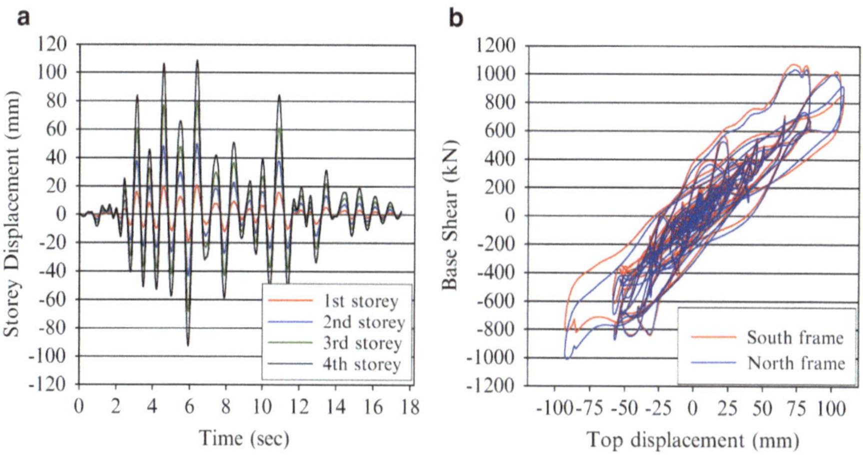

Fig. 17.8 Experiment at 0.25g acceleration (**a**) Variation of the storey displacements with time, (**b**) Base shear versus top storey displacement for the north and south frames

opened at the ground beam of the foundation at the base of the wall and by a lap-splice failure that appeared in the outer column at the east side of the south frame. It should be noted that the presence of the CFRP on the bounding columns of the wall prevented a similar failure, thus allowing completion of the test.

In examining Figs. 17.9 and 17.10 it can be observed that there is a steady decrease of the inter-storey shear from about 1,000 kN at the 1st storey to about 400 kN at the top storey. What can be also observed is that while the interstorey drift at the 1st storey is about 20 mm, the interstorey drift for the upper three floors is on the order of 30 mm, which shows the influence of the larger stiffness of the RC

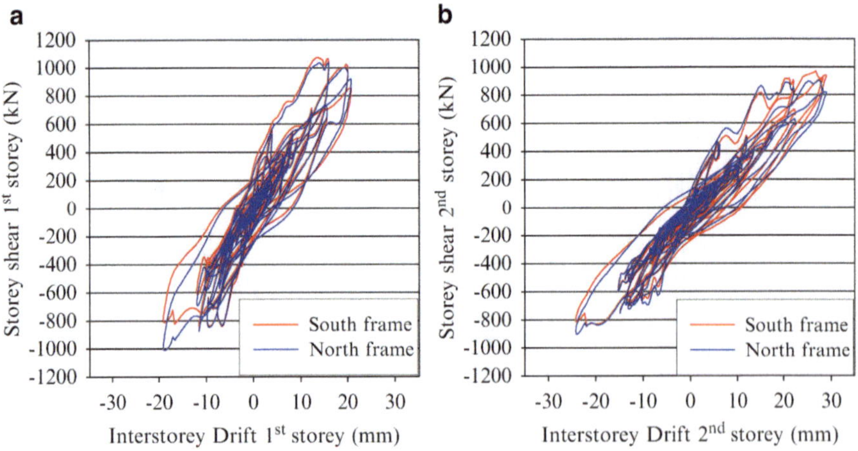

Fig. 17.9 Test at 0.25g acceleration (**a**) 1st storey interstorey drift, (**b**) 2nd storey interstorey drift

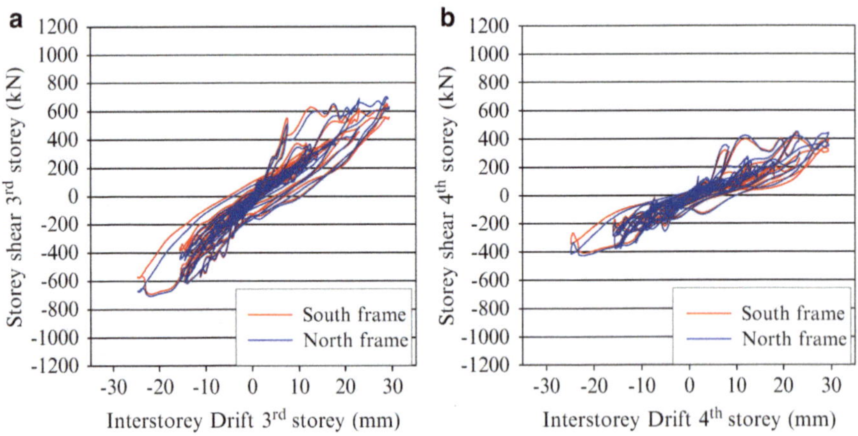

Fig. 17.10 Test at 0.25g acceleration (**a**) 3rd storey interstorey drift, (**b**) 4th storey interstorey drift

infill wall connection at the base. It can be also observed that the hysteresis loops are stable and provide some energy dissipation.

Figure 17.11a, b show the variation of the storey forces with time for the South and North frame, respectively, between the 4th and 7th seconds of the tests. From both figures it is obvious that there is at least a 2nd mode effect since the peak force in the 4th floor decreases, while the ones in the 2nd and 3rd increase. This effect is more pronounced for the South frame at about 5.6 s where the forces in the 2nd and 3rd floor are larger than that of the 4th floor, which is equal to the one of the 1st floor. At the same instance, the storey-forces in the North frame are almost equal indicating a rectangular distribution of forces. In addition, at about 6.3 s the force at

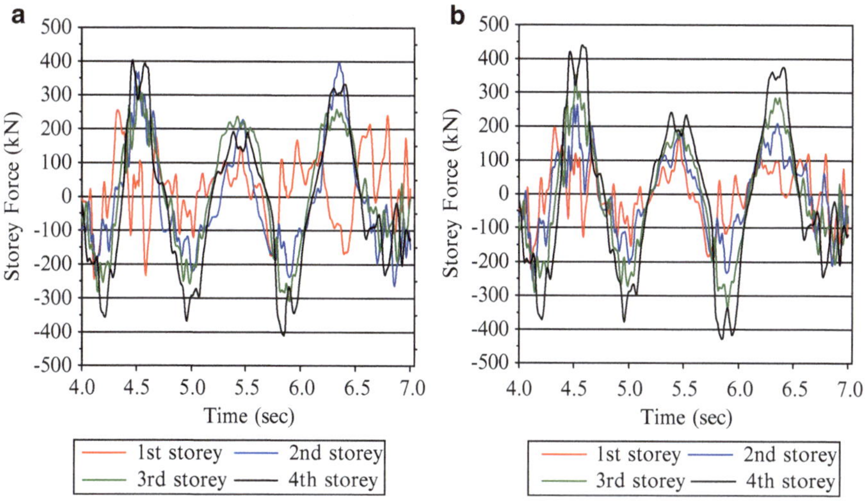

Fig. 17.11 Experiment at 0.25g acceleration (**a**) Variation of storey forces of south frame with time, (**b**) Variation of storey forces of north frame with time

the 1st floor of the South frame is negative while the ones at the rest of the floors are positive, indicating a load reversal.

Concerning the general behavior of the specimen during the two PsD tests, its performance was in accordance with the damage expected for the retrofit design corresponding to a life-safety limit-state for the 0.25g earthquake. There were no visible diagonal cracks on the walls, confirming that the wall responded in flexure. In nearly all the corner columns and at all floors, a horizontal crack appeared at a height of 0.55 m, corresponding to the limit of the lap-splice; in some cases spalling of the concrete cover was observed. Some vertical cracks appeared in the beams close to the beam/column interface, but no severe damage was observed, despite the fact that there were no ductile connections in the structure. In general, the stronger north frame had an overall better behavior compared to the south frame; nevertheless the differences between the two frames were minor.

During the final cyclic test a displacement history was imposed at the top storey (92, −92, 89, −125, 37, 0 mm). The objective of the test was to obtain a 20 % reduction of the peak strength of the specimen, so as to complete the global force-displacement envelope of the specimen. The base shear versus the top storey displacement of the cyclic test is shown in Fig. 17.12. As it can be observed in the first cycle the structure was able to reach 92 mm in both directions. In the second cycle the objective was to study the response of the specimen at 125 mm. However, while in the negative direction a displacement of 125 mm could be reached, in the positive direction the test could not go beyond 89 mm due to saturation of the second floor actuator in the south frame. This was due to the fact that the force in the second storey actuator of the south frame was saturated in a direction opposite to the displacement of the structure in order to keep the same top storey displacement

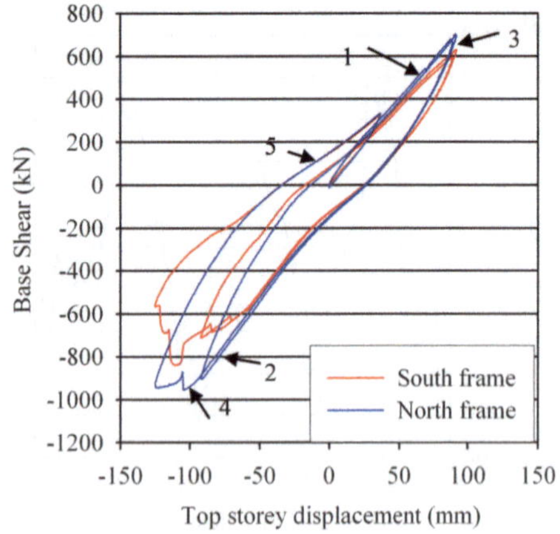

Fig. 17.12 Base shear versus top storey displacement for the south and north wall for the cyclic test. The numbers show the cycles

in the two frames, while maintaining a lower base shear in the south, weaker-frame. The force-displacement envelope in the negative direction shows that the strength of the south frame dropped from −838 kN at −110 mm to −553 kN at a displacement of −125 mm. This amounts to a drop in strength of 34 %, larger than the target of 20 %. After that, the top displacement was reduced to 37 mm in the positive direction and from there to zero.

It is worth noting also that for the same level of displacements, in the negative direction the structure develops a higher shear than in the positive direction. One reason for this difference is that the cyclic test was performed after an earthquake test, where the structure may have experienced larger levels of damage in the negative direction. A better reason may be that the east-end column of the south frame suffered a lap-splice failure, leading to lower stiffness of the south frame when the structure was pushed away from the reaction wall (towards the west).

17.7 Local Results

Invaluable results have been obtained for the local behavior of the RC infilled frame by monitoring the local behavior of the walls and the bounding beams and columns. Figure 17.13 shows the instrumentation that has been used to monitor the local behavior. The numbers in circles indicate the number of each sensor, which will be used for the discussion of the results, while the letters in squares indicate the west (W) and east (E) sides of the frames. Figures 17.14 and 17.15 show the strains of the ground-floor columns of the south and north frames, respectively. For the columns next to the walls two transducers covering a length of 1,250 and 1,500 mm were used, and for the two edge columns one transducer covering a length of 2,750 mm. In order to have comparable results, the displacements of the two transducers of the columns next to the walls were added and the strain was obtained

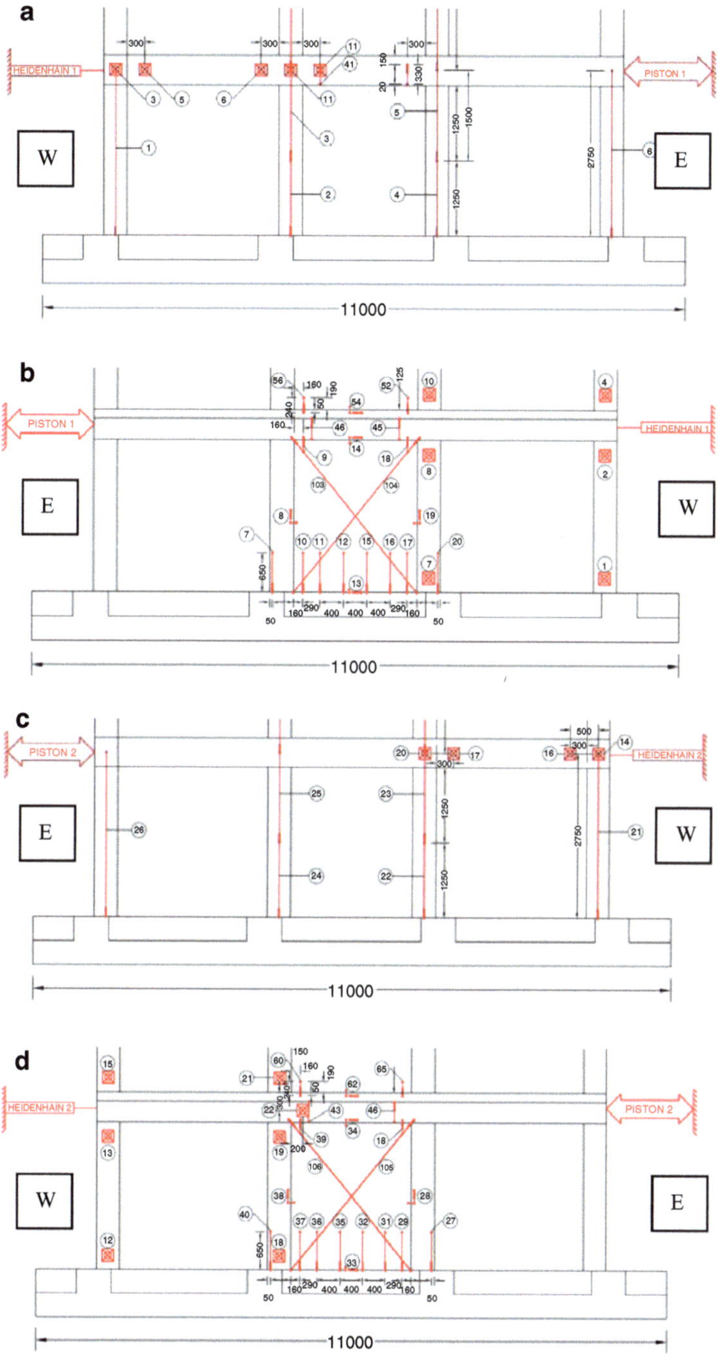

Fig. 17.13 (**a**) Instrumentation of south frame (outside face), (**b**) Instrumentation of south frame (inside face), (**c**) Instrumentation of north frame (outside face), (**d**) Instrumentation of north frame (inside face)

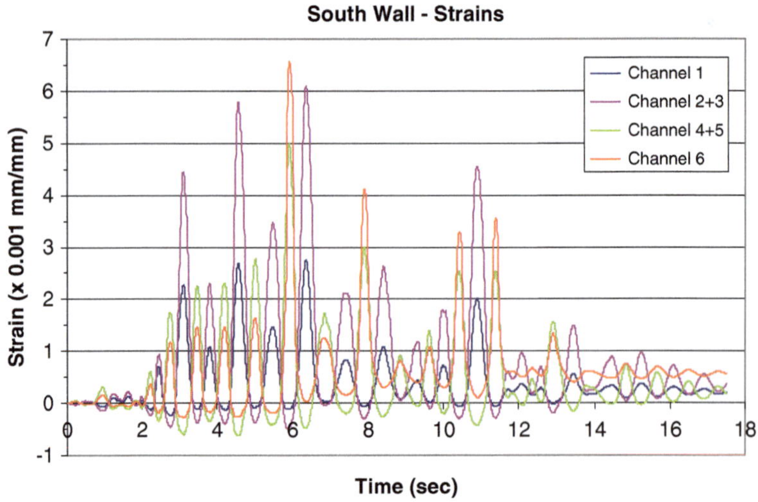

Fig. 17.14 Strain in ground-floor columns of south frame

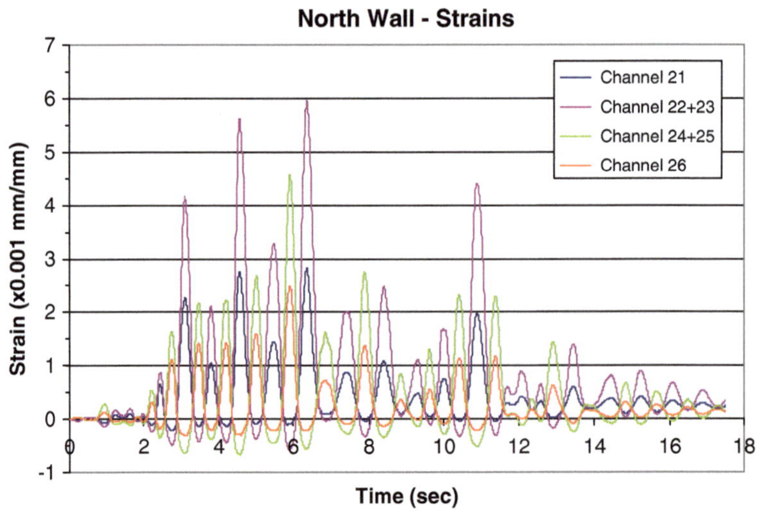

Fig. 17.15 Strain in ground-floor columns of north frame

dividing by a length of 2,750 mm. This gives average strains for the total length of the columns, which may be misleading, since the displacements at the lower part of the column (channels 2 & 4 for the south frame, and channels 22 & 24 for the north frame) may be larger than those of the upper part of the columns (channels 3 & 5 and 23 & 25 for the south and north frames, respectively).

On examining Fig. 17.14 it can be observed that, as expected, the columns next to the wall (Channel 2 + 3 and 4 + 5) had larger strains than the outer ones.

This changed at about 5.95 s, when the strain in the east column (Channel 6) increased suddenly to 0.0065 mm/mm, and it continued to be larger than the east bounding-column of the south wall (Channel 4 + 5), in most of the cases, and remained in tension for the rest of the experiment. This was associated with the sudden failure of the lap-splice of this column and the formation of a crack which increased the recorded displacement resulting in a permanent strain of 0.00055 mm/mm or a displacement of 1.51 mm, which is noticeable in Fig. 17.14. Therefore, although the graph shows an increase in the strain of the outer column, this is not true since it is due to the formation of the crack.

A comparison between Figs. 17.14 and 17.15 shows that the columns next to the wall had larger strains for the whole experiment, which indicates that most probably there was no splice failure in the columns of the north frame at the ground floor, or if there was, it was unnoticeable. Another observation is that, unlike the east column of the south frame, the corresponding one of the north frame (channel 26) had the same permanent deformation as the rest of the columns.

It can be also observed that the strains in both frames are more or less the same for the whole duration of the experiment, except for the tensile strains of the east outer- and east bounding-columns of the south frame (channels 6 and 4 + 5), which at 5.95 s become larger than the corresponding ones of the north frame (channels 26 and 24 + 25, respectively). The largest difference of 0.004 mm/mm appears in the east outer-column that has suffered lap-splice failure. A maximum tensile strain difference of about 0.0005 mm/mm at 5.95 s appears at the east bounding-column, which can be attributed to a small movement of the lap-splice, but which at a later stage almost diminishes. On the other hand, the tensile strains of the east outer-column of the south frame remain larger than those of the corresponding column of the north frame, for the rest of the experiment.

A similar picture appears on the compression side, where the waveforms are the same for both frames, with the exception of that of the east-outer column of the south frames after 5.95 s. The maximum compressive strains appear in the east-bounding columns of the frames at 4.55 s with that of the north frame reaching 0.00068 mm/mm and the one for the south frame 0.00063 mm/mm.

Figures 17.16 and 17.17 show the distribution of strain along the base of the south and north wall, respectively. The strains were monitored with eight transducers covering the width of the wall horizontally (2.8 m out of a total of 2.9 m since the first and the last were positioned at 50 mm from the edge of the wall-column) and extending 650 mm in the vertical direction (Fig. 17.13b, d). This height was selected so as to be longer than the end-point of the lap-splices of both the column reinforcement (550 mm) and that of the wall web-reinforcement (600 mm), in order to be able to capture a crack that would possibly form at that level. The lines plotted correspond to the points shown in Table 17.2. It should be noted that the lines plotted for the maximum compressive strain at west edge and for the maximum tensile strain at east edge coincide; therefore only three lines appear on both Figs. 17.16 and 17.17.

Examination of the two figures indicates that distribution of strains is more or less linear, except for the tension at the west side of both walls. This can be

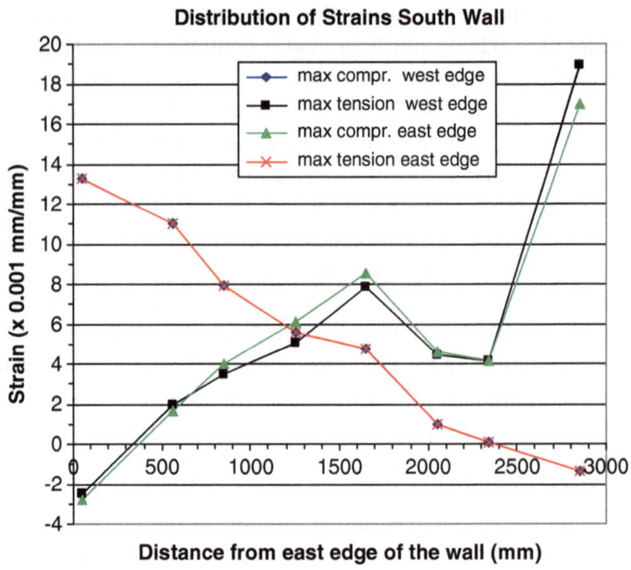

Fig. 17.16 Strain distribution at the base of the south wall

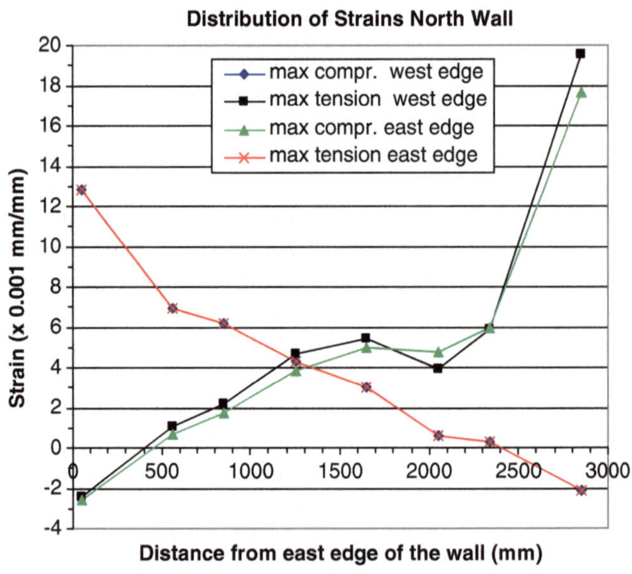

Fig. 17.17 Strain distribution at the base of the north wall

attributed to a crack that formed in the ground beam at that location, on which the transducers were anchored (Fig. 17.18). While this did not have any effect on the measurements in the compression phase, the recorded measurements in the tension phase were affected (smaller measurements) due to the opening of the crack.

Table 17.2 Maximum tensile and compressive strains in the south and north walls

	Time(s)	South wall ($\times 10^{-3}$)	North wall ($\times 10^{-3}$)
Max compr. west edge	5.9	−1.4	−2.1
Max tension west edge	6.4	19.0	19.6
Max compr. east edge	4.6	−2.8	−2.6
Max tension east edge	5.9	13.3	12.8

Fig. 17.18 Crack in the ground beam where the transducers were anchored

The overall maximum compressive strain in the south wall was 0.0028 at the east edge (0.0026 for the North wall). The overall maximum tensile strain was about 0.019 at west side and 0.013 at east side for both the south and north walls, well beyond the yield strain of 0.0023. The neutral axis was at about 400 mm from the two edges of both walls (total length of wall including the two bounding columns is 2,900 mm). This positions the neutral axis at the interface between the bounding column and the wall (width of columns is 400 mm). This is the limit specified by EC8 beyond which additional confinement reinforcement is required. For this experiment, it seems that the provision of the CFRP at the edges of the wall provided the necessary confinement that prevented the lap-splice failure and at the same time allowed the walls to sustain loads close to their ultimate capacity.

Figure 17.19 shows the strain distribution along the bounding columns of the south wall. Both in the east and west side the strains in the ground floor are considerably larger compared to those at the 1st floor; the ones for the 2nd and 3rd floors are negligible. The same applies to the north wall with the only difference that the tensile strains are slightly smaller, but the compressive strains slightly larger. Note that, since these strains are averaged for the total length of the column, much smaller values are obtained compared to those in Figs. 17.16 and 17.17.

Figure 17.20 shows the slip displacement between the wall and the bounding beams of the ground floor, for both the south and north walls. The maximum displacements are of the order of 0.8 mm for both the south and north walls. In the south wall the displacements are larger at the interface between the wall and the ground beam for the whole duration of the test, while for the north wall they are larger at the interface between the wall and the top beam in the positive

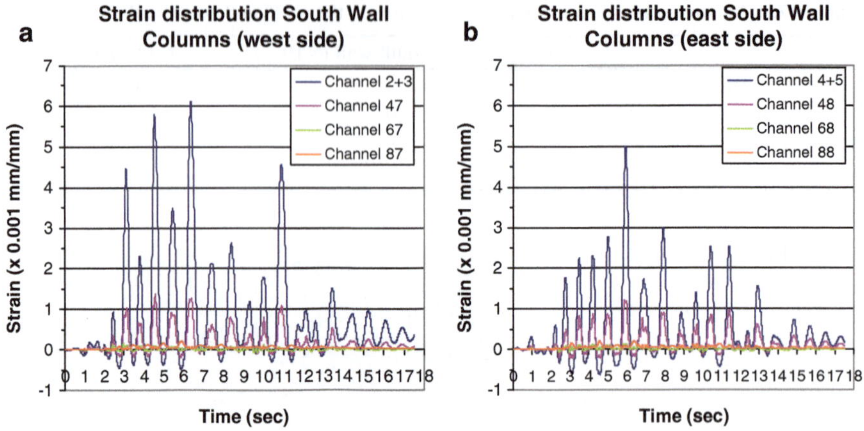

Fig. 17.19 (**a**) Strain distribution along the west side bounding-columns of the south wall (**b**) Strain distribution along the east side bounding-columns of the south wall

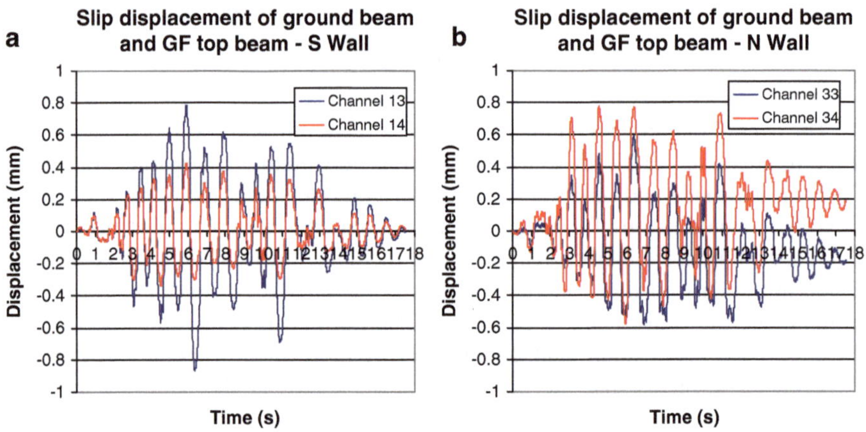

Fig. 17.20 (**a**) Slip displacement at the ground beam and ground floor top beam for the south wall (**b**) Slip displacement at the ground beam and ground floor top beam for the north wall

direction for the whole duration, while after 6.5 s the ones at the bottom become larger in the negative direction, mainly due to the shifting of the slip values that takes place. This shifting is not observed in the south wall. At the bottom interface both the starter bars and the dowels contribute to the doweling action, while at the top beam the dowels double as bending and shear reinforcement. Dowels are considered to produce their full capacity at a slip of 10 % of their diameter. For both walls the dowels have a diameter of 20 mm, therefore they have not been fully engaged. The starter bars in the south wall have a diameter of 10 mm while those of the north wall a diameter of 12 mm. Both of these diameters are close to the 0.8 mm slip displacements that were observed, especially the ones of the south wall.

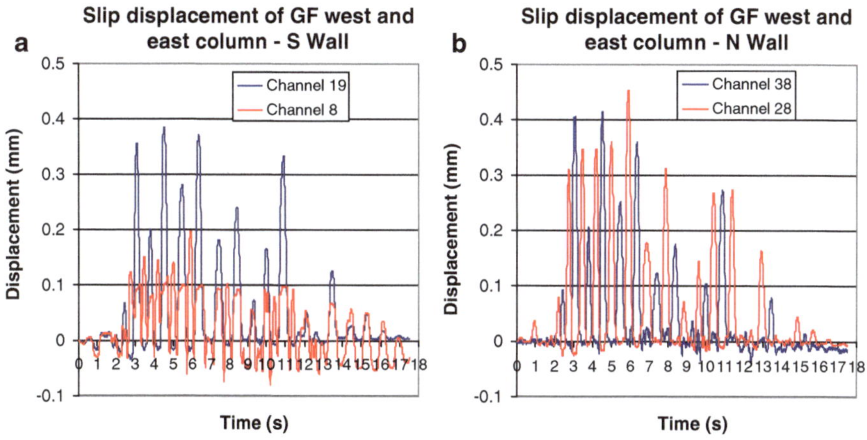

Fig. 17.21 (**a**) Slip displacement at the west and east GF columns for the south wall (**b**) Slip displacement at the west and east GF columns for the north wall

No solid conclusions can be drawn regarding the effectiveness of the two schemes used, since they do not give consistent results. Similar behavior at the wall-beam interfaces were observed for the rest of the stories with the slip-displacements being of the order of 0.3 mm.

The slip-displacements between the wall and the west and east columns of the south and north frame are shown in Fig. 17.21. The west column is connected to the wall with both starter bars and dowels while the east column with dowels only. The maximum slip displacements are of the order of 0.4 mm for both walls, which is much smaller than the one required for the full activation of either the starter bars or, especially, the dowels. Again no solid conclusions can be drawn on the effectiveness of the two types of connection, since the results do not show a repeated pattern. For the higher floors the slip-displacements are even smaller and of the order of 0.1–0.2 mm.

In Fig. 17.22, the opening and closing displacements of the bottom face of both the south and north walls at the 1st floor are shown. In this case, the displacements show exactly the same pattern, both in tension and compression, indicating nonlinear behavior depicted by the shifting of the curves. The peak tensile-value for the east edge of the south wall is about 1.3 mm, resulting in a strain of 0.0054 mm/mm while that of the north wall is about 1.5 mm, resulting in a strain of 0.0063 mm/mm. For the west edge the tensile displacement for the south wall is about 1 mm while that of the north wall is about 1.2 mm, resulting in strains of 0.0042 and 0.0050, respectively. All of the above strains are larger than the yield strain of the reinforcement used and shows that the reinforcement of the wall at the 1st floor has also yielded.

On the compression side, the displacements for both walls are about the same and of the order of 0.25 mm for the east edge and 0.20 mm for the west edge resulting in strains of the order of 0.0010 and 0.00083, respectively. These strains are considerably smaller that the ultimate strain of concrete.

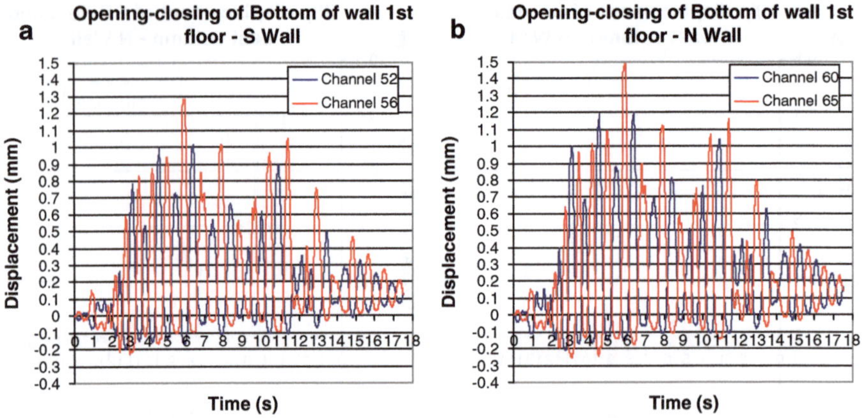

Fig. 17.22 (**a**) Opening and closing of the interface at the bottom of the south wall at the 1st floor (**b**) Opening and closing of the interface at the bottom of the north wall at the 1st floor

17.8 Conclusion

The effectiveness of seismic retrofitting of multi-storey multi-bay RC-frame buildings by converting selected bays into new walls through infilling with RC was studied experimentally on a full scale specimen at the ELSA facility of the Joint Research Centre at Ispra. The main parameters of the mock-up were the connection between the RC infill and the surrounding RC frame and the percentage of reinforcement in the RC infill. The effect of these parameters was studied during the experiment, by using different connection details and reinforcement percentages for the two infilled frames. Some findings regarding the behavior of the structure are: (a) the structure managed to sustain an earthquake of 0.25g without significant damage, (b) some column lap-splices failed with concrete spalling, but the structure continued to carry load, (c) the three-sided CFRPs protected the wall bounding columns at the 1st floor and prevented lap-splice failure, (d) The "weak" south frame behaved equally well as the "strong" north frame, (e) the slip-displacement at the horizontal interfaces of the ground-floor walls were of the order of 0.8 mm, which is very close to the full engagement of the starter bars but not of the dowels, (f) the slip-displacements between the wall and the bounding columns of the ground-floor was of the order of 0.4 mm, (g) the behavior of the wall was mainly flexural; yielding took place at both the ground-floor and the 1st floor wall, (h) the distribution of strains along the bounding columns of the walls show that those of the ground floor are much larger than those of the 1st floor, while those of the 2nd and 3rd floors are negligible, (i) the two connection arrangements used behaved satisfactorily, but no solid conclusions can be drawn regarding the advantages of one over the other, (j) higher mode effects appeared in the response of the structure, (k) some vertical cracks appeared at the connection of the beams to both the exterior and the wall columns, and (l) a horizontal crack appeared at the ground-beam of the walls, which was the main cause for the loss of strength of the south frame.

It was demonstrated that this is a viable method for retrofitting existing deficient structures. The recorded global and local behavior of the structure provides data for the development of numerical models, to facilitate the proposal of design guidelines for such a retrofitting method.

Acknowledgments The research leading to these results has received funding from the European Community's Seventh Framework Program [FP7/2007-2012] for access to the European Laboratory for Structural Assessment of the European Commission – Joint Research Centre under grant agreement no. 227887 [SERIES]. The authors would also like to acknowledge Prof. Michael Fardis for his invaluable suggestions for the setting-up of the mock-up and testing campaign, as well as Artur Pinto, Georges Magonette and all the staff of the ELSA laboratory for their contribution in constructing and testing the test specimen.

References

Altin S, Ersoy U, Tankut T (1992) Hysteretic response of reinforced-concrete infilled frames. ASCE J Struct Eng 118(8):2133–2150

Anil O, Altin S (2007) An experimental study on reinforced concrete partially infilled frames. Eng Struct 29:449–460. doi:10.1016j.engstruct.2006.05.011

BS8110 (1983) Structural use of concrete. Code of practice for design and construction, Part 1. The Council for Codes of Practice, British Standard Institution, London

BSI (1957) CP114-The structural use of reinforced concrete in buildings. The Council for Codes of Practice, British Standards Institution, London

BSI (1972) CP110-The structural use of concrete, Part 1. The Council for Codes of Practice, British Standards Institution, London

Canbay E, Ersoy U, Ozcebe G (2003) Contribution of reinforced concrete infills to seismic behavior of structural systems. ACI Struct J 100(5):637–643

CEN (2004) EN 1992-1-1 Eurocode 2: Design of concrete structures – Part 1–1: General rules and rules for buildings, Comité Européen De Normalisation, Brussels

Eibl J, Keintzel E (1988) Seismic shear forces in cantilever shear walls. In: Proceedings of the 9th world conference in earthquake engineering, Japan Association for Earthquake Disaster Prevention, Tokyo

Erdem I, Akyuz U, Ersoy U, Ozcebe G (2004) Experimental and analytical studies on the strengthening of RC frames. In: 13th world conference on earthquake engineering, Vancouver

Erdem I, Akyuz U, Ersoy U, Ozcebe G (2006) An experimental study of two different strengthening techniques for RC frames. Eng Struct 28:1843–1851

Hayashi T, Niwa H, Fukuhara M (1980) The strengthening methods of the existing reinforced concrete buildings. In: Proceedings of the 7th world conference on earthquake engineering, vol IV. Turkish National Committee on Earthquake Engineering, Ankara, pp 89–96

Higashibata Y et al (1978) Studies on seismic rehabilitation of low- to medium-rise reinforcement concrete buildings, Part 1 and 2. Summary of technical papers of annual meeting, vol 9. Architectural Institute of Japan, Tokyo

KANEPE (2012) Code for intervention in reinforced concrete buildings. Earthquake Planning and Protection Organization (OASP), Athens

Kara ME, Altin S (2006) Behavior of reinforced concrete frames with reinforced concrete partial infills. ACI Struct J 103(5):701–709

Kato D, Katsumata H, Aoyama H (1984) Study on evaluation of the strength of post-cast shear walls without opening. J Struct Constr Eng Archit Inst Jpn 337(3):81–89

Keintzel E (1990) Seismic design shear forces in reinforced concrete cantilever shear wall structures. Eur J Earthq Eng 3(1):7–16

Masuo K, Inoue T, Otani T, Uemetsu T (1998) Test on strengthening of RC frames with CFRP jacketing and infill wall. GBRC Rep 90(4):38–51

Molina FJ, Magonette G, Pegon P, Zapico B (2011) Monitoring damping in pseudo-dynamic tests. J Earthq Eng 15(6):877–900

Molina FJ, Bosi A, Pegon P, Magonette G, Pinto A (2012) Error study of a hybrid testing system of structures through a state-space model. In: Proceedings of the 15th world conference on earthquake engineering, Lisbon

Pegon P, Molina FJ, Magonette G (2008) Continuous pseudo-dynamic testing at ELSA. In: Saouma VE, Sivaselvan MV (eds) Hybrid simulation; theory, implementation and applications. Taylor & Francis/Balkema Publishers, London, pp 79–88

Shiohara H, Hosokawa Y, Nakamura T, Aoyama H (1984) Tests on the construction method of seismic rehabilitation of existing reinforced concrete buildings. In: Proceedings, vol 6. Japan Concrete Institute, Tokyo, pp 405–408

Sonuvar M, Ozcebe G, Ersoy U (2004) Rehabilitation of reinforced concrete frames with reinforced concrete infills. ACI Struct J 101(4):494–500

Sugano S (ed) (2007) Seismic rehabilitation of concrete structures, International Publication Series IPS-2. American Concrete Institute, Farmington Hills

Sugimoto T, Masuo K, Komiya T, Ueda M (1999) Experimental study on RC infill walls which used adhesive anchor with specification for practice. Summary of technical papers of annual meeting, vol 9. Architectural Institute of Japan, Tokyo, pp 219–222

Takeyama H, Satoh A, Minai M, Sometani T (1998) Construction and performance test of precast concrete infills walls. Prestress Concr 40(4):219–222

Teymur P, Yuksel E, Pala S (2008) Wet-mixed shotcrete walls to retrofit low ductile RC frames. In: Proceedings of the 14th world conference on earthquake engineering, Beijing, paper 0146

Turk M, Ersoy U, Ozcebe G (2003) Retrofitting of reinforced concrete frames with reinforced concrete infills walls. In: Proceedings of the *fib* symposium, concrete structures in seismic regions, Athens

Chapter 18
Hybrid Control of a 3-D Structure by Using Semi-Active Dampers

Gürsoy Turan

Abstract A base isolated three storey 3-D building is semi-actively controlled not to exceed the maximum allowable base displacement. Large displacements are likely to cause failure in the isolation system, and hence, failure in the superstructure is expected. If a base isolated structure is positioned next to a very long fault line, such as the North Anatolian Fault, the structure will mostly undergo far field type excitations. Near field effects will be seen less occasionally, but design considerations should be made to account for both types of excitations. In case of nearby seismic action, the isolated building should be smart enough to modify its isolation impedance to resist against large ground displacement and velocities. For this study, an isolated three storey building model together with four dampers, which are all placed at the base level, is considered. The dampers have controllable orifices (damping coefficients) and the magnitudes of these damping coefficients are assigned by using a linear quadratic regulator (LQR). During an earthquake excitation, the storey displacements and velocities are used as feedback in the calculation of the optimal control force that is producible by viscous dampers, at each time step. This force, however, is applied only at times when critical displacements and/or velocities occur. The performance of the set of controllers is presented via time simulations of the system for three recorded earthquakes. In addition, these records are time shifted five folds to see the effect of near field action. The results indicate that the control effectively reduces the maximum displacements of the isolation system, while maintaining a reasonable isolation to the superstructure.

G. Turan (✉)
Department of Civil Engineering, Izmir Institute of Technology, Izmir, Turkey
e-mail: gursoyturan@iyte.edu.tr

A. Ilki and M.N. Fardis (eds.), *Seismic Evaluation and Rehabilitation of Structures*, Geotechnical, Geological and Earthquake Engineering 26,
DOI 10.1007/978-3-319-00458-7_18,

18.1 Introduction

Semi-active dampers are foolproof control devices being widely accepted in structural control. Dampers are utilized to absorb energy from the structure. Thus, the larger the damping, the less will be the relative velocity and displacement. The accelerations, however, will increase. For building-type structures, the only case with semi-active control via dampers is the case at which buildings are seismically isolated. The role of the dampers in this type of structures is to limit the displacement of the dampers so that they don't rupture. The presence of a damper in parallel to a base isolation system obviously decreases the effectiveness of the structure's earthquake isolation. Nevertheless, it will keep the elastomeric bearings from being driven into large displacements, thus securing the base isolation system.

Extensive research has been conducted to model and implement variable orifice dampers. Kurata et al. (1999) designed a full scale building that is controlled by semi-active dampers. The damper used in his design is capable of producing a 1,000 kN damping force, while only 70 W electric energy is consumed for this purpose. Wongprasert and Symans (2005) used variable-orifice fluid dampers to enhance the response of a base isolated 1:4 scale three storey frame model. They simulated the response of the system both with software and on an earthquake simulator. Aldemir and Bakioğlu (2000) designed a time varying controller for a damper in a single degree of freedom system. They showed that the maximum displacement of the controlled response is about 18 % less than the passive response. Çetin et al. (2009) worked on a six storey building that was to be controlled via a Magneto rheological damper at the floor level. Although the device is different from a variable orifice damper, the principle remains the same. They modeled the structure as a single degree of freedom system and designed a robust Hinf controller.

In this study, a set of linear quadratic regulator based controllers are designed for various damping levels in the isolated structure. These damping levels arise due to the orifice settings of the dampers. These various settings comprise the control which is smartly applied. An upper controller selects the controller that corresponds to the system with the set damping value, and decides if it should apply the optimum damping. This last phenomenon is crucial, because a maximum damping level would normally be the choice of the optimal control. Large damping in the base level, on the other hand, is not beneficial for the superstructure. Here, one needs to design for an acceptable maximum isolator displacement and inter-storey drift values. To overcome this economic balance problem an upper controller is designed and simulation results are presented.

18.2 Three Storey 3-D Building Model

A three storey building model is considered for the control effectiveness evaluation. Elastomeric base isolators are used at the base and four dampers are connected to the two opposite corners of the building. Figure 18.1 shows the three dimensional

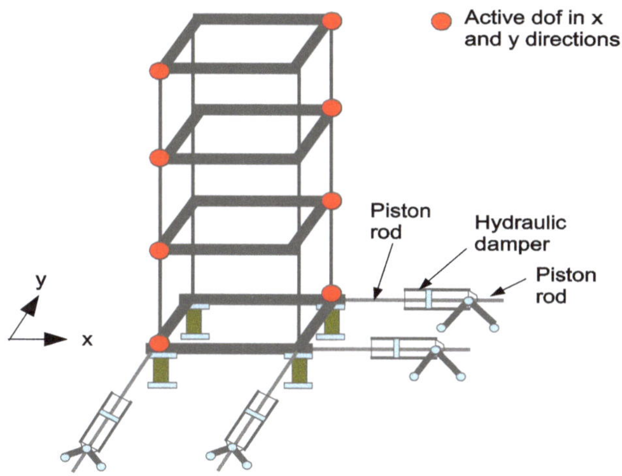

Fig. 18.1 3-D building model

view of this structure. A total number of four hydraulic dampers are connected in-between the structure base and the ground; two in the x-direction and two in the y-direction. The building is modeled by using 3-D steel beam elements (columns: 17.5 mm × 17.5 mm, beams: 90 mm × 90 mm × 5 mm). The storey heights are 80 cm, the structures cross sectional dimensions are 100 cm (x-dir) by 60 cm (y-dir). Each storey, including the base, has a total mass of 200 kg. The structure is constraint at the ground level and in the vertical direction. The remaining degrees of freedom (dof) except for the lateral dof at and above the damper connections are statically condensed, and finally a damping ratio of 0.006 is assigned to all modes. This number is based on the calculated damping ratios of a similar structure that exists in the IYTE Structural Mechanics Laboratory (Turan and Aydin 2010).

The resulting system is a second order differential equation with 12 dof for the fixed base building, and 16 dof for the isolated building. Table 18.1 displays the major modes of vibration in which the fixed building has periods denoted by T0, and the periods with isolators are denoted by Ti. The first three modes of the building occur mostly in the base, which are the isolation modes. Modes 13 through 16 of the building have high frequencies that correspond to a skew deformation in the denoted storey level only. All modes are preserved in the simulation model, because the added dampers cause non-proportional damping.

The fixed building has a fundamental period of 0.67 s, which is indicated as a vertical line in Fig. 18.2. The figure shows the influence of the chosen earthquakes onto the building. In order to isolate the building from the effect of these earthquakes, the elastomeric bearing stiffness is appropriately chosen as 1,200 N/m (for comparison purposes, the columns have a stiffness of 36,600 N/m). Thus, the fundamental period of the isolated building is increased to 3.19 s. This change is beneficial for far field earthquakes as it can be seen on Fig. 18.2a. Here, the expected absolute acceleration of the isolated building is significantly reduced. For near field type

Table 18.1 Vibration modes

	Fixed		Isolated	
Mode #	Mode	T_0 (s)	Mode	T_i (s)
1	Trans – x	0.674	Trans – x	3.198
2	Trans – y	0.666	Trans – y	3.111
3	Twist – xy	0.593	Twist – xy	2.897
4	Trans – x	0.239	Trans – x	0.369
5	Trans – y	0.236	Trans – y	0.000
6	Twist – xy	0.210	Twist – xy	0.000
7	Trans – x	0.000	Trans – x	0.199
8	Trans – y	0.162	Trans – y	0.000
9	Twist – xy	0.144	Twist – xy	0.173
10	Trans – x	0.008	Trans – x	0.000
11	Trans – y	0.0076	Trans – y	0.1500
12	Twist – xy	0.0074	Twist – xy	0.1330
13	–	–	skew – B	0.0110
14	–	–	skew – 2	0.0109
15	–	–	skew – 1	0.0103
16	–	–	skew – 3	0.0089

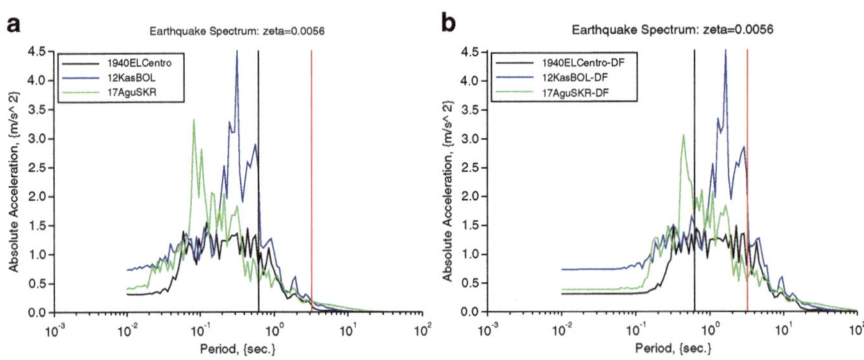

Fig. 18.2 Earthquake spectra and the influence on the first vibration mode of the fixed and isolated buildings, respectively. (**a**) Original earthquakes. (**b**) Modified (generic) earthquakes to simulate near field seismic action

earthquakes, on the other hand, Fig. 18.2b shows that the expected absolute acceleration may increase or decrease for various types of earthquakes as the fundamental structural period is increased. This unknown trend in the structural response shows the importance of an appropriate control mechanism.

18.3 Semi-Active Damper

A hydraulic piston is modified by a pipe that interconnects the two chambers, and a stepper motor controlled valve is placed in series with this pipe. Figure 18.3 shows the modified piston.

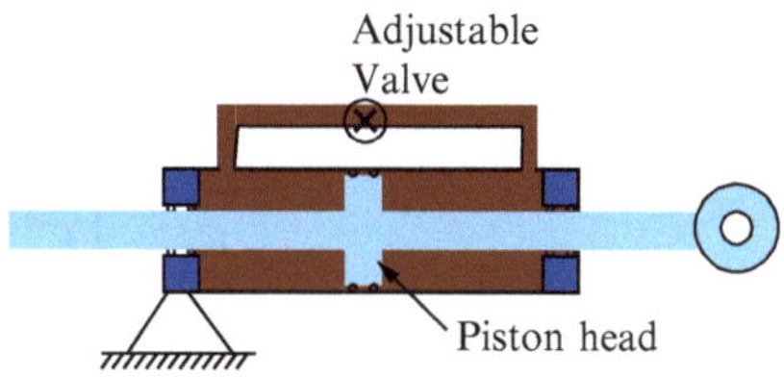

Fig. 18.3 Cross section of the semi-active damper

The force exerted onto the damper is assumed to have a linear relation with the piston head's velocity as shown in Eq. 18.1.

$$F_d = -c_d \dot{x}_d \tag{18.1}$$

F_d is the damper force, $\dot{x}_d$ the damper velocity and c_d the damping coefficient. The damper constant is evaluated to be in the range of 5,000–25,000 Ns/m. The upper limit is selected so that the piston's capacity of 5,000 N is not exceeded, whereas the lower limit corresponds to the valve being completely open.

18.4 Control Design

A hybrid control method, namely Gain Scheduling, is utilized in this study. The isolated building model is constructed with each damper valve opening possibility. The damping constants of the four dampers are each varied by 5,000 Ns/m increments, resulting in 5 possible damping positions for each damper, and 625 damping positions for the structure with four dampers. Feedback gains are designed for each of these possible configurations by using the linear quadratic regulator (LQR) scheme. During an earthquake simulation, the required force is calculated, and the closest damping constants for all four devices are selected such that the dampers are able to produce the required control forces.

18.4.1 Linear Quadratic Regulator (LQR) Design

Equation 18.2 shows the differential equation of the isolated building with damping control forces and earthquake effect.

$$M\ddot{x} + C\dot{x} + Kx = -M\Gamma_{eq}\ddot{x}_g + \Gamma F_d \tag{18.2}$$

Here, M, C, and K are the structural mass, damping and stiffness matrices, respectively, and x is the structural displacement with respect to the ground. $\ddot{x}_g$ is the ground acceleration, Γ_{eq} is the ground acceleration application matrix, Γ is the damper force application matrix, and F_d is the damping force vector of the four dampers, which is constructed as follows

$$F_d = \left\{F_d^{(1)}, F_d^{(2)}, F_d^{(3)}, F_d^{(4)}\right\}^T \tag{18.3}$$

where $F_d^{(i)}$ is the i'th damper force. Since F_d is a force vector of passive devices and is linearly related to the structural velocity, as indicated in Eq. 18.1, it can be moved to the left hand side of Eq. 18.2.

$$M\ddot{x} + \left(C + \Gamma C_d \Gamma^T\right)\dot{x} + Kx = -M\Gamma_{eq}\ddot{x}_g \tag{18.4}$$

Equation 18.4 is transformed to a first order differential equation by introducing a variable transformation of $q = \{x, \dot{x}\}^T$, which is the system state.

$$\dot{q} = Aq + B_1\ddot{x}_g + B_2 u \tag{18.5}$$

Matrices $A, B_1,$ and B_2 are defined below, and the variable u is the optimal control force vector to be evaluated.

$$A = \begin{bmatrix} 0 & I \\ -M^{-1}K & -M^{-1}\left(C + \Gamma C_d \Gamma^T\right) \end{bmatrix}, \; B_1 = \begin{bmatrix} 0 \\ \Gamma_{eq} \end{bmatrix}, \; B_2 = \begin{bmatrix} 0 \\ M^{-1}\Gamma \end{bmatrix} \tag{18.6}$$

The aim, is to design a controller so that the base displacements in the x and y directions are minimized. This is established by making use of the linear quadratic regulator formulation in which the cost function to be minimized is as follows

$$V_q = \int_0^\infty q^T Q q + u^T R u dt \tag{18.7}$$

where Q and R are positive semi definite weighting matrices. Q is arranged to be a diagonal matrix with values of unity corresponding to the base displacements and zero for all other states. The purpose of this setting is to make the base return to the zero state at times when the controller is active. The matrix R is taken as an identity matrix (same weights for all dampers) with a common multiplier of 1e-8. This common multiplier is the relative weight among the matrices Q and R. The optimal control effort that minimizes Eq. 18.7 requires that

$$u^o = -R^{-1}B_2^T\bar{P}q^o = K_c q^o \tag{18.8}$$

where K_c is the feedback gain matrix, q^o and u^o are the optimum results of the state and control force, respectively. $\bar{P}$ is the symmetric matrix that is the solution to the Riccati equation

$$A^T\bar{P} + \bar{P}A + Q - \bar{P}B_2R^{-1}B_2^T\bar{P} = 0 \tag{18.9}$$

Equations 18.6, 18.7, 18.8 and 18.9 are evaluated for all 625 damping constant possibilities, resulting in 625 K_c matrices. These matrices are stored as a hyper matrix (3 dimensional) and they are recalled during the simulation when needed.

18.4.2 Upper Controller (Gain Scheduling)

An upper controller is designed to switch between the 625 feedback control gains during earthquake simulations. At each time step, the optimal control force is calculated based on the feedback gain for the system with damping constants that are calculated in the previous step. The force that is required for the i'th device is divided by the i'th dampers velocity to obtain the optimum damping constant (see Eq. 18.1). Then the closest damping constant within [5,000–25,000 Ns/m at increments of 5,000] is selected for the next time step.

A passive device, as is the case for dampers, may only absorb energy from the system. That is why the damping force can only act in the opposite direction of its velocity. Hence, if the calculated optimum damping constant has a negative sign, the required force will not be producible. In this case, the damping constant will take its minimum value of 5,000 Ns/m. In addition a numerical precaution is taken to prevent a "divide by zero" error. During the calculation of the optimum damper constant, the smallest absolute damper velocity is limited to 1 mm/s. This does not have a detrimental effect to the structural response, since the worst case causes a force of 25 N only.

The upper controller also decides when the optimum control forces should be applied. The control should only take effect when the isolators are in danger; where "danger" in this study is defined as an isolator displacement of 15 mm or more. Once an isolator exceeds this value, the controller is activated until a minimum or maximum displacement instance is reached that is less than 15 mm. Figure 18.4 shows a schematic representation of the working principle of this upper controller.

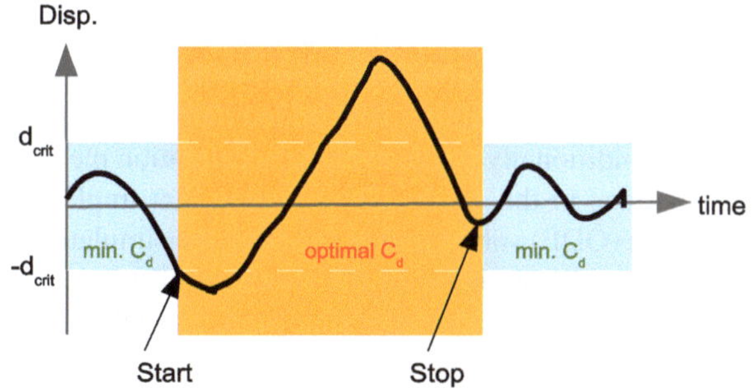

Fig. 18.4 Schematic working principle of the upper controller

Table 18.2 Earthquake characteristics

Earthquake			Date	A_{max} (g)	V_{max} (m/s)	D_{max} (m)
1	1940ElCentro	I-ELC-180	1940	0.296	0.236	0.130
		I-ELC-270	1940	0.179	0.279	0.221
2	12KasBOL	Bolu-000	1999	0.728	0.405	0.230
		Bolu-090	1999	0.822	0.620	0.132
3	17AguSKR	SKR-090	1999	0.317	0.284	0.442
4	1940ElCentro-DF	I-ELC-180-DF	–	0.304	1.18	3.25
		I-ELC-270-DF	–	0.182	1.39	5.51
5	12KasBOL-DF	Bolu-000-DF	–	0.733	2.03	5.74
		Bolu-090-DF	–	0.830	3.10	3.30
6	17AguSKR-DF	SKR-090-DF	–	0.337	1.42	11.0

18.5 Simulations

Six different earthquakes are selected for the simulation of the hybrid controlled isolated 3-D building. The first three are the 19-05-1940 Imperial Valley (El Centro Station), 12-11-1999 Düzce (Bolu Station) and the 17-08-1999 Kocaeli (Sakarya Station) earthquake acceleration records (http://peer.berkeley.edu). The North-South and the East-West components of the earthquake records are applied to the x and y-directions of the structure, respectively. Unfortunately, the North-south component of the Sakarya Station is not available; therefore the East-West components of this earthquake are used in both directions. In order to obtain near fault seismic action, these three earthquake records are modified to obtain earthquakes with high period responses. This is done by simply extending the sampling period to fivefold of the original sampling time ($dt = 0.01$ s. $\rightarrow dt = 0.05$ s). Thereafter, cubic spline interpolation is carried out to obtain data with $dt = 0.01$ s. The response characteristics of the earthquakes used in this study are presented in Table 18.2.

The generated earthquake records in Table 18.2 have much larger velocity and displacement values due to the increased time step. The use of spline interpolation also introduced slightly larger maximum accelerations than the original data. This fact is expected from the numerical procedure and it can be neglected.

The simulations are carried out for the first 20 s of the first three, and the first 100 s of the last three earthquakes. The major response is seen in this time frame, and it also allows for more detail in the illustrations. A direct integration method for the solution of the equation of motion in Eq. 18.2 is used. Superposition of modal responses is not possible for systems with non-proportional damping, as is the case with the current structure with added dampers. The Newmark β method (by using the unconditionally stable average acceleration method) is used as the solver for all simulations in this study. The function that implements this ordinary differential equation (ODE) solver makes sure that the simulation time step is 20 times smaller than the smallest period of the structure. If this is not the case, it interpolates the excitation data for a smaller time step, and later outputs the response at a 0.01 s. In this work, the building type structure has a minimum period of 0.108 s. Thus, the simulation takes place at $0.108/20 = 0.0054$ s.

Figures 18.5, 18.6, 18.7, 18.8, 18.9 and 18.10 show simulation responses of the isolated building subjected to the selected earthquake records (see Table 18.2) with the upper controller. Each figure shows three responses on a single plot together with a digital indicator at the bottom. The three plots are the responses of the isolated building with damping at minimum stage, damping at maximum stage, and optimally controlled damping by using the upper controller. The digital indicator at the bottom shows if the upper controller is activated or not. The plotted responses are in the x-direction (North-South) of the building. The y-directional responses have smaller amplitudes, and hence are not shown.

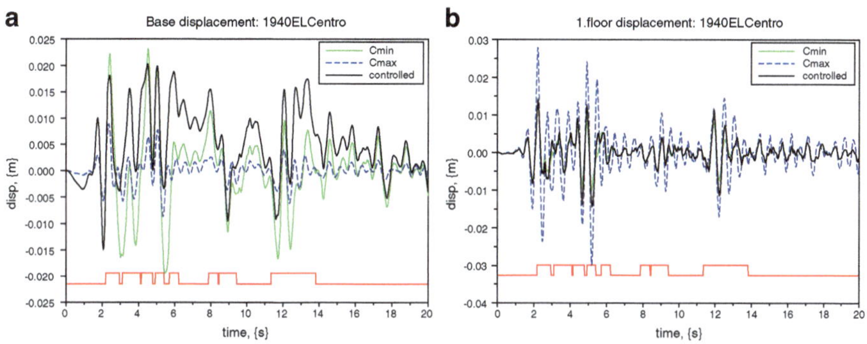

Fig. 18.5 El Centro simulation response. (**a**) Base displacement (node 3, x-dir.). (**b**) 1. Floor displacement (node 5, x-dir.)

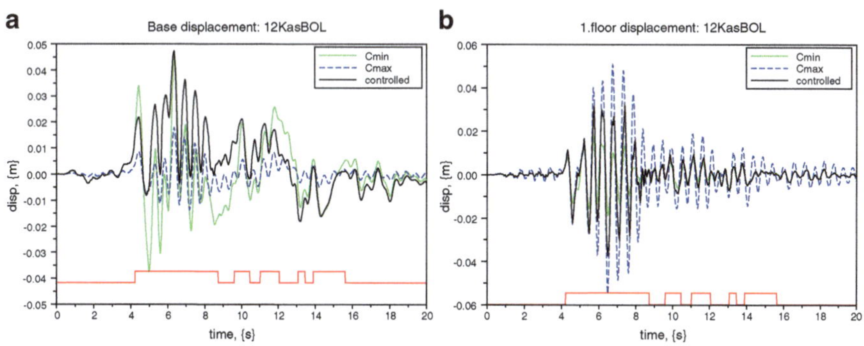

Fig. 18.6 12KasBOL simulation response. (**a**) Base displacement (node 3, x-dir.). (**b**) 1. Floor displacement (node 5, x-dir.)

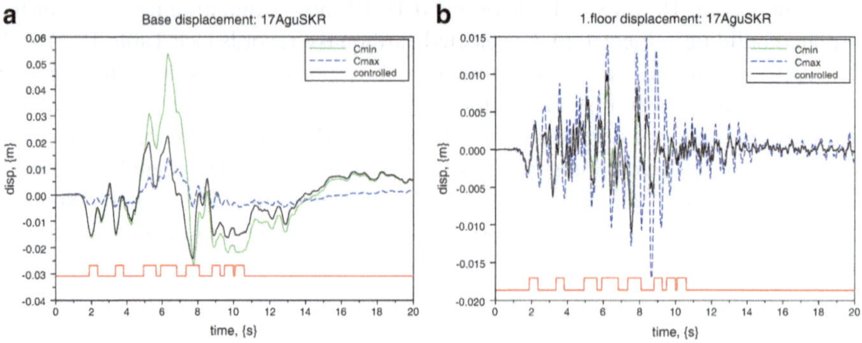

Fig. 18.7 17AguSKR simulation response. (**a**) Base displacement (node 3, x-dir.). (**b**) 1. Floor displacement (node 5, x-dir.)

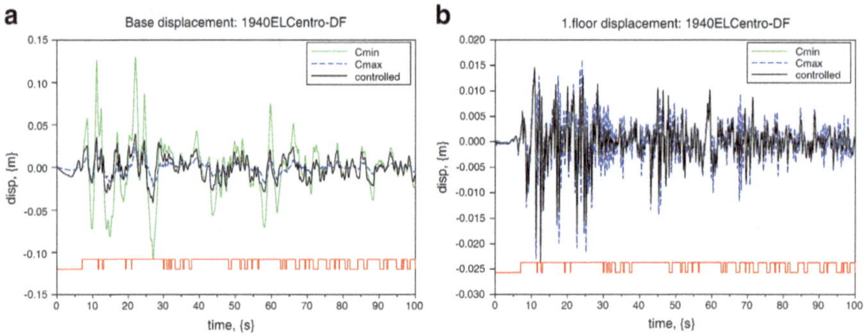

Fig. 18.8 El Centro-DF simulation response. (**a**) Base displacement (node 3, x-dir.). (**b**) 1. Floor displacement (node 5, x-dir.)

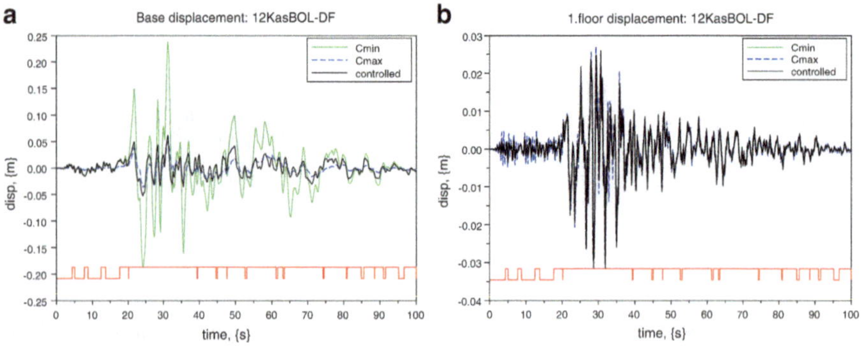

Fig. 18.9 12KasBOL-DF simulation response. (**a**) Base displacement (node 3, x-dir.). (**b**) 1. Floor displacement (node 5, x-dir.)

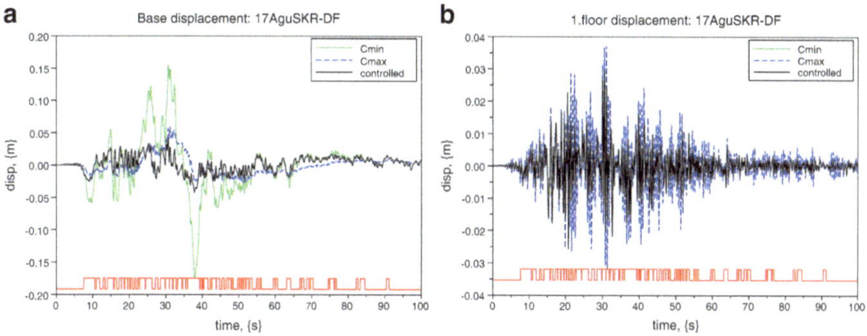

Fig. 18.10 17AguSKR-DF simulation response. (**a**) Base displacement (node 3, x-dir.). (**b**) 1. Floor displacement (node 5, x-dir.)

It is expected that the controlled response of the semi-active seismically isolated building stays in between the responses of the isolated building with passive minimum and maximum damping levels. Figures 18.5a, 18.6, 18.7, 18.8, 18.9 and 18.10a show the base displacement response for the earthquakes under consideration in which the expected response behavior cannot be seen directly in these figures. The reason seems to arise from the over-damping effect, which hinders the base from returning to the origin at some instances. If these specific instances are neglected, the controlled response obeys the above mentioned rule. Figures 18.5b, 18.6, 18.7, 18.8, 18.9 and 18.10b show the first floor response of the controlled simulation and these are always in-between the lower and upper bounds of the pure minimum and pure maximum damping levels, as expected.

The major benefit of the control mechanism can be seen in the comparison of the far field with the near field earthquake responses. The maximum base displacement of the structure with minimum damping is below 0.05 m for the far field earthquakes according to Figs. 18.5a, 18.6 and 18.7a. These maximum values rise to about 0.25 m for the near field earthquakes (Figs. 18.8a, 18.9 and 18.10a). The controlled base displacement is affected much less. It is below 0.05 m for both near field and far field earthquakes.

In Figs. 18.5b, 18.6 and 18.7b the controlled first floor response approaches the minimum damping behavior, while in Figs. 18.8b, 18.9 and 18.10b, the controlled first floor response approaches the maximum damping behavior. In other words, the structure reaction is soft for far field type of ground excitations, and stiff for near field type of ground excitations. This behavior can also be verified by the fact that the upper controller is active for a much longer time period during the near field type earthquakes (see the red digital line at the bottom of these figures).

18.6 Conclusions

The aim of seismic isolation is to decrease the drift in the superstructure, and the aim of the present control design is to avoid damage in the isolators and hence the superstructure. From the superstructure's point of view the case with minimum damping, or even further, no damping at the base level appears to be the most feasible control solution for far-field earthquake excitations. From the isolation system's point of view, the highest damping case would be the preferred one since the base displacements will be small. On the contrary, in the case of near field earthquakes, the most feasible solution for both the isolators and the superstructure is to have large damping. A balance is established by using the upper controller. It can smoothly switch between minimum and maximum damping values and thereby reduce the structural response based on any given excitation type.

A damper with a fixed damping coefficient will not be able to achieve a similar performance as the presented control method. The best response in far field type earthquakes is established by minimum damping with an extra control force at some time instances; near field type earthquakes, on the other hand, appear to be best handled with a fixed base. Therefore, in either case a fixed valued damper will not suffice to produce a desirable structural response. At last, a smartly controlled semi-active damper is able to protect a seismically isolated building from both near and far field type earthquakes.

References

Aldemir Ü, Bakioğlu M (2000) Semiactive control of earthquake-excited structures. Turk J Eng Environ Sci 24:237–246

Çetin S, Sivrioğlu S, Zergeroğlu E, Yüksek İ (2009) Semi-active Hinf robust control of six degree of freedom structural system using MR damper. Otomatik Kontrol Türk Milli Komitesi Otomatik Kontrol Ulusal Toplantısı, YTÜ, İstanbul

Kurata N, Kobori T, Takahashi M, Niwa N, Midorikawa H (1999) Actual seismic response controlled building with semi-active damper system. Earthq Eng Struct Dyn 28:1427–1447

Turan G, Aydin E (2010) Degisken sönümleme katsayili amortisörlerin deprem simülasyonu ile uc katli bir yapiya olan etkisinin degerlendirilmesi, Rapor No 107M353 (in Turkish). Tübitak

Wongprasert N, Symans MD (2005) Experimental evaluation of adaptive elastomeric base-isolated structures using variable-orifice fluid dampers. ASCE J Struct Eng 131(6):867–877

Chapter 19
Substructure Pseudo-Dynamic Tests on Seismic Response Control of Soft-First-Story Buildings

Hideto Kanno, Tetsuya Nishida, and Jun Kobayashi

Abstract Soft-first-story structures, such as piloti buildings, are known as vulnerable structures against earthquakes. In this chapter, a simple scheme for reducing the structural damage of such buildings is proposed. Its effectiveness is experimentally examined through substructure pseudo-dynamic tests. In the proposed method, low yield strength steel devices are applied as elasto-plastic dampers at the first story of the buildings to reduce the seismic response and damage. A six-story single-span piloti model, with or without steel dampers, are the subject of the test. The behavior of the two exterior columns at the first story and the steel damper are tested. The substructure pseudo-dynamic tests are successfully performed to investigate the elasto-plastic behavior of the damper and the reinforced concrete columns at the soft-first-story, as well as the overall structural performance. The experimental results show that the seismic damage of piloti buildings can be reduced with steel dampers, which have been found to work as effectively as expected.

19.1 Introduction

Soft-first-story structures, such as piloti buildings, are vulnerable against earthquake (Naeim and Lew 2000). In the 1995 Kobe earthquake, some piloti buildings designed with modern design codes, as well as many older buildings, suffered serious structural damage. Most of the damage was concentrated at the first story, owing to the change in lateral stiffness and strength compared to the upper part of the structure. One way to reduce the seismic damage of such structures is to increase the column section at the first story, in order to reduce the stiffness discontinuity (Kaushik et al. 2009 and Lu et al. 1999). However, this causes construction

H. Kanno (✉) • T. Nishida • J. Kobayashi
Department of Architecture and Environment Systems, Faculty of Systems Science and Technology, Akita Prefectural University, 84-4 Ebinokuchi, Tsuchiya, Yurihonjo, Akita 015-0055, Japan
e-mail: kanno@akita-pu.ac.jp; tetsuya_nishida@akita-pu.ac.jp; jun_kobayashi@akita-pu.ac.jp

A. Ilki and M.N. Fardis (eds.), *Seismic Evaluation and Rehabilitation of Structures*, Geotechnical, Geological and Earthquake Engineering 26,
DOI 10.1007/978-3-319-00458-7_19,

problems, due to the discontinuity of the main bars of the columns. Another way is to dissipate seismic energy in the soft-first-story using dynamic dampers, thus taking advantage of the concentration of structural deformations in the soft story (Iqbal 2006, Mezzi and Parducci 2005 and Todorovska 2008). In this chapter a simple scheme is proposed in order to reduce structural damage in such piloti buildings, using low-yield-strength-steel devices as elasto-plastic dampers in the first story. The detailed response of such buildings depends on the precise elasto-plastic behavior of the steel dampers. The shape of the hysteresis loop of the steel damper, including fatigue behavior, is particularly important. A full size test seems to be almost impossible; therefore a scaled model and the substructure pseudo-dynamic test method were conceived for this purpose.

In this study, six-story piloti reinforced concrete (RC) frames are experimentally tested, in order to examine the effectiveness of steel dampers in reducing the structural damage. Two sets of structural models are compared to each other through substructure pseudo-dynamic tests.

19.2 Outline of the Tests

19.2.1 Building Model

Figure 19.1 shows the prototype of the six-story single-span piloti frame. Its transverse span is 9 m. The height of the first story is 3.5 m and that of the upper floors is 3.1 m. There are four spans in the longitudinal direction of the building. The central single frame in the transverse direction is considered under one-directional in-plane loading. The steel damper is placed between the lower and the upper beam at the first story of the frame. The connections of the damper are assumed to be rigid, and their deformations are neglected.

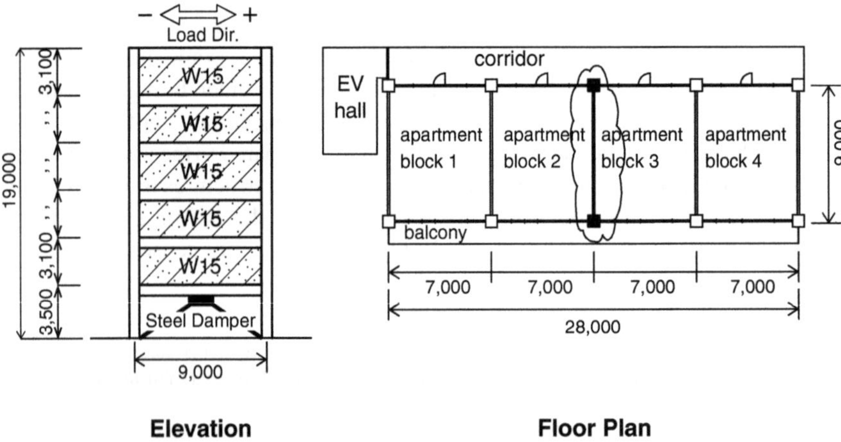

Fig. 19.1 Model frames (*dimensions in mm*)

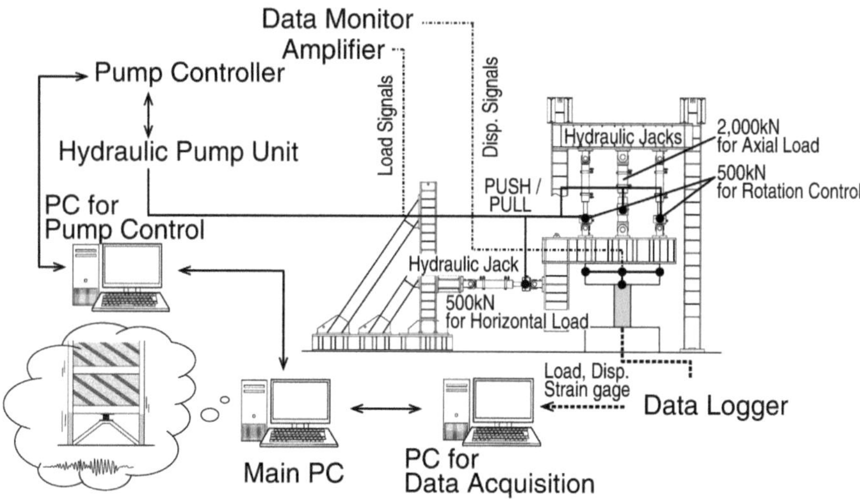

Fig. 19.2 Schematic diagram of testing system

19.2.2 Testing System

In order to conduct the substructure pseudo-dynamic test, the prototype structure is modelled in two parts. One is the actual specimen which is experimentally tested so that the elasto-plastic behavior of the critical part of the structure is realistically represented. The behavior of the other part, which is expected not to critically affect the total response of the structure, is numerically calculated using a conventional nonlinear computational approach. The substructure pseudo-dynamic test is developed on a basis of pseudo-dynamic testing, which is an experimental technique for simulating the seismic response of the tested structure or component.

In the present tests, two exterior columns and a steel damper on the first floor are the tested part, which are represented by three separate specimens. Figure 19.2 shows the scheme of pseudo-dynamic testing developed at the Akita Prefectural University (Teramoto et al. 2008). The system consists of the main management part (Main PC), the data acquisition and the hydraulic pump control. The main management part controls the system and executes the numerical analysis during the tests. The main PC sends command signals, such as the target displacement for each tested part, to the hydraulic pump control unit, so that the target displacement can be reached. Then the measured information, such as the restoring forces and moments of the tested parts, is sent back to the Main PC together with analytical results of the numerical models to calculate the total structural behavior. The loading conditions for the tested part, the deformations, axial loads and rotation angles of beam-column connections, are calculated step by step prior to the next step of the loading. Loading is carried out with an accuracy of 0.01 mm using servo-controlled hydraulic jacks. Loads in the two directions, horizontal and vertical, and the moment at the column top, are measured simultaneously.

The integration method using the operator-splitting (OS) method was applied in the tests. This method has been used in many Substructure pseudo-dynamic experiments to calculate the seismic response considering the interaction between the specimen and the whole frame (Nakashima et al. 1990). The formulation of this method is as follows:

$$Ma_{\mathrm{n+1}} + Cv_{\mathrm{n+1}} + K^{I}d_{\mathrm{n+1}} + K^{E}_{n+1}\,d_{\mathrm{n+1}} = P_{\mathrm{n+1}} \tag{19.1}$$

$$\tilde{d}_{\mathrm{n+1}} = d_{\mathrm{n}} + v_{\mathrm{n}}\,\Delta t + a_{\mathrm{n}}\left(\frac{\Delta t}{2}\right)^{2} \tag{19.2}$$

$$d_{\mathrm{n+1}} = \tilde{d}_{\mathrm{n+1}} + a_{\mathrm{n+1}}\left(\frac{\Delta t}{2}\right)^{2} \tag{19.3}$$

$$v_{\mathrm{n+1}} = v_{\mathrm{n}} + (a_{\mathrm{n}} + a_{\mathrm{n+1}})\left(\frac{\Delta t}{2}\right) \tag{19.4}$$

In Eqs. (19.1), (19.2), (19.3) and (19.4) K^{I} and $K^{E}_{\mathrm{n+1}}$ are the linear and the non-linear stiffness matrices, M and C are the mass and viscous damping matrices, $\tilde{d}$ and d are the predictor and corrector displacement vectors, v and a are the velocity and acceleration vectors, and Δt is the integration time interval, respectively. The main characteristic of this method is the division of the stiffness of the whole structure into a linear and a non-linear stiffness. For the non-linear tested part, the explicit predictor-corrector method is used.

The Newmark's β method (linear stiffness integration method) is also applied to take account of the non-linear part of the whole structure. If the linear stiffness is much larger than the non-linear stiffness, the integration method is unconditionally stable. In this work, the equations were transformed into an incremental form. It was confirmed that even when the incremental form of the equations is applied, the condition where the linear stiffness is larger than the non-linear tangent stiffness ensures that the integration method gives a stable solution.

The procedure for the substructure pseudo-dynamic tests in this study is as follows:

1. By using the integration method (OS method), the target predictor displacement at the next step is calculated. The Main PC (shown in Fig. 19.2) is used in this calculation.
2. The main PC sends the target displacement (horizontal displacement and rotation angle at the top of the column and steel damper) to the PCs for the Pump Control Unit of each specimen (the two RC columns and the steel damper).

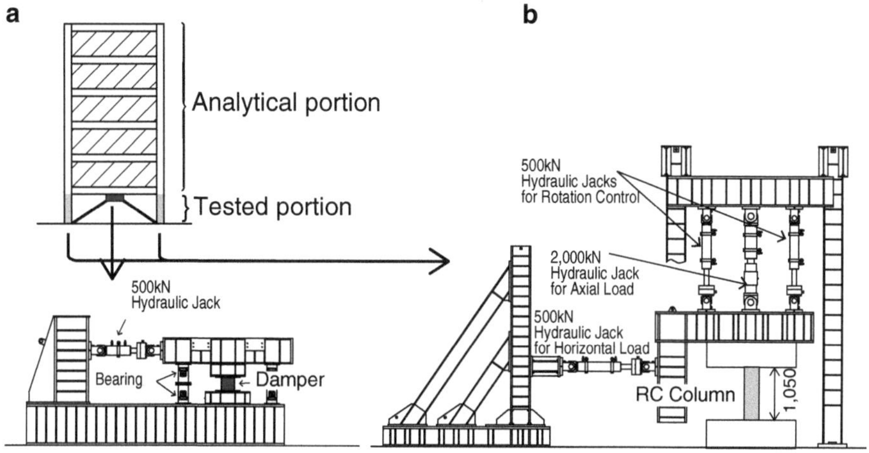

Fig. 19.3 Set up for tested parts. (**a**) Test setup for the damper. (**b**) Test setup for RC columns

3. Until the current displacement reaches the target value, the specimens are loaded by the four jacks (one jack in the case of the damper) under the control.
4. The restoring forces (horizontal load and moment at the top of the column) are measured when the displacements reach the target value by the load cells attached to the jacks, and the values are fed back to the Main PC.
5. After the restoring forces are fed back to the Main PC, the latter sends a command to the PC for Data Acquisition.
6. After the data acquisition is completed, step 1 is carried out using the data on the restoring force and the input acceleration.

19.2.3 Tested Part

The test setups are shown in Fig. 19.3. In this study, two RC specimens and one steel damper on the first floor are tested individually but concurrently. Therefore two kinds of loading systems are provided. Each specimen must be loaded simultaneously with the analytical calculated loads coming from the upper part of the framed structure. Therefore, four static hydraulic actuators are connected to each column specimen, so that loading with three degree of freedom can be performed. One actuator is connected to the damper and loading takes place with one degree of freedom. In the case of RC column, as shown in Fig. 19.3b, the middle actuator in the vertical direction has a loading capacity of 2,000 kN; the force capacity of the other two vertical actuators and one horizontal actuator is 500 kN. The force capacity of the horizontal actuator for the damper is 500 kN.

The specimens for the RC columns are shown in Fig. 19.4. The size of the column is 300 × 300 mm and its clear height 1.05 m, at a 3:8 scale of the prototype

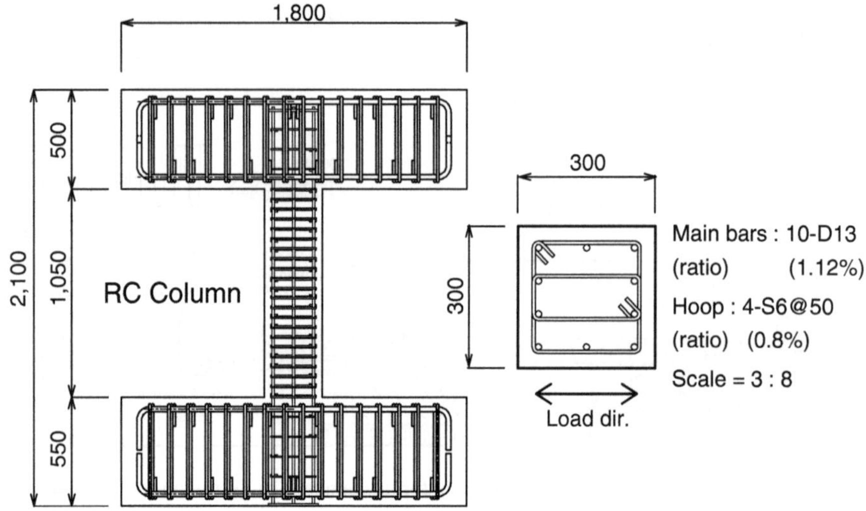

Fig. 19.4 Details of the RC column

building. The main reinforcement consists of ribbed (deformed) 13 mm bars with an average yield strength of 377 N/mm^2. The shear reinforcement consists of ribbed (deformed) 6 mm high strength bars, with an average strength of 1,100 N/mm^2. The compressive strength of the concrete is 38.8 N/mm^2.

The steel damper is designed so that about 10 % of the critical damping can be provided to the building model. Figure 19.5 shows details of the low-yield-strength steel damper.

The steel damper is panel-shaped, with dimensions of 262 mm by 262 mm and 6 mm thickness. A full-scale model is used for the damper, to provide realistic elasto-plastic hysteretic behavior. The low yield strength steel has an average yield strength of 154 N/mm^2.

19.2.4 Analytical Models

In the analytical part, shear walls in the upper part of the building are modeled as elastic axial springs and elasto-plastic shear springs. Multi-springs (MS) are added to both end sections of shear walls, as illustrated in Fig. 19.6. The MS model consists of five layers of steel and concrete springs at selected locations. These springs are assumed as individual axial springs, representing the stiffness of the longitudinal reinforcement and the concrete. A shear spring is used to represent the shear hysteretic behavior of the shear wall. In this study, the beams connecting to shear walls are treated as rigid at both ends of the shear wall.

The hysteretic characteristics of the reinforcement and the concrete are shown in Fig. 19.7. The force-deformation relationship with a bilinear skeleton curve for

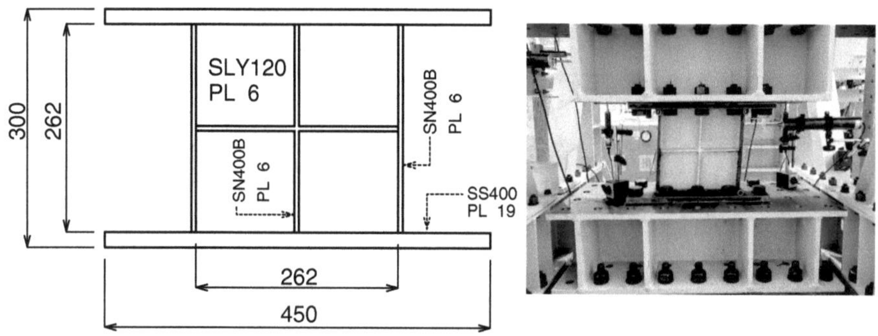

Fig. 19.5 Details of the low-yield-strength steel damper

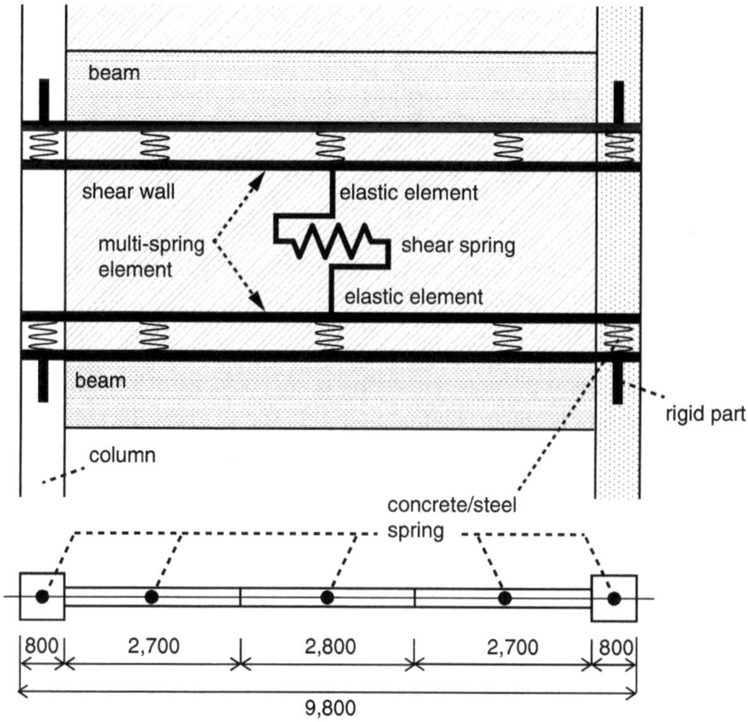

Fig. 19.6 Analytical model for shear walls

steel springs (the reinforcement) is shown in Fig. 19.7a, where f_s and d_s are axial force and deformation of the steel spring, respectively, and f_{sy}, d_y are the yield strength and yielding deformation. The unloading stiffness is assumed equal to the initial stiffness of the steel spring.

The hysteretic relationship for the concrete spring is assumed for simplicity as a tri-linear skeleton curve, as shown in Fig. 19.7b, where f_c and d_c are axial force and axial deformation of the concrete spring respectively; f_{cy} is the compressive

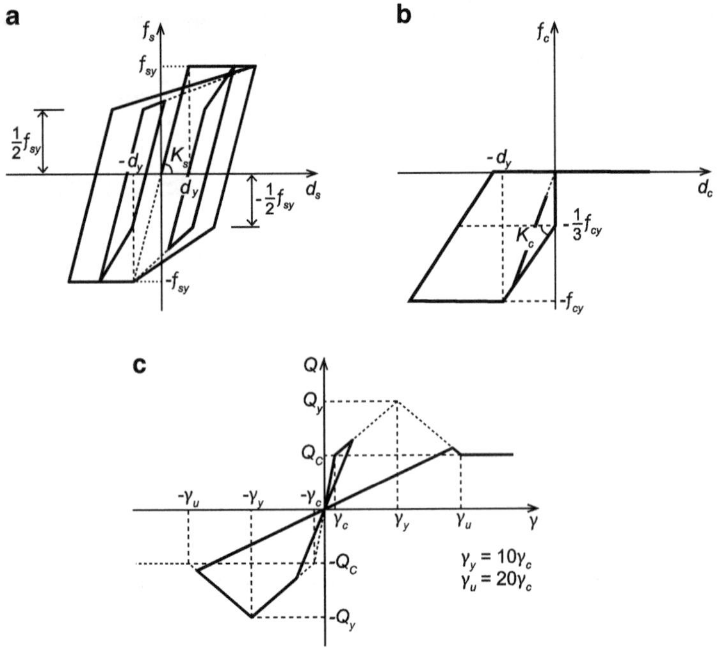

Fig. 19.7 Hysteretic model for axial (steel and concrete) and shear springs. (**a**) Steel spring. (**b**) Concrete spring. (**c**) Shear spring

strength of the concrete spring. The tensile resistance of concrete is neglected. The post-peak degradation of the concrete spring is not considered. The initial stiffness of the concrete spring is taken as infinitely rigid. The stiffness degrading behavior after the initial stiffness level is taken into account, when the axial stress is larger than $f_{cy}/3$.

Regarding the shear springs, it is assumed that shear has no effects on the axial force and bending interaction. The hysteretic model for the shear spring is assumed to be multi-linear model, as shown in Fig. 19.7c, where Q is shear force and γ the deflection of the shear wall due to shear deformation. The value of the shear force and deflection at cracking and yielding can be calculated according to the formulations proposed by the Architectural Institute of Japan.

Two series of substructure pseudo-dynamic tests were conducted. One on a piloti frame with the steel damper placed in the soft first story; the other on the piloti frame only without the steel damper, to confirm the damper effect. In these tests, the integration time interval was 0.01 s. Rayleigh damping was applied and the viscous damping ratio was set at 3 %. A simulated earthquake ground motion as shown in Fig. 19.8 was used, based on the phase characteristics of the El Centro 1940 record. The target spectral characteristic was based on the design spectrum specified in the Japanese building design code. Both series of tests, with the damper or without, were divided in three stages, from weak elastic response (RUN1) to the strong inelastic level (RUN3). After RUN3 (the maximum test level), the second stage level was carried out again as RUN4 to study the aftershock response.

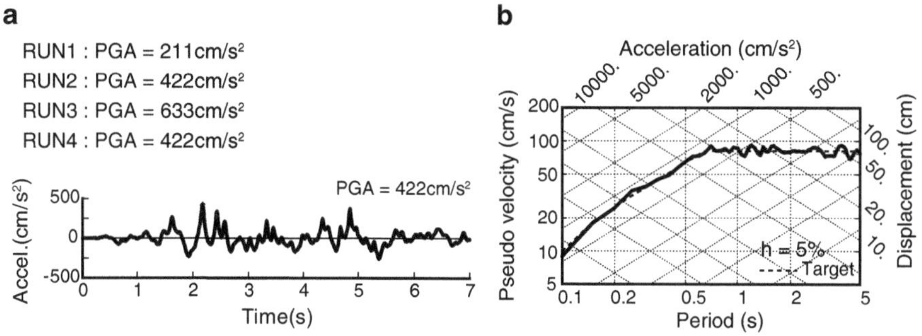

Fig. 19.8 Input motion. (a) Input wave. (b) Response spectrum

19.3 Results of Tests

19.3.1 Outline of Test Results

Figure 19.9 shows base shear-story drift hysteresis loops obtained from the tests. Figure 19.10 depicts the shear force-deformation relationships of the tested members during RUN3. The upper part of each figure refers to the piloti frame without a damper and the bottom one to the case with the damper. It can be seen that when the damper is used in the soft-first-story the performance of the piloti frame is good, especially at larger excitation levels.

During RUN1 both models (with or without a damper) show almost elastic behavior. In the model without a damper, some main bars yielded at the bottom of the RC column during RUN2. In the subsequent loading RUN3 all reinforcing bars yielded and the story drift angle reached about 12.5 mrad. By contrast, in the model with the damper, only some of the main bars yield during RUN3 and the peak displacement does not exceed half of that of the model without damper. During RUN4, which represents the behavior in an aftershock, both models show larger response than RUN2; however, the model with a damper has much lower displacement response than the one without a damper.

The peak lateral load distributions in both tests are shown in Fig. 19.11. Some studies (e.g. Lu et al. 1999) suggest that, in piloti structures the maximum lateral load on each floor is uniformly distributed, because most of the displacement tends to concentrate in the first story. The results of the present research confirm those findings, even when there is a damper at the soft-first-story.

19.3.2 Energy Response

The seismic energy response of soft-first-story frames is discussed based on the results of the substructure pseudo-dynamic tests. The total input energy, E_I, and

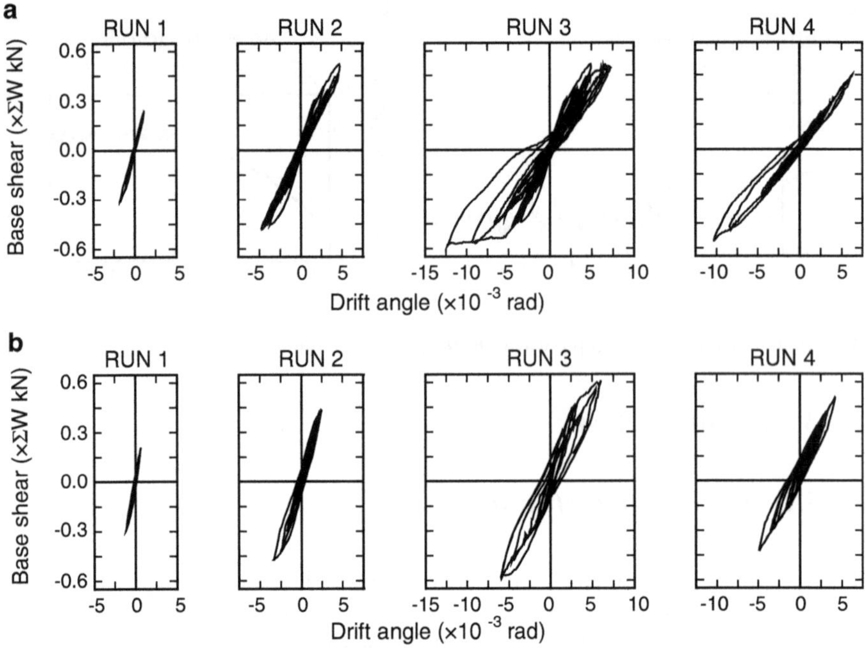

Fig. 19.9 Base shear-drift angle relationships (test results). (**a**) w/o damper. (**b**) With damper

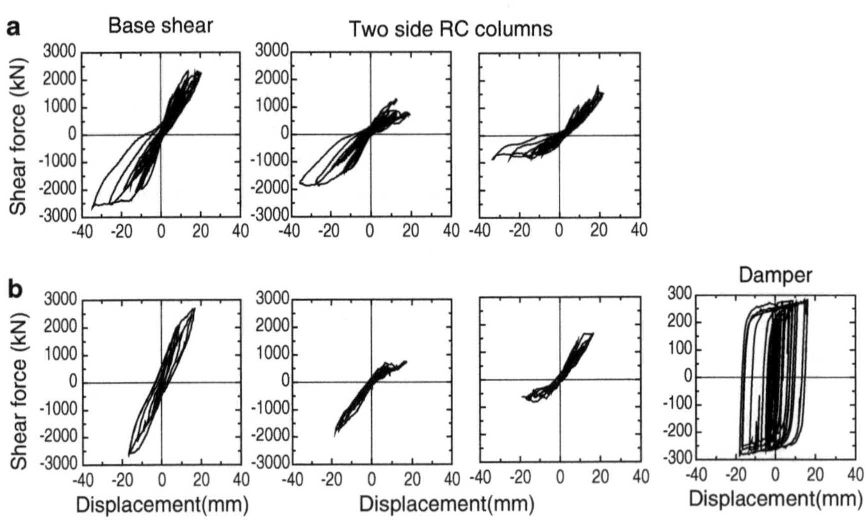

Fig. 19.10 Shear force-deformation relationships of tested members during RUN3. (**a**) w/o damper. (**b**) With damper

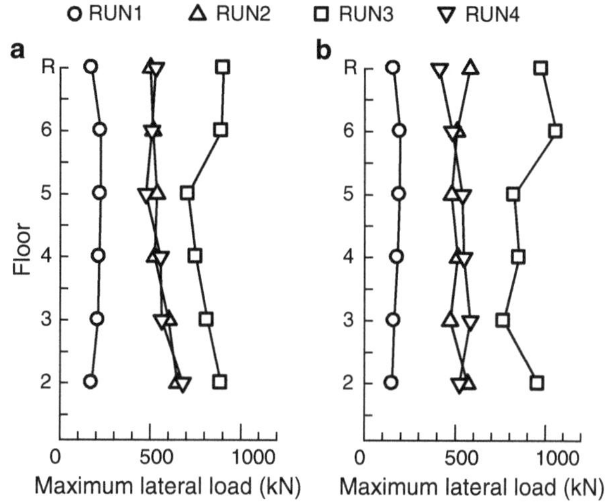

Fig. 19.11 Maximum lateral load distributions in both tests. (**a**) w/o damper. (**b**) With damper

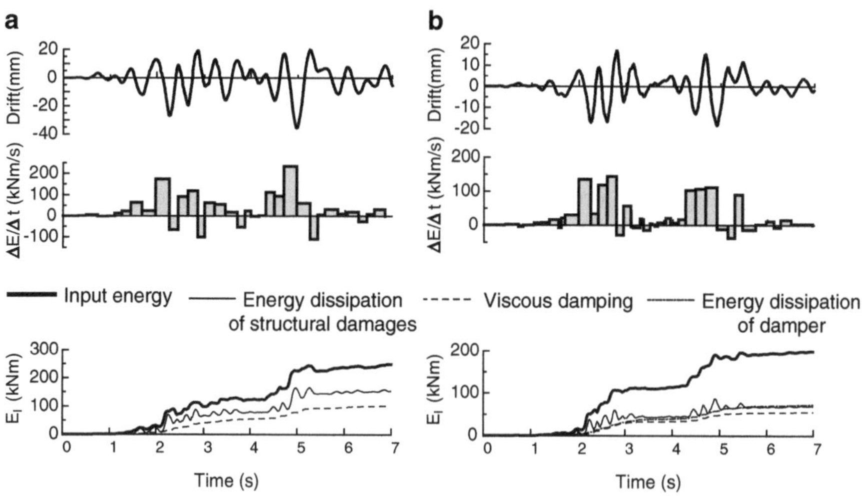

Fig. 19.12 Time history of energy response during RUN3. (**a**) w/o damper. (**b**) With damper

the momentary input energy, ΔE, are discussed dividing the input energy into hysteretic energy dissipation in the RC columns (in the lateral and the axial direction) of the soft-first-story and in the shear walls of the upper story, viscous damping and energy dissipation in the dampers of the first story. By momentary input energy is meant the increment of seismic input energy in half a cycle of the response (Nakamura et al. 1998). Figure 19.12 shows the time-histories of displacement and seismic response energy, ΔE and E_I, during RUN3, which is the maximum level of loading in the tests. It can be seen that, in both tests, ΔE is large when the displacement is increased. Figure 19.13 shows the relationships between the

Fig. 19.13 Maximum moment input energy-maximum story drift relationships

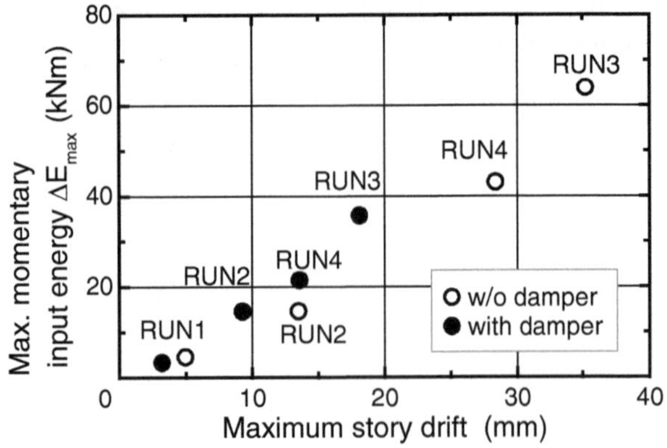

maximum momentary input energy $\Delta E_{\max}$ and the maximum drift of the first story. It suggests that $\Delta E_{\max}$ reflects the intensity of the energy input, related to the peak response displacement during the transient response.

Figure 19.14 shows the ratio of dissipated energy to seismic input in each stage of testing. In the cases without a damper, the energy dissipated by the RC columns is about 40–50 % of the input energy; the viscous damping energy is about 30–40 %. This tendency is observed in all the stages and input levels. When using the damper, the energy dissipation by the damper is very large (about 40 % of the total energy). As a result, the energy dissipation due to column damage and viscous damping is less than the one without the damper. More specifically, the energy dissipation ratio during RUN4 shows that the steel damper worked adequately, even for comparatively large aftershocks.

19.4 Conclusion

Substructure pseudo-dynamic tests were successfully performed to investigate the dynamic behavior of the reinforced concrete columns of the soft-first-story and of the damper placed in that story, as well as the total structural performance. Experimental results indicate that the seismic damage of piloti buildings can be reduced with steel dampers, as these worked as effectively as expected. More specifically, the steel dampers provide good performance during the largest ground motion and large aftershocks. The maximum response displacement has a stronger correlation with the maximum momentary input energy, $\Delta E_{\max}$, than with the amount of total input energy, because of the characteristics of the inelastic behavior of RC structures during earthquakes.

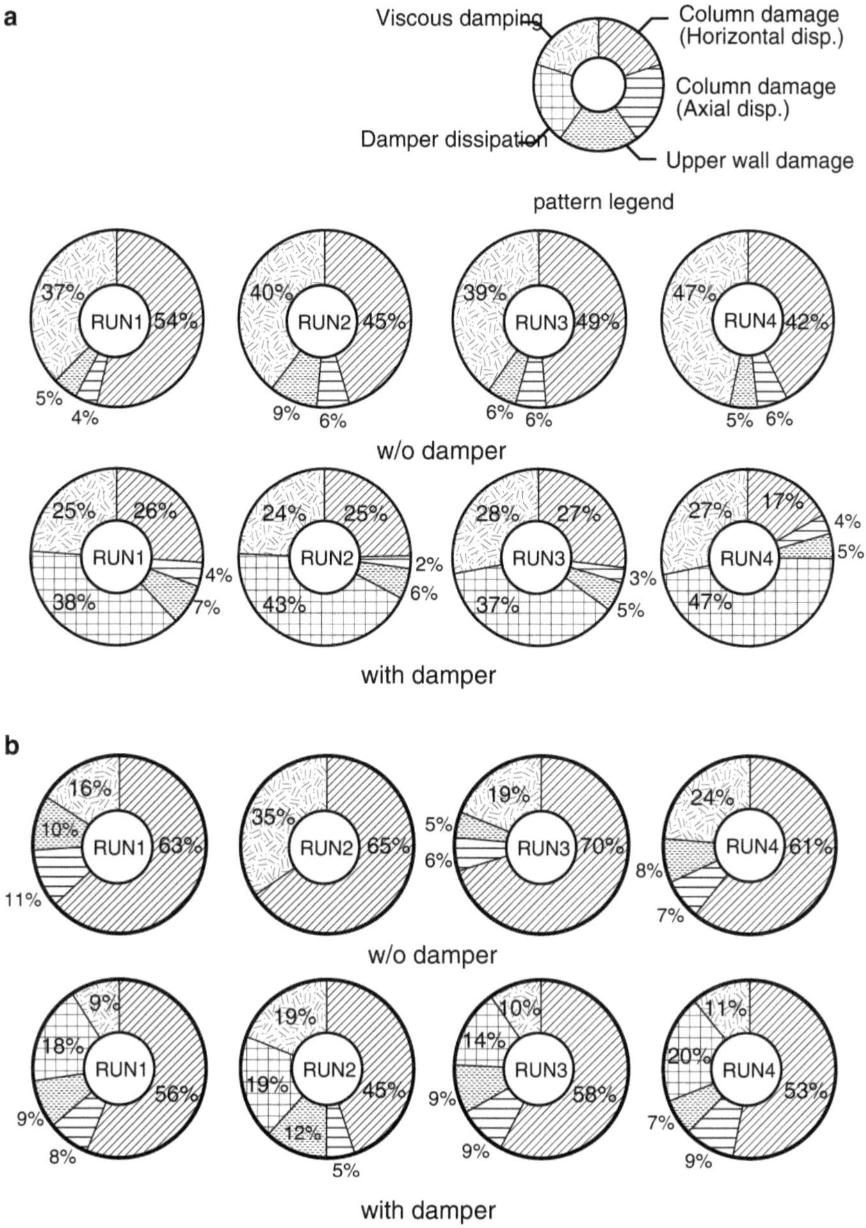

Fig. 19.14 Energy dissipation ratio. (**a**) Total input energy. (**b**) Maximum momentary input energy

References

Iqbal A (2006) Soft first storey with seismic isolation system. In: NZSEE conference, Napier, CDROM paperID 36

Kaushik HB, Rai DC, Jain SK (2009) Effectiveness of some strengthening options for masonry-infilled RC frames with open first storey. J Struct Eng ASCE 135(8):925–937

Lu Y, Tassios TP, Zhang G-F, Vintzileou E (1999) Seismic response of reinforced concrete frames with strength and stiffness irregularities. ACI Struct J 96(2):221–229

Mezzi M, Parducci A (2005) Preservation of existing soft-first-storey configurations by improving the seismic performance. In: 3rd specialty conference on the conceptual approach to structural design

Naeim F, Lew M (2000) The 1999 earthquake disasters worldwide: how many times do we have to re-learn the fundamentals of seismic engineering? Struct Des Tall Build 9(2):161–182

Nakamura T, Hori N, Inoue N (1998) Evaluation of damaging properties of ground motions and estimation of maximum displacement based on momentary input energy. J Struct Constr Eng (in Japanese) 513:65–72

Nakashima M, Kaminosono T, Ishida M, Ando K (1990) Integration techniques for substructure pseudo-dynamic test. In: 4th US national conference on earthquake engineering, vol 2, pp 512–524

Teramoto N, Nishida T, Kobayashi J (2008) Structural performance of side columns subjected to varying axial load. In: 14th world conference on earthquake engineering, paper-ID 05-03-0234

Todorovska MI (1999) Base isolation by a soft first storey with inclined columns. J Eng Mech ASCE 1254:448–457

Chapter 20
Towards Robust Behavioral Modeling of Reinforced Concrete Members

Kutay Orakcal

Abstract Performance–based seismic design and assessment guidelines promote nonlinear response history analysis of structures using analytical models that can realistically represent the behavioral characteristics of the structural members. For implementation and improvement of such performance–based methodologies, reliable analytical modeling approaches must be used to represent the cyclic nonlinear behavior of the individual structural members in the building, as well as their interaction in the structural system. In this chapter, examples of novel analytical modeling approaches are presented, for simulating the nonlinear response of reinforced concrete structural components, under reversed cyclic loading conditions. Accuracy of the presented modeling approaches are demonstrated via comparison of the model predictions with test results conducted on reinforced concrete wall, column, and panel specimens, with various geometries, material strengths, and reinforcement detailing characteristics. Continuing research efforts on further improvement of the models are also outlined.

20.1 Introduction

Modern codes and guidelines on performance-based seismic design and assessment require nonlinear response analysis of structures, using analytical modeling approaches which can realistically represent the behavioral characteristics of the structural components at both global and local response levels. In code-compliant seismic performance evaluation methods for existing buildings, the performance of the building is dictated by the intensity of damage occurring in the structural members, which in turn depends on the level of nonlinear deformation (e.g., plastic rotations, strains on concrete and reinforcing steel, nonlinear inter-story drifts) on the structural

K. Orakcal (✉)
Department of Civil Engineering, Bogazici University, 34342 Bebek, Istanbul, Turkey
e-mail: kutay.orakcal@boun.edu.tr

A. Ilki and M.N. Fardis (eds.), *Seismic Evaluation and Rehabilitation of Structures*, Geotechnical, Geological and Earthquake Engineering 26,
DOI 10.1007/978-3-319-00458-7_20,

members, under multiple earthquake excitation scenarios. For implementation and improvement of such performance assessment methodologies, reliable analytical modeling approaches must be used to represent the cyclic nonlinear behavior of the individual structural members in the building, as well as their interaction in the structural system.

For this purpose, examples of novel analytical modeling approaches are presented herein, for simulating the nonlinear response of reinforced concrete structural components, under reversed cyclic loading conditions. The emphasis is on simulation of nonlinear flexural, shear, and bond-slip responses in reinforced concrete walls, columns, and panel elements. The modeling methodologies presented are shown to be behaviorally-robust in capturing the critical response attributes of walls and columns; in terms of capacity, stiffness, ductility, and reversed cyclic response characteristics, at both global (force-displacement) and local (inelastic rotations, flexural and shear deformation components, stresses and strains in concrete and reinforcing steel) response levels. Accuracy of the presented modeling approaches are demonstrated via comparison of the model predictions with test results conducted on densely-instrumented reinforced concrete wall, column, and panel specimens, with various geometries, material strengths, and reinforcement detailing characteristics. Continuing research efforts on further improvement of the models are also outlined.

20.2 Modeling of Nonlinear Flexural Responses in Slender Walls

Various phenomenological macroscopic models have been proposed for simulating the nonlinear inelastic flexural response of slender walls under reversed cyclic loading conditions. Upon extensive studies, the multiple-vertical-line-element model (MVLEM) proposed originally by Vulcano et al. (1988) has been shown to successfully balance the simplicity of a macroscopic model and the refinements of a microscopic model. The MVLEM captures important behavioral features (e.g., migration of the neutral axis, coupling of axial and flexural responses, nonlinear shear behavior), which are commonly ignored in simple models, and offers the flexibility to incorporate various hysteretic constitutive models for a refined representation of material behavior and confinement effects. A description of the model formulation, detailed hysteretic constitutive relationships implemented in the model, and the sensitivity of the model predictions to both model and material parameters were presented by Orakcal et al. (2004). The accuracy and limitations of the model were discussed by Orakcal and Wallace (2006), upon comparison of the model predictions with experimental results at both global and local response levels.

The model in Fig. 20.1a is an implementation of the generic MVLEM for structural walls. A horizontal spring placed at the element center of rotation (at relative height ch) simulates the shear response of the wall element. Flexural and

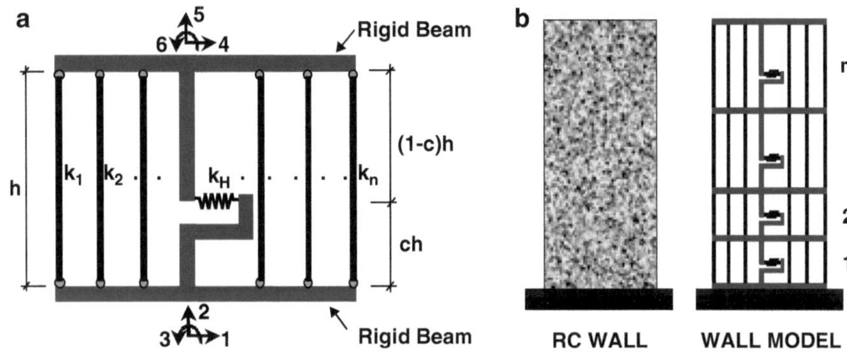

Fig. 20.1 Multiple vertical line element model: (a) MVLEM element, (b) model of a wall

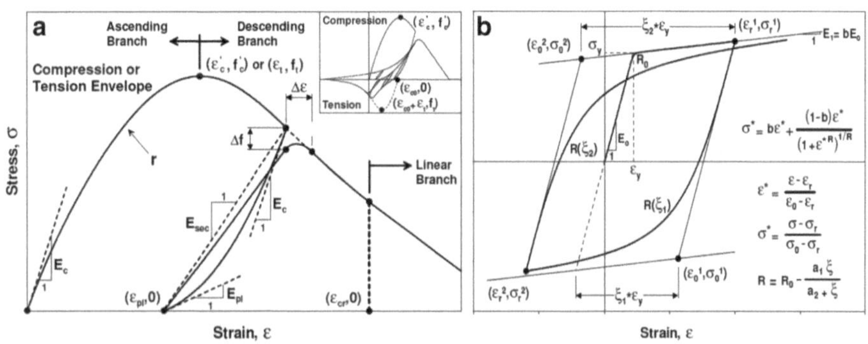

Fig. 20.2 Hysteretic material constitutive models: (a) concrete, (b) reinforcing steel

shear modes of deformation of the wall element are uncoupled (i.e., flexural deformations do not affect shear strength or deformation), which is a very commonly-used, yet arguable modeling assumption. A structural wall is modeled as a stack of m elements, which are placed one upon the other (Fig. 20.1b). The flexural response is simulated by a series of n uniaxial elements connected to infinitely rigid beams at the top and bottom levels. The primary simplification of the model involves applying the plane-sections-remain-plane assumption in calculating the strain level in each uniaxial element.

The stiffness properties k_i and force-displacement relationships of the uniaxial elements are defined according to constitutive stress–strain relationships implemented in the model for concrete and steel (Fig. 20.2) and the tributary area assigned to each uniaxial element. The reinforcing steel stress–strain behavior implemented in the wall model is the well-known nonlinear relationship of Menegotto and Pinto (1973) (Fig. 20.2b). The hysteretic constitutive relation developed by Chang and Mander (1994) (Fig. 20.2a) is used as the basis for the relation implemented for concrete because it is a general model that provides the

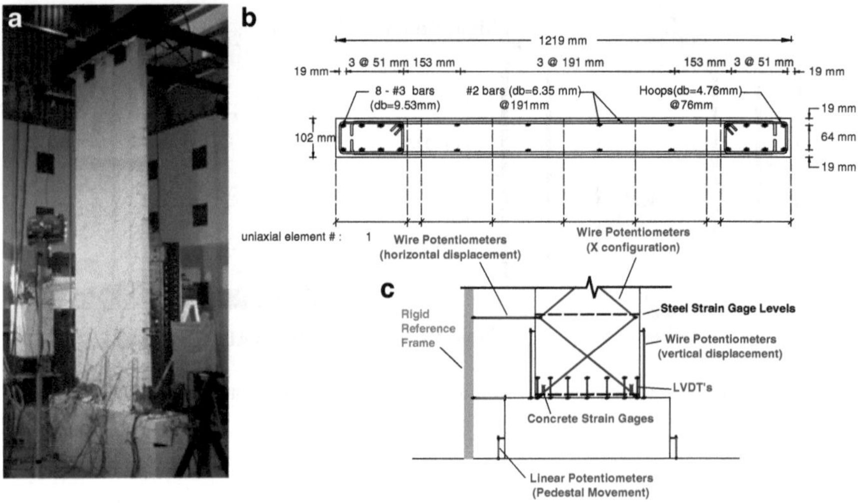

Fig. 20.3 Wall Specimen RW2: (**a**) test setup, (**b**) discretization of the cross-section, (**c**) instrumentation in the plastic hinge region

flexibility to model the hysteretic behavior of confined and unconfined concrete in both cyclic compression and tension, with particular emphasis paid to the transition between crack opening and closure.

For experimental validation of the model, test results obtained for two, approximately quarter-scale wall specimens tested by Thomsen and Wallace (2004): one specimen with a rectangular cross section (Specimen RW2, Fig. 20.3) and another specimen with a T-shaped cross section (Specimen TW2) were used. The walls were 3.66 m tall and 102 mm thick, with web and flange lengths of 1.22 m. The walls were relatively slender, with an aspect ratio of 3. Design of the specimens was based on a prototype building, with strength requirements satisfying the UBC94, and detailing requirements following a displacement-based evaluation. The specimens were subjected to constant axial load, corresponding to approximately 7 % of their axial load carrying capacity, via two hydraulic jacks attached to post-tensioning cables anchored in the strong floor (Fig. 20.3a). Drift-controlled cyclic lateral loads were applied at the top of the wall specimens via a hydraulic actuator mounted horizontally to a reaction wall.

Extensive instrumentation was provided in the tests to measure wall responses at various locations (Fig. 20.3c). Wall lateral displacements at story levels were measured by wire potentiometers connected to a rigid reference frame. Wire potentiometers were used vertically at the boundaries of the walls to obtain boundary vertical displacement and average wall rotation measurements within the first story of the walls. Wire potentiometers were also used in an X configuration along the bottom two stories of the specimens to measure shear deformations. Average strain measurements at the base of the wall were taken through seven linear transducers (LVDTs) along the length of the wall, and through concrete strain gages embedded in the boundary zones of the specimens. Steel strain gauges were provided on both longitudinal and transverse steel at various locations including wall base and first story levels. Linear potentiometers were also installed on the specimens to measure the rotation and sliding of the pedestals.

For comparison of the test results with the model predictions, the model was calibrated to represent measured material properties, as well as the geometry and reinforcement characteristics of the test specimens. Figure 20.3b depicts discretization of the wall cross section for the analytical model, with eight uniaxial elements defined along the length of the wall ($n = 8$) for Specimen RW2. The analytical model was discretized along the wall height to allow consistent strain comparisons between model and experimental results at all locations where LVDTs, concrete strain gauges, and wire potentiometers were provided. The height of the model elements used for all local strain comparisons were set equal to the gauge length of the instruments used. Accordingly, 16 MVLEM elements were used for the modeled walls ($m = 16$) with eight elements along the first-story height: four elements along the second story, and two elements along the third and fourth stories each. A value of 0.4 was selected for the parameter c defining the center of relative rotation for each wall element, based on findings of previous studies (Vulcano et al. 1988).

The monotonic parameters of the constitutive material models were calibrated to represent the results of uniaxial tests conducted on concrete cylinder specimens and rebar coupon samples. The compression envelope for confined concrete was calibrated using the empirical relations proposed by Chang and Mander (1994). The cyclic material parameters were calibrated per the empirical relationships recommended originally by Chang and Mander (1994) and Menegotto and Pinto (1973). The tensile yield strength and strain-hardening parameters were modified according to the empirical relationships proposed by Belarbi and Hsu (1994) to include the effect of tension stiffening on steel bars embedded in concrete. The relationships by Belarbi and Hsu (1994) were also used to calibrate the tensile strength of concrete, as well as the monotonic stress–strain envelope for concrete in tension.

The analytical model was implemented in Matlab for comparison of experimental and analytical results. A displacement-controlled nonlinear analysis strategy was incorporated to correlate the model results with results of the drift-controlled cyclic tests subjected to prescribed lateral displacement histories at the top of the walls. Before analysis, the lateral top displacement history applied during testing and the measured lateral story displacement histories for each specimen were processed to remove displacement contributions resulting from shear and pedestal movement to allow for a direct comparison of the measured and predicted flexural responses. The measured axial load histories applied on the wall specimens during testing, as measured by load cells during testing, were applied to the analytical models (on average, approximately 7 % of the axial load capacity for specimen RW2). Representative comparisons between model predictions of the flexural responses and test results for the rectangular wall specimens RW2 are presented in the following. Various other response comparisons, for both the rectangular and the T-shaped specimen, were presented by Orakcal and Wallace (2006).

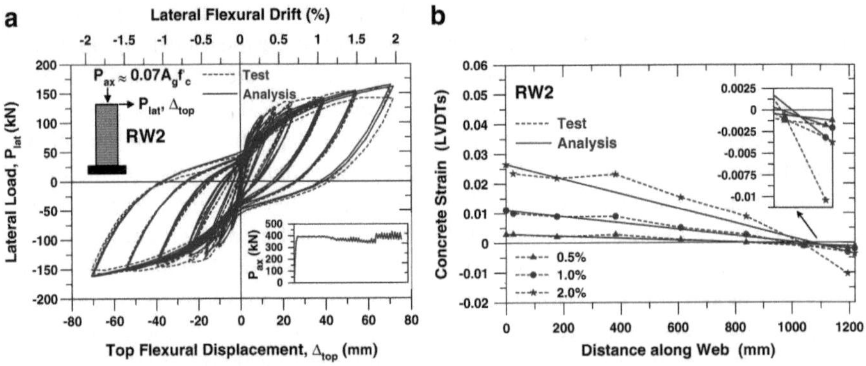

Fig. 20.4 Comparison of test results with model predictions for Specimen RW2: (**a**) lateral load vs. displacement response, (**b**) longitudinal concrete strain distribution at wall base under increasing drift levels

Figure 20.4a compares the measured and predicted lateral load vs. top flexural displacement responses for the rectangular wall Specimen RW2. The analytical model captures the measured response with reasonable accuracy. The lateral load capacity and the lateral stiffness of the wall are well-represented for most of the lateral drift levels. Cyclic properties of the response, including stiffness degradation, hysteretic shape, plastic (residual) displacements, and pinching behavior are all well represented in the analytical results.

A representative local response comparison is presented in Fig. 20.4b, which plots the average concrete strains measured by the seven LVDTs over a 229 mm gauge length at the base of the wall (Fig. 20.3c), at applied peak positive top displacement (top displacement reversal) data points, for selected drift levels applied during testing. Similar trends were observed in the results for other drift levels and also for peak negative top displacement data points. Results shown in Fig. 20.4b illustrate that the analytical model predicts the tensile strain profile reasonably well, but significantly underestimates the compressive strains. The compressive strain predictions at the location of the outmost LVDT correspond to 63, 55, and 30 % of the measured values for wall nominal drift levels of 0.5, 1, and 2 %, respectively. The accuracy in the prediction of the position of the neutral axis is reasonably good; the analytically predicted and experimentally obtained neutral axis positions vary by no more than 5 % of the wall length. As well, the model provides reasonable predictions of the average inelastic rotations (related to the slope of the longitudinal strain distribution) developing at the base of the wall specimens.

The larger measured compressive strains in concrete at the base of the wall may be due to stress concentrations induced at the wall-pedestal interface (the bottom of the LVDTs were mounted on the pedestal) due to the abrupt change in geometry. The concrete strain measurements may also have been influenced by the attachment of the LVDTs to the wall specimen, as the LVDTs were affixed to blocks glued to the wall surface; the measurements may not represent average compressive strains experienced along the thickness of the wall specimen when spalling is observed. Finally, the larger compressive strains may be partially due to the nonlinear shear response that the wall specimen experienced within the first-story height. Preliminary analysis results using a modified MVLEM with an implemented methodology for coupling shear and flexural displacements (described in a later section of this chapter), based on biaxial constitutive relationships for concrete with compression softening, yields compressive strains larger than those predicted with the flexural model used herein.

Overall, other than underestimation of compressive strains, the MVLEM proves to be an effective modeling approach for the flexural response prediction of slender RC walls, as the model provides good predictions of the experimentally observed global and local responses, including wall lateral load capacity and lateral stiffness at varying drift levels, yield point, cyclic properties of the load–displacement response, rotations (average over the region of inelastic deformations), position of the neutral axis and tensile strains.

20.3 Modeling of Bond Slip Responses in Columns with Inadequate Lap Splices

Most commonly-used macroscopic modeling approaches for simulating the behavior of columns with lap splices involve defining a zero-length moment vs. slip rotation spring (with predefined unloading and reloading rules) at the location of the lap splice. Although this modeling approach can provide an accurate prediction of the global (lateral load–displacement) response of a column, using predefined unloading/reloading for the rotational bond-slip spring introduces incompatibility between flexural deformations (strains) and bond slip deformations, developing in the lap splice region during cyclic loading. A more robust macroscopic model formulation was proposed by Chowdhury and Orakcal (2012), for simulating the cyclic bond slip responses of reinforced concrete columns with deficient lap splices or anchorage conditions. This model incorporates constitutive bond slip behavior in the formulation of the MVLEM described in the previous section, which allows distribution and kinematic coupling of flexural and bond slip deformations, and monitoring of local deformations on a column under reversed cyclic loading conditions. Several bond stress vs. slip constitutive relationships can be implemented in the model, depending on the type of reinforcing bar (deformed or plain) used, and possible failure modes (pullout or splitting). The flexible formulation of the model also allows considering the influence of strain penetration effects and the presence of 180-degree hooks, on the lateral load behavior of a column. Correlation studies conducted between model predictions and test results verified that the model can effectively reflect the global response characteristics and failure modes of various column configurations incorporating deficient lap splices or anchorage deficiencies.

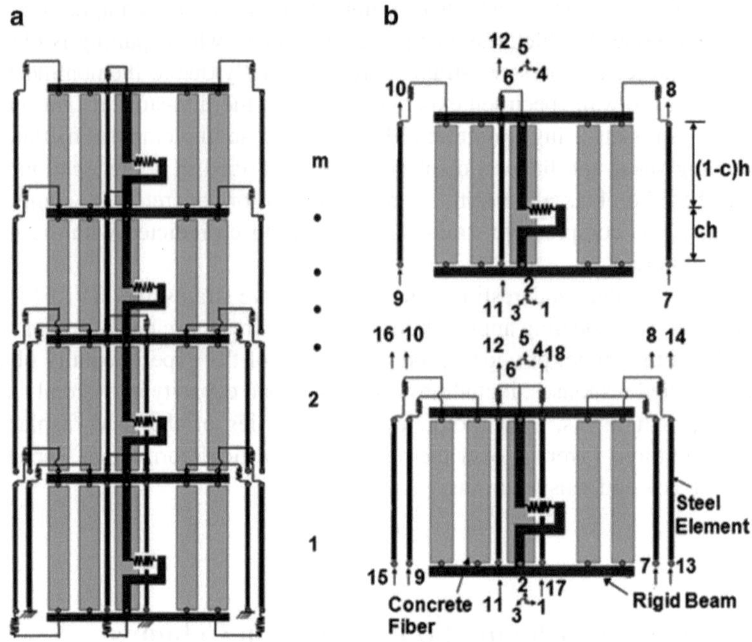

Fig. 20.5 Bond slip model: (**a**) model assembly, (**b**) model elements and degrees of freedom

The analytical model presented here is an extension of the MVLEM, which was shown to be effective in simulating nonlinear flexural responses. In the bond slip model formulation, in the present model formulation, the macro-fibers connecting the rigid beams at the top and bottom represent the uniaxial behavior of concrete only, whereas the uniaxial steel elements are connected to the rigid beams (and therefore to concrete) through uniaxial bond slip springs at the top level of each model element. The plane-sections-remain-plane assumption is therefore imposed on concrete only, and steel strains deviate from the concrete strains due to bond slip deformations developing between the two materials. The model element shown in the lower part of Fig. 20.5b (with 18 degrees of freedom) is used along the lap splice region (to represent spliced bars), and the element shown in the upper part (with 12 degrees of freedom) is used above the lap splice region of a column. In the model formulation, bond slip and flexural modes of deformation are kinematically coupled, whereas shear and flexural modes of deformation are uncoupled. The column is modeled as a stack of m model elements, which are placed upon one another, as shown in Fig. 20.5a.

In addition to the refined constitutive models adapted for concrete and reinforcing steel, advanced constitutive models are also implemented in the model formulation to define the bond stress vs. slip behavior of the bond slip springs (Fig. 20.6). For deformed (ribbed) reinforcing bars, the bond stress vs. slip model by Harajli (2009) is adopted to represent the splitting-type bond slip behavior expected under inadequate clear cover conditions for unconfined concrete

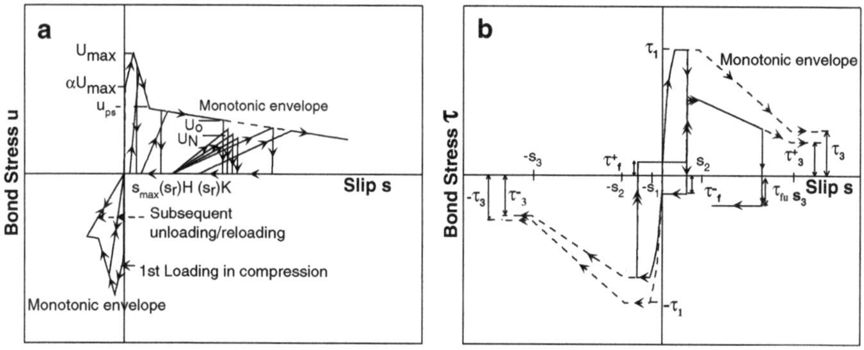

Fig. 20.6 Constitutive bond stress vs. slip deformation models for bond slip springs: (**a**) splitting-type behavior, (**b**) pullout-type behavior

(Fig. 20.6a), whereas the constitutive model by Eligehausen et al. (1983) is implemented to represent the pull-out type bond slip behavior expected under adequate clear cover and adequate confinement conditions (Fig. 20.6b). For plain (smooth) bars, the bond stress vs. slip constitutive model by Verderame et al. (2009) is incorporated, whereas the bar axial stress vs. hook end slip deformation relationship by Fabbrocino et al. (2004) is adopted to represent the behavior of 180-degree hooks.

The six full-scale lap-splice deficient column specimens tested by Melek and Wallace (2004) were first used to calibrate and validate the analytical model. The specimens consisted of cantilever columns with foundation blocks attached to a strong floor and loaded laterally at the top, representing the moment distribution (single curvature) along half the height of an interior column in a building. Specimen heights were 1.52, 1.68 and 1.83 m, and each specimen had 457 × 457 mm square cross section. Reinforcement configurations, with eight 25 mm nominal diameter vertical bars, and 9.5 mm diameter hoops with 90-degree hooks spaced at 305 mm on center, were based on a review of typical reinforcing details in older buildings (pre-1970 construction) in the United States. A lap splice length of $20d_b$ (20 longitudinal bar diameters) was used in the construction of specimens, corresponding to 60 % of the required lap splice length per ACI318-08. The test-day compressive strength of concrete used in the construction of the specimens was approximately 36 MPa, whereas yield strengths of 510 and 480 MPa were measured for the longitudinal and transverse reinforcement, respectively. The axial load levels applied on the specimens correspond to 10, 20, and 30 % of their axial load capacities. The lateral displacement history imposed during 5 of the 6 tests was fairly typical (standard); consisting of three cycles at each lateral displacement level, with monotonically increasing drift levels. The test setup used and the typical damage observed on the specimens (bond slip followed by crushing of concrete) are presented in Fig. 20.7a, b, respectively, whereas Fig. 20.7c illustrates the reinforcement configuration of the specimens. The specimens were instrumented with a large number of displacement transducers

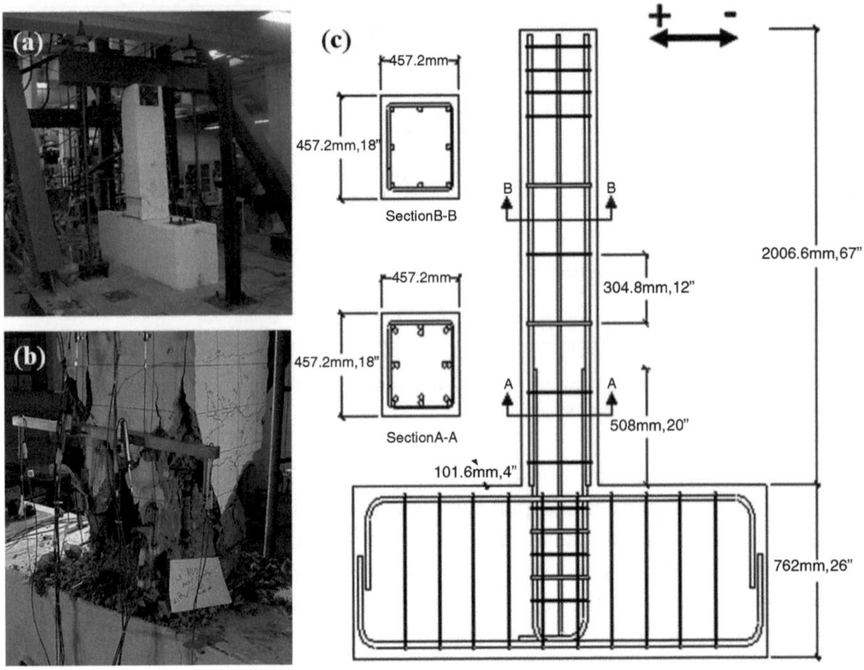

Fig. 20.7 Lap-spliced column specimens: (**a**) test setup, (**b**) observed damage, (**c**) reinforcement configuration

for measuring of average flexural, shear, and bond slip deformations, at various locations on each specimen, which allowed comparison of test results with the analytical model predictions also at local (deformation) response levels.

The model was calibrated to represent measured material properties as well as the geometry and reinforcement characteristics of the test specimens. Both monotonic and cyclic parameters of the bond stress vs. slip relationships used for the bond slip springs were calibrated as defined originally in the constitutive bond slip model formulations by Harajli (2009) and Eligehausen et al. (1983). The extended model formulation was used for comparing the model results with the experimentally-obtained load–displacement responses. Response comparisons also included rotations, strains in concrete and reinforcing steel, neutral axis position, and average bond stresses along the lap splice region. Only selected comparisons are presented herein, whereas all of the comparisons are presented by Chowdhury (2011).

The experimentally-measured lateral load vs. top displacement response for column specimens 2S10M (subjected to an axial load level corresponding to 10 % of its axial load capacity) is compared with the analytical model prediction in Fig. 20.8. The comparison indicates that the model provides reasonably accurate load–displacement response predictions. Overall, a good level of agreement is achieved between the test data and model results in terms of column lateral load

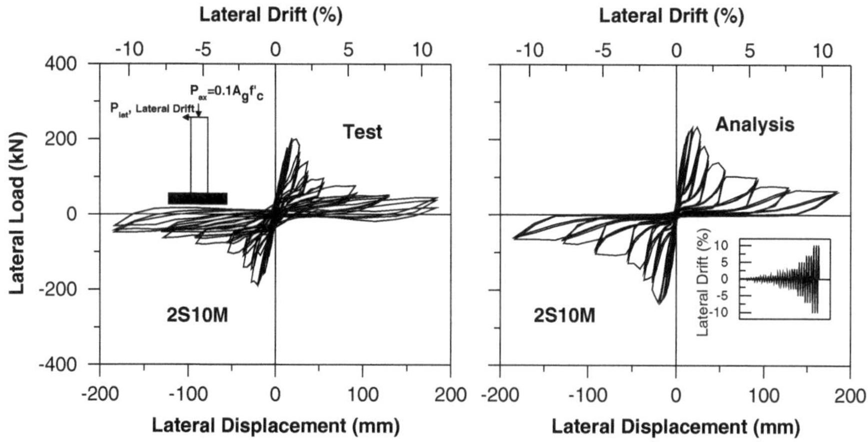

Fig. 20.8 Comparison of test results with model predictions for lateral load–displacement response for Specimen 2S10M

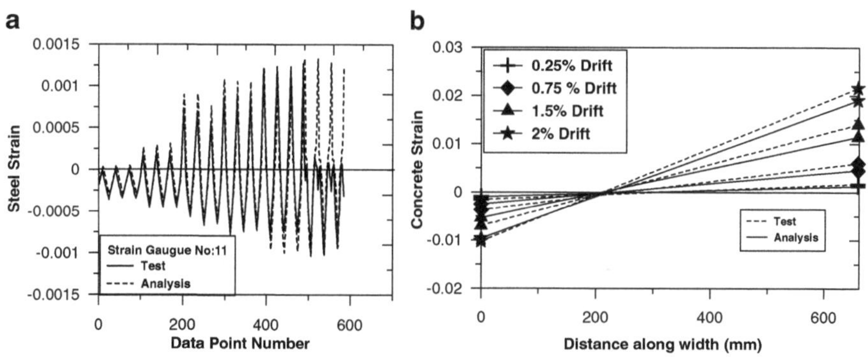

Fig. 20.9 Representative comparisons of measured and predicted local responses: (**a**) steel strain history on a starter bar in Specimen 2S20M, (**b**) average longitudinal concrete strain profile at the base of Specimen 2S10M

capacity, lateral stiffness, ductility, shape of the unloading/reloading loops, and pinching characteristics of the response. Furthermore, the behavior characteristics and failure modes observed during the tests, including bond slip initiation (formation and widening of longitudinal splitting cracks), yielding of reinforcement (widening of transverse flexural cracks), and crushing of concrete, were observed to be consistent with the analytically-predicted responses at varying drift levels.

Figure 20.9a compares the analytically-predicted steel strain histories with measurements of strain gauge No. 11, which was attached to one of the starter bars in Specimen 2S20M (subjected to an axial load level corresponding to 20 % of its axial load capacity). In general, the test measurements and model results show reasonable agreement, considering the typical scatter observed in strain gauge measurements during testing of reinforced concrete members, based on proximity of a strain gauge to a crack, up to a drift level of 1.5 % (data point 500), after which lateral load degradation starts with the initiation of bond deterioration along the splice length. Figure 20.9b compares the measured and predicted average longitudinal strain profiles (due to flexural deformation of concrete), as well as the neutral axis position (corresponding to zero strain), along the width of Specimen 2S10M, under increasing lateral drift levels. The concrete strain profiles shown consider average longitudinal strains over a distance of 330 mm from the column base, and were calculated using measurements from displacement transducers mounted vertically at the base of the column specimens, at peak positive top displacement (top displacement reversal) data points corresponding to selected drift levels. The results indicate that the analytical model is capable of providing reasonably accurate predictions of the average concrete strains (both compressive and tensile), rotations (related to the slope of the strain distribution), and the neutral axis depth position measured within the lap splice region of the specimens, particularly for lateral drift levels not exceeding 1.5 %. At larger drift levels, together with rapid degradation in lateral load due to splice failure, progressive crushing of concrete was observed at the base of the column specimens, which impaired the accuracy of the model in predicting the measured concrete strain profiles.

For further experimental verification of the model, analysis results were compared with numerous test results presented in the literature on columns with various configurations. Results of 30 cyclic loading tests conducted by 11 research groups were used for the comparisons. The column specimens investigated had varying geometries, material strengths, and reinforcement details and were tested under various axial load levels. The comparisons demonstrated that the model is effective in predicting various bond slip response characteristics, including strain penetration effects, bond-slip-governed responses, coupled-flexural-and-bond-slip responses, and responses of columns with smooth bars and hooks. Response comparisons for all 30 tests are available in the dissertation by Chowdhury (2011). Overall, the modeling approach was shown to be promising towards obtaining improved seismic response predictions for reinforced concrete buildings with splice-deficient columns.

20.4 Modeling of Nonlinear Shear Responses in Reinforced Concrete Panel Elements

For finite element modeling of reinforced concrete wall segments subjected to nonlinear shear actions (e.g., squat walls, wall piers, wall spandrels), although a number of cyclic constitutive models have been proposed for simulating the nonlinear responses of constitutive panel elements of the finite element model, most of these model formulations are not included in commonly-used structural analysis platforms due to complexities in their implementation. A new constitutive panel model (the Fixed-Strut-Angle-Panel Model, FSAM) was proposed by Orakcal et al. (2012) for simulating the behavior of RC panel elements under generalized, in-plane, reversed-cyclic loading conditions. The proposed constitutive model is presented as a feasible candidate for implementation into a two-dimensional finite-element analysis formulation, for efficient and practical response prediction for squat walls experiencing predominant shear or coupled flexural and shear responses, using the finite element modeling approach.

The Fixed Strut Angle Model (FSAM) is a constitutive model for simulating the behavior of RC panel elements under generalized, in-plane, reversed-cyclic loading conditions. As assumed by other RC panel models available in the literature, in the Fixed Strut Angle model, the strain field acting on concrete and reinforcing steel components of an RC panel is assumed to be equal to each other, implying perfect bond assumption between concrete and reinforcing steel bars. Further, reinforcing steel bars are assumed to develop zero shear stresses perpendicular to their longitudinal direction, implying no dowel action on reinforcement. While the reinforcing steel bars develop only uniaxial stresses under uniaxial strains in their longitudinal direction, the behavior of concrete is defined using stress–strain relationships in biaxial directions, and the orientation of those biaxial directions is governed by the state of cracking in concrete.

In the uncracked state of concrete, the stress–strain behavior of concrete is represented with a rotating strut approach (similar to the Modified Compression Field Theory, Vecchio and Collins (1986) and the Rotating Angle Strut and Tie Model, Pang and Hsu (1995)). The strain field imposed on concrete is transformed into principal strain directions, which are assumed to coincide with principal stress directions, and uniaxial stress–strain relationships for concrete are applied along the principal strain directions in order to obtain the principal stresses in concrete. Although the stress–strain relationships used for concrete in principal directions are fundamentally uniaxial in nature, they also incorporate biaxial softening effects including compression softening and biaxial damage. At this stage of the behavior, monotonic stress–strain relationships for concrete are used, since it is reasonable to assume that concrete behavior follows a monotonic (virgin) stress–strain relationship, prior to first cracking under a biaxial state of stress (Fig. 20.10a).

When the value of the principal tensile strain in concrete exceeds the monotonic cracking strain of concrete for the first time, the first crack is formed, and for following loading stages, the principal strain direction corresponding to first cracking in concrete is assigned as the first "Fixed Strut" direction for the panel. After formation of this first crack, while principal directions of the applied strain field continue to rotate based on the applied strain field, the principal stress directions in concrete are assumed to be along and perpendicular to the first Fixed Strut direction. The present model assumes that the first crack (or strut) direction coincides with the principal stress directions in concrete. This physically implies zero shear aggregate interlock along a crack, which was an inherent assumption of the original model formulation (Ulugtekin 2010). Since the direction of the first strut is fixed, a uniaxial hysteretic stress–strain relationship for concrete can now be applied in principal stress directions (parallel and perpendicular to the first strut), and history

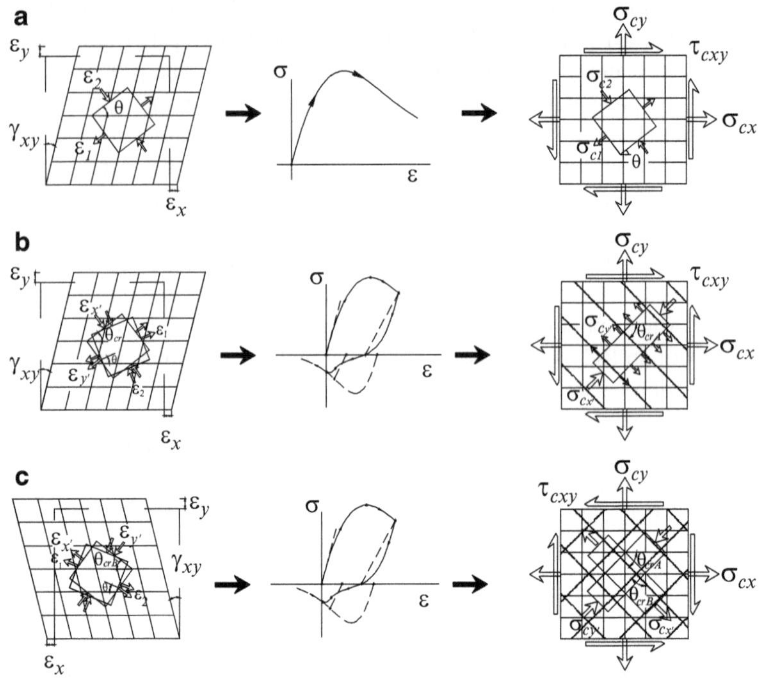

Fig. 20.10 Concrete biaxial behavior in the fixed strut angle model: (**a**) uncracked behavior, (**b**) behavior after formation of 1st crack, (**c**) behavior after formation of 2nd crack

variables in the concrete stress–strain relationship can be easily tracked and stored in the two fixed directions. For calculation of concrete stresses in principal directions, the applied strain field in concrete should be transformed into strain components that are parallel and perpendicular to the first fixed strut direction, instead of principal strain directions (Fig. 20.10b).

The analysis is continued in the form of a single fixed strut mechanism until the formation of the second crack, after which the second strut will develop in the panel model. During the first fixed strut stage of the analysis, the model tracks the concrete stress–strain behavior along the first strut direction, and when the strains along the first strut direction first exceed the cyclic cracking strain (which depends on both the monotonic cracking strain and the plastic strain upon reversal from a compressive stress state), the second crack is formed. In case of the zero aggregate interlock assumption, the second crack has to develop in perpendicular direction to the first crack, according to a stress-based cracking criterion, since the first strut direction is a principal stress direction and the concrete stress–strain relationship is assumed to be uniaxial along the first strut direction. After formation of this second crack, the second "Fixed Strut" will develop in the direction of the second crack (in perpendicular direction to the first strut), and for further loading stages, the concrete mechanism consists of two independent struts, working as interchanging compression and tension struts in the two Fixed Strut directions, based on the applied strain field. While principal directions of the applied strain field continue to rotate, the principal stress directions in concrete are assumed to be along the two Fixed Strut directions, again implying zero shear stresses (zero shear aggregate interlock) developing along the two cracks (Fig. 20.10c).

The inherent assumption of the model, which is principal concrete stress directions coinciding with crack directions, is based on interpretation of existing panel tests in the literature. Available test results in the literature (e.g., Stevens et al. 1991) typically indicate that for an RC panel, after formation of cracks, the principal stress direction in concrete does not change significantly with loading, although the principal strain direction on a panel may undergo significant variation. The principal stress directions in concrete being insensitive to loading may imply that after formation of cracks, the principal stress directions in concrete follow approximately the fixed crack directions, indicating that shear stresses along a crack (and thus shear aggregate interlocking along a crack) has marginal influence on the panel behavior. However, this assumption also allows the flexibility to incorporate a suitable cyclic shear aggregate interlock constitutive model (shear stress versus shear strain along a crack) in the FSAM, since the formulation of the model allows calculating shear strains along a crack. In the model formulation presented by Orakcal et al. (2012), a simple friction-based constitutive model was adopted to represent shear aggregate interlock effects. The proposed cyclic shear aggregate interlock model starts with linear loading/unloading behavior, relating the sliding shear strain along a crack to the shear stress, via a simple linear elastic relationship between the sliding shear strain and the resultant shear stress along the crack surface. However, the shear stress is restrained to zero value when the concrete normal stress perpendicular to the crack is tensile (crack open); and is bounded via the product of a friction coefficient and the concrete normal stress perpendicular to the crack, when the concrete normal stress is compressive (crack closed). The linear unloading/reloading slope of the shear stress vs. sliding strain relationship was taken as a fraction of the concrete elastic modulus (a value $0.4E_c$ was adopted, representing the elastic shear modulus of concrete), and a value of 0.2 was assumed for the friction coefficient. Under constant compressive stress in concrete perpendicular to the crack, this model yields an elasto–plastic aggregate interlock behavior under cyclic loading. In implementation of the material constitutive relationships in the model, the original formulation of the constitutive relationship by Chang and Mander (1994) was modified to represent behavioral features of concrete under biaxial loading; via inclusion of parameters representing compression softening (Vecchio and Collins 1993), hysteretic biaxial damage (Mansour et al. 2002), and tension stiffening effects (Belarbi and Hsu 1994). Details on the model formulation and constitutive parameters were described by Ulugtekin (2010).

Results of cyclic panel tests from two experimental programs were used by Orakcal et al. (2012) for experimental calibration and validation of the FSAM. The first of these two test programs, referred to by Stevens et al. (1991), was conducted using the “Shell Element Tester” facility at the University of Toronto (Fig. 20.11a); and the other referred to by Mansour and Hsu (2005), was performed using the “Universal Element Tester” facility at the University of Houston (Fig. 20.10b).

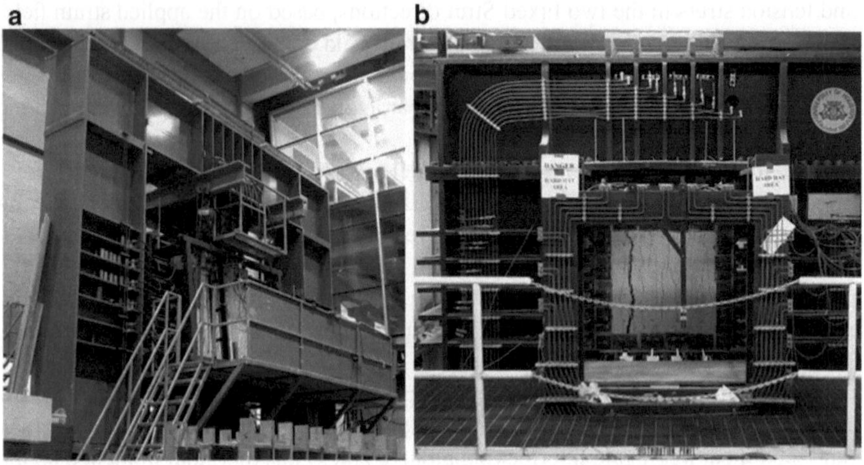

Fig. 20.11 Reinforced concrete panel test facilities: (a) shell element tester at University of Toronto, (b) universal element tested at University of Houston

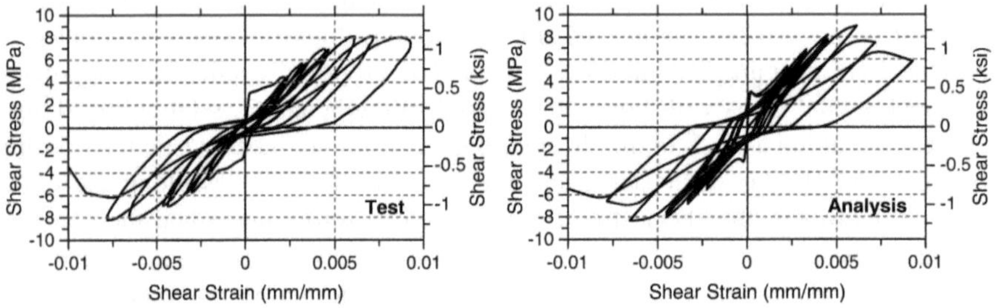

Fig. 20.12 Comparison of measured and predicted shear stress vs. shear strain responses for Specimen SE10

Only selected representative comparisons are presented here. The experimentally-measured shear stress vs. shear strain responses for specimen SE10 (Stevens et al. 1991) and specimen CA2 (Mansour and Hsu 2005) are compared with the analytical model predictions in Figs. 20.12 and 20.13, respectively. Specimen SE10 had dimensions of 1,625 × 1,625 mm, 285 mm thickness, concrete compressive strength of 34 MPa, reinforcing steel bars with yield strengths of 422 and 479 MPa, and uniformly distributed reinforcement ratios of 1 and 3 % in two orthogonal directions. The specimen was subjected to equivalent cyclic shear stresses (perpendicular to orthogonal directions of reinforcement), together with normal compressive stresses along reinforcement directions with magnitudes corresponding to one-third of the applied shear stress. Specimen CA2 had dimensions of 1,379 × 1,379 mm, 178 mm thickness, concrete compressive strength of 45 MPa, rebars with yield strength of 424 MPa, and equal distributed reinforcement ratios of 0.77 % in two orthogonal directions. The specimen was subjected to an equivalent

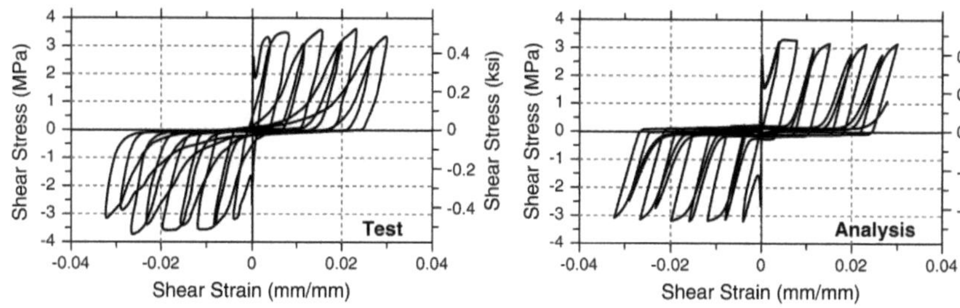

Fig. 20.13 Comparison of measured and predicted shear stress vs. shear strain responses for Specimen CA2

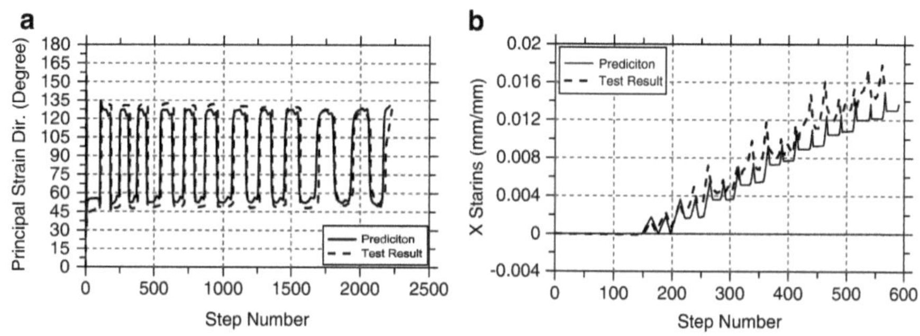

Fig. 20.14 Comparison of measured and predicted local responses: (**a**) principal strain direction history for Specimen SE10, (**b**) average normal strain history in the x direction for Specimen CA2

pure shear stress state, with shear stresses acting perpendicular to the orthogonal reinforcement directions.

The comparisons shown in Figs. 20.12 and 20.13 indicate that the model provides reasonably accurate shear stress vs. strain response predictions, for varying reinforcement ratios and loading conditions. Overall, a good level of agreement is achieved between the test data and model results in terms of shear stress capacity, stiffness, ductility, shape of the unloading/reloading loops, and pinching characteristics of the response. Test results were also compared with model predictions in terms of local response and deformation characteristics; including average normal strains in horizontal and vertical directions, principal strain directions, and principal stress directions. Experimentally-measured principal strain direction histories for specimen SE10 are compared with model predictions in Fig. 20.14a. The model predicts the variation in principal strain directions with reasonable accuracy, with slight underestimation of the principal strain directions at the beginning of the analysis (for relatively small shear strain values). Average normal strain history in the x direction (in one of the two orthogonal reinforcement directions) measured using displacement transducers attached to Specimen CA2, is compared with the model prediction in Fig. 20.14b. The model captures the average normal strain history with reasonable accuracy. The measured and predicted average normal strains on this specimen are large post–yield strains; which agrees with the reinforcement yielding failure mode observed during testing of this specimen, due to relatively low reinforcement ratios used in both directions. Overall, the FSAM was shown to capture the behavioral attributes of RC panels subjected to combined axial and shear stresses, and the model is expected to be a feasible candidate for implementation into a two-dimensional finite-element analysis formulation for efficient seismic response prediction of structural walls with various geometries, aspect ratios, and reinforcement details. Ongoing efforts focus on FEM implementation of the FSAM.

20.5 Modeling of Nonlinear Shear-Flexure Interaction Responses in Medium-Rise Walls

A majority of existing analytical models for predicting nonlinear behavior of RC walls do not incorporate shear–flexure interaction (SFI) and characterize the shear behavior of a wall using ad-hoc force-deformation rules defined independently from flexural modeling parameters. However, experimental evidence suggests that interaction between flexural and shear deformations exists even for relatively slender RC walls with aspect ratios of 3.0 and 4.0, with shear deformations contributing to lateral deformations by approximately 30 and 10 % of the first story and roof-level lateral displacement, respectively (e.g., Massone and Wallace (2004)). Experimental results have also shown that shear deformations for RC walls with aspect ratios of 1.5 and 2.0 contribute as much as 35 % of the wall top displacement (Tran and Wallace 2012). Modeling approaches that do capture this interaction are limited to ad-hoc or empirical approaches (e.g., Xu and Zhang 2011; Beyer et al. 2011) or models that are limited to monotonic responses only (e.g., Massone et al. 2006, 2009). In addition, none of the modeling approaches have been comprehensively validated, since the detailed data needed, especially for localized responses for cases with significant interaction, is not available.

An ongoing effort by Kolozvari et al. (2012) aims to develop a robust modeling approach for simulating the cyclic shear–flexure interaction behavior of structural walls, upon validation with recent tests conducted by Tran and Wallace (2012) on medium-rise walls. The analytical modeling approach proposed incorporates RC panel behavior into the Multiple-Vertical-Line-Element Model (MVLEM) for coupling of shear and flexural responses. The original formulation of the MVLEM, with uncoupled shear and flexural responses, is modified by replacing each uniaxial (macro-fiber) element with a RC panel element (Fig. 20.15). This enables coupling of axial and shear responses at the panel (fiber) level, which further allows coupling of flexural and shear responses at the model element level. The constitutive RC panel behavior under generalized reversed-cyclic loading conditions is described by the Fixed Strut Angle Model (FSAM) by Orakcal et al. (2012) described in the previous section. In the modified MVLEM formulation, the longitudinal normal (axial) strain ε_y and shear strain γ_{xy} values for the panel

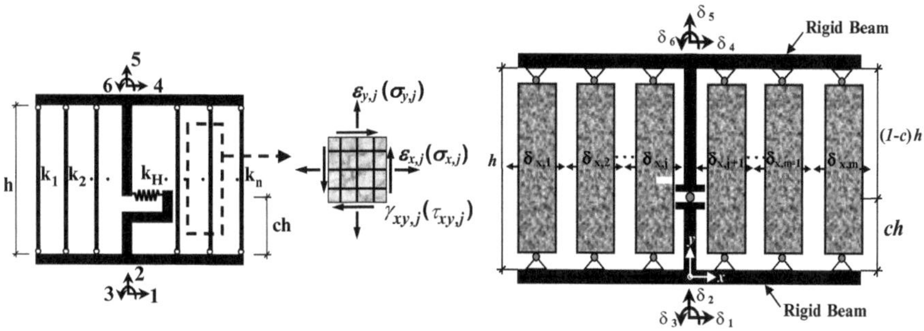

Fig. 20.15 Shear-flexure interaction model element (Kolozvari et al. 2012)

elements are first calculated based on the six prescribed degrees of freedom defined at the top and bottom rigid beams of the model element. Plane-sections-remain-plane condition is enforced and shear strain is assumed to be uniformly distributed along the wall section. Horizontal normal strain ε_x on each panel element (which is necessary to complete the strain field in each panel), is defined by use of additional degrees of freedom in the horizontal direction $\{\delta_x\} = \{\delta_{x,1}\ \delta_{x,2}\ldots\ \delta_{x,m}\}^T$ (where m is the number of panel elements in one model element), as shown in Fig. 20.13. These horizontal (extensional) degrees of freedom on the panel elements are assumed to be kinematically independent from the six nodal displacement degrees of freedom at the top and bottom of the element; therefore, the total degrees of freedom necessary to describe the deformation of one modified MVLE is increased, from 6 in the original formulation of the MVLEM, to 6 + m. For obtaining the displacements along these extensional degrees of freedom during the nonlinear analysis, the resultant transverse normal stress σ_x within each panel element (resultant of the contributions from concrete and reinforcing steel), is assumed to be zero, which is consistent with the boundary conditions at the sides of the wall. Previous studies performed by Massone et al. (2006) revealed that the zero resultant horizontal stress assumption is not capable of correctly reproducing experimental responses observed in walls with very low shear span-to-depth ratios (lower than approximately 0.5), underestimating the lateral load capacity of the wall by 13–40 %. This study, however, focuses on the application of the proposed modeling approach, with the implemented assumption of zero resultant horizontal stress along the length of the wall ($\sigma_x = 0$), to cantilever walls with aspect ratios larger than 1.0.

Experimental data used to validate the proposed analytical model was obtained from a recent test program (Tran and Wallace (2012)) designed particularly to investigate factors that influence the deformation capacity of moderate aspect ratio walls. The test program included five large-scale RC wall specimens, 183 or 244 cm tall, 122 cm long, and 15 cm thick; resulting aspect ratios are 1.5 and 2.0, respectively. The design compressive strength of concrete was 34.5 MPa, whereas the reinforcing steel bars were primarily of 420 MPa nominal yield strength. All test specimens were equipped with an extensive set of instrumentation, enabling the

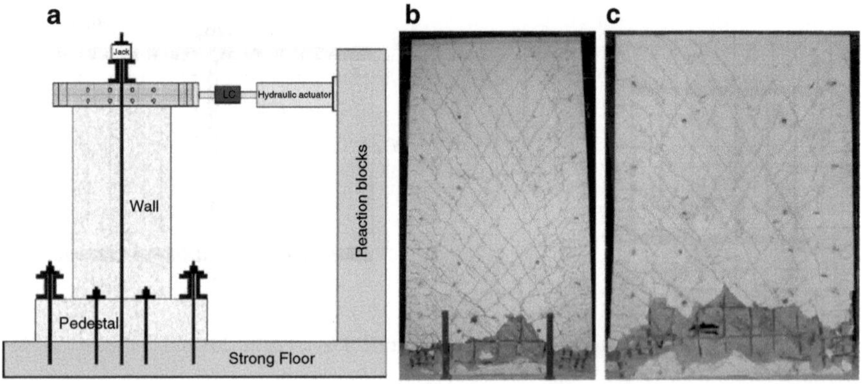

Fig. 20.16 Medium-rise wall tests by Tran and Wallace (2012): (**a**) test setup, (**b**) damage on Specimen RW-A20-P10-S63, (**c**) damage on Specimen RW-A15-P10-S78

collection of detailed experimental data, at both global and local (deformation) response levels. In addition to numerous strain gauges installed on the reinforcing bars, LVDTs mounted on the specimens were configured to measure the distribution of the average vertical, horizontal, and shear strains at multiple locations over the wall height, to measure the contributions of shear and flexural deformations to wall lateral displacements. Experimental data obtained for two of these wall specimens, specimen RW-A20-P10-S63 ($h/l = 2.0$) and RW-A15-P10-S78 ($h/l = 1.5$), were used to assess the ability of the proposed modeling approach to capture the measured global and local responses. Specimen RW-A20-P10-S63 incorporated a boundary reinforcement ratio of 7 % and a distributed web reinforcement ratio 0.6 %, whereas specimen RW-A15-P10-S78 had a boundary reinforcement ratio of 6 % and a distributed web reinforcement ratio 0.7 %. Both specimens were designed to yield in flexure prior to reaching nominal shear capacity, but with relatively higher average shear stress demand at flexural capacity. Both specimens were subjected to constant axial load corresponding to 10 % of their axial load capacity, and were tested in an upright position under drift-controlled cyclic lateral loads applied at the top of the specimens (Fig. 20.16a). Both specimens demonstrated nonlinear flexural and shear behaviors, with shear deformations contributing as much as 35 % to the overall lateral displacements measured at the top of specimens. Concrete crushing at the wall boundaries followed by buckling of longitudinal boundary reinforcement caused lateral strength degradation for both specimens (Fig. 20.16b, c), ultimately resulting in large sliding shear deformations observed at the base of the walls.

Comparison of analytically-predicted and experimentally-measured lateral load versus top displacement response for Specimen RW-A15-P10-S78 (aspect ratio of 1.5) is presented in Fig. 20.17a. The analytical model captures the experimentally-measured response of the specimen with reasonable accuracy. The lateral load prediction is in good agreement with the test results for most of the drift levels. Cyclic properties of the response, including degradation in the unloading/reloading

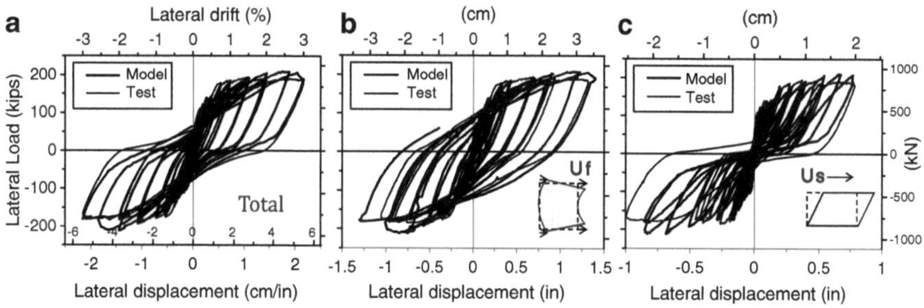

Fig. 20.17 Comparison of model predictions with test results for Specimen RW-A15-P10-S78: (**a**) lateral load vs. total displacement, (**b**) lateral load vs. flexural displacement, (**c**) lateral load vs. shear displacement

stiffness, residual (plastic) displacements, and pinching behavior are all represented reasonably well.

Model results for the lateral load vs. flexural and shear deformation components of top displacement are compared with test measurements in Fig. 20.17b, c, respectively. Again, the model provides reasonable predictions for the contributions of nonlinear flexural and shear deformations in the response of the wall, as well as the overall hysteretic shape of the lateral load vs. flexural displacement response and the pronounced pinching behavior observed in the lateral load vs. shear displacement response. Furthermore, the crack directions predicted by the model at various locations on the wall were consistent with the crack patterns observed during the test (Kolozvari et al. 2012). Overall, the proposed shear–flexure interaction modeling approach was shown to be promising towards simulating the coupled flexural and shear responses in medium-rise walls with moderate aspect ratios. Ongoing efforts focus on detailed local response comparisons (rotations, strains in concrete and reinforcing steel in longitudinal and transverse directions) as well as refinement of the modeling assumptions.

20.6 Summary and Conclusions

Examples of novel analytical modeling approaches were presented for simulating the nonlinear response of reinforced concrete structural components under reversed cyclic loading conditions. The emphasis was on simulation of nonlinear flexural, bond slip, shear, and shear–flexure interaction responses in reinforced concrete walls, columns, and panel elements. The modeling methodologies presented were shown to be behaviorally-robust in capturing the critical response attributes of walls and columns; in terms of capacity, stiffness, ductility, and cyclic response characteristics, at both global (force-deformation) and local (inelastic rotations, contributions of shear and flexural deformations, strains in concrete and reinforcing steel) response levels. Accuracy of the presented modeling approaches were demonstrated via comparison of the model predictions with test results conducted on reinforced concrete wall, column, and panel specimens. The modeling approaches presented are believed to be a significant improvement towards realistic representation of the nonlinear response of reinforced concrete walls and columns. Implementation of the model formulations into a computational structural analysis platform will allow improved seismic response predictions for existing buildings, for obtaining more reliable performance assessment results and arriving at more informed decisions on rehabilitation.

Acknowledgments This chapter summarizes findings of several collaborative research studies, with contributions of Prof. John W. Wallace and Kristijan Kolozvari from UCLA, Prof. Leonardo M. Massone from University of Chile, Prof. Sharmin R. Chowdhury from Ahsanullah University of Science and Technology, and Denizhan Ulugtekin from Istanbul Technical University. Contribution of these researchers is gratefully acknowledged. The author would also like to thank Prof. John W. Wallace and Thien Tran from UCLA, Dr. Murat Melek from Ove Arup and Partners Inc., John H. Thomsen from SGH Inc., and Prof. Mohamad Mansour from University of Houston for providing experimental data.

References

Belarbi A, Hsu TC (1994) Constitutive laws of concrete in tension and reinforcing bars stiffened by concrete. ACI Struct J 91:465–474

Beyer K, Dazio A, Priestley N (2011) Shear deformations of slender reinforced concrete walls under seismic loading. ACI Struct J 108:167–177

Chang GA, Mander JB (1994) Seismic energy based fatigue damage analysis of bridge columns: Part I – evaluation of seismic capacity. NCEER technical report, State University of New York

Chowdhury SR (2011) Analytical modeling of reinforced concrete columns with lap splices. Dissertation, Bogazici University

Chowdhury SR, Orakcal K (2012) An analytical model for reinforced concrete columns with lap splices. Eng Struct 43:180–193

Eligehausen R, Popov EP, Bertero VV (1983) Local bond stress-slip relationships of deformed bars under generalized excitations. EERC report, University of California, Berkeley

Fabbrocino G, Verderame GM, Manfredi G, Cosenza E (2004) Structural models of critical regions in old-type RC frames with smooth rebars. Eng Struct 26:2137–2148

Harajli MH (2009) Bond stress-slip model for steel bars in unconfined or steel, FRC, or FRP confined concrete under cyclic loading. ASCE J Struct Eng 135:509–518

Kolozvari K, Tran T, Wallace JW, Orakcal K (2012) Modeling of cyclic shear–flexure interaction in reinforced concrete structural walls. In: Proceedings of the 15th world conference on earthquake engineering, Lisbon

Mansour MY, Hsu TC (2005) Behavior of reinforced concrete elements under cyclic shear. ASCE J Struct Eng 131:44–53

Mansour MY, Hsu TC, Lee JY (2002) Pinching effect in hysteretic loops of RC shear elements. ACI Spec Publ 205:293–321

Massone LM, Wallace JW (2004) Load-deformation responses of slender reinforced concrete walls. ACI Struct J 101:103–113

Massone LM, Orakcal K, Wallace JW (2006) Shear-flexure interaction for structural walls. ACI Spec Publ 236:127–150

Massone LM, Orakcal K, Wallace JW (2009) Modeling of squat structural walls controlled by shear. ACI Struct J 106:646–655
Melek M, Wallace JW (2004) Cyclic behavior of columns with short lap splices. ACI Struct J 101:802–811
Menegotto M, Pinto E (1973) Method of analysis for cyclically loaded reinforced concrete plane frames. In: Proceeding of the IABSE symposium on resistance and ultimate deformability of structures acted on by well-defined repeated loads, Lisbon
Orakcal K, Wallace JW (2006) Flexural modeling of reinforced concrete walls – experimental verification. ACI Struct J 103:196–206
Orakcal K, Wallace JW, Conte JP (2004) Nonlinear modeling and analysis of reinforced concrete structural walls. ACI Struct J 101:688–698
Orakcal K, Ulugtekin D, Massone LM (2012) Constitutive modeling of reinforced concrete panel behavior under cyclic loading. In: Proceedings of the 15th world conference on earthquake engineering, Lisbon
Pang D, Hsu TC (1995) Behavior of reinforced concrete membrane elements in shear. ACI Struct J 92:665–677
Stevens NJ, Uzumeri M, Collins MP (1991) Reinforced concrete subjected to reversed cyclic shear – experiments and constitutive model. ACI Struct J 88:135–146
Thomsen JH IV, Wallace JW (2004) Experimental verification of displacement-based design procedures for slender reinforced concrete structural walls. ASCE J Struct Eng 130:618–630
Tran T, Wallace JW (2012) Experimental study of nonlinear flexural and shear deformations of reinforced concrete structural walls. In: Proceedings of the 15th world conference on earthquake engineering, Lisbon
Ulugtekin D (2010) Analytical modeling of reinforced concrete panel elements under reversed cyclic loadings. MS thesis, Bogazici University
Vecchio FJ, Collins MP (1986) The modified compression field theory for reinforced concrete elements subjected to shear. ACI Struct J 83:219–231
Vecchio FJ, Collins MP (1993) Compression response of cracked reinforced concrete. ASCE J Struct Eng 83:219–231
Verderame GM, Carlo GD, Ricci P, Fabbrocino G (2009) Cyclic bond behavior of plain bars, Part II: analytical investigation. Constr Build Mater 23:3512–3522
Vulcano A, Bertero VV, Colotti V (1988) Analytical modeling of RC structural walls. In: Proceedings of 9th world conference on earthquake engineering, Tokyo-Kyoto
Xu SY, Zhang J (2011) Hysteretic shear-flexure interaction model of reinforced concrete columns for seismic response assessment of bridges. Earthq Eng Struct Dyn 40:315–337

Chapter 21
Earthquake Engineering Experimental Facility for Research and Public Outreach

Ece Eseller-Bayat, Seda Gokyer, and Mishac K. Yegian

Abstract Over the past two decades, important advancements have been made in earthquake engineering practice aimed at reducing seismic risk to urban communities worldwide. Since earthquakes occur infrequently and often in unpredictable locations, the role of experimental research in these advancements has been invaluable. Experimental tests performed under controlled environment can provide high quality data that can advance fundamental knowledge of the behavior of geotechnical and structural elements, validate analytical models, and help explore development of innovative, cost-effective seismic mitigation technologies. It has become evident also that implementation of and investment in seismic risk reduction technologies to vulnerable urban communities requires heightened awareness at all levels of society of the earthquake risk. The authors have found that a shaking table facility while providing valuable research opportunities, is also ideally suited for educational and outreach activities tailored for regional communities and media to heighten their awareness of earthquake risk and demonstrate the important role engineers play in seismic mitigation. The shaking table utilized is uni-directional (1.5 m $\times$ 2 m) and has 254 mm peak to peak lateral displacement capacity. The shaking table has been crucial in many research projects in areas such as structural and soil isolation, dynamic interface properties of geosynthetics, seismic permanent deformations, and liquefaction mitigation. The facility is also utilized weekly, as well as at times of heightened public and media interest in earthquake damage, to present various

E. Eseller-Bayat (✉)
Civil Engineering Faculty, Istanbul Technical University, Maslak, Istanbul 34469, Turkey
e-mail: ebayat@itu.edu.tr

S. Gokyer • M.K. Yegian
Department of Civil and Environmental Engineering, Northeastern University, Boston, MA 02115, USA
e-mail: sedagokyer@gmail.com; m.yegian@neu.edu

A. Ilki and M.N. Fardis (eds.), *Seismic Evaluation and Rehabilitation of Structures*, Geotechnical, Geological and Earthquake Engineering 26,
DOI 10.1007/978-3-319-00458-7_21,

educational modules and carry out shaking table model demonstrations. This paper and presentation will focus on highlights of research conducted using the shaking table and how the facility has been integrated into formal and community-wide educational and outreach programs.

21.1 Introduction

Research performed to date in earthquake engineering has had tremendous effect on understanding earthquake motion, its modification through subsurface soil profiles and its devastating damage to infrastructures and to loss of lives. There is still need for the development of new practical and low-cost mitigation techniques to reduce the loss of lives and prevent collapse. Figure 21.1 demonstrates a collapsed building during the 2011 Van Earthquake in Turkey. The role of earthquake engineering experimental infrastructure can be significant in developing new technologies, and testing novel risk reduction ideas and measures. The Earthquake Engineering Experimental Facility at Northeastern University, Boston, MA provides an integrated infrastructure and instrumentation for several research and educational projects performed on seismic improvement of ground and civil structures, and recently on the development of a new mitigation technique against liquefaction. The experimental facility includes a one-directional servo-controlled shaking table, a data acquisition card and software, accelerometers, linear variable displacement

Fig. 21.1 Building collapsed during the 2011 Van Earthquake in Turkey (Courtesy of National Turk)

transducers (LVDTs), miniature pore water pressure transducers and bender elements and bending disks. Recently, a new mitigation technique was evaluated by performing cyclic shear strain tests on partially saturated sand specimens using the experimental facility. The facility also has had a great impact on increasing public awareness by performing open house demonstrations where students can build structural models and test them under real earthquake motions. In this paper, the details of a current research project that has been conducted using the facility is presented, along with the significant role of the facility in improving public awareness about earthquake risk.

21.2 Earthquake Engineering Experimental Facility at Northeastern University

The earthquake engineering experimental facility at Northeastern University provides large-scale dynamic testing of soil samples under uniform cyclic or earthquake motions. It includes a shaking table (Fig. 21.2), a data acquisition card (NI-DAQ) and software (LabVIEW), a set of instruments including accelerometers (Crossbow, 1g, 2g, 5g, Fig. 21.3a), linear variable displacement transducers (LVDT, RDP DCTH400AG, Fig. 21.3b), miniature pore pressure transducers (GE Druck PDCR 81, Fig. 21.3c), and multiple bender element and bending disk measurement equipment for S and P wave measurements (Fig. 21.3c–e).

The shaking table is one-dimensional, medium size (1.5 m × 2 m), light weight (made of 25.4 mm thick aluminum plate), and servo hydraulic controlled system. The table motion can be controlled from a control unit as well as from a computer. The control unit provides harmonic excitation options such as sinusoidal, step motion, triangular at various frequencies and amplitudes utilizing a plug-in function generator. On the other hand, earthquake records can be simulated from a computer

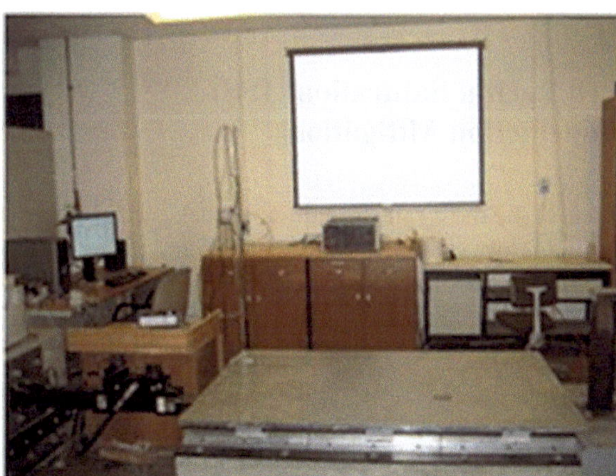

Fig. 21.2 1-D Shaking table used for research, education and outreach activities at Northeastern University

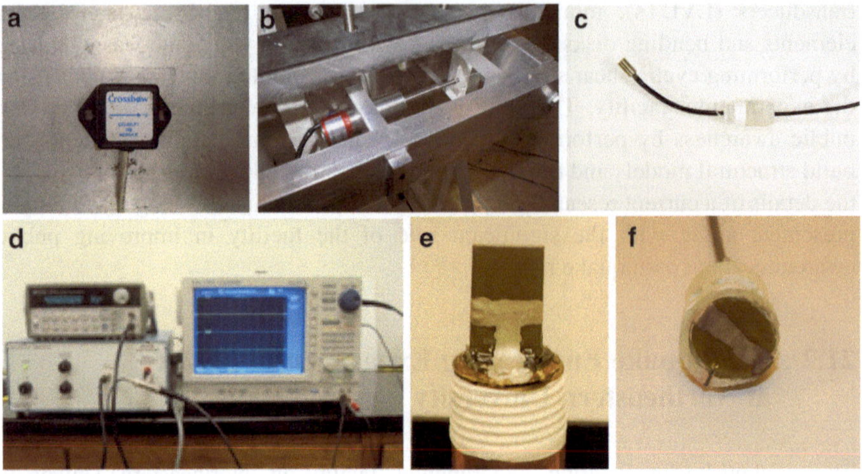

Fig. 21.3 Instruments used in Earthquake Engineering Experimental Facility at Northeastern University: (**a**) Crossbow Accelerometer; (**b**) LVDT (Linear Variable Displacement Transducer); (**c**) GE Druck PDCR81miniature pore pressure transducer; (**d**) HP Function generator, power amplifier and Yokogawa Oscilloscope; (**e**) Bender element; (**f**) Bending disk

utilizing a data acquisition card (NI-DAQ) and data processing software LabVIEW. The shaking table has a load capacity of 25 kN and can effectively perform at a frequency range of 0–50 Hz. It incorporates MTS Model 244.21 actuator with a 254 mm stroke capability ($\pm$127 mm lateral displacement) (Kadakal and Whelpley 1997).

Various research projects in areas of structural and soil isolation, dynamic interface properties of geosynthetics, seismic permanent deformations, and liquefaction mitigation have been conducted using this facility. A recent research project conducted on liquefaction mitigation will be presented in the following section.

21.3 Induced Partial Saturation (IPS) for Liquefaction Mitigation

Liquefaction is one of the most catastrophic earthquake-induced events that impacts infrastructure built on fully saturated loose sands. Using the earthquake engineering experimental facility at Northeastern University, the liquefaction strength of partially saturated sands has been investigated by performing cyclic simple shear tests on prepared specimens.

A special liquefaction box was designed and built in which fully and partially saturated sand specimens can be prepared and tested under uniform cyclic simple shear strains induced through a shaking table. The liquefaction box, named Cyclic Simple Shear Liquefaction Box (CSSLB), can accommodate an integrated set of instrumentation, minimize the sidewall boundary effects and induce uniform shear

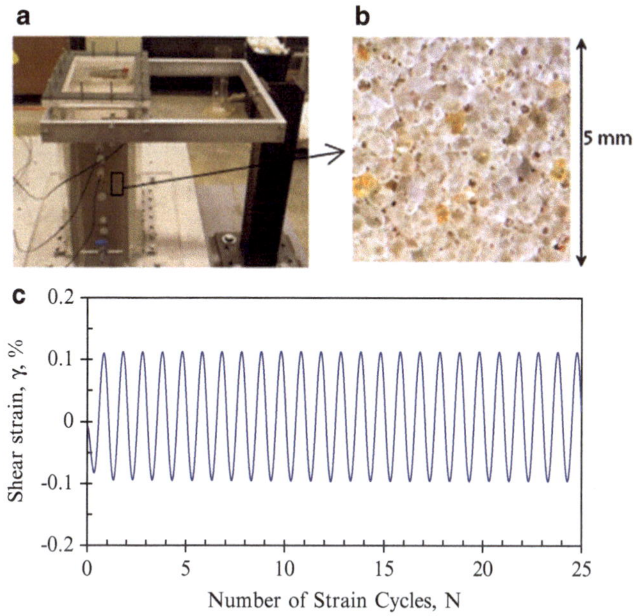

Fig. 21.4 (**a**) Partially saturated sand specimen tested in Cyclic Simple Shear Liquefaction Box (CSSLB); (**b**) A micro picture of air induced sand specimen (S = 80 %) (Gokyer 2009); (**c**) Simple shear strain record applied using the shaking table (for shear strain amplitude is 0.1 %)

strains on large sand specimens. Figure 21.4a demonstrates the side-view photograph of CSSLB which has inside plan dimensions of 190 mm × 305 mm (7.5" × 12") and a height of 490 mm. CSSLB consists of two fixed walls (fixed walls being in the direction of shaking) and two rotating walls which are hinged to the bottom plate, and also are connected to the two fixed walls and the bottom plate by a joint sealant called Sikaflex 15LM. The walls are made of plexiglass. The sealant makes the joints water tight yet flexible allowing movements along the joints by being compressed and elongated. An aluminum frame is attached to the top of the CSSLB at one end and to an outsider fixed beam next to the shaking table at the other end, through unthreaded bolts. Hence, when the shaking table is excited with cyclic displacements, controlled simple shear strains can be induced on the specimens through rotation around the bottom hinge as well as by means of compression and/or elongation of Sikaflex. The elastic compression and elongation capacity of Sikaflex in connections up to 5 mm leads to a maximum 1 % shear strain capacity on sand specimens. The details of the CSSLB design can be found at (Ortakci 2007).

Partially saturated sand specimens were prepared using a special chemical compound "sodium perborate", a main ingredient of dental product "Efferdent" which produces oxygen gases when reacted with water. Specimens were prepared by wet pluviation technique which basically consists in raining the powdered Efferdent-dry sand (Ottawa sand) mix in the partially filled with water CSSLB (keeping 30 mm of free water on top). The intended partial saturation level could be

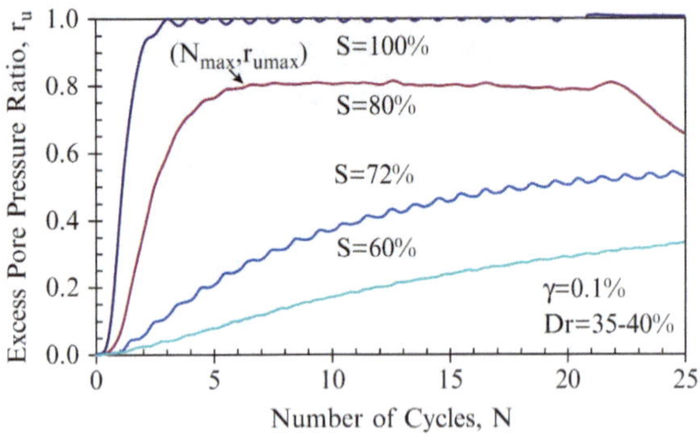

Fig. 21.5 Comparison of excess pore pressure ratio (r_u) generation in specimens at different degrees of saturation ($\gamma = 0.1$ % $\sigma'_v = 2.5$ kPa)

achieved in each specimen based on a correlation determined between the degrees of saturation induced in the specimen versus the Efferdent-sand mass ratios. The presence of oxygen gases and their distribution within the sand specimens were evaluated by taking micro pictures of the specimens with a professional camera as shown in Fig. 21.4b (Gokyer 2009).

A series of cyclic simple shear strain tests were performed on partially saturated sand specimens prepared at degrees of saturation ranging from S = 40 % to S = 90 %. The effects of the main parameters: "degree of saturation (S), relative density (D_r), shear strain (γ)", on the maximum excess pore water pressure generation were evaluated. Consequently, based on the test results, an empirical model was developed, which predicts maximum excess pore water pressure ratios in partially saturated sand specimens.

Figure 21.5 demonstrates the excess pore pressure ratio (r_u) generations in medium dense sand specimens with different degrees of saturation under the shear strain record shown in Fig. 21.4c. The results reveal that during cyclic loading, excess pore pressures also generate in partially saturated sands and can remain high as the degree of saturation increases. However, as the degree of saturation reduces, maximum excess pore pressure (r_{umax}) gets lower, while at the same time the number of cycles required to reach r_{umax} (N_{max}) gets higher.

A parametric study was performed on experimental results to estimate individual, as well as cooperate effects of S, D_r and γ on r_{umax}. Test results showed that r_{umax} depends significantly on the degree of saturation (S) and to a lesser extent on relative density (D_r) and amplitude of the cyclic shear strain (γ). Ultimately, a mathematical model was developed to predict maximum excess pore water pressure ratios (r_{umax}) in partially saturated sands. The r_{umax} model function (Eqs. 21.1, 21.2, 21.3, 21.4 and 21.5) was obtained by the product of a base function f_b and scaling factor functions F_D and F_γ. Details on the formulation of these functions and estimation of model parameters and their statistics are presented in (Eseller-Bayat 2009).

$$r_{u\max} = f(S, D_r, \gamma) \tag{21.1}$$

$$r_{u\max} = f_b(S, D_r = 20\,\%,\ \gamma = 0.1\,\%) F_D(S, D_r) F_\gamma(S, \gamma) \tag{21.2}$$

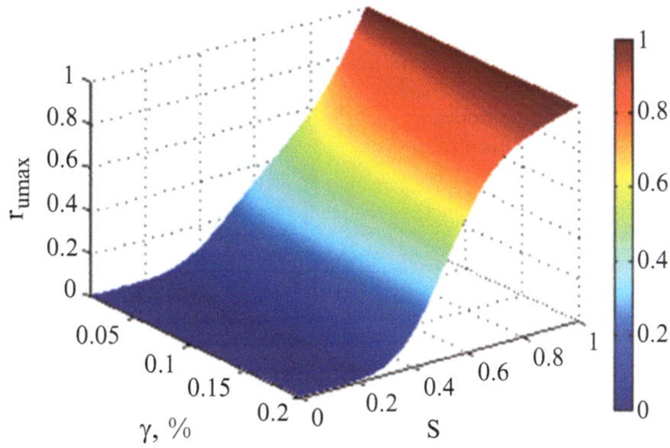

Fig. 21.6 Graphical presentation of the mathematical model presenting maximum excess pore pressure ratio (r_{umax}) generated in loose sands based on cyclic simple shear tests

where:

$$f_b = S^{0.5} e^{-\left[\frac{1-S}{0.54}\right]^4} \tag{21.3}$$

$$F_D = 1 - 8.75(D_r - 0.2)(1 - S)\exp\left[-\frac{1}{2}\left(\frac{1-S}{1-0.84\left(\frac{0.2}{D_r}\right)^{0.25}}\right)^2\right] \tag{21.4}$$

$$F_\gamma = 1 + 1.75\left(\log\frac{\gamma}{0.001}\right)(1 - S)\exp\left[-3.1(1 - S)^2\right] \tag{21.5}$$

Figure 21.6 demonstrates graphical representation of the r_{umax} model for loose sands ($D_r = 25$ %). The model depicts that the effect of shear strain on r_{umax} is more pronounced at degrees of saturation higher than 60 %. Below S = 60 %, regardless of the shear strain level applied, r_{umax} is smaller than about 0.4.

Finally, the shaking table tests on partially saturated sands allowed the authors to investigate the beneficial effect of partial saturation on liquefaction strength and led to the development of a new liquefaction mitigation technique.

21.4 Use of Earthquake Engineering Experimental Facility for Public Awareness and Education

The Northeastern University Earthquake Engineering Experimental Facility has had a great impact on attracting the interest of middle, high school and undergraduate students in civil and earthquake engineering. Open house activities are organized for middle and high school students to increase their interest in science, technology, engineering, and mathematics as well as to learn, hands-on, what engineering is about. In the open house activities, after seeing the basic principles

Fig. 21.7 Shaking table testing of toy building models designed by high school students under earthquake motion records

of how earthquakes can damage buildings, they are expected to design their own structure models, using plastic structural elements, which are in turn tested under real earthquake motions simulated on the shaking table (Fig. 21.7). These activities are very useful in improving the creativity and engineering skills of the students.

Graduate and undergraduate students are exposed to innovative experimental research aimed at mitigating earthquake-related damage to the built environment. Such exposure enhances their awareness of and interest in the field of earthquake engineering. Furthermore, class demonstrations help students understand essential concepts in earthquake engineering such as dynamic behavior of single and multi degree of freedom models, response spectra and earthquakes. Also, demonstrations performed in the Earthquake Engineering Experimental Facility at Northeastern University which are covered by the media raise public awareness about the risks of earthquakes to our society.

21.5 Conclusion

In this paper, the Earthquake Engineering Experimental Facility at Northeastern University was presented. The facility includes a uni-axial shaking table, a data acquisition card and software, accelerometers, LVDTs, miniature pore pressure transducers and bender element and bending disk measurement equipment. The facility has been used for various research projects including testing of partially saturated sands under earthquake motions. The liquefaction strength of partially saturated sands were confirmed through cyclic simple shear tests performed utilizing the shaking table facility. The results of the experimental study further led to the development of a new liquefaction mitigation technique: "Induced Partial Saturation" which is an ongoing project at Northeastern University. The facility also hosts outreach activities to attract the interest of middle, high school and also undergraduate students in earthquake engineering problems and risks.

In summary, earthquake engineering facilities have a significant role in seismic risk reduction because they

- provide high quality data that can advance fundamental knowledge of the behavior of geotechnical and structural elements,
- validate analytical models,
- help explore the development of innovative, cost-effective seismic mitigation technologies,
- encourage educational and outreach activities and increase awareness of earthquake risk
- demonstrate the important role engineers play in seismic mitigation.

Acknowledgements The research presented in Sect. 21.4 was funded by the National Science Foundation through Geoenvironmental Engineering and Geohazard Mitigation Program under Grant No: CMS-0509894. The support of NSF and Program Director Dr. Richard J. Fragaszy is greatly appreciated. The contributions to this research by former graduate students A. Emre Ortakci and Oguz Deniz and undergraduate civil engineering students Kaitlin Vacca and Meredith Washington are acknowledged. Special appreciation is expressed to Dr. David Whelpley for his valuable support to the design and manufacturing of the liquefaction box.

References

Eseller-Bayat EE (2009) Seismic response and prevention of liquefaction failure of sands partially saturated through introduction of gas bubbles. PhD thesis, Department of Civil and Environmental Engineering, Northeastern University, Boston

Gokyer S (2009) Inducing and imaging partial degree of saturation in laboratory sand specimens. MS thesis, Northeastern University, Boston

Kadakal U, Whelpley D (1997) System characteristics and performance of the shaking table setup at Northeastern University. Report, Northeastern University, Boston

Ortakci M (2007) Design and manufacturing of a cyclic simple shear liquefaction box (CSSLB). MS thesis, Northeastern University, Boston

In summary, earthquake engineering facilities have a significant role in seismic risk reduction because they:

[illegible]

Chapter 22
Physical Modeling for the Evaluation of the Seismic Behavior of Square Tunnels

Grigorios Tsinidis, Charles Heron, Kyriazis Pitilakis, and Gopal Madabhushi

Abstract The Chapter summarizes results from dynamic centrifuge tests performed on a rectangular tunnel model embedded in dry sand. The tests were carried out at the geotechnical centrifuge facility of the University of Cambridge, within the Transnational Access Task of the SERIES Research Project (Project: TUNNELSEIS). The experimental data is presented in terms of acceleration and displacement-time histories in the soil and on the tunnel, soil surface settlements, earth pressures on the side walls of the tunnel and internal forces of the tunnel lining. The goal of the experiment is twofold: to better understand the seismic behavior of these types of structures, and to use the high quality and perfectly constrained data to validate the numerical models which are commonly used for the design of rectangular embedded structures. The interpretation of the results reveals (i) rocking response of the tunnel model, (ii) existence of residual values on the earth pressures on the side walls and on the internal forces and (iii) important influence of the tunnel on the shear wave field. These issues are not well understood and are usually not taken into account in the simplified seismic analysis methods.

G. Tsinidis (✉) • K. Pitilakis
Department of Civil Engineering, Research Unit of Geotechnical Earthquake Engineering and Soil Dynamics, Aristotle University of Thessaloniki, PO BOX 424, 54124 Thessaloniki, Greece
e-mail: gtsinidi@civil.auth.gr; kpitilak@civil.auth.gr

C. Heron • G. Madabhushi
Department of Engineering, Schofield Centre, University of Cambridge, High Cross, Madingley Road, Cambridge CB3 0EL, UK
e-mail: cmh78@cam.ac.uk; mspg1@cam.ac.uk

A. Ilki and M.N. Fardis (eds.), *Seismic Evaluation and Rehabilitation of Structures*, Geotechnical, Geological and Earthquake Engineering 26,
DOI 10.1007/978-3-319-00458-7_22,

22.1 Introduction

Underground structures behave generally better than near surface or aboveground structures during strong earthquakes. However, several cases of extensive damage and even collapse are reported in the literature (i.e. Sharma and Judd 1991; Power et al. 1998). The most interesting case is that of the Daikai station in Kobe, Japan, that collapsed during the major Hyogoken-Nambu earthquake (1995) (Iida et al. 1996; Hashash et al. 2001). This is the first well-reported case of a total collapse of a large underground structure under seismic shaking. These recent failures revealed some important weaknesses in the current seismic design practices.

The seismic response of embedded structures to ground shaking and permanent ground deformations (due to ground failure) is very distinct with respect to the aboveground structures. The kinematic loading imposed on the structure from the surrounding soil is prevalent, while the inertial effects are of secondary importance. The soil-structure interaction effects, that are expected to be increased in cases of rectangular embedded structures, are closely related to two crucial parameters, namely (i) the relative flexibility of the structure and the ground and (ii) the interface characteristics between the structure and the surrounding soil. The exact effect of these parameters on structural seismic response is not well known.

For the evaluation of the seismic response of underground structures, several methods may be found in the literature based on different levels of complexity, ranging from uncoupled methods, simplified closed form solutions and equivalent static analysis schemes, to the most sophisticated full dynamic analysis of the coupled soil-structure system incorporating advanced numerical methods (i.e. Wang 1993; Penzien 2000; Hashash et al. 2001; ISO23469 2005; FWHA 2009 etc). The results of these methods may substantially deviate, even under the same assumptions, indicating the lack of knowledge regarding some very crucial issues that significantly affect the seismic response (Pitilakis and Tsinidis 2012). The lack of knowledge is even more pronounced for the non-circular structures. The seismic earth pressures on the side walls, the seismic shear stresses around the perimeter of the structure and the soil-structure interface characteristics, the complex deformation modes during the shaking (i.e. rocking for stiff structures or inward deformations for the flexible structures) and the impact of the soil-structure relative flexibility on the seismic response are, among others, issues that need further study.

To this end, a comprehensive set of tests have been performed at the University of Cambridge (UCAM) jointly with the Aristotle University of Thessaloniki (AUTH) on square tunnel models embedded in dry sand. The tests were carried out at the geotechnical centrifuge facility of the University of Cambridge (as part of the TA project TUNNELSEIS), within the Transnational Access Task of the EU-funded research project SERIES. The produced experimental data is primarily to be used to better understand the seismic behavior of rectangular embedded structures. The data, of high quality and control, will be also used to validate advanced numerical models and to improve the design methods. The experimental procedure and the setup for one of the tests are presented in the following along with some representative experimental results.

22.2 Test Setup and Model Preparation

22.2.1 *Centrifuge Facility*

The tests were performed at the "Turner beam centrifuge" of the University of Cambridge (Schofield Centre), under centrifuge acceleration of 50g (Fig. 22.1a). The 10-m beam-like structure that rotates around a central vertical axis accommodates a swinging platform with the model and the actuator on one end and a counterweight on the other (Schofield 1980).

Earthquake input motions are applied using the Stored Angular Momentum (SAM) actuator (Fig. 22.1b) (Madabhushi et al. 1998), which is designed to apply sinusoidal input motions at a maximum frequency up to 60 Hz and at a maximum amplitude of 20g (in model scale). The actuator, developed at the University of Cambridge, can operate under a centrifuge acceleration of 100g.

A large Equivalent Shear Box (ESB) is used as the container for the models (Fig. 22.2). The box consists of aluminum dural frames connected to each other through rubber layers. The inside dimensions of the model are 673 × 255 × 427 (mm). The box is designed to match the shear stiffness of the contained soil for the range of shear strains of interest, in order to minimize any soil-container interactions (Schofield and Zeng 1992).

22.2.2 *Materials*

The soil deposit is made of uniform Hostun HN31 sand having a relative density of about 90 %. The mechanical properties of the sand are tabulated in Table 22.1.

The tunnel model, manufactured from 6063A aluminum alloy, is 100 mm wide and 220 mm long, having a thickness of 2 mm. The aluminum alloy mechanical

a b

Fig. 22.1 (**a**) Turner beam centrifuge, (**b**) SAM actuator

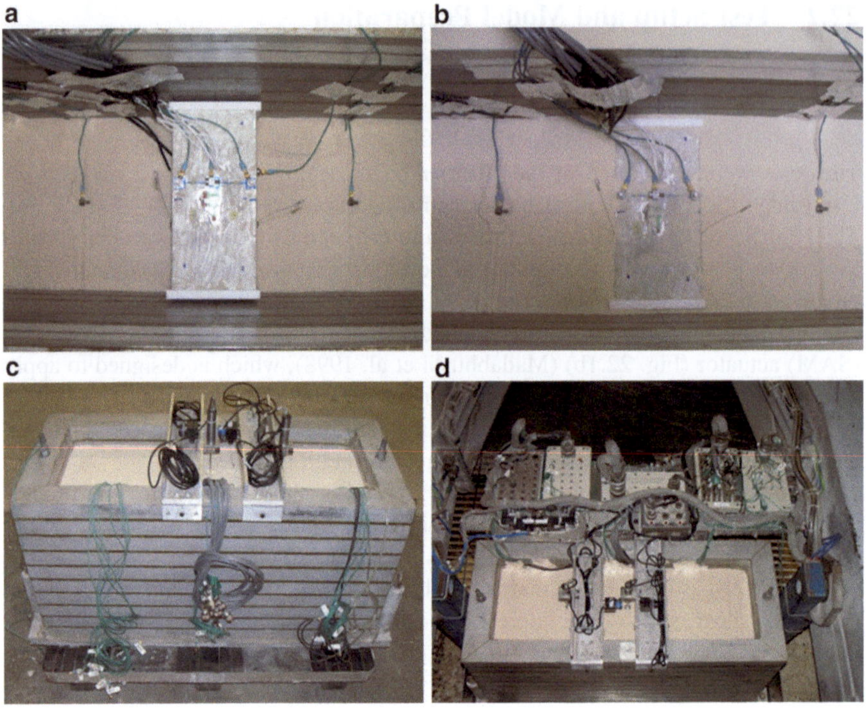

Fig. 22.2 (**a**), (**b**) Installation of the tunnel-model, (**c**) prepared box before loading, (**d**) loading of the box on the centrifuge

Table 22.1 Sand mechanical properties (After Mitrani 2006)

	ρ_s (g/cm^3)	e_{max}	e_{min}	d_{10} (mm)	d_{50} (mm)	d_{60} (mm)	φ_{crit} (°)
Hostun HN31	2.65	1.01	0.555	0.209	0.335	0.365	33

Table 22.2 Model mechanical properties

Unit weight, γ (kN/m^3)	Elastic modulus, E (GPa)	Poisson ratio ν	Tensile strength, f_{bk} (MPa)
2.7	69.5	0.33	220

properties are summarized in Table 22.2. According to the scale factor ($N = 50$), the model corresponds to a 5 × 5 × 11 (m) square tunnel having an equivalent concrete lining thickness equal to 0.13 m (assuming $E = 30$ GPa for concrete). This thickness is obviously unrealistic in practice, as the design analysis for the static loads will result in a much thicker lining. However, this selection was necessary to achieve measurements of the lining strains. To simulate more realistically the soil-structure interface, Hostun sand was stuck on the external face of the tunnel-model, creating a rough surface. Consequently, the shear stresses at the interface are expected to be increased.

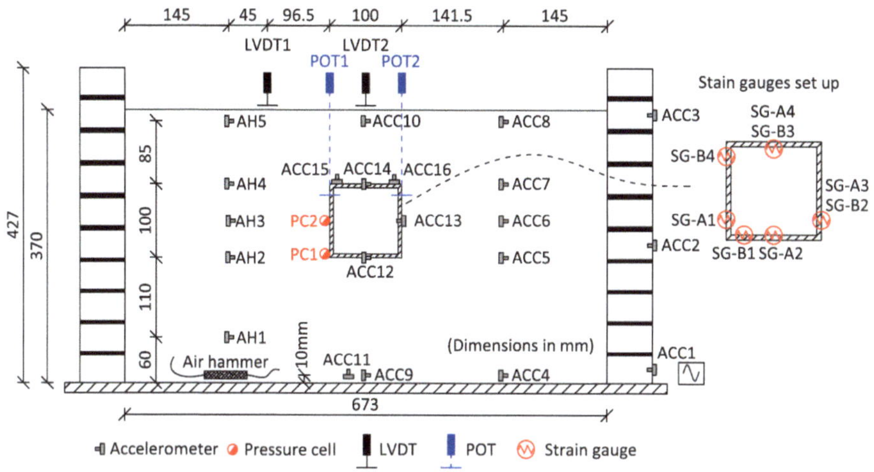

Fig. 22.3 Model layout – instrumentation scheme

22.2.3 Model Preparation

The sand pouring was performed with an automatic hopper system (Madabhushi et al. 2006; Zhao et al. 2006). The soil relative density is actually controlled by the sand drop height and the nozzle diameter of the hopper. A muzzle with two small sieves was also added under the nozzle in order to create a dense model. To achieve the desired soil density, trial pours were made to aid the selection of the aforementioned parameters.

During construction, the tunnel and all the embedded transducers were positioned in the model. To avoid any interaction of the tunnel with the ESB box, the tunnel was shorter than the box width. Two PVC rectangular plates were placed at both the tunnel ends to avoid the sand entrance into the tunnel-model. The plates, having larger dimensions than the model (110 mm × 110 mm × 10 mm), were connected to each other through a large screw running inside the model.

22.2.4 Model Layout – Instrumentation

Figure 22.3 presents the final model layout and the instrumentation scheme. Miniature accelerometers were used to measure the acceleration in the soil, on the tunnel and on the ESB box.

To estimate the soil shear wave velocity profile before each earthquake, air hammer tests were performed (Ghosh and Madabhushi 2002). A small air-hammer was introduced close to the base of the soil layer while a set of accelerometers were placed above it, forming an array, allowing a record of the arrival times of the waves emanating from the air-hammer.

The soil surface settlements were recorded at two locations using linear variable differential transformers (LVDTs), while two position sensors were attached on the upper side of the walls of the tunnel to capture the vertical displacement and the possible rocking of the tunnel model. Both the LVDTs and the POTs were attached on gantries running above the ESB box. Two miniature total earth pressure cells were attached to the left side wall of the tunnel, allowing the measurement of the soil earth pressures on the wall. Resistance strain gauges (model TML by Tokyo Sokki Kenkyujo) were glued on the inner and the outer face of the tunnel to measure the model strains at several locations. Four of them measured the axial strains at the walls and the slabs, while four others recorded the bending moment strains near the model corners and at the middle of the roof slab. One of the latter strain gauges (middle of the roof slab) was broken during testing. Both the bending moment and the axial force strain gauges were forming full Wheatstone bridge circuits to achieve maximum resolution. In the following results, positive values represent bending moment with tensile stress increments for the internal lining face and tensile axial force. The gauges cables were running from the inside face of the tunnel, to avoid any interactions with the soil-tunnel interface.

All the instruments were adequately calibrated before and after the test. Regarding the strain gauges, a special procedure for their calibration was followed. More specifically, the calibration factors for both the axial and the bending moment strain gauges were derived for simple static loading patterns. For each loading case, the model was incrementally loaded and unloaded by adding and reducing weights on it through a frame. Readings of each strain gauges voltage output were taken during each step. The support and loading configurations were changed, reversing the model, while the whole procedure was done twice so as to check the repeatability of the strain gauges' response. This procedure allowed multiple records for different static systems for each strain gauge to be collated.

The internal forces at each gauge position were computed through a simple static analysis of the model and plotted against the measured voltage change, to evaluate the gauge factor. The structural model was simulated in ABAQUS (ABAQUS 2010) with shell elements taking into account the exact supports and loading positions.

Figure 22.4 presents the static model configurations for the bending moment and the axial strain gauges, while a calibration curve is also depicted, as an example.

22.2.5 Experimental Procedure

After the preparation, the model was loaded on the centrifuge and all the instruments were plugged into the on-board acquisition systems. During each flight, the centrifuge was spun up in steps stopping at 10g, 30g and 50g and then the earthquakes were fired in a row, leaving some time between them to acquire the data. The data was recorded at sampling frequency of 4 Hz during the swing up and at 4 kHz during the earthquakes. The CDAQS (Centrifuge Data Acquisition System) was used for the data acquisition.

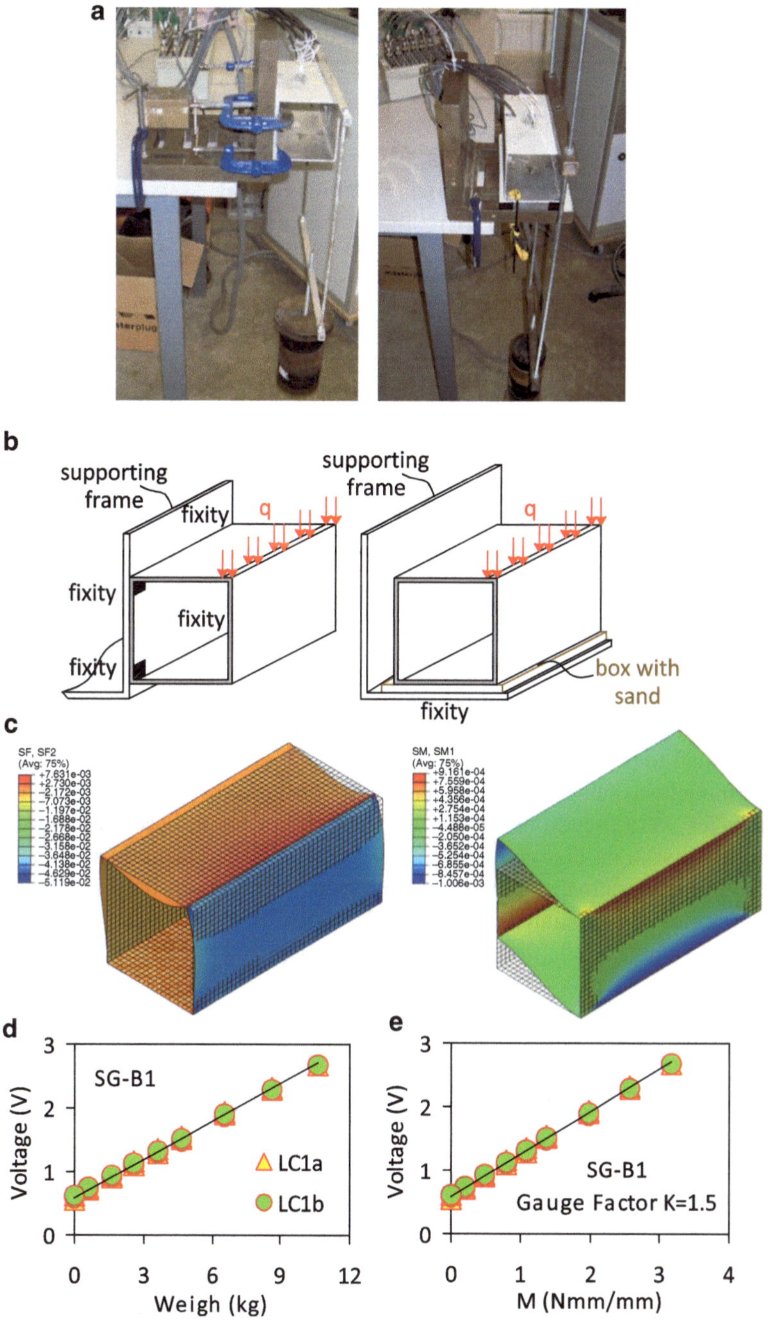

Fig. 22.4 Strain gauges calibration procedure (**a**) static loading for bending moment and axial force strain gauges, (**b**) static systems, (**c**) static analyses in ABAQUS-deformed shapes, (**d**) voltage-mass calibration curve derived for repetitions of the same loading scenario (**e**) voltage-bending moment calibration curve

Table 22.3 Input motion characteristics

	Input type	Frequency (Hz)	Amplitude (g)	Nominal duration (s)
Flight 1				
EQ1	Pseudo-harmonic	30 (0.6)	1.0 (0.02)	0.4 (20)
EQ2		45 (0.9)	4.0 (0.08)	0.4 (20)
EQ3		50 (1)	6.5 (0.13)	0.4 (20)
EQ4		50 (1)	12.0 (0.24)	0.4 (20)
EQ5	Sine sweep	60 (1.2)	12.0 (0.24)	3.0 (150)
Flight 2				
EQ6	Pseudo-harmonic	50 (1)	5.8 (0.116)	0.4 (20)
EQ7		50 (1)	6.0 (0.12)	0.6 (30)
EQ8		50 (1)	11.0 (0.22)	0.5 (25)

Thirty channels were used in total; 16 of them for the accelerometers and 14 for the other instruments (strain gauges, LVDTs, POTs and pressure cells).

Before each shake, air-hammer tests were conducted to evaluate the new (modified) shear wave velocity (V_s) profile of the soil deposit. The V_s profiles were estimated based on the travel times of the waves, between accelerometers that are placed at known distances apart. These distances did not change significantly after each shake, as the recorded soil settlements were small due to the high relative density of the soil deposit. The travel times were estimated in a simplified way from the arrivals of the waves, produced by air-hammer. To make sure that the arrival times were adequately recorded, the DasyLab software was used as the acquisition system for the air-hammer array of accelerometers, allowing for a sampling frequency equal to 50 kHz.

The model was subjected to a total of eight "earthquakes" during two flights. The earthquakes were pseudo-harmonic wavelets except for the last earthquake fired during the first flight that was a sine sweep. This signal is actually a sine wave of decreasing acceleration amplitude and frequency. The main characteristics of the input motions are tabulated in Table 22.3 both in model and prototype scale (bracketed values), while the time histories are depicted in Fig. 22.5.

22.3 Experimental Results

22.3.1 Air Hammer Testing

Figure 22.6 presents the shear wave velocity profiles as estimated for the first flight after each shake. Although the model is subjected to severe shakes, the computed profiles indicate minor changes of the mechanical properties of the soil deposit, expressed in terms of shear wave velocity at different depths. This is attributed to the high relative density and strength of the soil deposit. It is noted that this is a generic estimation of the velocity, as even small movements of the accelerometers,

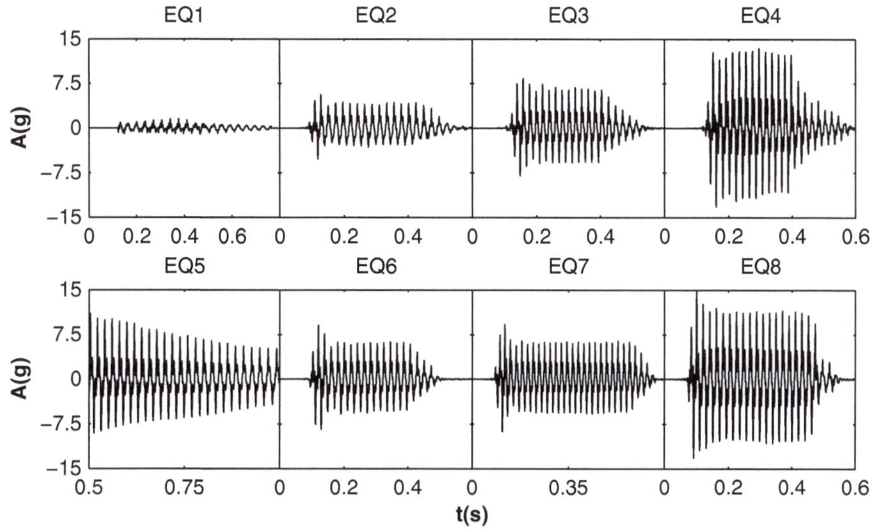

Fig. 22.5 Input motions

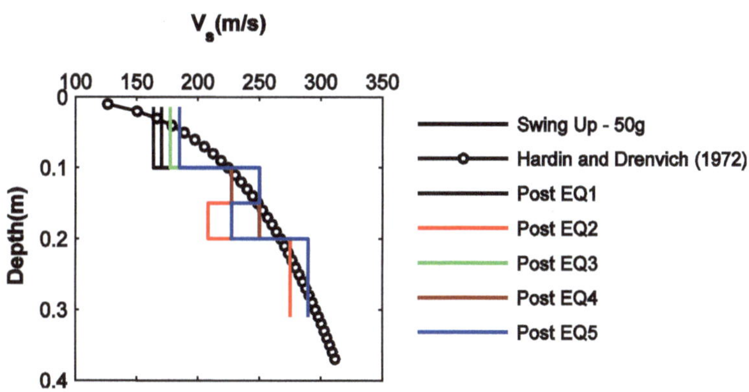

Fig. 22.6 V_s profiles as computed from air hammer testing after the swing up and the shakes during the first flight. Comparison with empirical formulation of Hardin and Drnevich (1972)

caused by the soil settlements, may significantly affect the results. The "initial" V_s profile (estimated after the completion of the swing up) is in relatively good agreement with the profile computed by the empirical formulation proposed by Hardin and Drnevich (1972).

22.3.2 Static Response

Indicative records of the model response during the first flight swing up are presented in terms of soil surface settlements (Fig. 22.7a), earth pressures on the side walls (Fig. 22.7b), bending moments (Fig. 22.7c), and axial forces of the

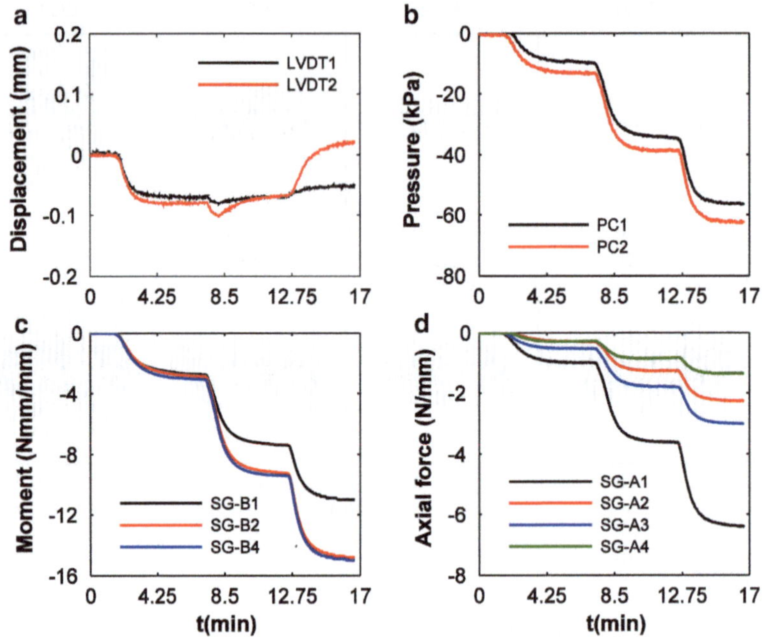

Fig. 22.7 Model response during first flight spin up (**a**) soil surface displacements, (**b**) earth pressures on the side wall, (**c**) bending moments, (**d**) axial forces

structure (Fig. 22.7d). The recorded settlements are small due to the high relative density of the soil. Moreover, the earth pressures measured at the middle height of the sidewall are increased with respect to the wall-slab joint.

The internal forces of the structure increased during the swing up procedure due to the increase of the model self weight. The bending moment strains are in relatively good agreement, as the gauges are located at the same distances from the joints. The axial forces are found to be larger for the invert slab with respect to the roof slab, due to the larger overburden stress at tunnel foundation. Finally, the differences between the axial forces of the two walls could be attributed to differences of the soil-model interface characteristics at these locations and also to differences of the calibration factors of the gauges.

22.3.3 Dynamic Response

In the following paragraphs the dynamic response of the soil-tunnel model is presented and discussed. To interpret the results, the data was windowed neglecting the parts of the signals before and after the main duration of the signals, while a filtering procedure was conducted in the frequency domain. More specifically, the acceleration-time histories were filtered between the frequencies of 10–400 Hz,

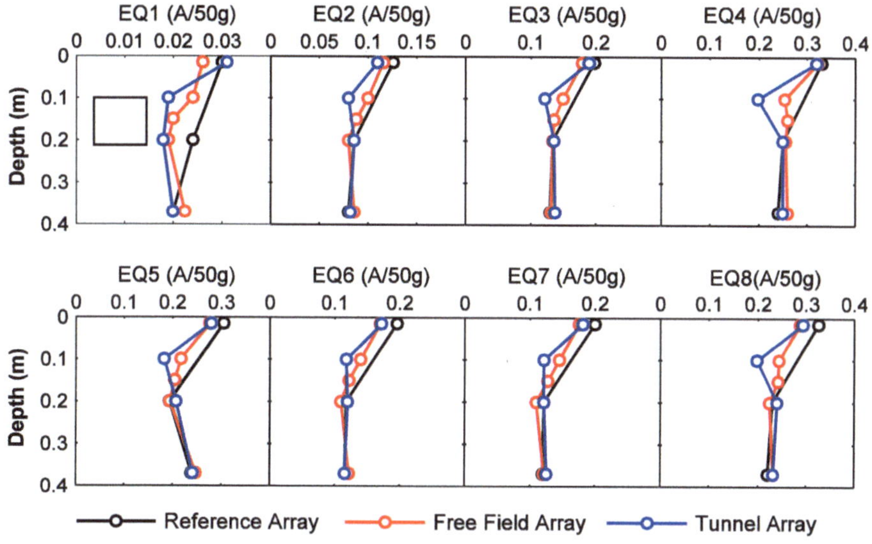

Fig. 22.8 Maximum accelerations along accelerometer arrays

using a band pass filter, subsequently the displacement-time histories were obtained through double integration. The band pass filter was selected to avoid drift of the accelerometer signals during integration. The other data (i.e. soil settlements, earth pressures on the wall, strain-time histories etc) were filtered using a low pass 8th order Butterworth type filter, for frequencies up to 400 Hz.

22.3.3.1 Acceleration**s**

Figure 22.8 presents the peak acceleration profiles with depth along the three main arrays of accelerometers, namely the reference array (on the ESB box), the "free field" array and the vertical array across the tunnel (tunnel array). The peak values are computed as the average semi-amplitude of the cycles in the processed acceleration-time histories. Generally, as expected, the peak acceleration decreases with depth. Moreover, the acceleration at the tunnel roof slab is found to be reduced, with respect to the free field for the same burial depth, for all the earthquakes. This indicates an influence of the tunnel to the wave propagation field. The differences are amplified with the intensity of the input motion.

Figure 22.9 plots the peak "surface" acceleration to peak "base" acceleration ratios as computed for each vertical array for both the flights. All values are larger than unity indicating an amplification of the acceleration from the base to the surface. The amplification ratio decreases with an increase of the input motion amplitude, due to the non-linear soil behavior. The response of the reference array (ESB box) is always larger than the soil. The amplification along the tunnel array is

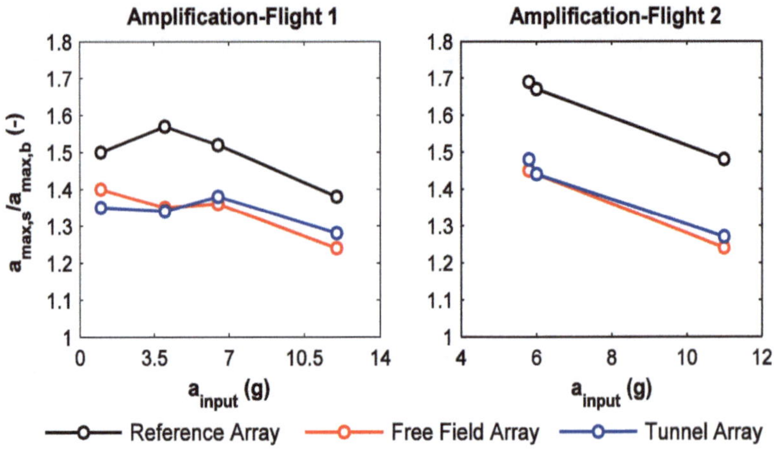

Fig. 22.9 Amplification of horizontal acceleration along arrays

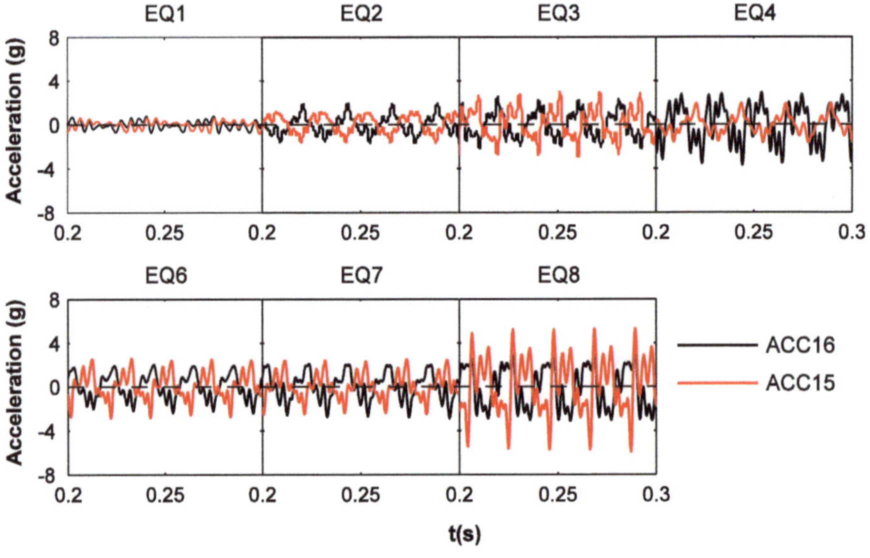

Fig. 22.10 Vertical acceleration time histories on model roof slab

slightly larger than the free field array for input motions larger than 4g (0.08g), while the opposite is observed for smaller earthquakes.

Vertical acceleration was recorded on the sides of the model roof slab. Figure 22.10 presents windows of the recorded accelerations. It is observed that the time histories are out of phase indicating a rocking mode of vibration for the model, in addition to the classical racking mode.

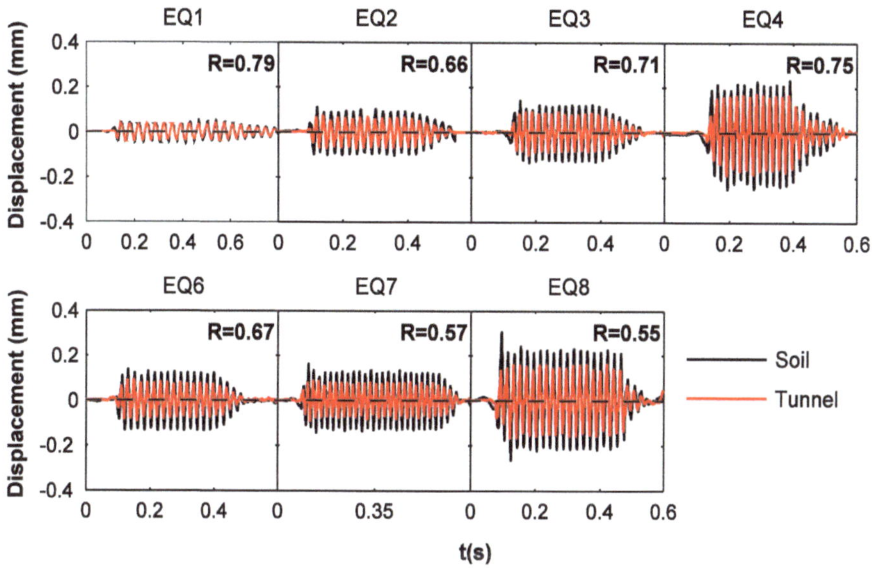

Fig. 22.11 Racking distortions computed from displacement time histories obtained from acceleration time histories through double integration. Racking ratios for each earthquake

22.3.3.2 Racking Deformations

The racking distortions of the tunnel and the soil were computed from the displacement-time histories, derived from acceleration-time histories through double integration (Fig. 22.11). The racking ratios (structural/soil distortions) were then computed, using the following formulation:

$$R = \frac{\Delta_{structure}}{\Delta_{ff}} = \frac{u_{ACC14} - u_{ACC12}}{u_{ACC7} - u_{ACC5}} \tag{22.1}$$

The results indicate that the tunnel behaves as a rigid structure with respect to the surrounding soil, as the structural distortions are decreased with respect to the soil. This behavior is not consistent with the results of the well known analytical procedure proposed by Wang (1993) for the estimation of the relative soil to structure flexibility. According to this methodology, the flexibility ratio is equal to 180, indicating a quite flexible structure. The analytical procedure precludes the structure's rocking vibration, assuming pure racking deformation as the prevailing deformation mode. Moreover, the soil deformations computed by the accelerations could be affected by the presence of the tunnel (accelerometers array close to the tunnel) and the interaction with the ESB box. These issues could explain to some extent this important observation. Further research is certainly needed to better constrain this complex behavior and to elaborate an improved methodology for the seismic design of rectangular shallow tunnels considering also the real soil-structure relative flexibility that will account for the complex modes of vibration.

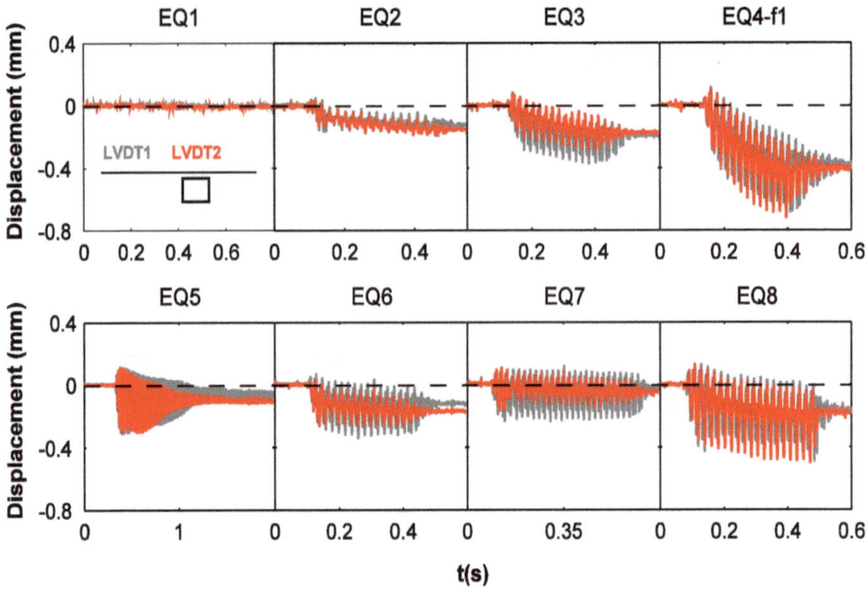

Fig. 22.12 Dynamic soil surface settlements recorded during each shake

22.3.3.3 Soil Settlements

Small settlements were recorded at the soil surface after each earthquake, due to the high relative density of the soil (Fig. 22.12). Moreover, the settlements at the free field and above the tunnel were found to be almost the same.

22.3.3.4 Earth Pressures

Figure 22.13 presents the dynamic earth pressures at the left side wall, as recorded during each shake at the middle of the wall and near the invert slab-wall corner. Three phases may be indentified, namely a transient stage, a steady state stage and finally a post-earthquake residual stage. During the first few cycles of loading, seismic earth pressures were building up. Then in the steady state stage the earth pressures were oscillating around a mean value. During this step, the pressures were shifting from active to passive state due to the tunnel oscillation. Finally, in the post-earthquake stage residual stresses were recorded on the tunnel lining. This behavior has been also reported during similar centrifuge tests (Cilingir and Madabhushi 2011). The dynamic increments of the pressures were found to be larger near the corner, while at the middle of the wall the increments were much smaller due to the flexibility of the wall at this location. In the majority of the cases the residual values were larger at the joint (corner) location, while it was not possible to find a clear connection between the residual values and the severity of an earthquake. This complex behavior is probably attributed to the soil plastic deformations and the small soil densification during shaking.

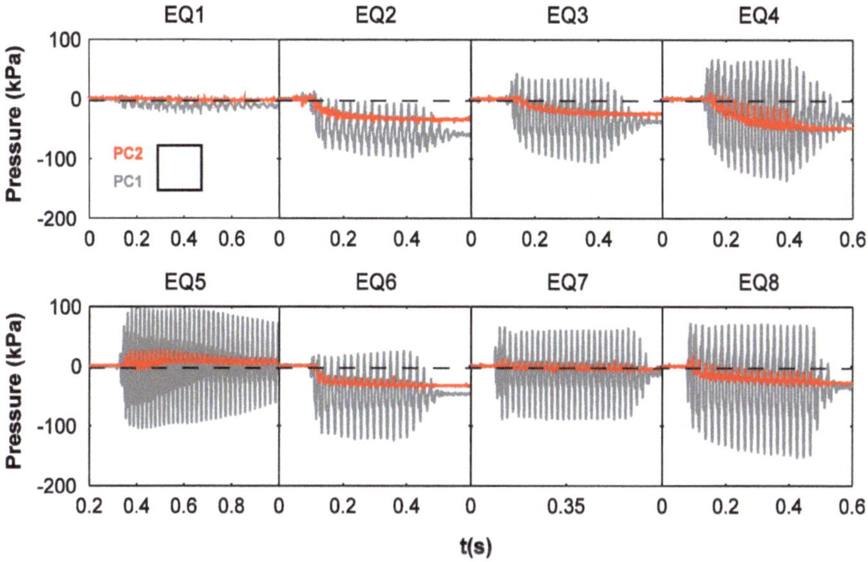

Fig. 22.13 Dynamic soil earth pressures on the left side wall of the structure

22.3.3.5 Bending Moment Time-Histories

Representative dynamic bending moment-time histories, as recorded for EQ3, are presented in Fig. 22.14. The recorded values are quite close to each other, as the strain gauges were actually located at equal distances from the corners. The bending moments were found to be higher after each shake. Actually, similar behavior to the earth pressures response was observed, as the three aforementioned stages were also reported in this case. The large residual values found after each shake, are attributed to stress redistributions in soil caused by the plastic soil deformations and to some extend to the small soil densification. This cumulative response has also been observed during dynamic centrifuge tests performed on circular tunnel models embedded in dry sand (Lanzano et al. 2012).

22.3.3.6 Axial Force-Time Histories

Residual values were also observed for the axial forces (Fig. 22.15). These residuals were generally smaller than for the bending moment and were found to be larger at the slabs. They can be attributed to the small soil densification as well as to the possible sliding effects on the soil-tunnel interface and to some extent to the soil non-linear behavior. It is important to note that the dynamic axial forces recorded on the sidewalls of the model were out of phase, indicating again the rocking mode of vibration of the tunnel, as mentioned above for the vertical accelerations.

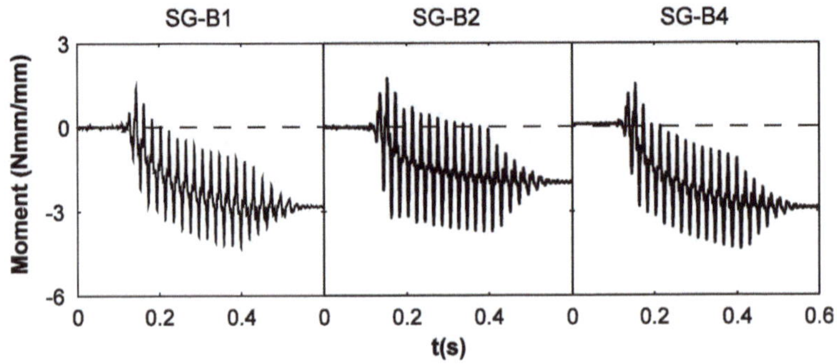

Fig. 22.14 Dynamic bending moments recorded during EQ3

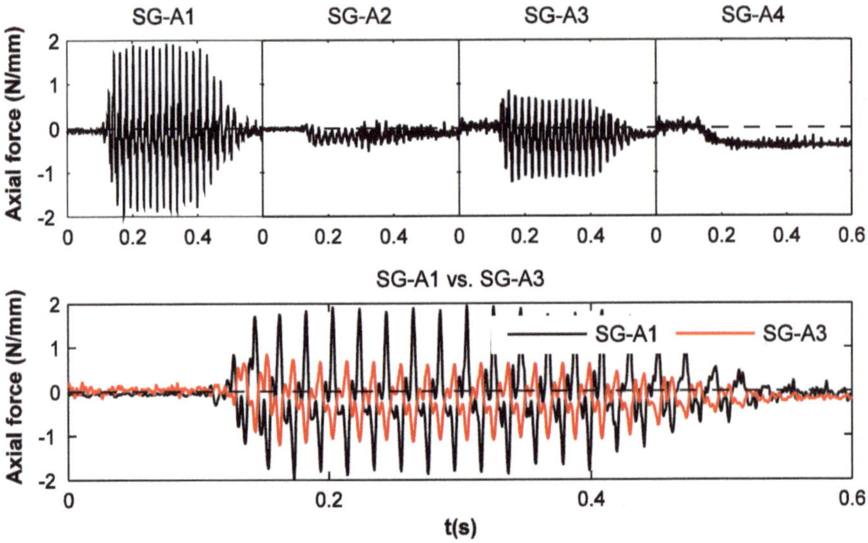

Fig. 22.15 Dynamic axial forces recorded during EQ3. Out of phase axial response of the side walls

22.4 Conclusions

The chapter presented the main experimental work conducted in a series of dynamic centrifuge tests on a square tunnel model embedded in dry sand. Representative experimental results have been presented and discussed. The main conclusions drawn by the interpretation of these results may be summarized as follows:

- During the spin up, the static response in terms of earth pressures on the structure and internal forces increased due to the increase of the self weight of the model. This response is found to be consistent with the theoretically expected behavior.

- The horizontal acceleration recorded at several locations is amplified towards the surface. The amplification ratio with respect to the base amplitude reduces with increased input motion amplitude due to the soil non-linear behavior. Moreover, the horizontal acceleration recorded at the roof slab of the tunnel is lower with respect to the acceleration at the same burial depth in free field. This difference increases with the increase of the input motion amplitude.
- Vertical acceleration-time histories recorded on the sides of the model's roof slab were out of phase indicating a rocking mode of vibration for the tunnel. Unfortunately, the position sensors did not manage to capture the exact evolution of this rotation. This rocking response may be slightly affected by the minor rocking response of the ESB box during shaking.
- According to the simplified procedure proposed by Wang (1993), the structure is classified as a flexible structure with respect to the surrounding soil. However, the tunnel's racking distortions, computed from the acceleration-time histories, indicate a rigid structure with respect to the surrounding soil. This inconsistency could be related to the complex vibration modes (i.e. rocking of the structure) that are not accounted for by the analytical procedure.
- Small ground settlements were recorded at the soil surface during swing up and shaking, due to the high relative density of the studied soil deposit.
- During shaking the dynamic increments of the pressures were higher for the slab-wall corner due to the larger rigidity of the model at this location compared to the middle of the wall. Residual values were observed after each shake; they are mainly attributed to plastic deformations of the soil and to a certain amount of soil densification during shaking that can cause stress redistribution.
- The dynamic bending moments of the structure responded in a similar way to the dynamic earth pressures. Large residual values are detected after each shake, as a result of cumulative strains during the shaking.
- Smaller residuals were observed for the dynamic axial forces after shaking. In addition to the aforementioned parameters that can cause these residual values, a small amount of sliding on the soil-tunnel interface can also affect this behavior.

Acknowledgments The research leading to these results has received funding from the European Community's Seventh Framework Programme [FP7/2007–2013] for access to the Turner Beam Centrifuge, Cambridge, UK, under grant agreement n° 227887 [SERIES]. The technical support received by the Technicians of the Schofield Centre is gratefully acknowledged.

References

ABAQUS (2010) Analysis user's manual, vols I–IV, Dassault Systèmes, SIMULIA Inc, USA

Cilingir U, Madabhushi SPG (2011) A model study on the effects of input motion on the seismic behavior of tunnels. J Soil Dyn Earthq Eng 31:452–462

FWHA (2009) Technical manual for design and construction of road tunnels – Civil elements. US Department of Transportation. Federal Highway Administration. Publication No. FHWA–NHI–09–010, p 629

Ghosh B, Madabhushi SPG (2002) An efficient tool for measuring shear wave velocity in the centrifuge. In: Phillips R, Guo PJ, Popescu R (eds) Proceedings of international conference on physical modelling in Geotechnics, St Johns, NF, Canada. AA Balkema Publishers, the Netherlands, pp 119–124

Hardin BO, Drnevich VP (1972) Shear modulus and damping in soils: design equations and curves. J Soil Mech Found Div 98(SM7):667–692

Hashash YMA, Hook JJ, Schmidt B, Yao JI-C (2001) Seismic design and analysis of underground structures. Tunn Undergr Sp Technol 16(2):247–293

Iida H, Hiroto T, Yoshida N, Iwafuji M (1996) Damage to Daikai subway station. Soils and Foundations, Special Issue on Geotechnical Aspects of the January 17 1995 Hyogoken-Nambu Earthquake. Japanese Geotechnical Society, pp 283–300

ISO 23469 (2005) Bases for design of structures – Seismic actions for designing geotechnical works. ISO International Standard. ISO TC 98/SC3/WG10

Lanzano G, Bilotta E, Russo G, Silvestri F, Madabhushi SPG (2012) Centrifuge modeling of seismic loading on tunnels in sand. Geotech Test J 35:1–15

Madabhushi SPG, Schofield AN, Lesley S (1998) A new stored angular momentum (SAM) based earthquake actuator. In: Kimura T, Kusakabe O, Takemura J (eds) Proceedings of the international conference Centrifuge '98, Tokyo, Japan. Balkema, Rotterdam, pp 111–116

Madabhushi SPG, Houghton NE, Haigh SK (2006) A new automatic sand pourer for model preparation at University of Cambridge. In: Proceedings of the 6th international conference on physical modelling in geotechnics, Hong Kong

Mitrani H (2006) Liquefaction remediation techniques for existing buildings. PhD thesis, University of Cambridge, Cambridge, UK

Penzien J (2000) Seismically induced racking of tunnel linings. Earthquake Eng Struct Dyn 29:683–691

Pitilakis K, Tsinidis G (2012) Performance and seismic design of underground structures. In: Proceedings of the II international conference on performance based design in earthquake geotechnical engineering, Taormina

Power M, Rosidi D, Kaneshiro J, Gilstrap S, Chiou SJ (1998) Summary and evaluation of procedures for the seismic design of tunnels. Final report for task 112-d-5.3(c). National Center for Earthquake Engineering Research, Buffalo, New York

Schofield AN (1980) Cambridge geotechnical centrifuge operations. Geotechnique 30(3):227–268

Schofield AN, Zeng X (1992) Design and performance of an equivalent shear beam (ESB) container for earthquake centrifuge modelling. Technical Report, CUED/D-Soils/I'R245, Cambridge University, UK

Sharma S, Judd WR (1991) Underground opening damage from earthquakes. Eng Geol 30:263–276

Wang JN (1993) Seismic design of tunnels – A simple state-of-the-art design approach. Monograph 7. Parsons Brinckerhoff, New York

Zhao Y, Gafar K, Elshafie MZEB, Deeks AD, Knappett JA, Madabhushi SPG (2006) Calibration and use of a new automatic sand pourer. In: Ng CWW, Zhang LM, Wang YH (eds) Physical modelling in geotechnics – 6th ICPMG '06. Taylor & Francis Group, London

Chapter 23
Susceptibility of Shallow Foundation to Rocking and Sliding Movements During Seismic Loading

Charles Heron, Stuart Haigh, and Gopal Madabhushi

Abstract Current design codes prevent the rocking and sliding of shallow foundations during seismic loading despite much research indicating the beneficial nature of allowing such movements. The primary benefit is the partial isolation of the structure from the soil beneath and subsequently the reduced ductility demands on the superstructure, saving money and reducing the risk of collapse. However, further research is required in order to be able to fully model and predict the behavior of the soil-foundation interface when sliding and rocking is permitted. The results presented in this chapter examine how several different parameters including structural stiffness, aspect ratio, soil relative density and earthquake magnitude affect the level of rotation and sliding experienced by the foundation. Six centrifuge tests were performed to examine how these parameters affected the response of the structure and high speed photography was used to track the movements of the foundation precisely. It was found that structures with a high centre of gravity slid more than structures with a low centre of gravity. Also, stiff structures were found to rotate more than flexible structures and structures located on dense sand rotate more than those located on loose sand.

23.1 Introduction

The design of shallow foundations located in seismically active zones is well defined in modern design codes. Generally, relative movement between the foundation and underlying soil is prohibited. This results in the foundation acting rigidly with the soil during an earthquake. Consequently the superstructure located on the foundation experiences the full magnitude of the earthquake.

C. Heron (✉) • S. Haigh • G. Madabhushi
Department of Engineering, Schofield Centre, University of Cambridge, High Cross, Madingley Road, Cambridge CB3 0EL, UK
e-mail: cmh78@cam.ac.uk; skh20@cam.ac.uk; mspg1@cam.ac.uk

A. Ilki and M.N. Fardis (eds.), *Seismic Evaluation and Rehabilitation of Structures*, Geotechnical, Geological and Earthquake Engineering 26,
DOI 10.1007/978-3-319-00458-7_23,

However, allowing movement between the foundation and soil would help to isolate the foundation and superstructure from the soil, hence reducing the transmission of energy into the superstructure (Kutter and Kunnath 2010; Anastasopoulos et al. 2009; Gajan and Saravanathiiban 2011; Gajan and Kutter 2008). To allow foundation-soil movement, the foundation will need to be smaller than that which would currently be specified by design codes and will therefore be cheaper and quicker to construct. In addition, with the superstructure being required to dissipate less energy, it too will not be required to be as strong reducing the cost of construction. Perhaps most importantly, a foundation which rocks will self-centre (Gajan and Kutter 2008). This means that plastic deformation can be allowed to occur in the soil without fear of overall collapse of the structure. It is clear therefore that there are many advantages to allowing relative foundation-soil movement. Despite these advantages, very few structures are currently being designed to utilize soil-structure interaction to reduce the ductility demands of the structure. This is due to the current difficulty in predicting fully how the soil-structure interface will act during an earthquake and hence how much energy will be transmitted into the superstructure. The current alternative isolation method, of using rubber bearings between the foundation and the structure, is considered more attractive due to the predictability and reliability of the materials being used, however it results in a significantly more expensive design. It is the aim of this chapter to start to address this lack in predictability of soil-structure interaction through an investigation of the entire soil-foundation-structure system.

A series of centrifuge tests was conducted at the University of Cambridge to examine soil-structure interaction and how the response changes when soil and structural parameters are altered. The effect of soil density, aspect ratio and structural stiffness on foundation settlement, rotation and horizontal sliding is presented in this chapter.

23.2 Methodology and Experimental Equipment

23.2.1 Soil Behavior and Centrifuge Modelling

Soil has highly non-linear stress–strain behavior and consequently soil stiffness is dependent on its stress state, as shown in Fig. 23.1. It can be seen that at low stress in a small scale model, a soil is significantly softer than would be the case for the same soil at higher stress in the prototype. It is therefore important when modelling soil to replicate the stress level of the prototype in the model. Without doing so, the soil stiffness would not be correct and hence test results would have no quantifiable relation to the prototype scenario. There are two possible methods available to ensure the stress state in the soil model is correct. The first method is to carry out tests at nearly full prototype scale proportions, hence accurately replicating the stress state in the soil. However this method is both time consuming and expensive.

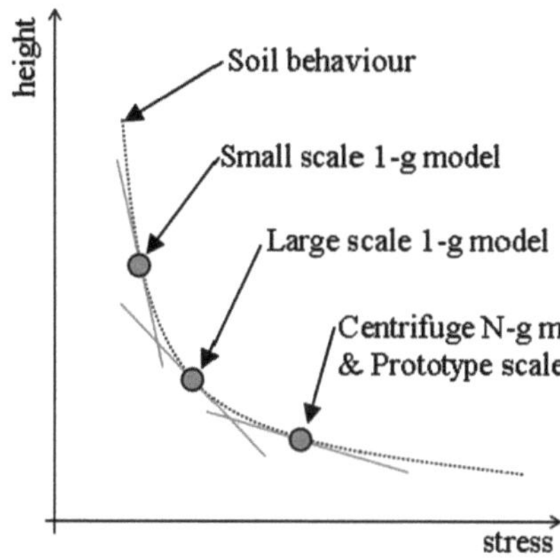

Fig. 23.1 Soil stiffness variability with stress state

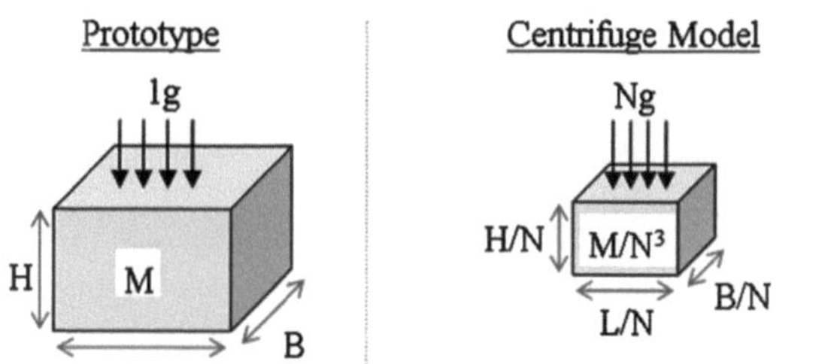

$$\sigma = \frac{Mg}{LB}, \quad \varepsilon = \frac{\delta L}{L} \qquad \sigma = \frac{\frac{M}{N^3} \times (N\gamma)}{\frac{L}{N} \times \frac{B}{N}} = \frac{M\gamma}{LB}, \quad \varepsilon = \frac{\frac{\delta L}{N}}{\frac{L}{N}} = \frac{\delta L}{L}$$

Fig. 23.2 Principles of centrifuge scaling

The second option is to create scaled models and then increase the g-level the model is subjected to. This is achieved through the use of a geotechnical centrifuge. If a model is created at 1/Nth scale of the prototype then an N times increase in g-level will result in the stress level in the soil being replicated accurately throughout the model (Fig. 23.2). A full set of scaling laws are shown in Table 23.1. Centrifuge modelling is relatively cheap to perform as well as being significantly quicker than conducting full scale tests. This allows a greater range of prototype scenarios to be tested while still obtaining quantitative results.

One potential issue with geotechnical centrifuge modelling is the fact that soil is a granular material, not a continuum. The soil grains used for the model cannot be scaled down, as to do so would result in grains being so small that it would alter the way in which they interact. It could be assumed that using the same soil grain size as the prototype would result in the N-g model representing a bed of rocks instead of soil grains due to the scaling effects detailed previously. However, the soil particles

Table 23.1 Centrifuge scaling laws

Parameter	Model/prototype	Dimensions
Length	$1/N$	L
Mass	$1/N^3$	M
Stress	1	$ML^{-1}\ T^{-2}$
Strain	1	1
Force	$1/N^2$	MLT^{-2}
Seepage velocity	N	LT^{-1}
Time (seepage)	$1/N^2$	T
Time (dynamic)	$1/N$	T
Frequency	N	T^{-1}
Acceleration	N	LT^{-2}
Velocity	1	LT^{-1}

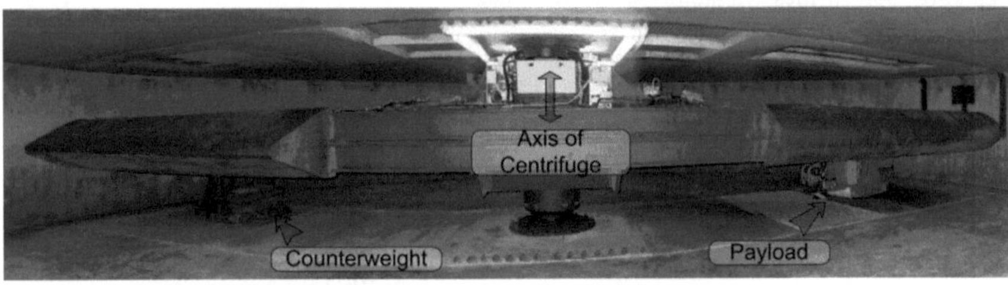

Fig. 23.3 Turner beam centrifuge

gain no increase in strength when subjected to an increased g-level and hence the particles have the same physical characteristics at N-g as they do in the prototype at 1-g. Many studies have been conducted to verify that soil can be considered a continuum when conducting centrifuge testing. However, when looking at soil-structure interaction effects it is important that the structure, whatever that may be, is at least approximately 35 times the mean diameter of the soil grain in order for the soil to be considered a continuum (Garnier and Gaudin 2007).

23.2.2 Testing Facilities

At the University of Cambridge there is a 10 m beam centrifuge on which both dynamic and static geotechnical problems can be investigated, Fig. 23.3. More information on the centrifuge can be found in Schofield (1980). To perform dynamic tests a secondary device, known as the SAM actuator, is added to the payload end of the centrifuge. The SAM actuator utilizes stored angular momentum energy to shake the test package as detailed in Fig. 23.4. The displacement and frequency of the shaking can be varied between earthquakes; however only simple sinusoidal (constant frequency, constant acceleration) or sine-sweep (decreasing

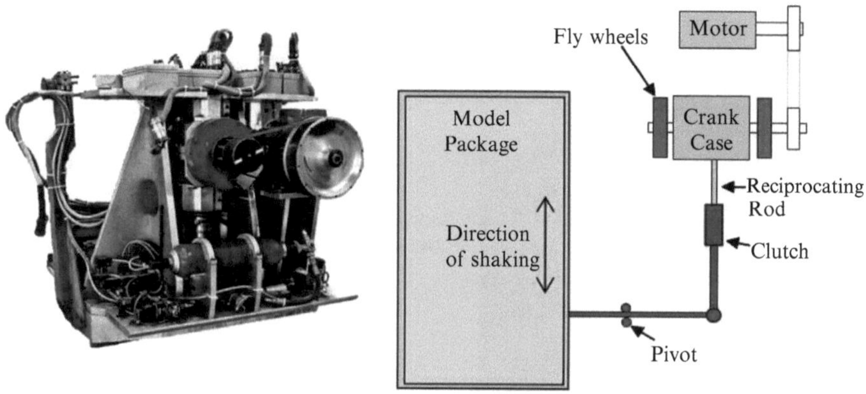

Fig. 23.4 Operation of SAM (earthquake) actuator

frequency, decreasing acceleration) traces can be triggered. Hence the input shaking motions are not truly representative of real earthquakes. Despite this, the results are still relevant, as different mechanisms in the model can be more easily isolated when the model is subjected to a simple single frequency input motion. More detailed information on the SAM actuator is presented by Madabhushi et al. (1998).

23.2.3 Centrifuge Model Setup

The results presented in this chapter were collected during a series of six centrifuge tests. The models consisted of a simple soil layer, prepared at a specific relative density (using a robotic sand pourer) on top of which a model structure was located. High speed photography techniques were used in order to accurately track the movement of both the model structures and the soil beneath. Due to scaling laws, model earthquakes last 1/N times as long as the prototype earthquake and at a frequency N times higher. The prototype input motions were chosen based on the typical duration and dominate frequency of real earthquakes; 25 s long at a frequency of 1 Hz, hence the model earthquakes were 0.5 s long at a frequency of 50 Hz. This very quick, high frequency shake requires photographs to be taken at up to 1,000 frames per second in order to fully capture the movements. A model container with a clear Perspex side allowed the movements of the structure and of the soil beneath to be observed. A view of a centrifuge package can be seen in Fig. 23.5. Dry Hostun sand (Flavigny et al. 1990) was used as the foundation soil for this series of tests the properties of which are shown in Table 23.2. Dry sand was used to remove the added complexity that saturation and consequently liquefaction would bring to the soil-structure-interaction problem.

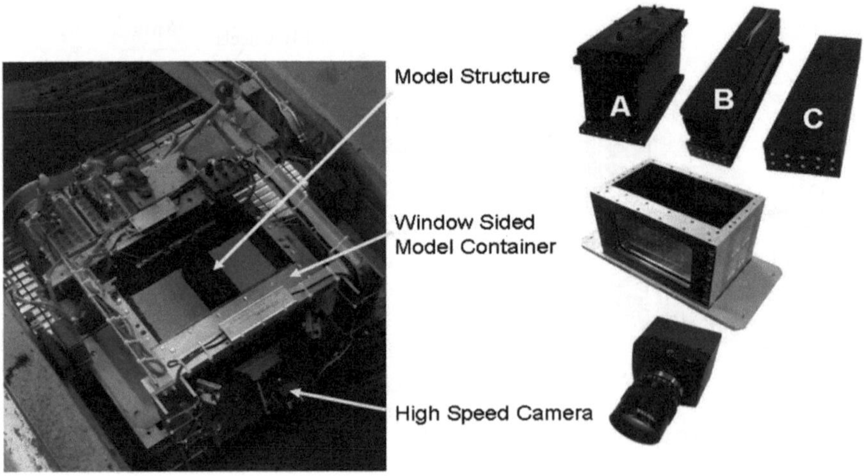

Fig. 23.5 Centrifuge package

Table 23.2 Properties of Hostun sand

Property	Value
e_{min} – Minimum voids ratio	0.555
e_{max} – Maximum voids ratio	1.067
d_{10} – 10 % of participles with diameter smaller	300 μm
d_{90} – 90 % of participles with diameter smaller	800 μm
I_d – Relative density	50 %//80 %
e – Voids ratio	0.811//0.657
ρ_d – Dry density	1,463 kg/m^3//1,599 kg/m^3
ρ_s – Saturated density	1,911 kg/m^3//1,996 kg/m^3

23.2.4 Model Structures

In order to explore the influence of a variety of structural properties, such as the structural stiffness and height of the centre of gravity, several model structures were constructed. Three distinct types of structure were tested. Type A structures were sway frame structures with adjustable web stiffness, foundation mass and superstructure mass, Fig. 23.5. This allowed any of the key structural properties to be adjusted as desired between tests. This type of structure was chosen initially as it was similar to the structures used in another series of tests conducted as discussed in Cilingir et al. (2011), allowing cross-comparisons to be made, ensuring the integrity of the PIV model configuration. To ensure the full soil deformation mechanism below the structure could be observed within the field of view of the camera, a smaller model structure was built, Type B, Fig. 23.5. These structures had a fixed configuration and hence the stiffness and mass could not be varied. However, like the Type A structures, they did have a superstructure mass in order to raise the centre of gravity to above the level of the foundation. No significant amount of sliding was apparent during the

Table 23.3 Properties of model structures (prototype scale)

Property	Type A{1}	Type A{2}	Type B	Type C
Bearing pressure (kPa)	100	100	100	100
Natural frequency (Hz)	1	400	400	–
Foundation dimensions (m)	4.2 × 10.5	4.6 × 10.5	2.5 × 10.7	2.5 × 10.5
Structure height	6.0	6.0	4.6	1.3
Height of centre of gravity	4.5	4.5	2.7	0.7

testing of the Type A and Type B structures, hence a final structure was fabricated, Type C, Fig. 23.5. This was a simple strip mass with no superstructure. Some of the key properties of the structures are shown in Table 23.3.

23.2.5 Geo-PIV Procedure

The images collected during the tests were processed using Particle Image Velocimetry (PIV) techniques. PIV was initially used by fluid mechanics to track fluid flows however it has been adapted for use investigating geotechnical problems (White et al. 2003). Geo-PIV allows patches of soil within an image to be tracked between successive images and hence a displacement vector field to be produced which shows the soils movement. This ability to visualize the deformations aids researchers in identifying failure mechanisms in a variety of soil-structure scenarios. The Geo-PIV techniques have been utilized in this series of tests to track the structural movements as well as the soil movements allowing the detailed rotation-settlement and sliding plots to be produced (discussed further in the next section).

23.3 Rotation-Settlement Behavior

Settlement-rotation plots have been widely used by researchers to explore the impact of different test parameters on the dynamic behavior of shallow foundations (Gajan et al. 2005; Gajan and Kutter 2008, 2009; Anastasopoulos et al. 2009; Kutter and Kunnath 2010). However, the majority of the available literature collates the data from numerical analysis, slow pseudo-dynamic testing or small-scale 1-g testing. Geo-PIV techniques allow settlement-rotation plots to be produced from true dynamic testing.

23.3.1 Relative Density Effect

A looser soil foundation will lead to increased settlements, however the impact on the rotations is perhaps less intuitive. The settlement-rotation traces for two tests with differing relative density, the same model structure (Type B) and the same

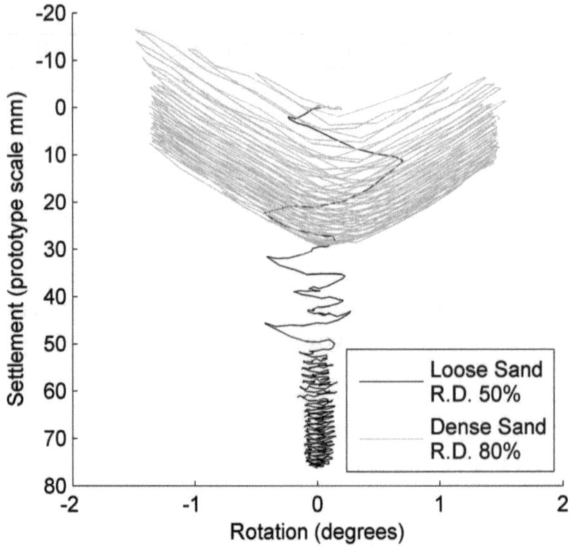

Fig. 23.6 Relative density effect on settlement-rotation behavior for Type B structure

earthquake (peak input acceleration of 0.2g) are shown in Fig. 23.6. From this figure the expected impact of relative density on cumulative settlement is apparent, with the structure located on the looser soil settling nearly three times more than that on the dense soil.

It is also clear from Fig. 23.6 that significantly greater rotation magnitude is experienced by the structure sited on the dense sand compared to that on the loose sand bed. There are three main modes in which a structure can react to seismic shaking; settlement into the soil, rotation on top of the soil and sliding across the soil surface. The bearing pressure and soil density have an important role in determining which mode will dominate. Any rotation of the foundation results in parts of the soil beneath the footing being subjected to an increased bearing pressure. In the case of the dense sand, the sand is capable of carrying the increase in bearing pressure and hence allows the structure to rock further. On the other hand, the loose sand cannot sustain the extra pressure and therefore compresses and deforms, resulting in settlement before any significant rotation can occur.

The variation in rotation magnitude and settlement rate during the earthquake is also of interest, with the structure on the dense sand maintaining an approximately constant level of rotation and settlement rate throughout the shaking. However, the structure on the loose sand decreased in rocking magnitude steadily until a point at which, within one cycle, it started to rock at a constant smaller magnitude and settle at a significantly slower and more constant rate. It is possible that at the transition point between the two phases the rotation magnitude and settlement rate has decreased to a level whereby the response transitions into a different mode of response, potentially as a result of lift-off no longer occurring between the foundation and the soil.

The observed changes in settlement-rotation behavior for varying relative density are independent of the type of structure being tested and are consistent across tests, as

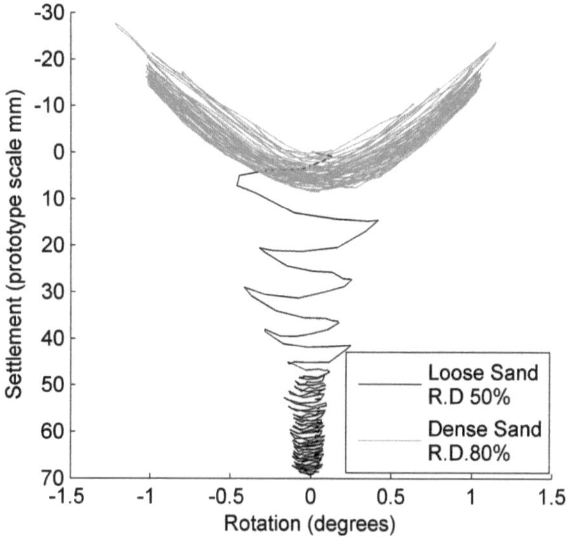

Fig. 23.7 Relative density effect on settlement-rotation behavior for Type A{2} structure

shown in Fig. 23.7 which shows the settlement-rotation plots produced from testing the Type A{2} structure. However, both Figs. 23.6 and 23.7 were produced from tests on stiff structures. As will be seen in the following section, structural stiffness further affects the settlement-rotation behavior.

23.3.2 Structural Stiffness Effect

Both of the previous figures were produced using data from stiff model structures. Stiff structures will not deform significantly internally and hence energy is dissipated through overall settlement, rotation and sliding of the model structure. Figure 23.8 shows a comparison of the settlement-rotation behavior between the Type A{1} and Type A{2} structures tested on dense sand and subjected to the same earthquake motion. It is clear that the stiff structure rotates significantly more than the flexible structure throughout the shaking. When the superstructure of a flexible structure is subjected to a sideways inertial force (d'Alembert's force) it will cause bending of the structural webs in addition to causing rotation of the foundation. The bending of the webs reduces the energy going towards the rocking mode and results in smaller rotations being observed. In addition, the earthquake frequency is matched to the natural frequency of the flexible structure, hence maximizing the shear deformation which occurs within the structure causing a particularly large reduction in rotation (Type A{1}, Fig. 23.8).

Overall settlement in both tests was a fraction of that observed with loose soil. The settlements did differ between the two structures with the stiff structure settling

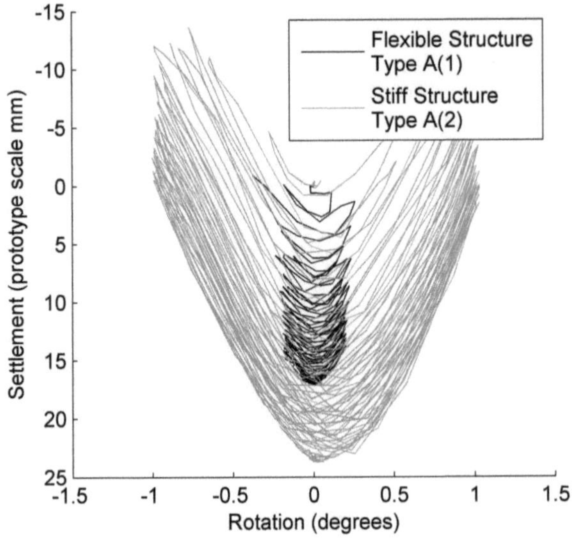

Fig. 23.8 Structural stiffness effect on settlement-rotation behavior on dense sand for Type A{1} and Type A{2} structures

a third more than the flexible structure. This difference can most likely be attributed to the problem with replicating identical relative densities in two different models. It would be expected that two models designated to be the same relative density could be up to 5 % different (Zhao et al. 2006). As discussed in the previous section relative density had a significant impact on settlement and hence even a small difference could lead to the variation in settlement observed in Fig. 23.8. In addition, larger rotations lead to localized increased bearing pressures. This will result in increased soil compaction leading to increased settlement as observed.

Given the sensitivity of the structural response to the foundation soil's relative density, it is important to also compare the response of model structures with varying stiffness situated on a loose sand deposit. Figure 23.9 shows the settlement-rotation behavior of the same two structures as Fig. 23.8 (Type A{1} and Type A{2}) subjected to the same earthquake but this time placed on loose foundation sand (50 % relative density instead of 80 %). The responses shown in Fig. 23.9 are more similar than for the response of the structures sited on the dense sand with the variation in rotation magnitude not being as large as that observed in Fig. 23.8. However, despite the stiff structure initially rocking with a larger rotation magnitude than the flexible structure, once the stiff structure transitions into a steady state (as discussed earlier) the rotation magnitude decreases to below the level of the flexible structure.

In terms of total settlement, it can be seen in Fig. 23.9 that both structures settled by the same amount. The minimal difference in rotation angle between the two tests resulted in similar bearing pressure regimes under both structures, therefore no differential settlement occurred.

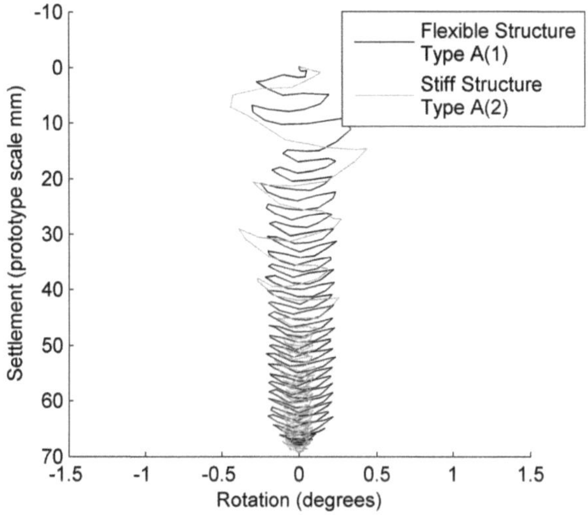

Fig. 23.9 Structural stiffness effect on settlement-rotation behavior on loose sand for Type A{1} and Type A{2} structures

To conclude, the effect of a change in structural stiffness depends on the relative density of the foundation soil, highlighting once again how important knowledge of the relative density is when predicting settlement-rotation behavior.

23.3.3 Aspect Ratio Effect

As the aspect ratio increases and the centre of gravity rises, the inertial loads that the structure experiences will apply a larger moment to the base of the structure. This results in a larger rotation and more lift-off of the foundation from the sand. Figure 23.10 shows the settlement-rotation behavior of the Type C structure, which was just a simple mass strip with no superstructure, and the Type B structure, which does have a superstructure mass. Both structures were situated on dense soil and subjected to the same magnitude of earthquake. The figure shows the expected response with near negligible amounts of rotation being experienced by the Type C structure. It is interesting to note that the Type B structure settled almost three times more than the Type C structure. The same differences are evident when examining the same two structures situated on loose soil however the Type B structure then settled nearly six times more. This increase in settlement is a result of increased localized bearing pressures under the rotating footing as discussed previously. As will be seen later, no significant amount of energy is dissipated through sliding of the Type C structure. It is clear therefore that the Type B structure dissipates significantly more energy through soil-structure interaction than the Type C structure. This observation is reinforced by examining the acceleration experienced by the structure's foundations during the earthquakes. Despite the input motions to the

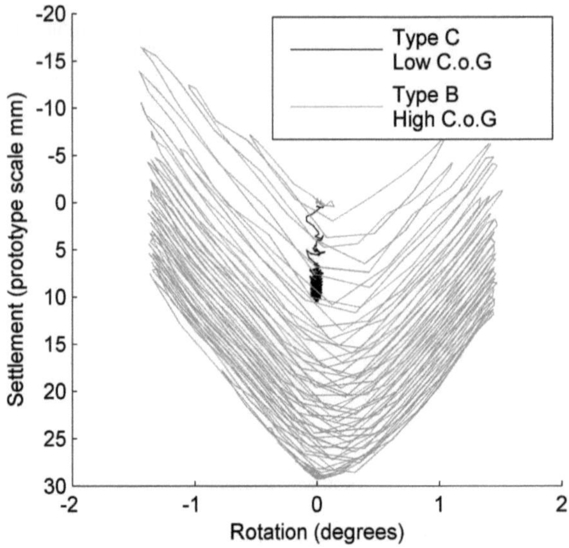

Fig. 23.10 Aspect ratio effect on settlement-rotation behavior for Type B and Type C structures located on dense sand

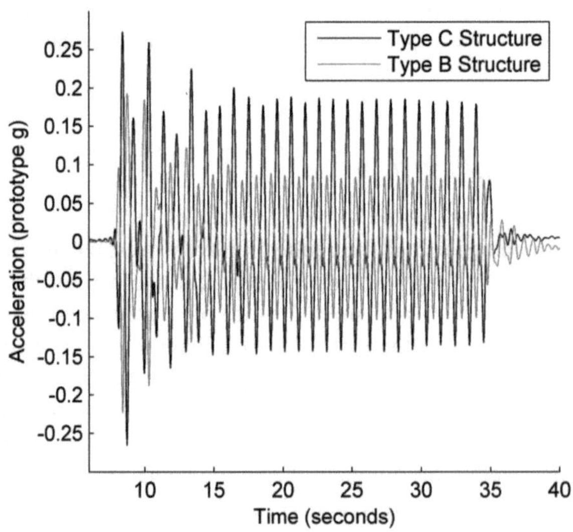

Fig. 23.11 Acceleration experienced by structure foundation

soil being the same (1 Hz, 0.14g peak input acceleration), the Type B structure would be expected to experience a smaller acceleration magnitude due to its significant rocking and settlement energy dissipation.

Figure 23.11 shows the acceleration magnitudes of the foundations for the same two tests as in Fig. 23.10. As predicted, the acceleration magnitude of the Type C structure is significantly larger than that of the Type B structure.

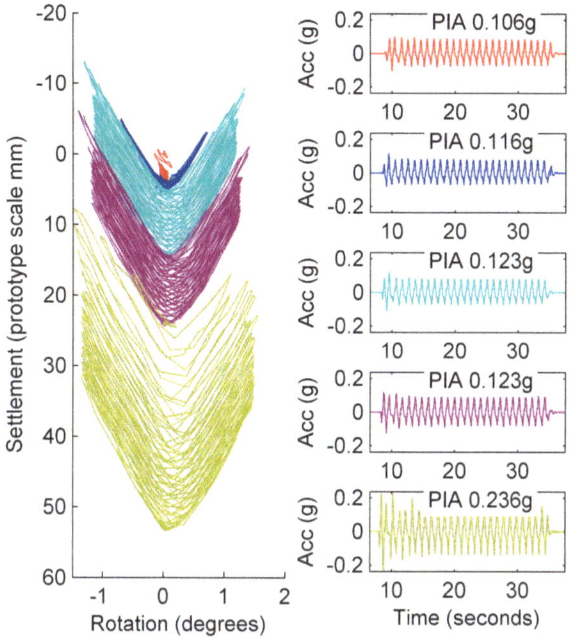

Fig. 23.12 Effect of earthquake magnitude on settlement-rotation behavior. *PIA* peak input acceleration

23.3.4 Earthquake Magnitude Effect

Intuitively a larger earthquake will result in the model structure undergoing larger settlements and experiencing larger rotations. Settlement-rotation behavior from five successive earthquakes of increasing magnitude and their corresponding input acceleration traces are shown in Fig. 23.12. The data was collected from a single centrifuge test performed with dense sand (80 % relative density), in order to minimize the impact of densification due to successive earthquakes, and with the Type B structure.

The initial earthquake with a peak input acceleration of 0.11g resulted in some initial settlements during the first 1.5 cycles; however minimal rotation occurred indicating the acceleration was not significant enough to cause rotation and lift-off of the foundation. The second earthquake with the larger 0.12g input acceleration did result in a significant rotation mode developing. The third and fourth earthquakes shown have identical input acceleration motions and the resulting settlement-rotation behavior is reassuringly very similar. The degree of rotation is larger than that experienced during the second earthquake, indicating that the larger acceleration has resulted in larger rotations, as would be expected, which in turn resulted in larger settlements. There is only 0.01g difference between the peak input acceleration magnitude of the second and third earthquakes and yet there is a significant difference in response. This is partly due to the average peak cyclic acceleration being larger during the third earthquake but also highlights an apparent sensitivity of the structural response to the earthquake magnitude.

To examine this apparent sensitivity to earthquake magnitude further, the model was finally subjected to a very large earthquake with a peak input acceleration of 0.24g. Despite being nearly twice the acceleration magnitude of the previous two earthquakes, the rotation magnitude is not significantly larger. It appears from Fig. 23.12 that either the structure will not rock significantly as per the first earthquake, or will rock at a fixed maximum rotation irrespective of increasing input acceleration. Therefore, instead of considering the structural response sensitive to earthquake magnitude, it is instead the case that the transition acceleration range between negligible rocking and maximum rocking is very small and hence the response could be construed as sensitive over this range only.

It is often suggested that allowing a structure to rock during earthquakes could result in toppling of the structure; however this result has shown that this will not occur. Despite attempts in some tests to cause the structure to rotate significantly, no large permanent rotations ever occurred.

23.4 Sliding Behavior

As discussed previously, in addition to rotation and settlement, a structure subjected to dynamic shaking can also dissipate energy through sliding across the soil-structure interface. The degree of sliding and the impact of several test parameters are discussed in this section.

23.4.1 Relative Density Effect

During each test the model structure is located on the sand surface with no embedment. This means that if the structure slides, there will not be passive resistance from the soil beside it. However, as the structure settles into the sand during the earthquake, passive soil wedges will form, resisting any horizontal movement. Figure 23.13 shows the sliding movement of the Type B structure subjected to the same earthquake but when located on two different densities of sand. The seismic amplitude of movement is very similar between the two tests; however the structure located on the dense sand accrues more residual horizontal movement compared to the structure located on the loose sand at the end of the shaking. This accumulation of horizontal movement is a result of the direction of the first cycle of shaking, which biases the structure to migrate more in one direction as observed. However, the structure on the loose sand migrated less

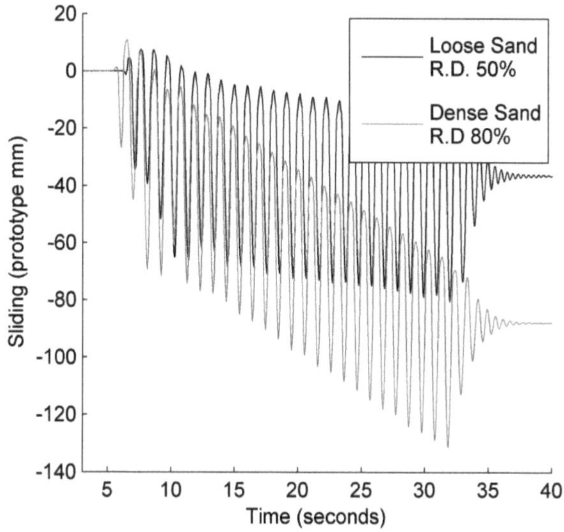

Fig. 23.13 Effect of relative density of magnitude on sliding behavior (Type B structure)

due to the passive sand wedges which form beside the structure as it settles into the sand. The same result can be seen in tests of the Type A structures. It is interesting that the passive wedges which form do not restrict the seismic component of the sliding, which is similar in both the tests shown in Fig. 23.13. It is likely that the magnitude of the force that the foundation applies to the sand wedges during the shaking is larger than that which the wedges can resist; hence the wedges simply get deformed as the foundation slides into them. However, as the structure attempts to migrate more in one direction, a larger passive wedge will develop which works to reduce the magnitude of the permanent movement as observed in Fig. 23.13.

23.4.2 Aspect Ratio Effect

The impact of aspect ratio on the settlement-rotation behavior was significant, with the Type C structure undergoing negligible rotation. It was anticipated that the Type C, with such a low centre of gravity, would cause it to experience a significant sliding mode. However, as shown in Fig. 23.14, this is not the case. In fact the Type B structure slides significantly more both in terms of co-seismic sliding (over 10 times) and accumulated movement (7 times) compared to the Type C structure. Similar to the earlier phenomenon whereby a structure which rotates more also settles more, a structure which rotates, will slide more also. As a foundation rotates, one side of the footing lifts off the sand decreasing the contact

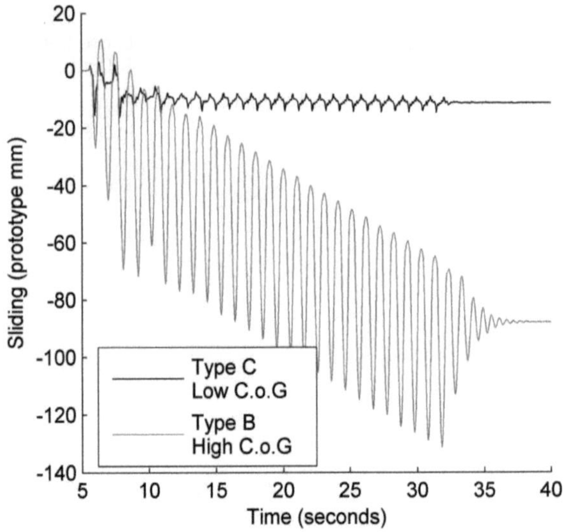

Fig. 23.14 Effect of aspect ratio on sliding behavior (dense soil model)

area. This increases the localized bearing pressure, which results in increased deformations under the corner of the structure still in contact. These deformations pave the way for the structure to slide more towards that corner. In addition, the sand wedge against the side of the foundation which has lifted off can collapse and flow below the edge of the footing which allows lager amounts of sliding movement. The lack of rotation experienced by the Type C structure means that the entire area of the foundation has to be sheared across the sand surface, in order for sliding to occur. This large interface friction, combined with the required deformation of any passive soil wedges that develop, results in the structure not sliding by a significant amount

23.4.3 Structural Stiffness Effect

The effect of structural stiffness on the sliding is also quite significant. However, once again it is related to how the stiffness affects rocking. As discussed previously, the stiff structure rocks more than the flexible structure on dense sand; therefore, the stiff structure also slides more, as shown in Fig. 23.15. The seismic component of the shaking is approximately five times greater for the stiff structure compared to the flexible structure. In addition the stiff structure also accrues more permanent horizontal movement during the shaking with the final horizontal movements of 15 mm, seven times greater than that of the flexible structure.

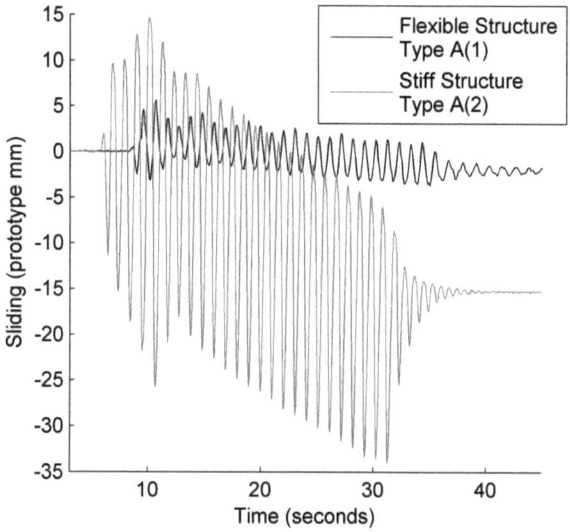

Fig. 23.15 Effect of structural stiffness on magnitude of sliding behavior (dense sand)

23.5 Conclusions

This chapter investigated the soil-structure interaction problem of a shallow foundation located on dry sand and subjected to seismic loading. Centrifuge tests and a detailed analysis were performed and allowed the rotation, settlement and sliding behavior of the foundation to be studied and understood. It was found that:

- Foundations located on loose sand experienced an increase in settlement and a decrease in rotation relative to a foundation on dense sand.
- The foundation of a flexible structure does not rotate as much as that of a stiff structure, due the ability of the flexible structure to dissipate energy internally.
- Counter-intuitively, a structure with a high centre of gravity slides more than a structure with a low centre of gravity. This is due to the larger rotations experienced by the structure with the high centre of gravity, which results in the edge of the foundation lifting off the sand surface and sliding over the top of passive soil wedges which form on either side. These passive soil wedges prevent structures with a low centre of gravity from experiencing any significant relative soil-foundation sliding.
- The rotation experienced by a foundation is not linearly dependent on input acceleration. Either the foundation will not rock, when the input motion is below a certain input acceleration threshold, or it will rock and rotate to a set maximum, when the input motion is just about the threshold.

Acknowledgments The research leading to these results has received funding from the European Community's Seventh Framework Programme [FP7/2007–2013] under grant agreement n° 227887 [SERIES]. The support of the other members of JRA3 in SERIES is acknowledged.

References

Anastasopoulos I et al (2009) Soil failure can be used for seismic protection of structures. Bull Earthq Eng 8(2):309–326

Cilingir U et al (2011) Cross-facility validation of dynamic centrifuge testing. In: Fardis MN, Rakicevic ZT (eds) Role of seismic testing facilities in performance-based earthquake engineering. Springer, Dordrecht/London, pp 83–98. doi: 10.1007/978-94-007-1977-4

Flavigny E, Desrues J, Palayer B (1990) Le sable d'Hostun. Géotechnique 53:67–70

Gajan S, Kutter BL (2008) Capacity, settlement, and energy dissipation of shallow footings subjected to rocking. J Geotech Geoenviron Eng 134(8):1129–1141

Gajan S, Kutter BL (2009) Effects of moment-to-shear ratio on combined cyclic load–displacement behavior of shallow foundations from centrifuge experiments. J Geotech Geoenviron Eng 135(8):1044

Gajan S, Saravanathiiban DS (2011) Modeling of energy dissipation in structural devices and foundation soil during seismic loading. Soil Dyn Earthq Eng 31(8):1106–1122

Gajan S et al (2005) Centrifuge modeling of load-deformation behavior of rocking shallow foundations. Soil Dyn Earthq Eng 25(7–10):773–783

Garnier J, Gaudin C (2007) Technical Committee TC2 – physical modelling in geotechnics catalogue of scaling laws and similitude questions in centrifuge modelling

Kutter BL, Kunnath SK (2010) Estimation of displacement demand for seismic design of bridges with rocking shallow foundations. In: 5th international conference on recent advances in geotechnical & earthquake engineering and soil dynamics, San Diego, p SOAP11

Madabhushi S, Schofield A, Lesley S (1998) A new stored angular momentum (SAM) based earthquake actuator. In: The international conference centrifuge '98, Tokyo, Japan, pp 111–116

Schofield AN (1980) Cambridge geotechnical centrifuge operations. Géotechnique 30(3):227–268

White DJ, Take WA, Bolton MD (2003) Soil deformation measurement using particle image velocimetry (PIV) and photogrammetry. Geotechnique 53(7):619–631

Zhao Y et al (2006) Calibration and use of a new automatic sand pourer. In: 6th international conference in physical modelling in geotechnics, Hong Kong, pp 265–270

Chapter 24
Centrifuge Modeling of Liquefaction Effects on Shallow Foundations

Andreia Sofia Pedroso da Silva Marques, Paulo Alexandre Lopes de Figueiredo Coelho, Stuart Haigh, and Gopal Madabhushi

Abstract Earthquake-induced liquefaction is a major concern for structures built on saturated cohesionless soils in seismically active regions, as it often causes failure of critical structures such as bridges and quay walls, which severely restricts post-earthquake emergency response and economic recovery. The destructive consequences of this phenomenon have remarkably increased since it was firstly identified in US and Japan in 1964. This paper describes an investigation on the performance of shallow foundations susceptible to seismic liquefaction, considering the particular vulnerability that this type of foundation has shown in the field during past earthquakes. The research program included three dynamic centrifuge experiments, conducted at the Schofield Centre, University of Cambridge, UK, as part of a SERIES' TNA Use Agreement focusing on the magnitude of liquefaction effects on shallow foundations, under different conditions, including interaction effects between adjacent structures, and on the assessment of the performance of innovative mitigation techniques, particularly narrow densified zones combined with selectively positioned high-capacity vertical drains.

24.1 Introduction

Earthquake-induced liquefaction is a major concern for structures built on saturated deposits of cohesionless soils in seismically active regions, the phenomenon being so complex that even its definition remains ambiguous (Boulanger 2005). Throughout

A.S.P. da Silva Marques (✉) • P.A.L. de Figueiredo Coelho
Department of Civil Engineering, University of Coimbra,
Rua Luís Reis Santos – Pólo II da Universidade, 3030 – 788 Coimbra, Portugal
e-mail: as.silvamarques@gmail.com; pac@dec.uc.pt

S. Haigh • G. Madabhushi
Department of Engineering, Schofield Centre, University of Cambridge,
High Cross, Madingley Road, Cambridge CB3 0EL, UK
e-mail: skh20@cam.ac.uk; mspg1@cam.ac.uk

A. Ilki and M.N. Fardis (eds.), *Seismic Evaluation and Rehabilitation of Structures*, Geotechnical, Geological and Earthquake Engineering 26,
DOI 10.1007/978-3-319-00458-7_24, © Springer International Publishing Switzerland 2014

the last 50 years, liquefaction has proven to be a major source of economic losses, often resulting in vast structural damage to buildings, infrastructures and even ground failure, which often severely restricts post-earthquake emergency response and economic recovery. Damage to shallow foundations can be particularly severe, mitigation measures being poorly understood (Mitchell 2003; Bardet et al. 1997).

The research previously carried out on the behavior of shallow foundations built on liquefiable ground undergoing seismic events identified major shortcomings affecting our fundamental understanding of the problem and our ability to model and to mitigate its effects. In view of the limitations of the data obtained from case histories and alternative experimental tools able to deal with such problems, centrifuge modeling was selected as the main research tool for this project.

Considering the key limitations of our current understanding of the problem and the critical requirements from practitioners working in this field, namely those employing sophisticated methodologies including performance-based design and risk analysis, the following objectives were defined for this research: establish the fundamental behavior of the soil-structure system through the analysis of a benchmark problem; evaluate the influence of the bearing pressure induced by the shallow foundation on the liquefiable ground on the performance of the system under seismic loading; assess the effectiveness of narrow densified zones with selectively positioned high-capacity vertical drains in the mitigation of earthquake-induced liquefaction effects. The results obtained through centrifuge modeling will be complemented with elementary tests carried out on the modeling sand and a judicious assessment of selected advanced numerical tools to perform realistic numerical predictions of the behavior of the shallow foundations affected by seismic loading inducing ground liquefaction. The results of the numerical simulations will be compared between various researchers and institutions to clarify the limitations and strengths of different numerical tools.

This paper describes the part of the research employing dynamic centrifuge modeling, which was made possible through the financial support obtained through a Transnational Access (TA) Use Agreement granted within a project funded by the European Commission (7th Framework Program): Seismic Engineering Research Infrastructures for European Synergies- SERIES. The international research team involved in the TA Research Project is led by the University of Coimbra and includes the University of Rome, the Polytechnic of Madrid and the Slovenian National Building and Civil Engineering Institute. The dynamic centrifuge experiments were carried out at the experimental facilities of the Schofield Centre – Cambridge University Engineering Department.

24.2 Characteristics of the Dynamic Centrifuge Models

A series of three centrifuge experiments were carried out on models similarly built and submitted to comparable earthquake simulations, to investigate the magnitude of liquefaction effects, under different conditions, and to assess the performance

Heavier Structure **Lighter Structure**

Fig. 24.1 Structures used in the centrifuge experiments

of innovative mitigation techniques, particularly combining densification and high-capacity vertical drains. The only factors varying in each test were the model structures and, where applicable, the depth of the densified zone created in the soil foundation under the footing and the length of the surrounding material simulating high-capacity vertical drains. Four different combinations for the improved zone were tested and compared to the behavior of the benchmark test where no ground improvement was carried out.

24.2.1 Model Structures

Considering the aim of the tests, structural models had a simple design, consisting of solid steel blocks with scale dimensions of 60×60 mm in plan view and a thickness of 24.5 and 15 mm, for the heavier (H) and lighter (L) model structures, respectively. In the tests under consideration, performed at 50-g centrifuge acceleration, the pressures transmitted through the foundations basis, at prototype scale, were 58 and 95 kPa for structures L and H, respectively. Figure 24.1 shows the model structures used to perform the centrifuge experiments.

24.2.2 Dynamic Centrifuge Model A (CT-A)

The series of dynamic centrifuge tests included an experiment (CT-A) intended to evaluate the fundamental behavior of the system through the analysis of a benchmark problem for further comparison with the subsequent tests. This test was also carried out to establish the influence of the bearing pressure induced by the shallow foundation on the liquefiable ground on the performance of the soil-structure system under seismic loading and to assess potential interaction effects between the structures placed in the same model. Figure 24.2 presents a scheme of the model

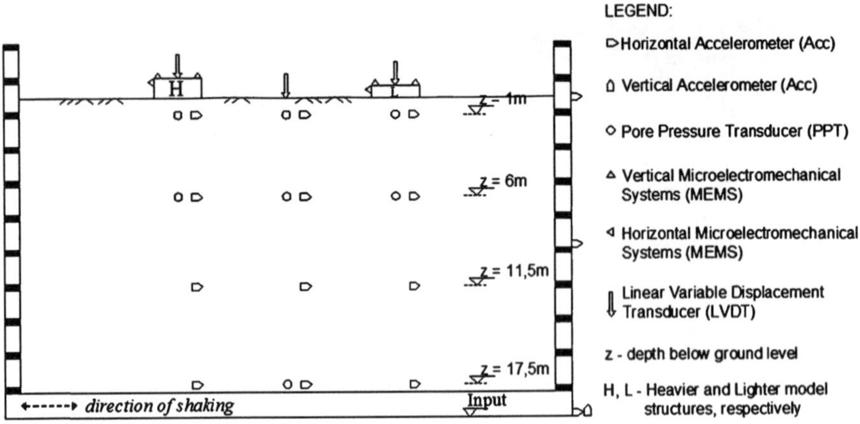

Fig. 24.2 Centrifuge model layout in experiment CT-A (cross section view)

experiment, where the units are at prototype scale. As the figure shows, no ground improvement was simulated in CT-A, which intends to establish the performance during the earthquake simulation of shallow foundations directly on liquefiable ground. CT-A was performed using two different structures placed in opposite sides of the model – structures H and L. The ground foundation represented a fully saturated liquefiable sandy layer of 18 m deep with relative density of 50 %.

To perform the subsequent tests, two different structures could be used to compare the results with CT-A – H or L structure. By using the heavier structure it might show the effects of the mitigation methods more clearly, while using the lighter structure might give a better representation of real structures built on liquefiable ground. Consequently, the footing chosen to perform subsequent tests was the H structure.

24.2.3 Dynamic Centrifuge Model B (CT-B)

Test CT-B was carried out with model structure H, to maximize liquefaction effects, and employing ground conditions similar to those used in benchmark test CT-A, except for the fact that some ground improvement was carried out under the footings. In this test, two equal shallow foundations were placed resting on narrow densified zones having the same depth as the layer of loose sand (Fig. 24.3). As the figure shows, one densified zone under one of the structures was also embedded by vertical drains extending to the bottom of the deposit, simulated through a particular geotextile. The geometry of the densified zone was established following the guidelines suggested by Coelho (2007). This centrifuge test intended to assess the effectiveness of narrow densified zones with selectively positioned high-capacity vertical drains in the mitigation of earthquake-induced liquefaction effects, as proposed by Coelho (2007).

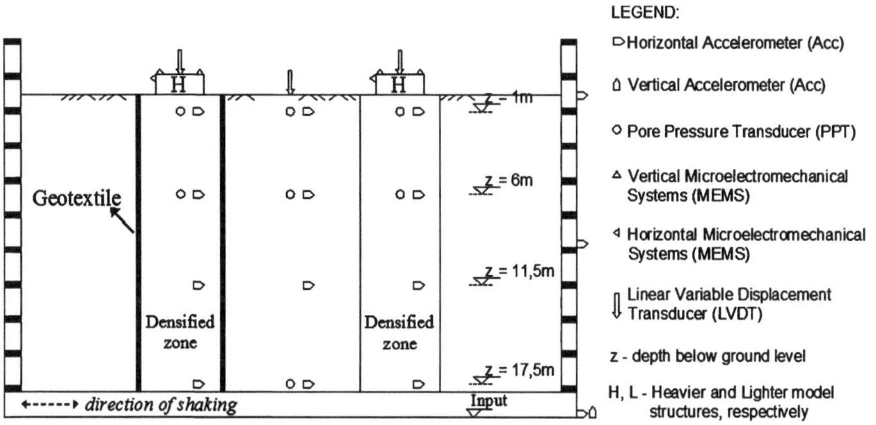

Fig. 24.3 Centrifuge model layout in experiment CT-B (cross section view)

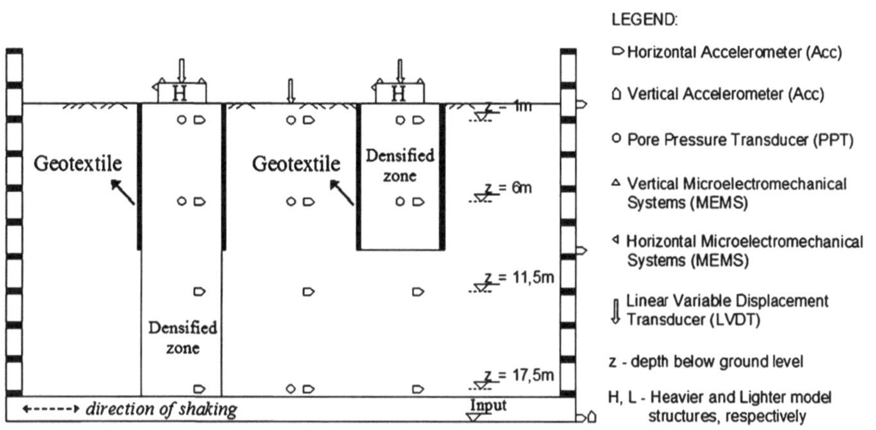

Fig. 24.4 Centrifuge model layout in experiment CT-C (cross section view)

24.2.4 Dynamic Centrifuge Model C (CT-C)

Centrifuge model CT-C (Fig. 24.4) tests different configurations for the hybrid resistance measure initially tested in model CT-B, to evaluate the influence on the performance. As the figure shows, the geotextile extends in both cases to half the depth of the liquefiable deposit (9 m) while the narrow densified column encased in the geotextile extends to the bottom of the liquefiable layer, in one case, and to the same depth as the geotextile in the other.

Table 24.1 Properties of Hostun sand (Stringer 2008)

Angle of repose (°)	D_{10} (mm)	D_{50} (mm)	$C_u = D_{60}/D_{10}$	e_{max}	e_{min}	G_s
33	0.286	0.424	1.59	1.067	0.555	2.65

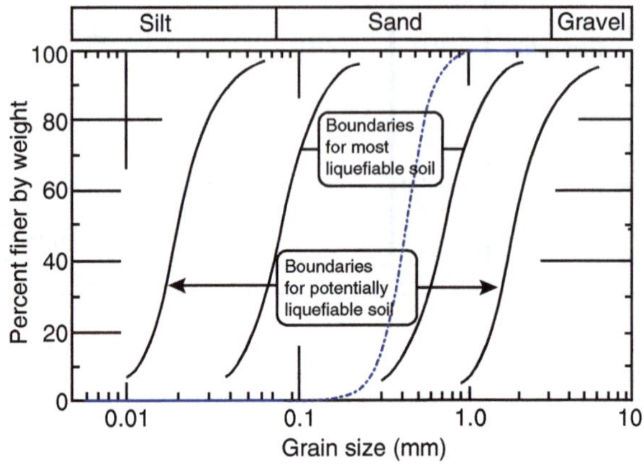

Fig. 24.5 PSD for Hostun sand, superimposed on liquefaction susceptibility curves (After Tsuchida (1970) – Adapted from Stringer (2008))

24.3 Experimental Techniques and Materials

24.3.1 Materials Used

Hostun sand, selected as the modeling sand for this experimental investigation, is a clean and uniform sand (Table 24.1) with a particle size distribution (PSD) curve that lies well within the bounds of soils most susceptible to liquefaction (Fig. 24.5). Its properties are described in detail by Flavigny et al. (1990).

To simulate the effect of drainage systems, a specific geotextile was chosen to be rigid enough to avoid squeezing of the vertical drainage paths once the final horizontal stresses were installed in the model (Fig. 24.6). The geotextile used in the tests as part of the proposed hybrid liquefaction resistance measure may eventually be materialized in the field by closely-spaced high-capacity vertical drains.

24.3.2 Equivalent Shear Beam Container

As the boundary conditions are extremely important in the results observed, the container where the models were prepared need to be carefully chosen. Thus, the models were prepared and tested within an Equivalent Shear Beam (ESB) container, described by Schofield and Zeng (1992), having flexible walls intended to replicate the soil dynamic behavior and minimize boundary effects. However, due to the large degradation of soil properties during liquefaction caused by pore pressure build-up and subsequent effective stress reduction, the container cannot exactly match the soil behavior at all times during the test.

Fig. 24.6 Geotextile used to simulate high-capacity vertical drains

24.3.3 Model Preparation Techniques

24.3.3.1 Sand Deposition

The soil deposit was prepared using an air dry pluviation technique, by means of an automatic sand pourer (Fig. 24.7). This equipment allows different sized nozzles to be placed at the bottom of the hopper in order to control the flow rate, while a selected drop height is kept constant through a computational program used to control the equipment. The relative density achieved using this technique, which depends on the flow rate and drop height, is usually up to ±5 % of the desired value. Sand is poured in pairs of steps, passing in each one of them along the model and along one axis in a single step. More details on the automatic sand pourer can be found in Madabhushi et al. (2006). Calibration was carried out before pouring the model to determine the correct drop height and nozzle size to achieve the desired relative density, varying from 50 to 80 % (Table 24.2).

In the cases where a densified block had to be created in the model, under the footing, a box made of thin metallic sheet was employed to temporarily support that zone while the model was built. This temporary formwork, which was removed during the pouring process, was not required when the densified zone was encased within a geotextile, in which case the dense sand could be placed directly inside the geotextile. Taking into account the disturbance that could be caused by the formwork removal, as well as the larger difficulty in creating a narrower denser zone with the automatic sand pourer, the geotextile and the formwork were 20 mm wider (at model scale), in each direction, than the width of the model structures. The relative density of the sand in the models was in every case about 50 and 80 % in the loose and densified zones, respectively.

24.3.3.2 Instrumentation

The instruments' position, represented for each test in Figs. 24.2, 24.3 and 24.4, were maintained in all the tests to ensure that results can be compared. A series of

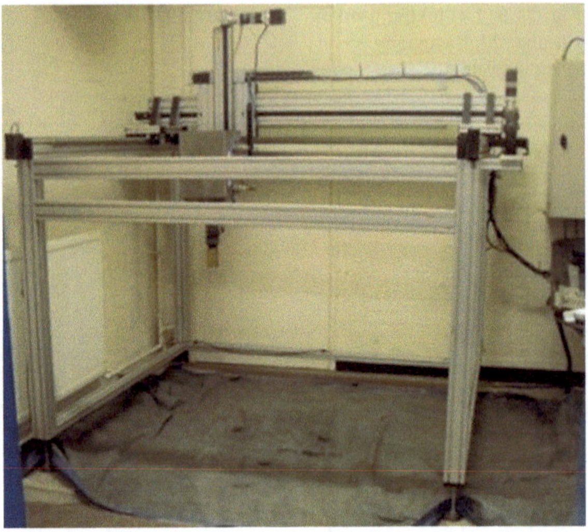

Fig. 24.7 Automatic sand pourer

Table 24.2 Drop height and nozzle diameter for each desired relative density

I_D (%)	Drop height (mm)	Ø Nozzle
50	500	7
80	580	9

PPTs, Acc and LVDTs were installed to assess the soil behavior during the centrifuge tests. A series of MEMS were carefully attached to the footings to measure the vertical and horizontal accelerations. Several Acc were installed at the bottom and walls of the container to measure the vertical and horizontal input motions and to evaluate the horizontal motion propagation through the ESB walls. Special arms were designed to place LVDTs over each footing to measure its vertical displacements.

The amount, distribution and placement of the instruments installed in the centrifuge models and the flexibility of the connecting cables are carefully chosen to minimize the disturbance on soil behavior. In general, large concentration of instruments was avoided and cables were positioned so that soil reinforcement and creation of preferential flow paths was hampered. The loading of the model was carefully performed to minimize the 1-g vibrations induced in the model, which could affect the instruments position.

24.3.3.3 Model Saturation

Model saturation with a viscous fluid is an essential part of a model preparation in centrifuge-based liquefaction research, requiring strict control for superior results. A solution of Hydroxypropyl Methylcellulose in water was used as the pore fluid with a viscosity of ≈50 times that of water, in order to achieve the so-called viscosity scaling at 50-g and overcome the conflict between time scaling in flow

Fig. 24.8 Model saturation

and dynamic phenomena. The model was placed under vacuum and de-aired viscous fluid was slowly introduced through the bottom of the sand, using small water pressure gradients. The saturation system (Fig. 24.8), which is controlled by a computational program – CAM-Sat – that ensures its smoothness, is described in detail by Stringer et al. (2009).

24.3.3.4 Footings Positioning

To minimize model disturbance during transportation to the beam, the structures were only installed after the model was loaded in the swing of the centrifuge beam. The positioning of the footings was very careful and involved verification of the evenness of each structure by means of a leveler.

24.3.4 Centrifuge Facilities

The three dynamic centrifuge experiments were conducted using the 10-m diameter Turner Beam Centrifuge at the Schofield Centre, University of Cambridge, UK, which is described in detail by Schofield (1980). The machine has a 150g-tonne capacity and it is capable of achieving a maximum centrifugal acceleration of approximately 120g at 4,125 radius. The actuator used in the centrifuge tests to generate seismic simulations is known as Stored Angular Momentum (SAM) actuator (Madabhushi et al. 1998), which is a simple and reliable mechanical actuator that uses the energy stored in a pair of flywheels to generate the input motion. Despite not being able to reproduce real seismic actions, it is able to generate nearly sinusoidal horizontal acceleration motions of chosen duration and amplitude, which is considered valuable for fundamental research on earthquake effects.

24.4 Earthquake Simulations

All the centrifuge models were submitted to a similar input seismic motion at the base, applied parallel to the long side of each model and designed to replicate a relatively strong real earthquake motion, planned to last about 25 s, have a predominant frequency of 1 Hz and impose maximum peak horizontal accelerations close to 0.3-g.

The time histories and FFTs of the earthquake simulations applied to the centrifuge models are depicted in Fig. 24.9, confirming that the planned loading was achieved in tests CT-B and CT-C. Although the input simulation in test CT-A was not measured, due to an instrument malfunction, there is no reason to believe this simulation was different. It should also be noted that the seismic simulation is not single-frequency, although as required the predominant frequency matches the desired value (1 Hz).

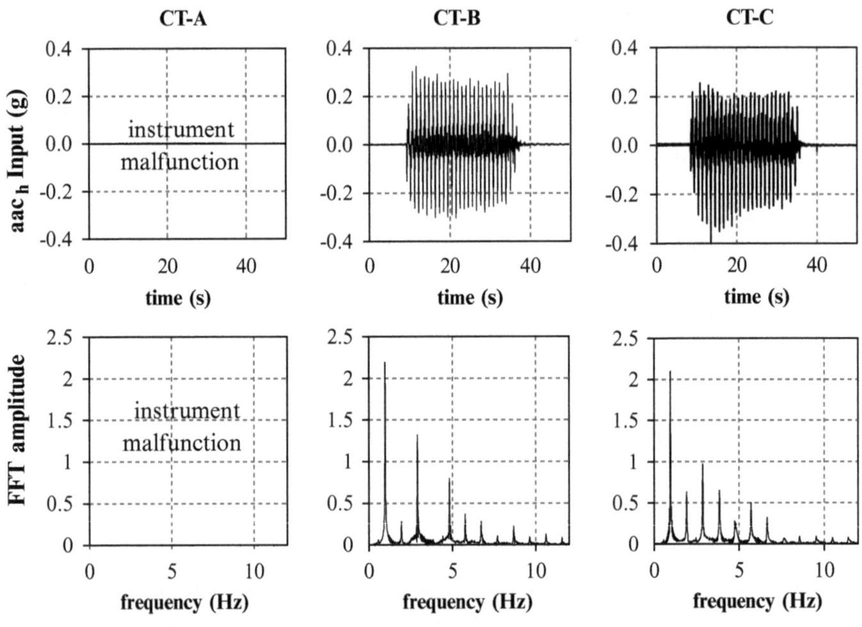

Fig. 24.9 Time histories of the seismic loading applied to all the models

24.5 Main Results

This section presents the data recorded in the centrifuge tests, the behavior of the models being successively compared in terms of the excess pore-pressure generation and dissipation in the granular ground, the horizontal accelerations induced on the structures and the footings settlements. In the data presented in the following, Foot.^{left} and Foot.^{right} refer to the structures placed on the left and right-hand sides of the model, respectively, as described in Sect. 24.2 for each model. The zone between the structures is regarded as free-field, although this hypothesis requires verification from analysis of the experimental data, since the structures and/or the ground improvement may influence the behavior at that location. A scheme of the particular configuration tested, as summarized in Table 24.3, is presented in the figures to facilitate the analysis.

24.5.1 Excess-Pore-Pressure Generation

Figure 24.10 presents the excess-pore-pressure (epp) measured in the liquefiable ground during shaking in all the tests, showing that, as the earthquake starts, the epp increases in the free-field (z = 17.5 m) from the first loading cycles. Under the footings, however, the epp generated during the seismic simulations strongly depend on the characteristics of the footing and the ground improvement carried out in the model, even if when the shaking ends, a more or less significant positive epp is observed in every location. The data also suggests that the epp developed in the so-called free-field location in the model may not be absolutely independent of the structure and/or ground improvement existing in the model. This is particularly true in model CT-B, where the epp generated in the free-field, at a depth of 17.5 m, seems to be restricted by the ground improvement extending to the full depth of the deposit under both left and right structures. The epp dissipation, in model CT-B, immediately after the end of shaking, also seems to be slightly accelerated by the surrounding liquefaction resistance measures.

Table 24.3 Different configurations tested in each centrifuge experiment

Centrifuge experiment	CT-A		CT-B		CT-C	
	H	L	H:(D + G)	H:D	H:(D + G/2)	H:(D/2 + G/2)
Left and right configurations	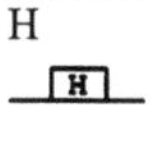	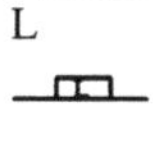	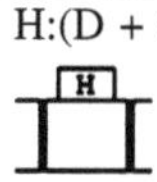	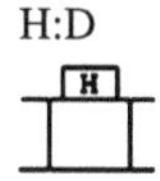	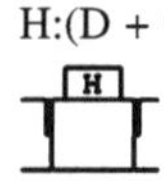	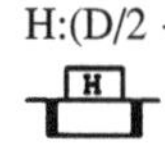

H heavy structure, *L* light structure, *D* full-depth Densified zone, *G* Geotextile encasing densified zone up to the full depth, *D/2* half-depth Densified zone, *G/2* Geotextile encasing densified zone up to the mid depth

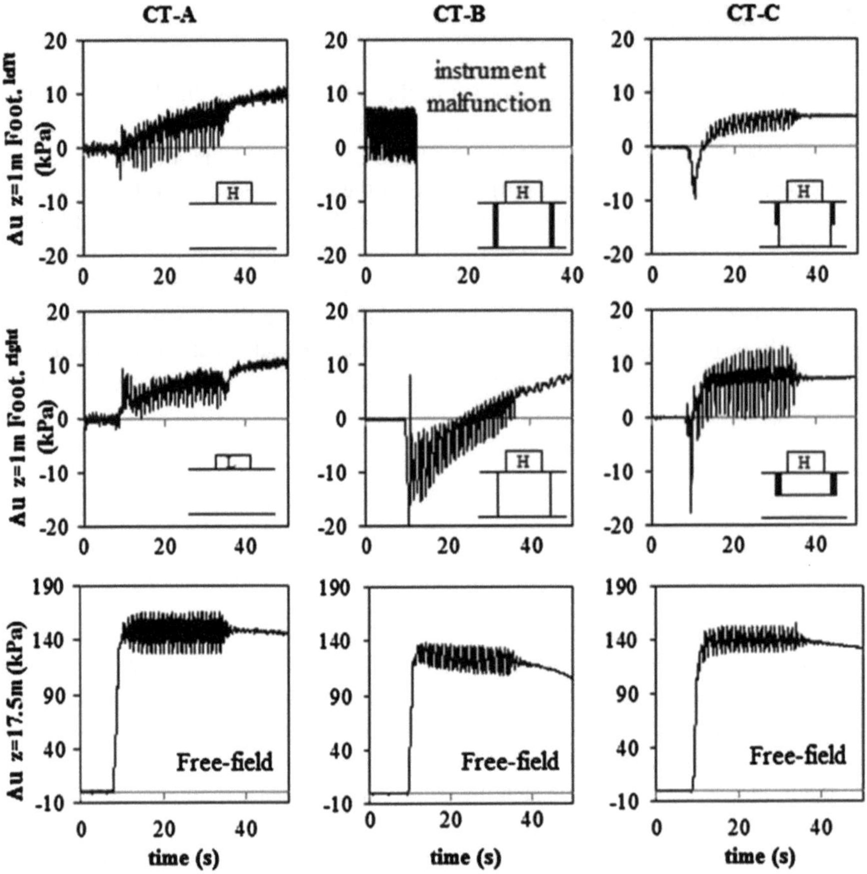

Fig. 24.10 Excess pore pressure at different depths under the footings during the earthquake simulation

When no ground improvement exists (CT-A), the final value for the epp at 1-m depth is slightly higher under H structure than under L structure. This probably results from the fact that footing H induces on the soil a higher initial effective stress, being able to develop a higher final epp. When liquefaction resistance measures are present (tests CT-B and CT-C), important negative epp are observed under the footing once the first couple of cycles take place. As permanent negative pore-pressure variation can only be induced by monotonic loading, this suggests that vertical stress concentration occurs under the footing at this stage. Also, the results show that the configuration H:D results in the longer period of negative epp induced under the footing, which continues far longer than the first loading cycles. Another main conclusion that can be drawn from the results is that, except when vertical drains were simulated with a geotextile, the epp tends to rise again immediately after the end of shaking (CT-A and CT-B's H:D). On the other hand, the data from CT-C shows that the epp reaches the maximum value after a

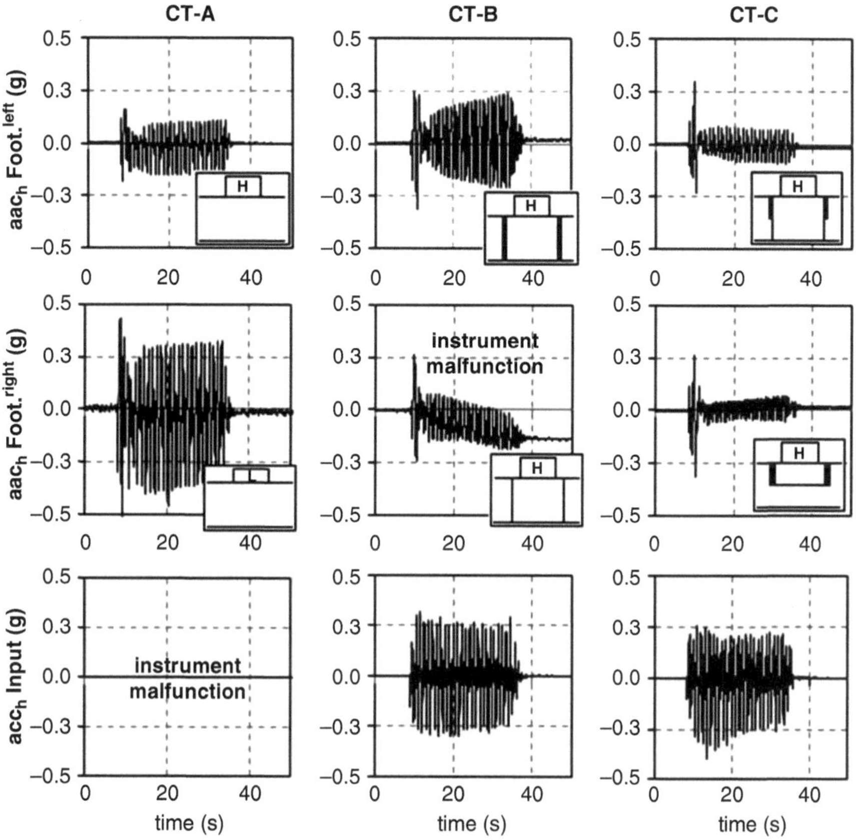

Fig. 24.11 Horizontal accelerations at the foundations and imposed to the base

few cycles once the seismic loadings starts. This suggests that hybrid techniques combining narrow densified zones with vertical drains mitigate faster the epp developed under footings during earthquake shaking.

24.5.2 Propagation of Accelerations

Figure 24.11 shows the horizontal motions measured at the footings for all the models tested. CT-A data shows that the peak horizontal accelerations measured in the lighter structure (L) are about twice as large as those in H, which may be eventually explained by Newton's 2nd Law of Motion, assuming that the horizontal force transmitted by the liquefied sand to footings L and H is similar (Marques et al. 2012). Figure 24.11 also shows that by increasing the soils' relative density under structure H, the peak horizontal acceleration induced on the structures is also increased. This phenomenon is even visible in situation H:D in CT-B, since the instrument only start to malfunction after the first loading cycles. In configuration

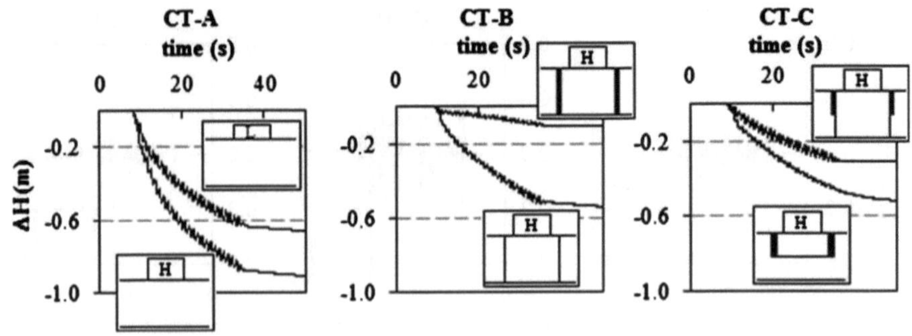

Fig. 24.12 Settlements of the footings measured during the period of the earthquake simulation

H:(D + G) (CT-B), there is no significant attenuation with the number of cycles, which is consistent with a less pronounced soil stiffness degradation that may result from the combination of a full-depth densified zone (H:D in Fig. 24.9) with vertical drains.

Observations in configurations H:(D + G) in CT-B, and H:(D + G/2) in CT-C, prove that by using a hybrid technique with narrow densified zones throughout the liquefiable soil layer and vertical drains until half of it results in almost half of the horizontal accelerations in the footing after the first loading cycles. As expected, by densifying the ground only up to half the full depth of the liquefiable deposit and encasing it with vertical drains (situation H:(D/2 + G/2) in CT-C), the horizontal accelerations measured in the footing are considerably smaller. However, the peak horizontal motions recorded in the first loading cycles still reach approximately the same high values irrespective of the depth the resistance measure employed.

24.5.3 *Footing Settlements*

Figure 24.12 shows the settlements measured in each case during the seismic simulations. The settlements of the footings built on densified ground with or without vertical drains are in all cases significantly reduced in comparison with any of the footings presented in CT-A, where no ground improvement was carried out. However, the footings suffer significantly different settlements during the earthquake, which range approximately from 0.10 to 0.50 m, for configurations H:(D + G) and H:D (CT-B); and from 0.40 to 0.30 m, for configurations H:(D/ 2 + G/2) and H:(D + G/2) (CT-C).

In configuration H:D the structure suffers a fairly non-even settlement with an average value of approximately 0.50 m, which is about 50 % smaller than the one measured with no improvement under structure H (≈0.90 m). However, this structure did not remain in its original position during the seismic loading, moving towards the looser sand. Consequently, the co-seismic settlement measured in this case would possibly be reduced if the structure had rested exclusively on the dense

Table 24.4 Total settlements of the footings

	CT-A		CT-B		CT-C	
	H	L	H:(D + G)	H:D	H:(D + G/2)	H:(D/2 + G/2)
ΔH Total (m)	−1.120	−0.868	−0.138	−0.736	−0.290	−0.443

zone. In contrast, in situation H:(D + G) in CT-B, the footing suffers an even settlement of approximately 0.10 m, which is about 90 % smaller than the one measured with no improvement. With respect to the experiments where hybrid mitigation techniques were used, Fig. 24.12 shows that the footing settles visibly more in situation H:(D/2 + G/2) in CT-C, as loose sand exhibits a much softer response to dynamic plus monotonic loading than dense sand.

The total settlements measured at the end of each centrifuge test for every footing is depicted in Table 24.4. Comparing the results obtained for the cases where mitigation techniques were used, it is clear that the total settlements in CT-C are noticeably larger than in H:(D + G) but much smaller than in H:D, the configurations tested in centrifuge experiment CT-B. However, the total settlements observed when liquefaction resistance measures are employed are always reduced in comparison to the benchmark case (H in CT-A), the reduction reaching almost 90 % when the tested hybrid technique is carried out up to the bottom of the deposit (Table 24.4).

24.6 Conclusions

A centrifuge-based research program was carried out at Cambridge University Engineering Department's Schofield Centre to characterize the behavior of shallow foundations built on saturated deposits of loose sand during seismic events, as part of a TA agreement established within the EU funded SERIES project. The performance of hybrid mitigation techniques, combining densification with vertical drainage was also assessed by considering different geometries for a narrow densified zone under the footings with and without geotextile surrounding it.

This paper describes the design of all the centrifuge tests, the test setup and the experimental techniques used to prepare the models, as well as the materials used. The experimental observations are also presented and analyzed, revealing the importance of centrifuge modeling in research aiming at highlighting the effects of earthquake-induced liquefaction. The results also prove the effectiveness of hybrid mitigation techniques combining vertical drains with narrow densified columns in mitigating liquefaction effects.

Acknowledgments The research leading to these results has received funding from the European Community's Seventh Framework Programme [FP7/2007–2013] for access to Turner Beam Centrifuge, Cambridge, UK, under Grant Agreement nº 227887 [SERIES]. The first author also would like to acknowledge the valuable assistance of the technicians at the Schofield Centre, as well as the research scholarship provided by FCT – Fundação para Ciências e a Tecnologia.

References

Bardet JP, Idriss IM, O'Rourke TD, Adachi N, Hamada M, Ishihara K (1997) Report on North America. In: Japan workshop on the geotechnical aspects of the Kobe, Loma Prieta, and Northridge earthquakes. Civil Engineering Department, University of Southern California, Los Angeles

Boulanger RW (2005) General report- earthquake related problems (TS-4b). In: International conference on soil mechanics and geotechnical engineering, Osaka, September 2005, pp 3117–3124

Coelho PALF (2007) In situ densification as a liquefaction resistance measure for bridge foundations. PhD thesis, Engineering Department, Cambridge University, UK

Flavigny E, Desrues J, Palayer B (1990) Le sable d'hostun. Geotechnique 53:67–70

Madabhushi SPG, Schofield AN, Lesley S (1998) A new stored angular momentum earthquake actuator. In: Proceedings of international conference centrifuge '98, Tokyo, pp 111–116

Madabhushi SPG, Houghton NE, Haigh SK (2006) A new automatic sand pourer for model preparation at University of Cambridge. In: 6th international conference on physical modelling in geotechnics, Hong Kong

Marques AS, Coelho PALF, Cilingir U, Haigh SK, Madabhushi SPG (2012) Centrifuge modeling of liquefaction-induced effects on shallow foundations with different bearing pressures. In: 2nd Eurofuge conference on physical modelling in geotechnics, Delft

Mitchell JK (2003) Performance of improved ground in earthquakes (Do the results justify the expense?). Seminar in Geotechnical Engineering at Cambridge University Engineering Department, Cambridge University, UK, February 2003

Schofield AN (1980) Cambridge geotechnical centrifuge operations. Geotechnique 30(3):227–268

Schofield AN, Zeng X (1992) Design and performance of an equivalent-shear-beam (ESB) container for earthquake centrifuge modeling. Technical report CUED/D-SOILS/TR245, University of Cambridge, UK

Stringer ME (2008) Pile shaft friction in liquefaction. First year report, University of Cambridge

Stringer ME, McMahon BT, Madabhushi SPG (2009) Cam-Sat: computer controlled saturation for geotechnical modelling. Technical report CUED/D-Soils/TR348, University of Cambridge, UK

Tsuchida H (1970) Prediction and countermeasure against the liquefaction in sand deposits. Abstracts of seminars in the Port and Harbours Research Institute, pp 3.10–3.33

Chapter 25
Stability Control of Rafted Pile Foundation Against Soil Liquefaction

Ahmed Mohammed Youssef Mohammed and Koichi Maekawa

Abstract Reinforced concrete (RC) piles are widely used to support structures on soft soil deposits with high liquefaction potential. While the lateral spreading of liquefied soil during earthquakes may cause severe damage to the RC piles, the authors propose using steel sheet pile walls (SSPW) to protect the RC piles from damage and control the overall stability of superstructures. The chapter deals with the nonlinear seismic response and damage evolution and control of multi-storey buildings supported by RC rafted piles on a liquefiable soil. The engineering focus is on the effect of using SSPW to protect existing multi-storey buildings which are supported by rafted pile foundation in a liquefiable soil. The effect of the SSPW embedment length is investigated. Drained and undrained conditions of the soft soil deposits are analytically considered. The results show that the sheet pile wall could improve the overall stability of the reinforced concrete superstructure, but it leads to a higher base shear on the structure.

25.1 Introduction

During the Alaska and Nigata 1964 earthquakes, serious structural damage due to soil liquefaction was reported. Likewise, severe damage and stability failure of multi-storey buildings were also reported during the Hygoken-Nanbu earthquake. Subsequently, research on the damage evolution and control has drawn increasing attention in the structural engineering field. Accordingly, efforts are being made towards enhanced damage control techniques and numerous experimental investigations with different scales have been carried out (Toshi 1986, Wilson et al. 2000). Nowadays, the most common countermeasures for damage control

A.M.Y. Mohammed (✉) • K. Maekawa
Concrete Lab, Civil Engineering Department, Faculty of Engineering,
The University of Tokyo, 7-3-1 Hongo, Bunkyo-ku, Tokyo 113-8656, Japan
e-mail: mohammed@concrete.t.u-tokyo.ac.jp; maekawa@concrete.t.u-tokyo.ac.jp

A. Ilki and M.N. Fardis (eds.), *Seismic Evaluation and Rehabilitation of Structures*, Geotechnical, Geological and Earthquake Engineering 26,
DOI 10.1007/978-3-319-00458-7_25,

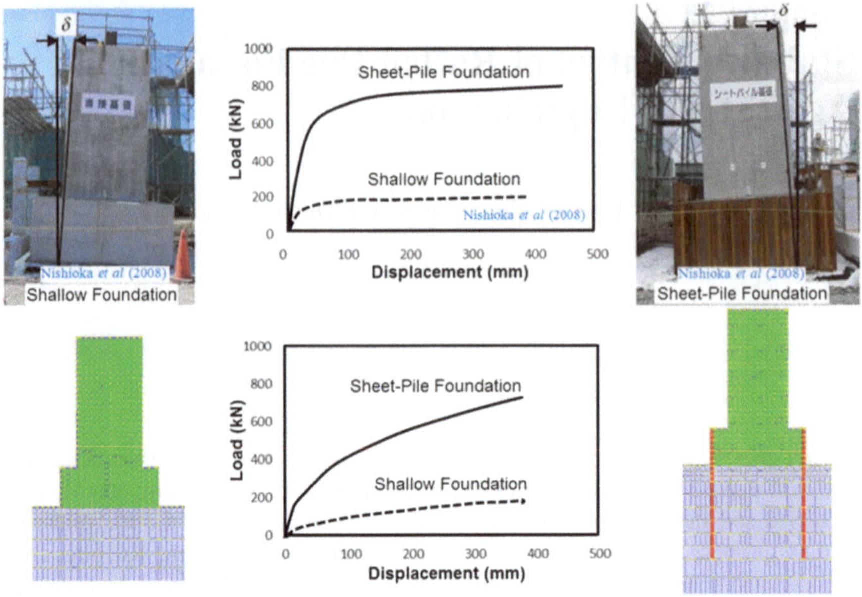

Fig. 25.1 Experimental and analytical full scale pushover test of two bridge piers resting on raft alone or on raft with SSPW

are soil densification, compaction grouting and dissipation of excess pore water pressure through gravel or pipe drains (Towhata 2008). These techniques are costly and practically difficult to use as a means to control the damage of multi-storey buildings. For improved horizontal seismic resistance, Nishioka et al. (2008) have proposed a new type of foundation, which combines a RC mat with steel sheet pile wall (SSPW) used for earth-retaining works during excavations. They conducted a series of both small and large scale bridge column experiments, with comparisons to analysis as depicted in Fig. 25.1. The results show a significant increase in the lateral seismic resistance of the RC mat foundation when it is combined with the steel sheet-pile wall. Yet, the effect of using SSPW on the superstructure is not discussed. The current chapter attempts to address this issue.

While a full scale experiment to investigate the seismic behavior of a real structure together with SSPW is hard to conduct, an analytical study with a well verified computational RC and soil model (Maekawa et al. 2003) can serve as an alternative to capture the overall efficiency of SSPW. In this study, the authors use a full three dimensional nonlinear finite element analysis of soil-structure-pore water systems. The applicability of the system was verified by shaking table experiments of top-heavy piles embedded in a liquefiable model foundation (Maki et al. 2004, 2005, Mohammed and Maekawa 2012; Mohammed et al. 2012a). Using this analytical platform, the effects of SSPW as a seismic countermeasure for the rafted pile foundation is discussed with regard to the base shear induced in the RC multi-storey superstructure and its overall stability. Both cases of drained and undrained soils are considered in the analysis to clarify the potential effect of soil liquefaction.

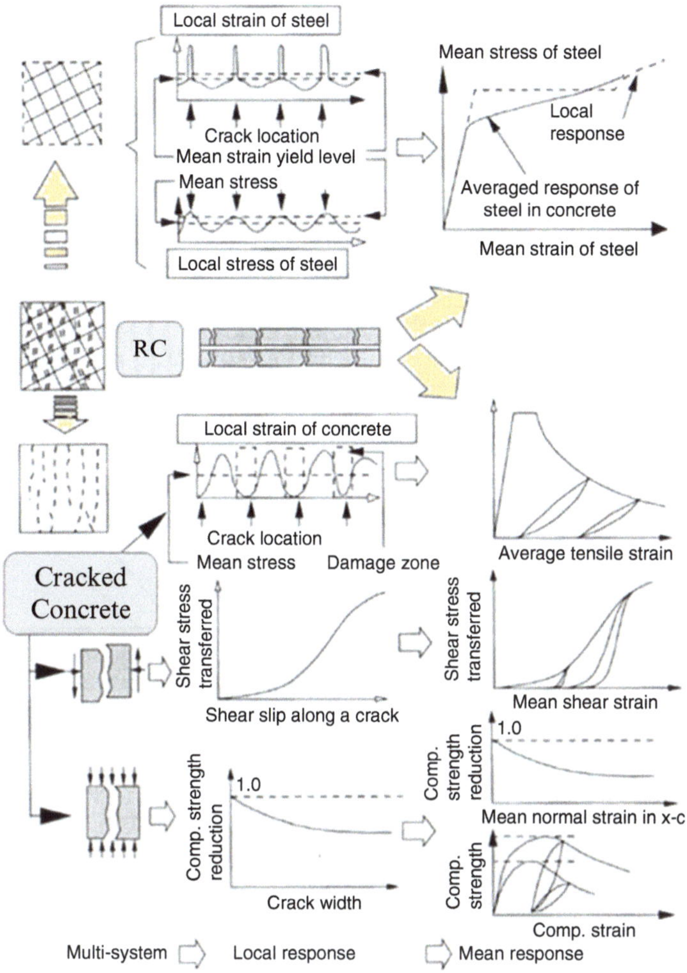

Fig. 25.2 RC constitutive laws

25.2 Constitutive Modeling

A reinforced concrete (RC) material model has been constructed by combining constitutive laws for cracked concrete and those for reinforcement. The fixed multi-directional smeared crack constitutive equations (Maekawa et al. 2003) are used, as summarized in Fig. 25.2.

Crack spacing and diameters of reinforcing bars are implicitly taken into account in smeared and joint interface elements, no matter how large they are. The constitutive equations of structural concrete satisfy uniqueness for compression, tension and shear transfer along crack planes. The bond between concrete and reinforcing bars is taken into account in the form of a tension-stiffening model; the space-averaged

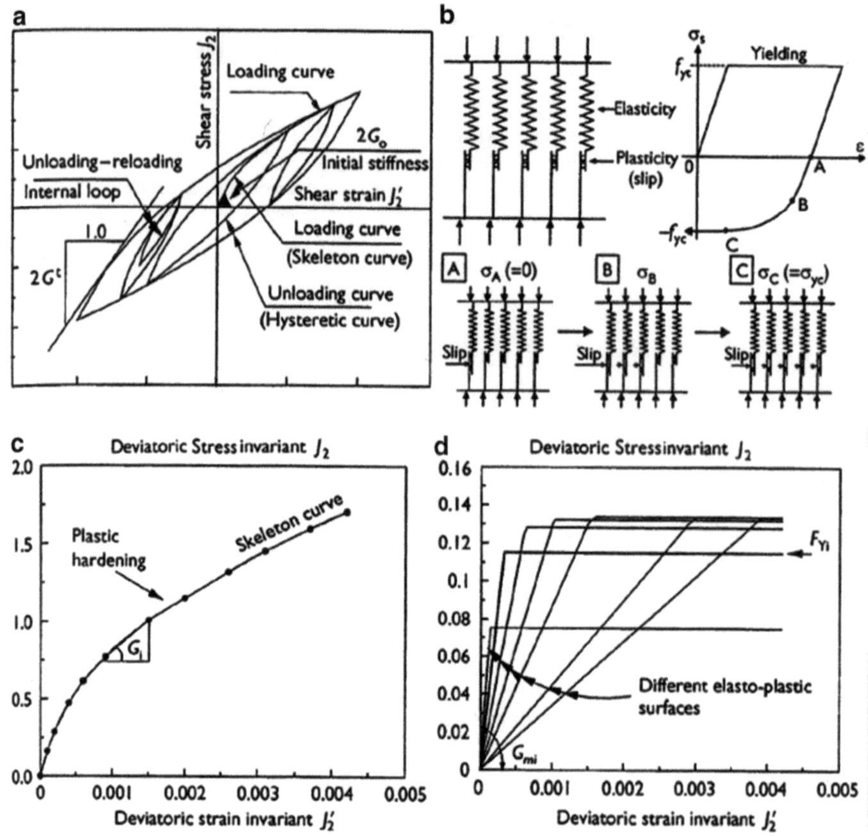

Fig. 25.3 Constitutive modelling for soil (**a**) Stress-strain in deviatoric mode, (**b**) Multi-phase yield modelling, (**c**) Division of envelope into *n* broken straight parts, (**d**) Decomposition into some elasto-perfectly plastic surface with different yield strength and stiffness

stress-strain relation of reinforcement is assumed to represent the localized plasticity of steel around concrete cracks. The hysteresis rule of reinforcement is formulated based on the Kato (1979) model for a single bar under reversed cyclic load. This 2D constitutive modeling of RC has been verified by member-based and structure-oriented experiments. For details of the RC material modeling reference is made to Maekawa et al. (2003).

A nonlinear path-dependent constitutive model for the soil mainly depends on the shear stress-shear strain relationship, which is extended to three-dimensional generic conditions and assumed to follow Masing's rule for the soil hysteresis. The soil is idealized as an assembly of a finite number of elasto-perfectly plastic elements connected in parallel as shown in Fig. 25.3 (Okhovat et al. 2009, Mohammed and Maekawa 2012; Mohammed et al. 2012a). The nonlinear behavior of the soil system in liquefaction is assumed as in undrained state, since its drainage takes much longer than the duration of an earthquake (Towhata 2008). The soil undrained behavior is shown in Fig. 25.4.

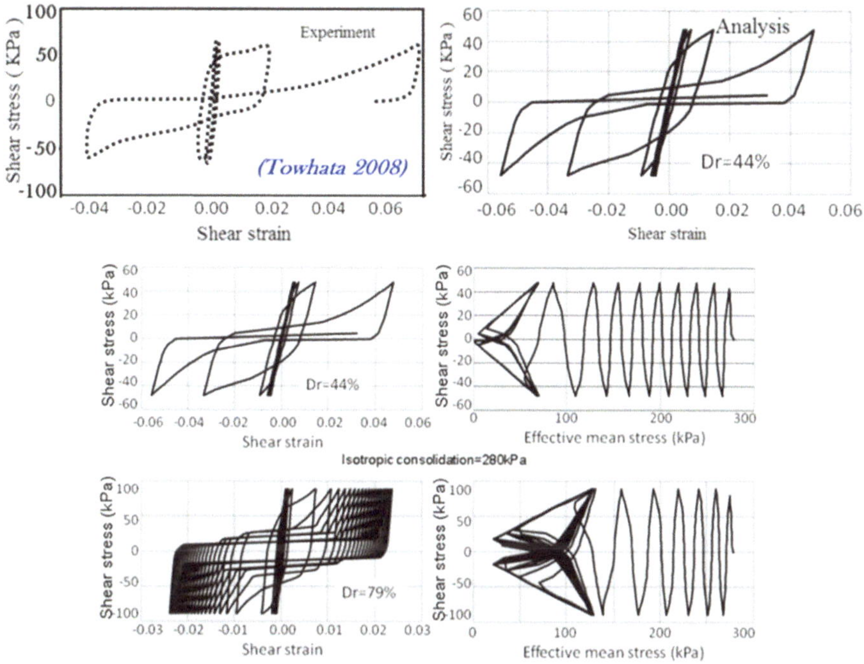

Fig. 25.4 Experimental and analytical confinement dependent soil model under undrained condition

25.3 Finite Element Model

For the purposes of the analytical investigation, a seven-storey building supported by nine columns on a soft sandy soil foundation is considered. The building is 12 m wide and 24.5 m tall and is supported on nine square columns (0.7 m × 0.7 m). A RC rafted pile foundation is considered for supporting these columns on 16.5 m thick of soft sandy soil to bear on a very dense base soil.

The soft soil is taken to consist of five layers, from a soft one at the surface to a compacted layer at the bottom and assumed to overlie a dense sandy soil that acts as the engineering bedrock where the ground acceleration is defined. The soil layer properties are described in Table 25.1. For simplicity, all concrete slabs and the foundation mat are considered to have a uniform thickness of 0.5 m and are modelled as 3D elastic solid elements with unit weight 25 kN/m^3, elastic modulus 28,000 MPa and Poisson's ratio 0.2. The finite element discretization was carried out using 3D-nonlinear solid element for the soil and 3D-elastic solid element for RC slabs and the mat foundation, as depicted in Fig. 25.5.

For modeling RC piles and columns, prismatic Timoshenko frame elements are used. The cross-section of members is divided into component cells according to the shape and the reinforcement layout (Fig. 25.6); in each cell,

Table 25.1 Soil properties used in analysis

Soil layer	Gs (MPa)	Relative density	Density (kg/cm^2)	Friction angle φ
A	740	100	2.2	45
B	150	42	1.8	37
C	90	40	1.8	35
D	70	36	1.7	34
E	50	30	1.7	31
F	30	20	1.6	25

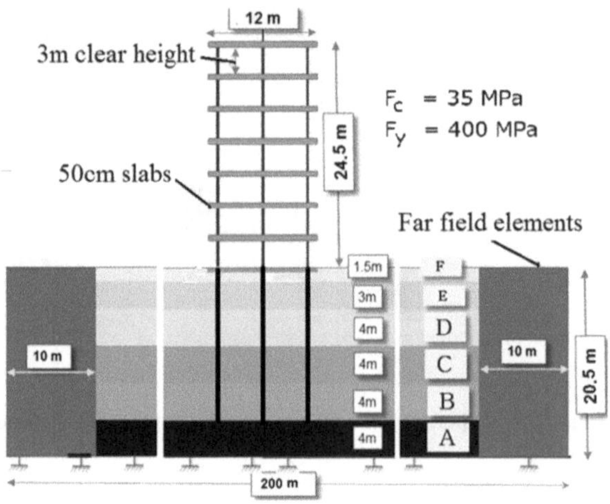

Fig. 25.5 Model of multi-storey building

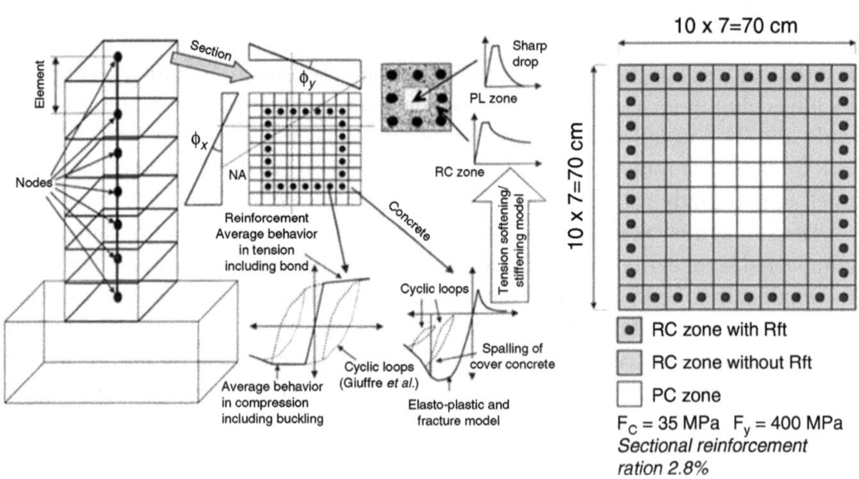

Fig. 25.6 Fibre technique, uniaxial constitutive models (Maekawa et al. 2003), and cross section modeling of RC column/pile

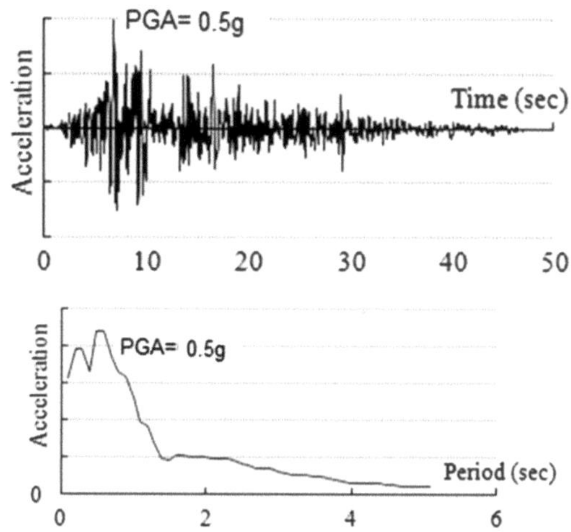

Fig. 25.7 Kobe earthquake record scaled to a PGA of 0.5g

the nonlinear tension-compression constitutive model of cracked reinforced concrete is applied and the zoning procedure is implemented to take into account the crack dispersion due to bond between reinforcement and concrete. By means of the section discretization, the nonlinear relation of flexural moment versus space-averaged curvature is internally computed in the finite element solution, using the constitutive modeling of materials allocated in each cell as shown in Fig. 25.6 (Maekawa et al. 2003). The compressive strength of concrete and the yield strength of reinforcing bars are assumed to be 35 and 400 MPa, respectively. A scaled Kobe earthquake with a PGA of 0.5g is used in the analysis, as shown in Fig. 25.7.

25.4 Stability Control of Multi-storey Buildings

Urban areas built on potentially liquefiable soft soil foundation are highly vulnerable to seismic actions. According to the history of soft saturated soils with liquefaction potential, most superstructures do not suffer mainly structural damage, but from vertical settlement, horizontal irrecoverable drift and tilting, as reported in the Niigata earthquake. This chapter studies the effect of using steel sheep pile walls to control the overall stability of existing RC multi-storey buildings constructed on soil having a rather high liquefaction potential. In order to investigate the effect of SSPW as a countermeasure against soil liquefaction, four analyses are carried out as shown in Fig. 25.8.

The steel sheet pile wall is assumed to be made of mild steel with a Young modulus of 200,000 MPa and a Poison's ratio of 0.3. It is modeled using elastic elements. For instance, the cross section and the properties of a sheet pile used in practice are shown in Fig. 25.9. The thickness of the steel finite elements is assumed to be 0.2 m, giving bending stiffness equivalent to the real one. Since the thickness of

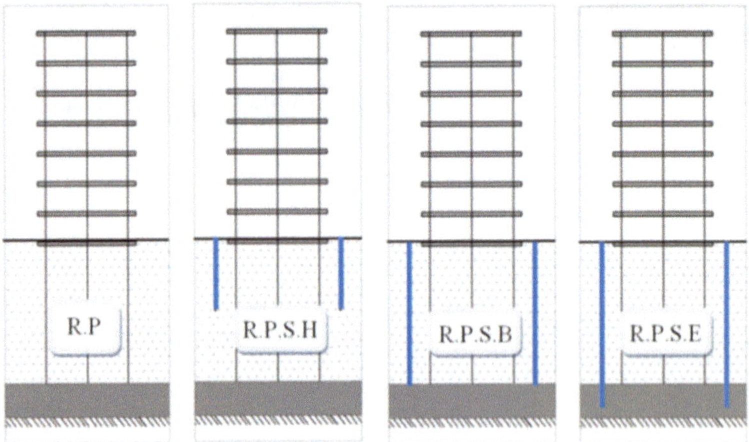

Fig. 25.8 Four models considered in the analysis

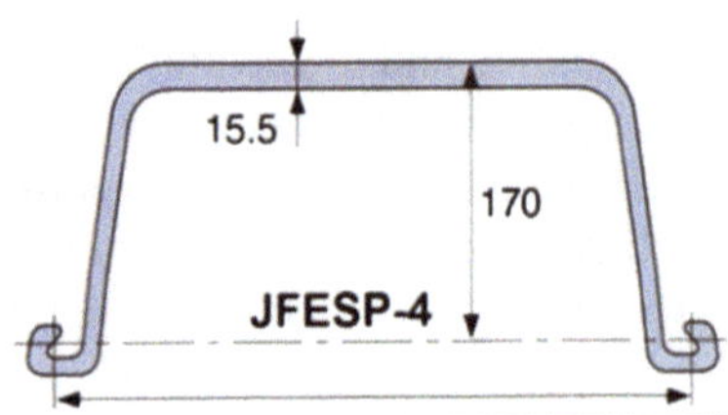

Fig. 25.9 Cross-section and properties of steel sheet-pile

Properties per 1 m of Wall			
Sectional Area	Mass	Moment of Inertia	Section Modulus
242.5 cm^2/m	190.0 kg/m	12090 cm^4/m	1175 cm^3/m

the liquefiable soil is 16.5 m as mentioned before, these three sheet pile lengths (8.5, 16.5 and 18.5 m) correspond respectively (see Fig. 25.8):

- half-length of the liquefiable layer thickness (R.P.S.H),
- just bearing on the non-liquefiable soil surface (R.P.S.B), and
- embedded by 2 m in the non-liquefiable soil (R.P.S.E).

The damage is illustrated by the magnitude of the base shear force transmitted to RC building superstructures by the soil foundation. The stability of the superstructure is defined by the absolute peak values of vertical subsidence, lateral drift motion and tilting angle. The relative performance of each foundation type is presented in the subsequent subsections.

25.4.1 Lateral Displacement

The structural lateral displacement (drift) during earthquake is an important criterion to evaluate whether a steel sheet pile wall could help to control stability. The

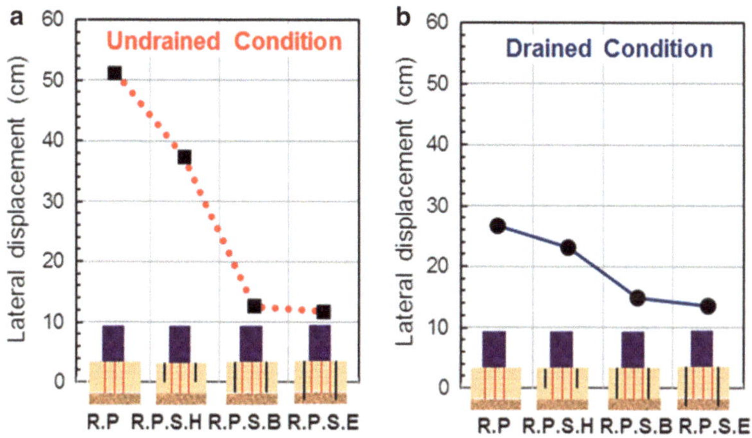

Fig. 25.10 SSPW impact on superstructure lateral displacement (**a**) undrained condition (**b**) drained condition

maximum lateral displacement is 0.51 m when there is no SSPW (R.P) in the case of undrained soil or 0.27 m for drained soil, reflecting the significant impact of soil liquefaction on the stability of the RC superstructure. SSPW with a length of 8.5 m (Half embedded, R.P.S.H) slightly increases the lateral resistance of the superstructure in both cases of drained and undrained soil. Accordingly, this solution is not recommended for controlling the lateral movement of the superstructure against soil lateral spreading. SSPW with a length of 16.5 m (bearing on the non-liquefiable soil surface, R.P.S.B) significantly increases the lateral resistance of the superstructure in both drained and undrained soils, as shown in Fig. 25.10. SSPW with a length of 18.5 m (embedded 2-m deep in the non-liquefiable soil, R.P.S.E) slightly increases the lateral resistance compared to the R.P.S.B case. Therefore, the steel sheet pile wall is really needed to minimize the lateral movement of superstructure in case of soil liquefaction.

25.4.2 Subsidence of On-Ground RC Framed Superstructure

Liquefaction of a soft soil deposit during earthquake may cause flotation of underground structures and subsidence of on-ground reinforced concrete structures. The experience from past earthquake provides strong evidence of these phenomena; experimental and analytical investigations (Towhata 2008, Okhovat et al. 2009, Mohammed et al. 2012b) provide further support. While the main function of piles is to support the superstructure safely, the subsidence of the rafted pile foundation without using SSPW is not very: the vertical settlement of cases R.P, R.P.S.H, R.P.S. B, and R.P.S.E are less than 30 mm for both drained and undrained conditions, as shown in Fig. 25.11. It is clear that the pile foundation has a sufficient resistance against subsidence in a liquefiable soil.

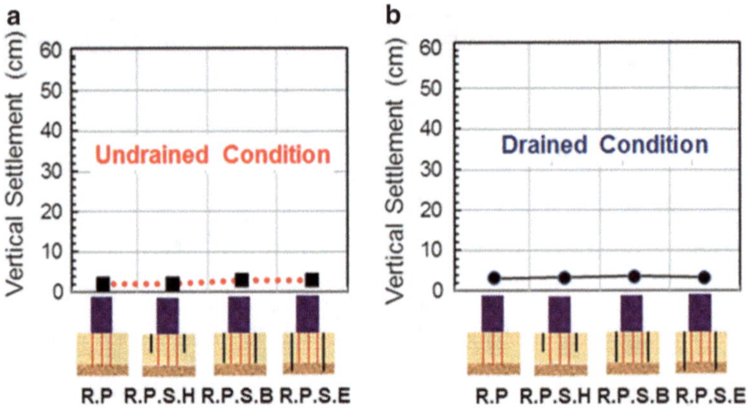

Fig. 25.11 SSPW impact on vertical settlement of superstructure (**a**) undrained condition (**b**) drained condition

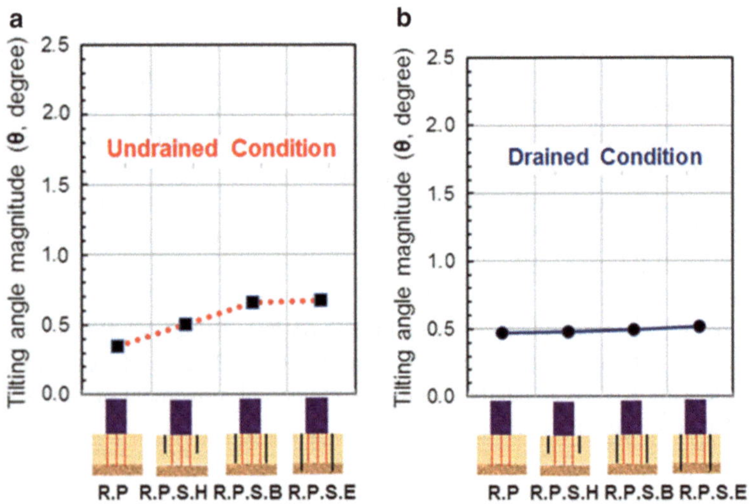

Fig. 25.12 SSPW impact on tilt of superstructure (**a**) undrained condition (**b**) drained condition

25.4.3 Tilt Angle

The peak values of tilt angle in the eight cases considered here are calculated and illustrated in Fig. 25.12. As shown in Fig. 25.12a, the tilt angle in the case of rafted pile foundation (R.P) is about 0.35-deg. SSPW with a length of 8.5 m (half embedded) increases the tilt angle to 0.5-deg in the case of rafted pile foundation (R.P.S.H); SSPW with a length of 16.5 m increases the tilt to 0.66-deg in the case of rafted pile foundation (R.P.S.B); SSPW with a length of 18.5 m increases the tilt to 0.67-deg in the case of rafted pile foundation (R.P.S.E). Accordingly, in the case of

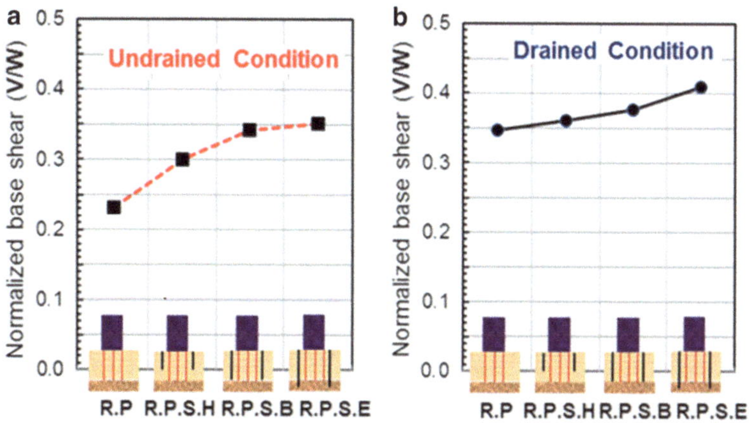

Fig. 25.13 SSPW impact on base-shear transmitted to superstructure (**a**) undrained condition (**b**) drained condition

undrained soil and rafted pile foundations, SSPW increases the tilt angle of the superstructure, as its embedment length increases due to the higher confinement of the soil foundation system which may lead to a higher rocking mode of the superstructure. This confirms the positive impact of the rafted pile foundation, which gives the smallest tilt angle.

Figure 25.12b shows the tilt angle variation for the drained condition.

25.4.4 Damage and Overall Stability of Superstructure

The peak storey shear forces, V, are calculated and normalized by the superstructure weight $W = 7{,}500$ kN, as shown in Fig. 25.13. The storey shear in the ground floor is the base shear transmitted to the superstructure by the foundation system. Among the four cases, the transmitted base shear is the lowest when only the rafted pile foundation, without SSPW, is used. This is because liquefied soil acts as a damping medium due to the reduction of the shear stiffness after liquefaction starts. The steel sheet pile wall confines the soil beneath the superstructure and may prevent soil liquefaction. Thus, it reduces the damping mechanism that may occur due to soil liquefaction. As a result, in the case of raft foundation, the longer the SSPW length, the more the normalized transmitted base shear (V/W) increases: from 0.23 (R.P), to 0.30, to 0.34 to 0.35 in cases R.P.S.H, R.P.S.B and R.P.S.E, respectively. Generally, the SSPW length increase causes more damage to the RC superstructure when used with the rafted pile foundation. Soil liquefaction is less hazardous in terms of human life, because the base shear is about 60 % of that from the dry soil case, as shown in Fig. 25.13.

Figure 25.14 presents the overall stability factor of the superstructure vs. the structural damage, as reflected by the transmitted base shear; it clearly shows

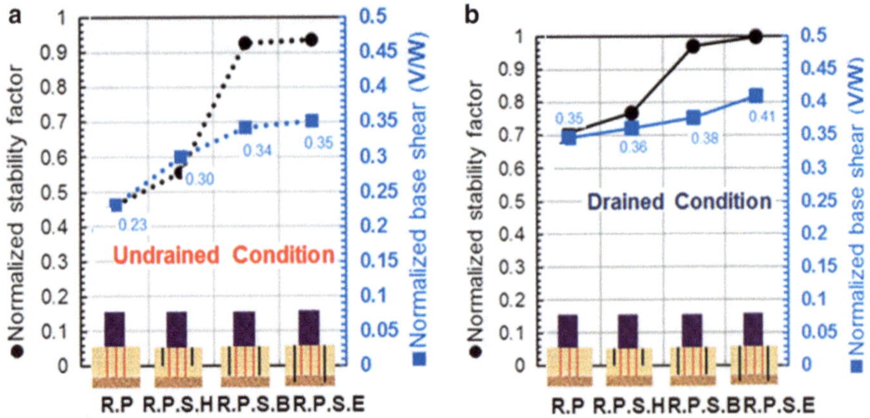

Fig. 25.14 SSPW impact on superstructure stability and damage (**a**) undrained condition (**b**) drained condition

the significant effect of increasing the overall stability of the on-ground structure against liquefaction due to the embedment length increase of SSPW. It also shows that the stiffer foundation system leads not only to higher stability but also to the largest structural damage.

25.5 Conclusions

The superstructure-soil interaction for a liquefiable soil is investigated by considering different kinds of foundations. The positive and negative effects of using steel sheet pile wall as a counter measure for control of overall stability and damage of the superstructure are elaborated. The steel sheet pile wall improves the overall stability of the multi-storey building, but induces more damage to the superstructure, which should be considered in its seismic design.

For existing multi-storey buildings supported through rafted pile foundations on a liquefiable soil, the steel sheet pile wall bearing on the non-liquefiable soil surface seems to be an optimum solution for protecting the building from soil liquefaction.

References

Kato B (1979) Mechanical properties of steel under load cycles idealizing seismic action. CEB Bull D'Inform 131:7–27

Maekawa K, Pmanmas A, Okamura H (2003) Nonlinear mechanics of reinforced concrete. Spon Press, London

Maki T, Maekawa K, Nakarai K, Hirano K (2004) Nonlinear response of RC pile foundations in liquefying soil. Japan Geotechnical Society

Maki T, Maekawa K, Matsuyoshi H (2005) RC pile-soil interaction analysis using a 3D-finite element method with fiber theory-based beam elements. Earthq Eng Struct Dyn 35(13):1587–1607
Mohammed AMY, Maekawa K (2012) Global and local impacts of soil confinement on RC pile nonlinearity. J Adv Concr Technol 10:375–388
Mohammed AMY, Okhovat MR, Maekawa K (2012a) Damage evolution of underground structural reinforced concrete: small-scale static-loading experiments. Int J World Acad Sci Eng Technol 61:696–703
Mohammed AMY, Okhovat MR, Maekawa K (2012b) Numerical investigation on damage evolution of piles inside liquefied soil foundation-dynamic-loading experiments. Int J World Acad Sci Eng Technol 71:1796–1803
Nishioka H, Koda M, Hirao J, Higuchi S (2008) Development of sheet-pile wall foundation that combines footing with sheet piles. Quart Rep Rail Tech Res Inst 49(2):73–78
Okhovat MR, Shang F, Maekawa K (2009) Nonlinear seismic response and damage of reinforced concrete ducts in liquefiable soils. Adv Concr Technol 7(3):439–454
Toshi I (1986) Soil liquefaction studies in Japan: state-of-the-art. Soil Dyn Earthq Eng 5(1):2–68
Towhata I (2008) Geotechnical earthquake engineering. Springer, Berlin
Wilson DW, Boulanger RW, Kutter BL (2000) Observed seismic lateral resistance of liquefying sand. J Geotech Geoenviron Eng 126(10):898–906

Chapter 26
Experimental Assessment of Seismic Pile-Soil Interaction

Armando L. Simonelli, Luigi Di Sarno, Maria Giovanna Durante, Stefania Sica, Subhamoy Bhattacharya, Matt S. Dietz, Luiza Dihoru, Colin A. Taylor, Roberto Cairo, Andrea Chidichimo, Giovanni Dente, Arezou Modaressi, Luìs A. Todo Bom, Amir M. Kaynia, George Anoyatis, and George Mylonakis

Abstract Physical modeling has long been established as a powerful tool for studying seismic pile-soil-superstructure interaction. This chapter presents a series of 1-g shaking table tests aiming at clarifying fundamental aspects of kinematic and inertial interaction effects on pile-supported systems. Pile models in layered sand deposits were built in the laboratory and subjected to a wide set of earthquake motions. The piles were densely instrumented with accelerometers and strain

A.L. Simonelli (✉) • L. Di Sarno • M.G. Durante • S. Sica
Department of Engineering, University of Sannio, Piazza Roma 21, 82100 Benevento, Italy
e-mail: alsimone@unisannio.it; ldisarno@unisannio.it; mgdurante@unisannio.it; stefsica@unisannio.it

S. Bhattacharya • M.S. Dietz • L. Dihoru • C.A. Taylor
Department of Civil Engineering, University of Bristol, Queen's Building, University Walk, Bristol BS8 1TR, UK
e-mail: s.bhattacharya@bristol.ac.uk; m.dietz@bristol.ac.uk; Luiza.Dihoru@bristol.ac.uk; colin.taylor@bristol.ac.uk

R. Cairo • A. Chidichimo • G. Dente
Department of Civil Engineering, University of Calabria, Ponte Pietro Bucci cubo 44b, 87036 Rende, CS, Italy
e-mail: rcairo@unical.it; achidichi@gmail.com; dente@dds.unical.it

A. Modaressi • L.A. Todo Bom
Laboratoire MSS-Mat CNRS UMR 8579, Ecole Centrale Paris, 92290 Chatenay-Malabry, France
e-mail: arezou.modaressi@ecp.fr; luis-andre.todo-bom@student.ecp.fr

A.M. Kaynia
Computational Geomechanics Section, Norwegian Geotechnical Institute (NGI), Sognsveien 72, N-0855 Oslo, Norway
e-mail: amir.m.kaynia@ngi.no

G. Anoyatis
Department of Civil Engineering, Geotechnical Engineering Laboratory, University of Patras, 26504 Patras, Greece
e-mail: anoyatis@upatras.gr

A. Ilki and M.N. Fardis (eds.), *Seismic Evaluation and Rehabilitation of Structures*, Geotechnical, Geological and Earthquake Engineering 26,
DOI 10.1007/978-3-319-00458-7_26,

gauges; therefore, earthquake response, including bending strains along their length, could be measured directly. Certain broad conclusions on kinematic and inertial SSI effects on this type of systems are drawn.

26.1 Background

Pile foundations are typically used when the soil under a structure cannot carry the vertical loads imposed by the superstructure. Piles often penetrate shallow loose deposits with their tip resting on competent bearing soils. Post-earthquake reconnaissance works (Mexico City 1985, Kobe 1995, Bhuj 2001 earthquakes) have shown that a large number of pile-supported buildings in layered soils suffered significant settlements and rotations; in certain cases pile damage was observed close to interfaces separating soil layers with significantly different shear moduli (i.e. Mizuno 1987; Bhattacharya and Madabhushi 2008). It is widely acknowledged that piles are affected by both the oscillations of the superstructure, inducing so-called "inertial" loads, and the movement of the surrounding soil, inducing so-called "kinematic" loads (Fig. 26.1). Recent building codes, such as Eurocode 8 (2004), include pile design provisions that account for the combined effect of both mechanisms.

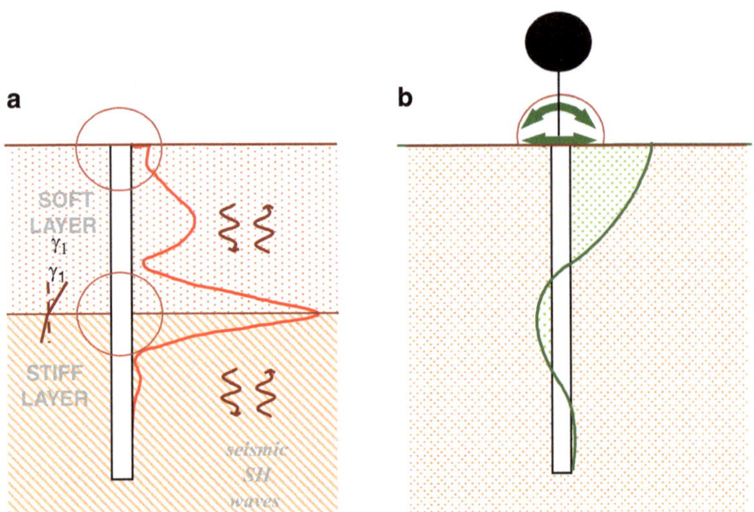

Fig. 26.1 Kinematic (**a**) and inertial (**b**) bending moments

G. Mylonakis
Department of Civil Engineering, Geotechnical Engineering Laboratory, University of Patras, 26504 Patras, Greece

Department of Civil Engineering, University of Bristol, Queen's Building, University Walk, Bristol BS8 1TR, UK
e-mail: mylo@upatras.gr; g.mylonakis@bristol.ac.uk

Assessment of the dynamic behavior of pile foundations under earthquake loading is decisive for evaluating the seismic performance of flexibly-supported structures, especially when the foundation rests on soft soil. The effects of soil-structure interaction (SSI) on foundations consisting of embedded piles include: (i) the variation between the free-field ground motion and the motion at the base of the superstructure, (ii) kinematic bending, axial and shear stresses applied along the piles – even in the absence of a superstructure. The kinematic bending moments may be significant especially for piles embedded in soft soils with high stiffness contrasts between consecutive layers, as depicted in Fig. 26.1a.

The scope of the present work is to examine the complex soil-pile-structure interaction problem by interpreting the results of a series of high-quality shaking table tests on model piles. The experimental program was performed at the Bristol Laboratory for Advanced Dynamics Engineering (BLADE), within the framework of the Seismic Engineering Research Infrastructures for European Synergies (SERIES). Tests were carried out on different pile group configurations, with and without pile caps, superstructures, subjected to both lateral and vertical earthquake shaking. The loading conditions include different input motions such as white noise, sine dwells and recorded earthquakes. The tests aimed at investigating various aspects of seismic Soil-Pile-Structure-Interaction (SPSI) such as the natural frequency of the systems (in both horizontal and vertical direction), the natural frequency and damping of embedded piles, the horizontal and vertical soil-pile kinematic interaction and foundation-structure interaction.

26.2 Rationale Behind the Model Tests

Ad-hoc laboratory investigations are essential for studying complex soil-structure interaction, as actual field data are rare, expensive to conduct and hard to interpret. While strain gauge data from instrumented piles under buildings of different vibrational characteristics subjected to actual earthquake motions would be ideal, such data are rare due to high cost and the unpredictable nature of earthquake occurrence. Therefore, well-controlled laboratory investigation on model piles alongside with analytical and numerical simulations are pivotal for understanding the seismic response of both single piles and pile groups (Muir Wood et al. 2002).

Kinematic and inertial bending moments constitute complementary aspects of the same phenomenon known as Soil-Pile-Structure Interaction (SPSI). Numerous technical reviews have been published, e.g. Novak 1991; Pender 1993; Gazetas and Mylonakis 1998, among others. As the kinematic and inertial interactions are not synchronous, a pertinent summation rule of the corresponding effects should be defined.

In seismic design practice it is common to neglect the influence of soil-pile interaction on the motion applied to the superstructure (i.e. the earthquake excitation is assumed to be equal to the free-field ground motion), except for important structures (NEHRP 2003; Eurocode 8 2004; Norme Tecniche per le Costruzioni 2008; Greek National Code, EAK 2000). It is worth discussing the relevant clauses from Eurocode 8 (EN 1998–5, 2004). The code prescribes that kinematic effects should be taken into account when all the following conditions simultaneously exist:

1. Seismicity of the area is moderate to high; it is being characterized by a normalized Peak Ground Acceleration $a_gS > 0.1g$, where a_g is the design ground acceleration on type A subsoil (essentially rock) and S is the soil factor
2. Subsoil type is D or worse, characterized by sharply different shear moduli between consecutive layers
3. The importance of the superstructure is of III or IV class.

The above factors have been echoed in the recent Italian building code (Norme Tecniche per le Costruzioni 2008).

Whereas the codes prescribe to design against kinematic action, they do not specify the methods for assessing the design loads. Therefore, one of the challenges faced by the engineers lies on the prediction of the maximum internal forces along piles, especially in the vicinity of interfaces having sharp stiffness contrasts.

Kinematic pile-soil interaction has been studied by several researchers, see for example Margason and Halloway (1977), Kuhlemeyer (1979), Krishnan et al. (1983), Gazetas et al. (1992), Kaynia and Mahzooni (1996), Wu and Finn (1997), Zhang et al. (2000), Padron et al. (2008), Chau et al. (2008), Mylonakis et al. (1997), Mylonakis (2001), Nikolaou et al. (2001), Saitoh (2005), Cairo and Dente (2007), Sica et al. (2007, 2011), Di Laora et al. (2012, 2013) and others; further, several systematic experimental investigations have been carried out (e.g. Meymand 1998; Moccia 2009; Dihoru et al. 2009, 2010a, b, among others).

Main issues to investigate include:

1. Under what conditions (i.e. soil stiffness and geometry, intensity and frequency of shaking) kinematic interactions may govern the pile design. Conversely, under which conditions could these effects be neglected
2. Effect of superstructure response on pile bending moments
3. Method of analysis of kinematic interaction
4. Importance of kinematic interaction in design and construction cost.

26.3 Scaling Laws and Modelling Issues

Derivation of the correct scaling laws constitutes a necessary first step in an experimental study. The similitude relationships are essential for interpreting the experimental data and also for scaling up the results to real prototypes. There are two ways to scale up the model test results as shown in Fig. 26.2. The first is to use standard tables (look-up approach) for scaling the model observations by pertinent factors to predict the prototype response. The alternative is to study the underlying mechanics/physics of the problem based on the model tests recognizing that not all the interaction mechanisms can be scaled accurately in a particular test. Once the mechanics/physics of the problem are identified in terms of pertinent dimensionless

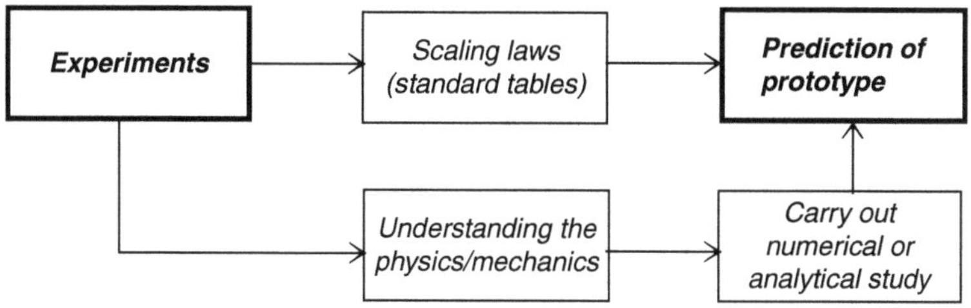

Fig. 26.2 Usefulness of scaling law

ratios, the prototype response can be predicted through analytical and/or numerical modeling in which the physics/mechanics discovered are implemented in a suitable way (problem-specific scaling relationships).

The second method is particularly useful for studying a complex dynamic foundation-structure interaction configuration as none of the physical modeling techniques can simultaneously satisfy all the interactions to the appropriate scale. It is recognized that not all physical mechanisms can be modeled adequately and, therefore, special considerations are needed when interpreting the test results.

Every physical process may be expressed in terms of dimensionless groups and the fundamental aspects of the governing physical processes must be preserved in the design of model tests. Derivations of the scaling laws for these experiments are beyond the scope of the paper and will be discussed elsewhere. However, in this work the following physical mechanisms and parameters are considered important: (i) ratio of soil stiffness in the two layers, (ii) characteristics of the Single-Degree-of-Freedom (SDOF) structure supported by the pile, (iii) frequency and intensity scaling of the earthquake record, (iv) ratio of strains and accelerations in the soil and the pile.

26.4 Experimental Procedure

26.4.1 Method and Materials

The earthquake response of pile groups was explored by means of carefully executed 1-g shaking table tests. Such laboratory tests were aimed at assessing the effects of both kinematic and inertial effects on the piles. The test campaign consisted of two series of tests: preliminary tests, carried out in November 2010, and a more comprehensive series of tests, including earthquake loading, carried out in June 2011. The 6-degree-of-freedom earthquake simulator of BLADE and the equivalent shear beam (ESB) laminar container were utilized to this end. The ESB is shown in Fig. 26.3: it consists of eight rectangular aluminium rings, which are stacked alternately with rubber sections to create a hollow yet flexible box of inner dimensions *1.190* m long by *0.550* m wide and *0.814* m deep (Crewe et al. 1995). The rings are made of aluminium box section to minimize inertia while providing sufficient constraint for

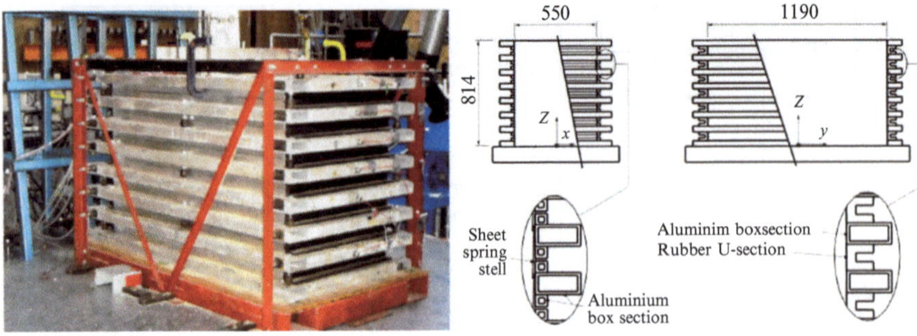

Fig. 26.3 Equivalent shear beam container (shear stack)

the K_0 condition. The stack is secured to the shaking table by its base and shaken horizontally lengthways (y direction). Its floor is roughened by sand-grain adhesion to improve the shear wave transfer; the internal end walls are similarly treated to enable complementary shear stresses. Internal side walls are lubricated with silicon grease and covered with latex membrane to ensure plane strain conditions.

This type of containers should be ideally designed to match the shear stiffness of the inner soil. However, the shear stiffness of the soil varies during shaking depending on strain level. Therefore, the matching between the end wall and the soil stiffness would be possible only at a particular strain level. The ESB of BLADE is designed considering a value of strain in the soil close to failure conditions. It is thus more flexible than the soil deposit at lower strain amplitudes and, consequently, the soil will always dictate the overall behavior of the container (Bhattacharya et al. 2011). Indeed, the resonant frequency and damping of the empty container in the first shear mode in the long direction were measured prior to testing as 5.7 Hz and 27 %, respectively, which are sufficiently different from the soil material properties.

26.4.2 Model Geometry

The sample test model consists of five piles embedded in a bi-layer soil (Fig. 26.4). The piles consist of an alloy aluminium tube (commercial model 6063-T6) with thickness $t = 0.71$ mm, outer diameter $D = 22.23$ mm and length $L = 750$ mm. The main properties of the aluminium tube are: unit weight $\gamma = 27$ kN/m^3, Young's modulus $E_p = 70$ GPa, Poisson's ratio $\nu = 0.3$, leading to bending stiffness $E_pI_p = 1.95 \times 10^8$ Nmm^2. Pile 3, 4 and 5 are closer to each other with a relative spacing $s = 70$ mm ($s/D \approx 3$); pile 1 and 2 are placed at a distance of 140 mm.

To model the superstructure, five different oscillators were used. Two types of columns (aluminium, steel) were selected in order to vary pier stiffness. All columns had a rectangular cross section, 3 × 12 mm. The aluminium and steel piers have a flexural stiffness, EI, of 1.89×10^{-3} kNm^2 and 5.67×10^{-3} kNm^2, respectively.

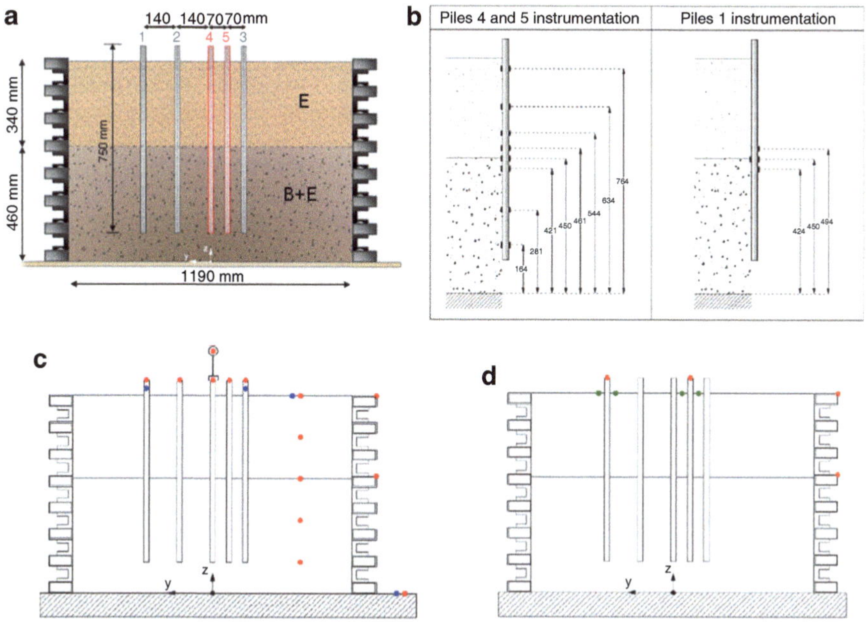

Fig. 26.4 Model setup: (**a**) subsoil configuration, (**b**) strain gauge, (**c**) accelerometer and (**d**) LVDT location

Accelerometers were used to monitor the accelerations of the shaking table, the shear stack, the soil along a vertical array, the pile heads and the superstructure. The LVDT transducers were employed to monitor the displacements of the pile in the horizontal and vertical direction. To evaluate the bending response along the piles, 8 strain gauge pairs have been attached on the shafts of pile 4 and 5; additionally, 4 strain gauges are placed on the shaft of pile 1 close to the layer interface. Overall, 63 data channels were employed.

26.4.3 Soil Material Properties

A two layer soil profile was deposited by pluviation. The top layer is made of Leighton Buzzard sand (LB) fraction E, deposited through a 40 mm diameter nozzle to achieve a mass density of 1.39 Mg/m^3. The bottom layer is a mix between LB fractions B and E (85 and 15 %, respectively) pluviated through a 12 mm diameter nozzle to achieve a mass density of 1.78 Mg/m^3. The free surface of the soil deposit is 800 mm above the base of the shear stack. The LB sand here adopted has been extensively used in the experimental research activity carried out at the BLADE. Numerous density and stiffness data can be found in previous

Table 26.1 LB sand index properties

Materials	Type	G_s [Mg/m^3]	e_{min}	e_{max}	D_{10}	D_{50}	References
LB fraction E	Sand BS 881-131	2.647	0.613	1.014	0.095	0.14	Tan (1990)
LB fraction B	Sand BS 881-132	2.647	0.486	0.78		0.82	Ling and Dietz (2004)
LB fraction E + B		2.647	0.289	0.614			Moccia (2009)

Table 26.2 Soil layer properties

Soil layers	Thickness H (mm)	Void ratio e	Relative density Dr (%)	Dry unit weight γ_d (kN/m^3)	Shear wave velocity Vs (m/s) Nov 2010	June 2011	Vs_2/Vs_1 Nov 2010	June 2011
Top LB(E)	340	0.9	28	13.63	51	54	1.59	1.57
Bottom LB (E + B)	460	0.48	41	17.46	81	85		

experimental studies (Stroud 1971; Cavallaro et al. 2001; Lings and Dietz 2004; Moccia 2009).

Table 26.1 outlines the sand index properties obtained from different previous studies, while in Table 26.2 the properties of the two soil layers of the present sample test models are listed.

The shear wave velocity, V_s, was derived from the white noise tests carried out before sine dwell and earthquake tests, for each stage of the experimental activity. As can be observed from Table 26.2, the shear wave velocity contrasts between the bottom and top layer are quite close for the two stages of tests (around 1.6), corresponding to a 2.5 times stiffer bottom layer.

26.4.4 Pile Model Configurations

Seven different model configurations were tested starting from the model setup of Fig. 26.4. The characteristics of all configurations are shown in Fig. 26.5. In configuration 1 (FHP), all pile heads are free to rotate and there is no oscillator attached. Configurations 2, 3 and 4 are characterized by free-head piles and one oscillator placed on Pile 1, 4 and 5, respectively. Configuration 5 (SC) has a small

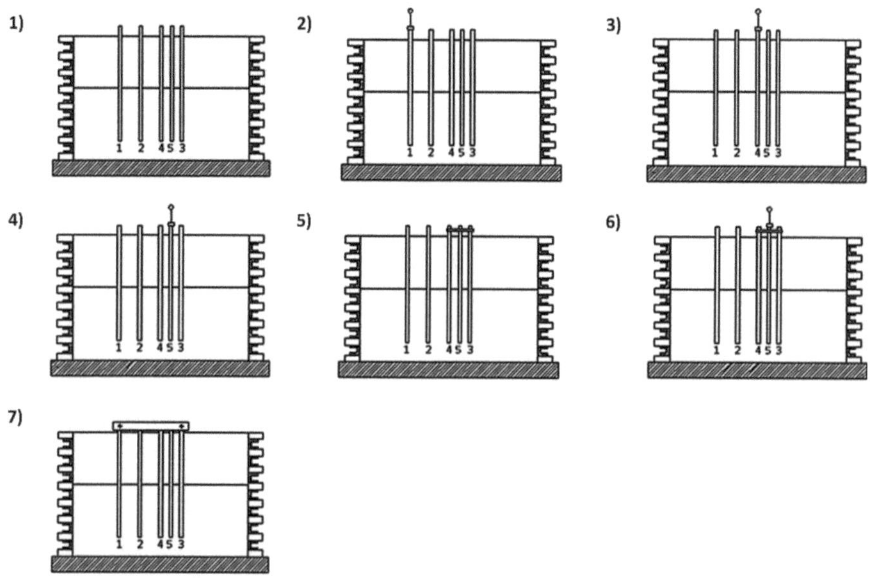

Fig. 26.5 Details of model configurations (*1*) Free head pile, (*2*) SDOF on pile 1, (*3*) SDOF on pile 4, (*4*) SDOF on pile 5, (*5*) Short cap connecting piles 3, 4, 5, (*6*) Short cap with SDOF on pile 5, (*7*) Long cap from pile 1 to pile 3

cap connecting piles 4, 5 and 3, without any oscillator. Configuration 6 (SC-O) is characterized by the small cap as before with an oscillator mounted. Finally, configuration 7 (LC) has a large cap (connecting all the piles) without an oscillator. Some pictures of model details and installation procedures are shown in Fig. 26.6.

26.4.5 Experimental Procedure

The main testing procedure was comprised of the following three input motions:

(a) *White noise excitation*: random noise signal of bandwidth 1–100 Hz and peak ground acceleration varying between 0.01g and 0.10g.
(b) *Harmonic excitation*: sine dwell acceleration time-histories with amplitudes varying between 0.01 ÷ 0.18g and frequencies varying from 5 to 45 Hz.
(c) *Earthquake excitation:* three earthquake records from the Italian database were selected: Tolmezzo (Friuli 1976), Sturno (Irpinia 1980) and Norcia (Umbria-Marche 1997). The earthquake motions were modified by a frequency scaling factor of 5 or 12; acceleration amplitude was varied between 0.043g and 0.577g.

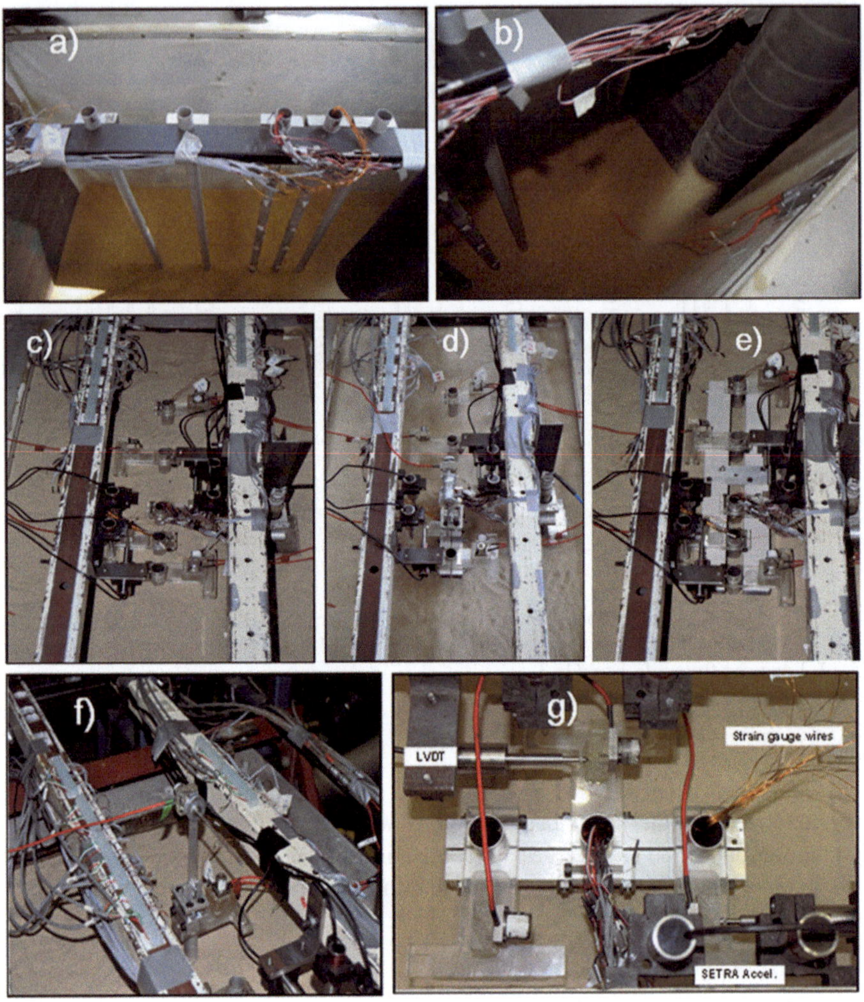

Fig. 26.6 Details of experimental setup: (**a**) and (**b**) soil deposition; (**c**) free-head piles; (**d**) short cap + SDOF; (**e**) long cap; (**f**) free-head piles + SDOF; (**g**) details of instrumentation on short cap configuration

26.5 Test Results

The results of typical tests of the comprehensive laboratory experimental program carried out on the BLADE shaking table are discussed hereafter, with the aim of investigating some particular aspects of seismic SPSI.

26.5.1 Effect of PGA on Free-Field Response and Kinematic Bending Moments

The results reported herein refer to tests performed during the first stage of the experimental research. The test configuration is the FHP one. Results for sine dwell input motion with three increasing levels of peak ground acceleration (*PGA*) applied on the shaking table, namely $PGA = 0.008$g, 0.027g and 0.069g and frequency of 30 Hz, are here discussed. In Fig. 26.7 the acceleration time-histories of the soil free-field response are shown. Naturally, free-field response increases with input acceleration level. The recorded accelerations are shown in the diagrams in the lower part of Fig. 26.7, where the peak acceleration profile in the soil (quoted as a_{max}) is plotted.

The diagrams of Fig. 26.7 are also included in Fig. 26.8 to compare these data with the counterparts derived from the tests characterized by $PGA = 0.013$g and 0.041g. The overall comparison in Fig. 26.8 shows that the profiles of a_{max} exhibit quite similar shapes. As regards the amount of amplification, it is believed that the effect of loading history on soil properties may play a key role and should be accounted for in a realistic manner in numerical simulations.

With reference to the same tests, pile response is investigated by means of time-histories of strains, measured by the strain gauges along pile shaft, and derived bending moments. Figures 26.9a and 26.10a show strain time-histories along pile 4 for sine dwells having a PGA at the shaking table of 0.027g and 0.069g, respectively.

The pile bending moment *M* and axial force *N* may be obtained from strains by as follows:

$$M = \frac{\varepsilon_1 - \varepsilon_2}{2} \cdot E_p I_p \cdot \left(\frac{2}{D}\right) \tag{26.1}$$

$$N = \frac{\varepsilon_1 + \varepsilon_2}{2} \cdot E_p A_p \tag{26.2}$$

where ε_1 and ε_2 are outer fiber strains in the opposite sides of the pile, *D* is pile diameter and A_p is the area of the (hollow) pile cross-section.

Computed bending moments for the two input motion (PGA = 0.027g and PGA = 0.069g) are presented in Figs. 26.9b and 26.10b respectively. A significant amount of kinematic interaction is observed close to the layer interface (z = 440 mm), where high values of strain amplitude are detected.

These two sets of experimental data have been selected as they represent typical responses of the sample pile group. In the first case (Fig. 26.9) the strains appear almost symmetric with respect to the horizontal axis and in opposite phase, which suggests that the pile is mainly subjected to bending under zero axial force.

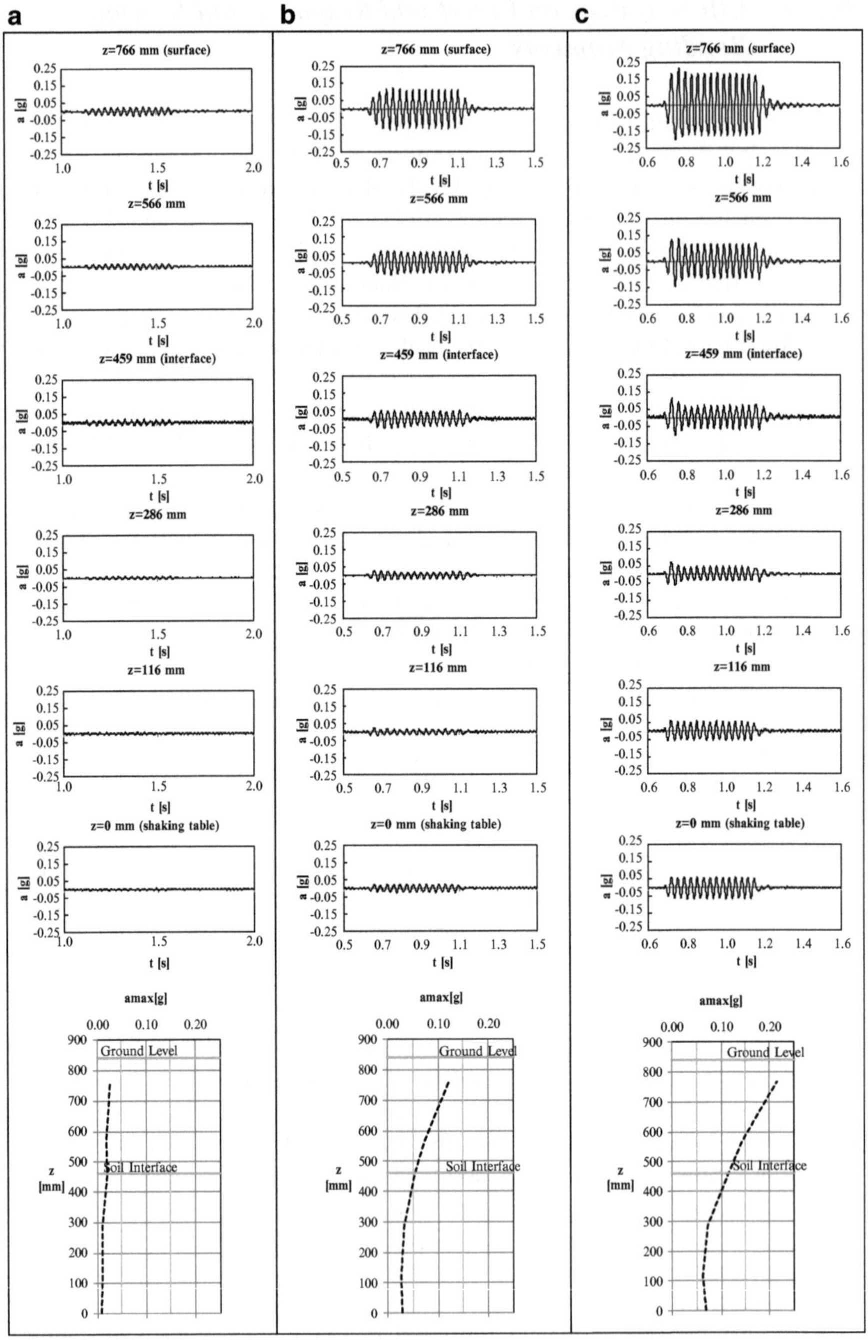

Fig. 26.7 Free-field accelerations (for the FHP configuration) for different amplitudes of the input acceleration at the shaking table: (**a**) PGA = 0.008g; (**b**) PGA = 0.027g; (**c**) PGA = 0.069g

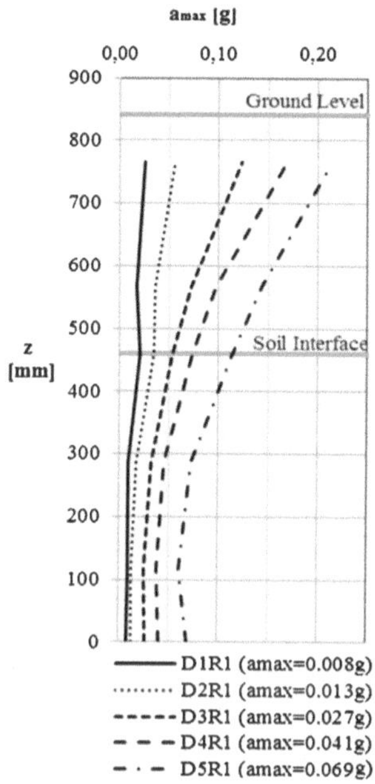

Fig. 26.8 Peak acceleration vs. z for different amplitudes of input acceleration

Additionally, when the excitation stops, the bending strains return to zero which indicates elasticity in both soil and pile.

In the second case (Fig. 26.10), strain time histories on the opposite sides show an offset, indicating a residual deformed configuration of the section, with a residual bending moment. This moment is slightly lower than the maximum moment induced during excitation.

The comparison of the different pile response is further presented in Fig. 26.11, where the time-history of normal stress is also plotted for both cases. It is evident that in the second case (PGA = 0.069g) the pile section at z = 440 mm experiences a residual normal stress. The strain-gauge time histories are, indeed, not symmetric with respect to the horizontal axis.

By taking the absolute maxima from the time-histories of bending moments, the envelopes of moment profiles with depth were plotted for piles 4 and 5, for the FHP configuration and the five tests with input PGA increasing from 0.008g to 0.069g (Fig. 26.12). The bending moments increase with input motion level, with peaks located at the layer interface; such bending moments are characteristic of kinematic interactions.

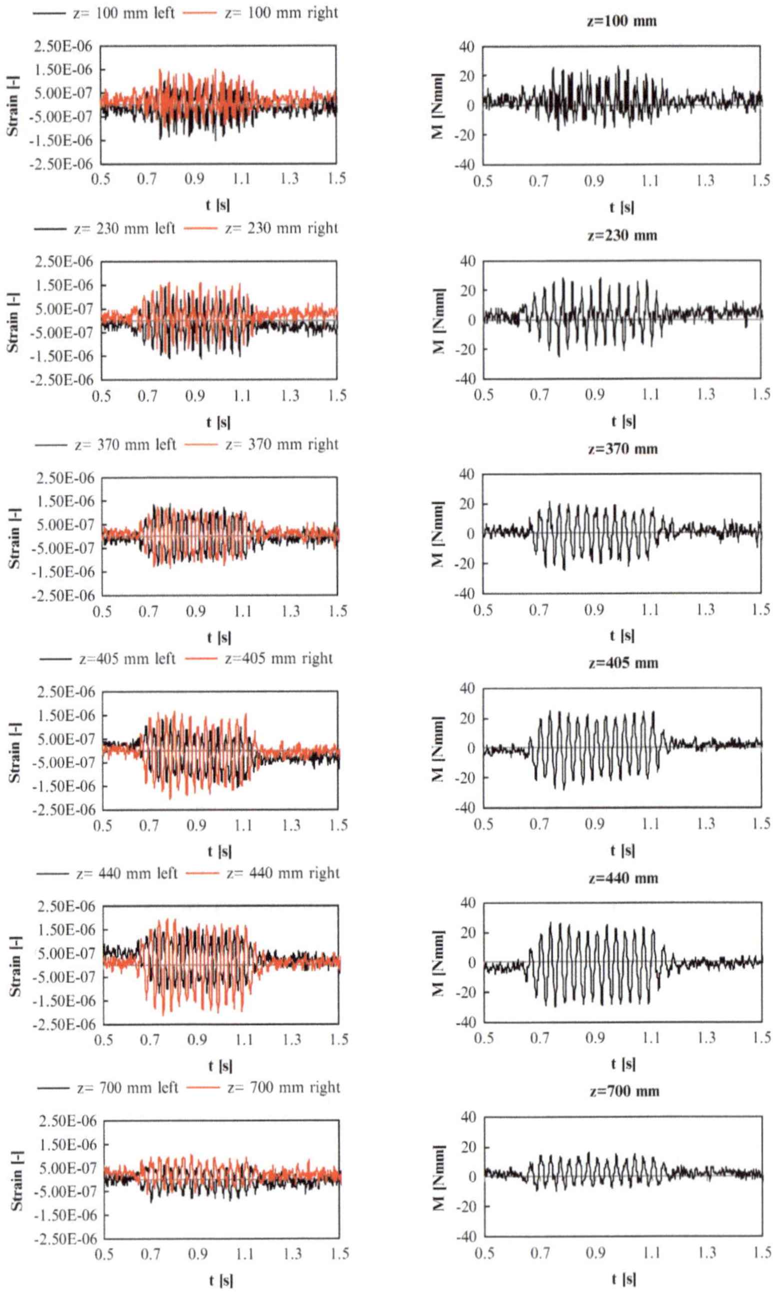

Fig. 26.9 Time histories of strains (**a**) and bending moments (**b**) at different elevation along pile 4 for the FHP configuration (PGA = 0.027g) (test n. 101115_D3R1)

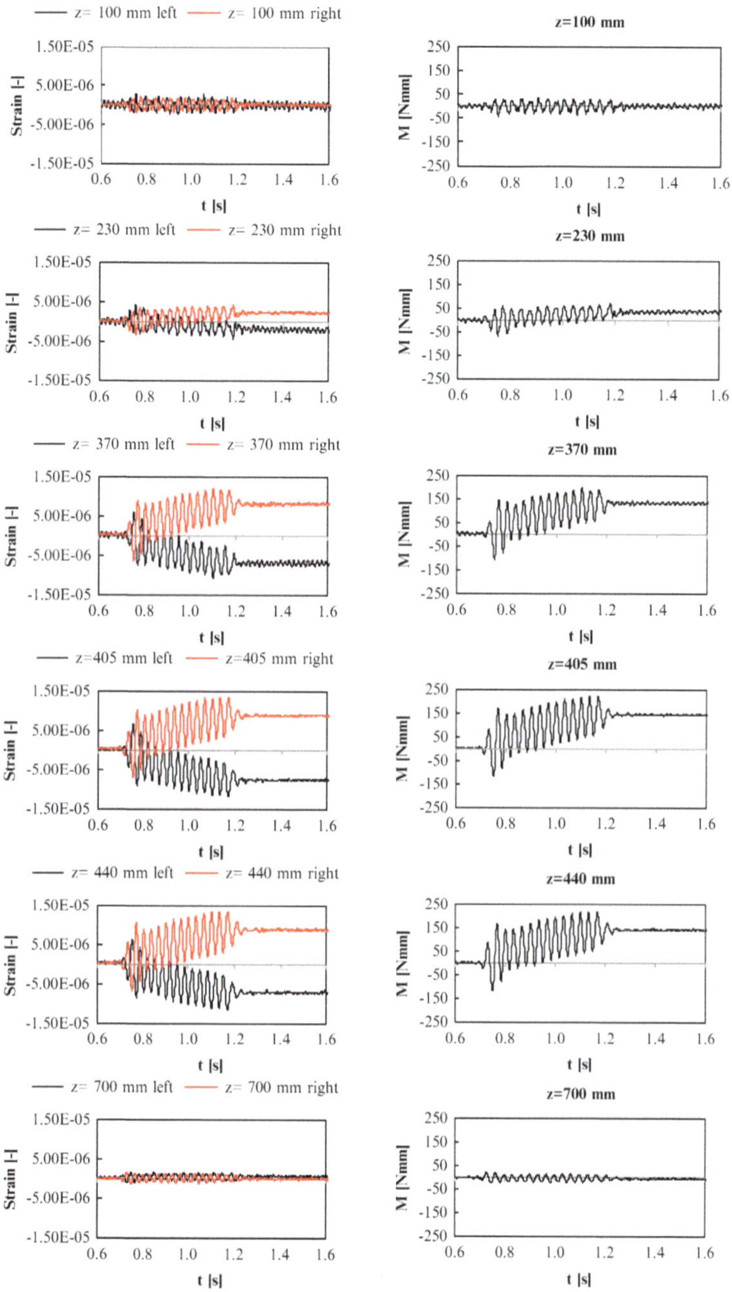

Fig. 26.10 Time histories of strains (**a**) and bending moments (**b**) at different elevation along pile 4 for the FHP configuration (PGA = 0.069g) (test n. 101115_D5R1)

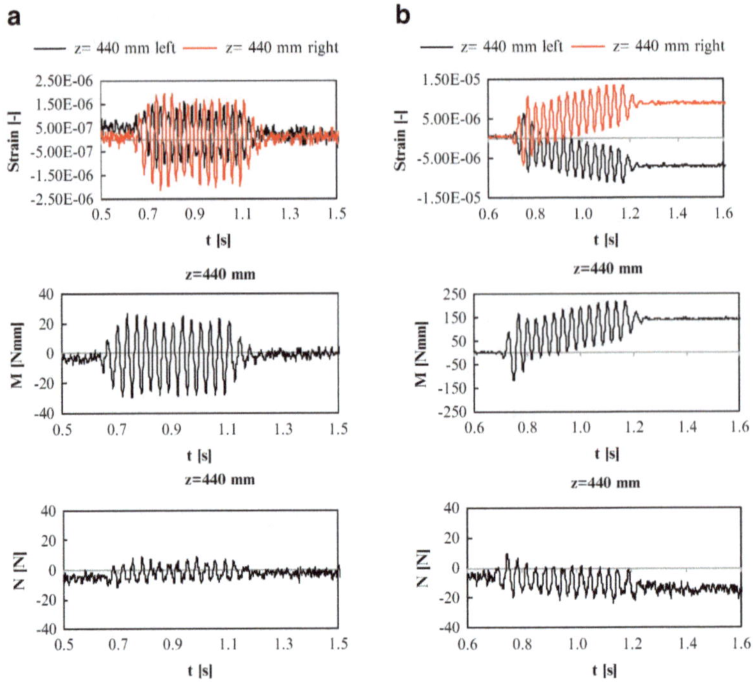

Fig. 26.11 Strains and bending moment at a given pile elevation (z = 440 mm). Results refer to the test with PGA = 0.027g (**a**) and PGA = 0.069g (**b**) at the shaking table

26.5.2 Configuration Effects

Time histories of free-field soil accelerations measured in three different configurations, namely free-head piles (FHP, test n. 101115_D16R1), free-head piles with an oscillator on pile 5 (FHP + SDOF, test n. 101125_D2R1) and short cap with an oscillator (SC + SDOF, test n. 101118_D2R1) have been considered. In all cases, a sine dwell with PGA around 0.125g and frequency of 7.5 Hz is applied on the shaking table. The free-field response measured by means of the accelerometer vertical array located quite far from piles does not significantly vary with the configuration layout. However, a minor difference in the free-field motion amplification can be detected from Fig. 26.13, where the peak acceleration, a_{max}, is plotted versus depth. This effect may be due to the progressive change in soil properties with loading history.

By taking the absolute maxima from the time histories of bending moments, the envelopes of moments versus depth for pile 4 and 5 in different test configurations were obtained (Fig. 26.14).

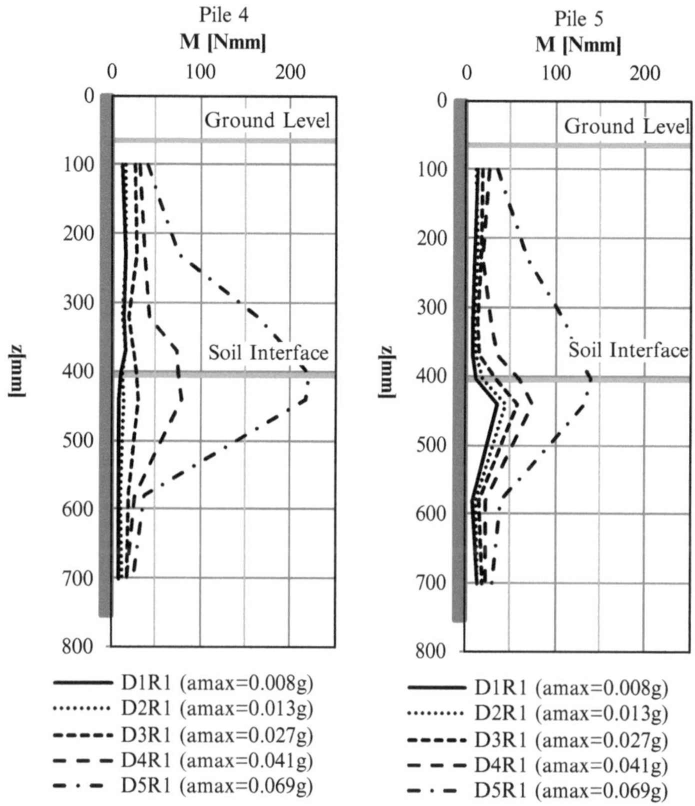

Fig. 26.12 Envelope of bending moments vs. depth for pile 4 and pile 5 in different model configurations

In the FHP configuration both pile 4 and pile 5 exhibit almost null moments near the top and bottom; bending moments induced along the piles are quite low.

For the FHP + SDOF configuration, the presence of the oscillator on pile 5 significantly increases the bending moment on the pile head, while the moment of (free-head) pile 4 at the same elevation remains zero. For this configuration slightly higher bending effects are induced at the interface, mainly due to kinematic interaction, the moments being almost equal for both the loaded and the unloaded pile. It is worthwhile noting that for the SC + SDOF configuration, the presence of the device (small cap) connecting the three piles (4, 5 and 3) produces, as expected, a redistribution at the pile heads of the bending actions induced by the superstructure.

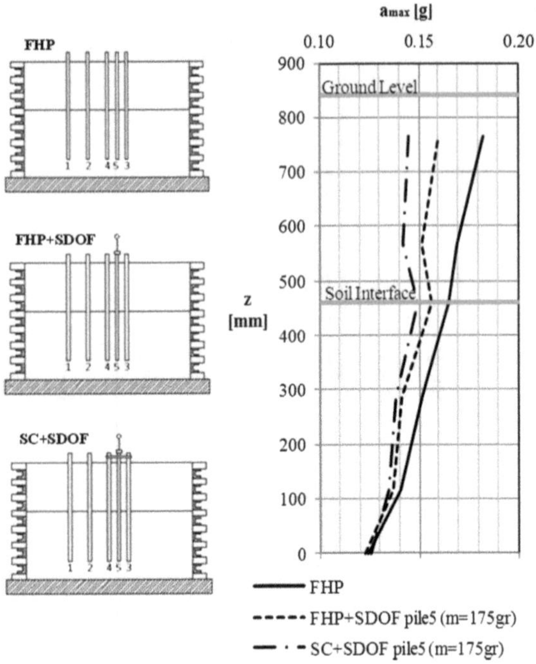

Fig. 26.13 Peak acceleration vs. z for different model configurations

26.6 Concluding Remarks

The main conclusions obtained from this study are:

1. Significant kinematic interactions were observed in the tests
2. The maximum kinematic bending moment generally occurs at the proximity of the soil layer interface
3. Complex seismic pile-soil interaction can be modeled in a shaking table for better understanding of their behavior leading to efficient design; the stiffness contrast of granular materials can be replicated by innovative ways, such as mixing Fraction B and E sand to obtain greater packing density
4. In presence of the superstructure (SDOF) it has been observed that the pile continues to deform even after the end of the input motion, due to the free vibration of the SDOF (inertial interaction)
5. Residual pile bending moments were observed due to inelastic action in the soil; the magnitude of these moments was comparable to the maximum one recorded during shaking.

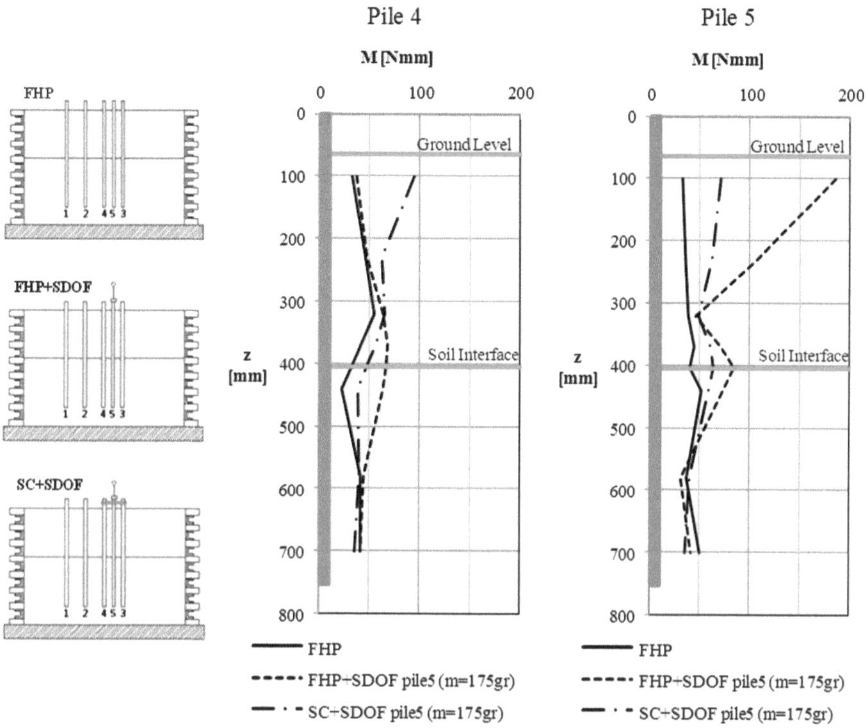

Fig. 26.14 Envelope of bending moments vs. depth for pile 4 and pile 5 in different model configuration

Acknowledgments The research leading to these results has received funding from the European Union Seventh Framework Programme (FP7/2007–2013) for access to the Bristol Laboratory for Advanced Dynamics Engineering (BLADE), University of Bristol, UK under grant agreement n° 227887 [SERIES]. The financial support provided by the project ReLUIS (Task MT2) funded by the Italian Civil Protection is also appreciated. Further, George Anoyatis was partially supported by the University of Patras through a Caratheodory Grant (#C.580).

References

Bhattacharya S, Madabhushi SPG (2008) A critical review of the methods for pile design in seismically liquefiable soils. Bull Earthq Eng (Springer) 6:407–446

Bhattacharya S, Lombardi D, Dihoru L, Dietz M, Crewe AJ, Taylor CA (2011) Container design for soil-structure interaction studies. In: Fardis MN, Rakicevic ZT (eds) Role of seismic testing facilities in performance-based earthquake engineering. Springer, Dordrecht/New York. ISBN 978-94-007-1976-7

Cairo R, Dente G (2007) Kinematic interaction analysis of piles in layered soils. In: 14th European conference on soil mechanics and geotechnical engineering, ISSMGE-ERTC 12 Workshop Geotechnical Aspects of EC8, Madrid (Spain), Pàtron Editore, Bologna, cd-rom, paper n. 13

Cavallaro A, Maugeri M, Mazzarella R (2001) Static and dynamic properties of Leighton Buzzard sand from laboratory tests. In: 4th international conference on recent advances in geotechnical earthquake engineering and soil dynamics, San Diego, Paper 1.13

Chau KT, Shen CY, Guo X (2008) Non-linear seismic soil-pile-structure interaction: shaking table tests and FEM analyses. Soil Dyn Earthq Eng 29(2):300–310

Crewe AJ, Lings ML, Taylor CA, Yeung AK, Andrighetto R (1995) Development of a large flexible shear stack for testing dry sand and simple direct foundations on a shaking table. In: Elnashai AS (ed) European seismic design practice. Balkema, Rotterdam

Di Laora R, Mandolini A, Mylonakis G (2012) Insight on kinematic bending of flexible piles in layered soil. Soil Dyn Earthq Eng 43:309–322

Di Laora R, Mylonakis G, Mandolini A (2013) Pile-head kinematic bending in layered soil. Earthq Eng Struct Dyn 42:319–337

Dihoru L, Bhattacharya S, Taylor CA, Muir Wood D, Moccia F, Simonelli AL Mylonakis G (2009) Experimental modeling of kinematic bending moments of piles in layered soils. IS-Tokyo

Dihoru L, Bhattacharya S, Taylor CA, Muir Wood D. Moccia F, Simonelli AL Mylonakis G (2010a) Physical and numerical modelling of kinematic pile-soil interaction under seismic conditions. In: Physical modelling conference, Zurich

Dihoru L, Bhattacharya S, Taylor CA, Muir Wood D, Moccia F, Simonelli AL Mylonakis G (2010b) Stiffness design for granular materials – theoretical and experimental considerations. In: Physical modelling conference, Zurich

EAK 2000 (2003) Greek Seismic Code. Earthquake Planning and Protection Organization, Athens

Eurocode 8 (2004) Design of structures for earthquake resistance, Part 5: foundations, retaining structures and geotechnical aspects. CEN EC for Standardization, Brussels

Gazetas G, Fan K, Tazoh T, Shimizu K, Kavvadas M, Makris N (1992) Seismic pile-group-structure interaction. Piles under dynamic loads, Geotechnical Special Publication No 34, ASCE, pp 56–93

Gazetas G, Mylonakis G (1998) Seismic soil-structure interaction: new evidence and emerging issues. In: Dakoulas P, Yegian MK, Holtz RD (eds) Geotechnical earthquake engineering and soil dynamics III, Geo-Institute ASCE conference, Seattle, vol II, pp 1119–1174

Kaynia AM, Mahzooni S (1996) Forces in pile foundations under seismic loading. J Eng Mech ASCE 122:46–53

Krishnan R, Gazetas G, Velez A (1983) Static and dynamic lateral deflection of piles in non-homogeneous soil stratum. Geotechnique 33:307–326

Kuhlemeyer RL (1979) Static and dynamic laterally loaded floating piles. J Geotech Eng ASCE 105:289–304

Norme Tecniche per le Costruzioni (2008) Italian Building Code, DM 14 Jan, GU n 29, 4 Feb, n. 30

Lings ML, Dietz MS (2004) An improved direct shear apparatus for sand. Geotechnique 54 (4):245–256

Margason E, Halloway DM (1977) Pile bending during earthquakes. In: Proceedings VI world conference on earthquake engineering, Meerut, pp 1690–1696

Meymand PJ (1998) Shaking table scale model tests of nonlinear soil-pile-superstructure interaction in soft clay. PhD thesis, University of California, Berkeley

Mizuno H (1987) Pile damage during earthquakes in Japan (1923–1983). In: Proceedings of the session on dynamic response of pile damage. Nogami T (ed) Special publication of ASCE, Atlantic City, 27 April, pp 53–77

Moccia F (2009) Seismic soil pile interaction: experimental evidence. PhD dissertation, Università degli Studi di Napoli Federico II, Napoli

Muir Wood D, Crewe A, Taylor CA (2002) Shaking table testing of geotechnical models. UPMG-Int J Phys Model Geotech 2:1–13

Mylonakis G (2001) Simplified model for seismic pile bending at soil layer interfaces. Soils Found 41:47–58

Mylonakis G, Nikolaou A, Gazetas G (1997) Soil-pile-bridge seismic interaction: kinematic and inertial effects. Part I: soft soil. Earthq Eng Struct Dyn 26:337–359

NEHRP (2003) Recommended provisions for seismic regulations for new buildings and other structures (FEMA 450). Building Seismic Safety Council, Washington, DC

Nikolaou S, Mylonakis G, Gazetas G, Tazoh T (2001) Kinematic pile bending during earthquakes: analysis and field measurements. Géotechnique 51(5):425–440

Novak M (1991) Piles under dynamic loads: State of the art. In: Proceedings of 2nd international conference on recent advances in geotechnical earthquake engineering and soil dynamics, St. Louis, vol 3, pp 2433–2456

Padron LA, Aznarez JJ, Maero O (2008) Dynamic analysis of piled foundations in stratified soils by a BEM-FEM model. Soil Dyn Earthq Eng 5:333–346

Pender M (1993) Seismic pile foundation design analysis. Bull New Zeal Natl Soc Earthq Eng 26 (1):49–160

Saitoh M (2005) Fixed-head pile bending by kinematic interaction and criteria for its minimization at optimal pile radius. J Geotech Geoenviron Eng 131(10):1243–1251

Sica S, Mylonakis G, Simonelli AL (2007) Kinematic bending of piles: analysis vs code provisions. In: Proceedings of the 4th international conference on earthquake geotechnical engineering, Thessaloniki

Sica S, Mylonakis G, Simonelli AL (2011) Transient kinematic pile bending in two-layer soil. Soil Dyn Earthq Eng 31(7):891–905. doi:10.1016/j.soildyn.2011.02.001

Stroud MA (1971) The behaviour of sand at low stress levels in the simple shear apparatus. PhD thesis, University of Cambridge, UK

Tan FSC (1990) Centrifuge and theoretical modelling of conical footings on sand. PhD Dissertation, University of Cambrige, UK

Wu G, Finn W (1997) Dynamic elastic analysis of pile foundations using finite element method in the time domain. Can Geotech J 34:44–52

Zhang F, Kimura M, Nakai T, Hoshikawa T (2000) Mechanical behavior of pile foundations subjected to cyclic lateral loading up to the ultimate state. Soils Found 40(5):1–17

[illegible]

Chapter 27
Experimental Investigation of Dynamic Behavior of Cantilever Retaining Walls

Panos Kloukinas, Augusto Penna, Anna Scotto di Santolo, Subhamoy Bhattacharya, Matt S. Dietz, Luiza Dihoru, Aldo Evangelista, Armando L. Simonelli, Colin A. Taylor, and George Mylonakis

Abstract The dynamic behavior of cantilever retaining walls under earthquake action is explored by means of 1-g shaking table testing, carried out on scaled models at the Bristol Laboratory for Advanced Dynamics Engineering (BLADE), University of Bristol, UK. The experimental program encompasses different combinations of

P. Kloukinas (✉)
Department of Civil Engineering, Geotechnical Engineering Laboratory,
University of Patras, 26504 Patras, Greece
e-mail: pkloukin@upatras.gr

A. Penna • A.L. Simonelli
Department of Engineering, University of Sannio,
Piazza Roma 21, 82100 Benevento, Italy
e-mail: aupenna@unina.it; alsimone@unisannio.it

A. Scotto di Santolo
Universita Telematica Pegaso, Piazza Trieste e Trento 48, 80132 Naples, Italy
e-mail: anscotto@unina.it

S. Bhattacharya • M.S. Dietz • L. Dihoru • C.A. Taylor
Department of Civil Engineering, University of Bristol,
Queen's Building, University Walk, Bristol BS8 1TR, UK
e-mail: s.bhattacharya@bristol.ac.uk; m.dietz@bristol.ac.uk; Luiza.Dihoru@bristol.ac.uk; colin.taylor@bristol.ac.uk

A. Evangelista
Department of Civil and Environmental Engineering,
University of Naples Federico II, via Claudio 21, 80125 Naples, Italy
e-mail: evangeli@unina.it

G. Mylonakis
Department of Civil Engineering, Geotechnical Engineering Laboratory,
University of Patras, 26504 Patras, Greece

Department of Civil Engineering, University of Bristol,
Queen's Building, University Walk, Bristol BS8 1TR, UK
e-mail: mylo@upatras.gr; g.mylonakis@bristol.ac.uk

A. Ilki and M.N. Fardis (eds.), *Seismic Evaluation and Rehabilitation of Structures*, Geotechnical, Geological and Earthquake Engineering 26,
DOI 10.1007/978-3-319-00458-7_27,

retaining wall geometries, soil configurations and input ground motions. The response analysis of the systems at hand aimed at shedding light onto the salient features of the problem, such as: (1) the magnitude of the soil thrust and its point of application; (2) the relative sliding as opposed to rocking of the wall base and the corresponding failure mode; (3) the importance/interplay between soil stiffness, wall dimensions, and excitation characteristics, as affecting the above. The results of the experimental investigations were in good agreement with the theoretical models used for the analysis and are expected to be useful for the better understanding and the optimization of earthquake design of this particular type of retaining structure.

27.1 Introduction

Reinforced concrete cantilever walls represent a popular type of retaining system. It is widely considered as advantageous over conventional gravity walls as it combines economy and ease in construction and installation. The concept is deemed particularly rational, as it exploits the stabilizing action of the soil weight over the footing slab against both sliding and overturning, thus allowing construction of walls of considerable height.

The traditional approach for the analysis of cantilever walls is based on the well-known limit equilibrium analysis, in conjunction with a conceptual vertical surface AD (Fig. 27.1b, c) passing through the innermost point of the wall base (vertical virtual back approach). A contradictory issue in the literature relates to the calculation of active thrust acting on the virtual wall back, under a certain mobilized obliquity ranging naturally from 0 (a perfectly smooth plane) to φ (a perfectly rough plane). Efforts have been made by numerous investigators to establish the proper roughness for the analysis and design or this type of structures as reported by Evangelista et al. (2009, 2010) and Kloukinas and Mylonakis (2011). Nevertheless, the issue of seismic behavior remains little explored. In fact many modern Codes, including the Eurocodes (CEN 2004) and the Italian Building Code (NTC 2008), do not explicitly refer to cantilever walls. The current Greek Seismic Code (EAK 2003) addresses the case of cantilever retaining walls by adopting the virtual back approach in the context of a pseudo-static analysis under the assumption of gravitational infinite slope conditions (Rankine 1857) and various geometric constraints.

Recent theoretical findings obtained by means of stress limit analysis (Evangelista et al. 2010; Evangelista and Scotto di Santolo 2011; Kloukinas and Mylonakis 2011) for the seismic problem of Fig. 27.1a indicate that a uniform Rankine stress field can develop in the backfill, when the wall heel is sufficiently long and the stress characteristics do not intersect the stem of the wall ($\omega_\beta < \omega_{wall}$). Given that the inclination of the stress characteristics depends on acceleration level, a Rankine condition is valid for the vast majority of cantilever wall configurations under strong seismic action. This is applicable even to short heel walls, with an error of about 5 % (Huntington 1957; Greco 2001).

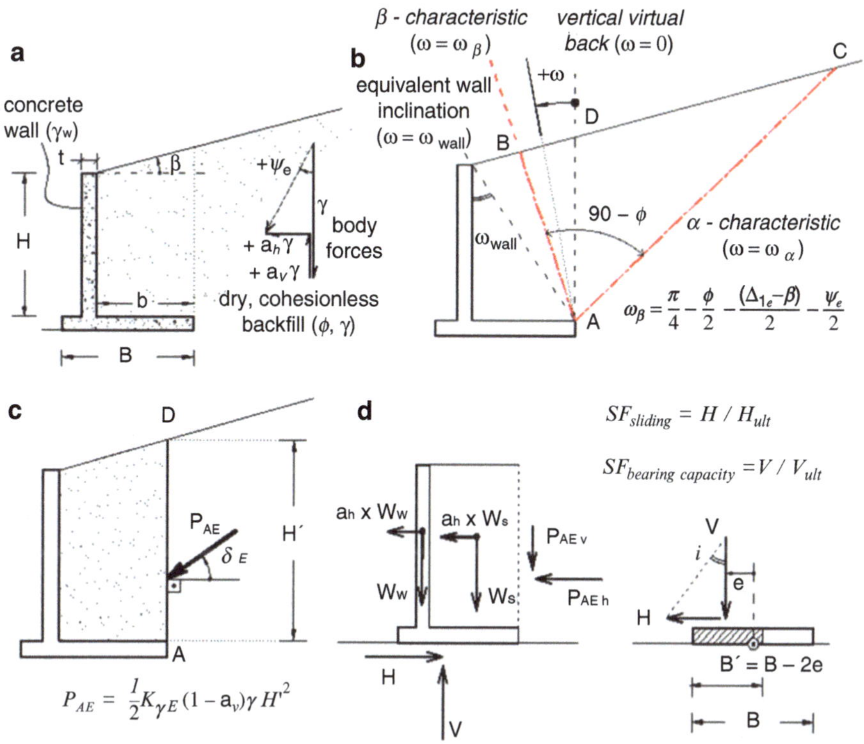

Fig. 27.1 Pseudo-dynamic analysis of cantilever retaining walls for seismic loading: (**a**) Problem under consideration, (**b**) seismic Rankine stress field characteristics, (**c**) Modified virtual back approach by stress limit analysis, (**d**) Stability analysis according to Meyerhof (CEN 2003)

Following the above stress limit analysis studies, closed-form expressions are derived for both the pseudodynamic earth pressure coefficient K_{AE} and the resultant thrust inclination, δ_E (Fig. 27.1c), given by (Kloukinas and Mylonakis 2011):

$$\delta_E = \tan^{-1}\left[\frac{\sin\varphi\ \sin(\Delta_{1e} - \beta + \psi_e)}{1 - \sin\varphi\ \cos(\Delta_{1e} - \beta + \psi_e)}\right] \tag{27.1}$$

where $\Delta_{1e} = \sin^{-1}[\sin(\beta + \psi_e)/\sin\varphi]$ and $\psi_e = \tan^{-1}[a_h/(1-a_v)]$ are the so-called Caquot angle and the inclination of the overall body force in the backfill. The same result for δ_E has been derived, though in a different form, by Evangelista et al. (2009, 2010). In the case of gravitational loading ($\psi_e = 0$), the inclination δ_E equals the slope angle β, coinciding to the classical Rankine analysis. In presence of a horizontal body force component, δ_E is always greater than β, increasing with ψ_e up to the maximum value of φ, improving consequently wall stability. The robustness of the above stress limit analysis becomes evident since under Rankine conditions and mobilized inclination δ_E, the stress limit analysis and the Mononobe-Okabe formula results coincide (Lancellotta 2007; Mylonakis et al. 2007). These findings have been confirmed by numerical analysis results (Evangelista et al. 2010; Evangelista and Scotto di Santolo 2011).

A second key issue in the design of the particular type of retaining systems deals with the stability analysis shown in Fig. 27.1d. Traditionally, stability control of retaining walls is based on safety factors against bearing capacity, sliding and overturning. Of these, only the first two are known to be rationally defined, whereas the safety factor against overturning is known to be misleading, lacking a physical basis (Greco 1997; NTC 2008; Kloukinas and Mylonakis 2011). It is important to note that the total gravitational and seismic actions on the retaining wall are resisted upon the external reactions H and V acting on the foundation slab. The combination of these two actions, together with the resulting eccentricity e, determines the bearing capacity of the wall foundation, based on classical limit analysis procedures for a strip footing subjected to an eccentric inclined load (e.g. EC7, EC8). This suggests that the wall stability analysis is actually a footing problem and from this point of view, understanding the role of the soil mass above the foundation slab and the soil-wall interaction is of paramount importance. The above observations provided the initial motivation for the herein-reported work.

27.2 Shaking Table Experimental Investigation

The dynamic behavior of L-shaped cantilever walls was explored by means of 1-g shaking table testing. The aim of the experimental investigation is to better understand the soil-wall dynamic interaction problem, the relationship between design parameters, stability safety factors and failure mechanisms, and the validation of the seismic Rankine theoretical model. The test series were conducted to the Bristol Laboratory for Advanced Dynamics Engineering (BLADE), University of Bristol, UK. Details on the experimental hardware, materials, configurations and procedure are provided in the ensuing.

27.2.1 Experimental Hardware

27.2.1.1 Earthquake Simulator (ES)

The 6DOF ES consists of a 3 m × 3 m cast aluminum platform weighing 3.8 tonnes, with payload capacity of 15 tonnes maximum and operating frequency range of 1–100 Hz. The platform sits inside an isolated, reinforced concrete seismic block having a mass of 300 tonnes and is attached to it by eight 70 kN servo hydraulic actuators of 0.3 m stroke length. Hydraulic power for the ES is provided by a set of six shared variable volume hydraulic pumps providing up to *900 l/min* at a working pressure of *205 bar,* with maximum flow capacity of around *1,200 l/min* for up to 16 s at times of peak demand with the addition of extra hydraulic accumulators.

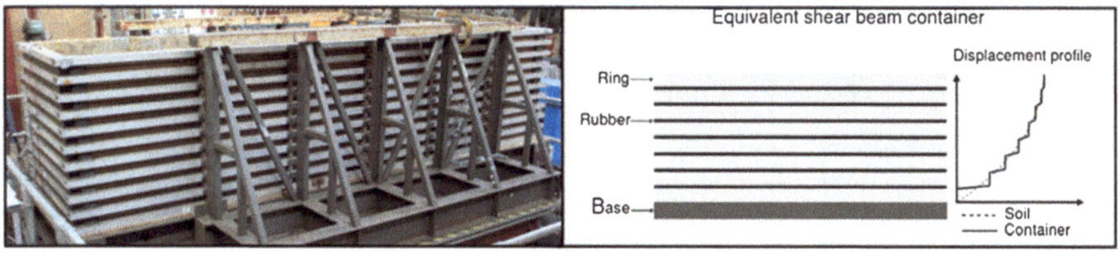

Fig. 27.2 Equivalent shear beam container ("shear stack")

27.2.1.2 Equivalent Shear Beam (ESB) Container

The apparatus, shown in Fig. 27.2a, consists of 11 rectangular aluminum rings, which are stacked alternately with rubber sections to create a flexible box of inner dimensions 4.80 m long by 1 m wide and 1.15 m deep (Crewe et al. 1995). The rings are constructed from aluminum box section to minimize inertia while providing sufficient constraint for the Ko condition. The stack is secured to the shaking table by its base and shaken horizontally lengthways (in the x direction). Its floor is roughened by sand-grain adhesion to aid the transmission of shear waves; the internal end walls are similarly treated to enable complementary shear stresses. Internal side walls were lubricated with silicon grease and covered with latex membrane to ensure plane strain conditions.

This type of containers should be ideally designed to match the shear stiffness of the soil contained in it, as shown in Fig. 27.2b. However, the shear stiffness of the soil varies during shaking depending on the strain level. Therefore the matching of the two stiffnesses (end-wall and soil) is possible only at a particular strain level. The "shear stack" at the University of Bristol is designed considering a value of strains in the soil close to failure (0.01–1 %). Therefore it is much more flexible than the soil deposit at lower strain amplitudes; as a consequence, the soil will always dictate the overall behavior of the container (Bhattacharya et al. 2012). Indeed the shear stack resonant frequency and damping in the first shear mode in the long direction when empty were measured prior to testing as 5.7 Hz and 27 % respectively, sufficiently different from the soil material properties.

27.2.1.3 Instrumentation

As seen in Fig. 27.3, 21 1-D accelerometers have been used to monitor the shaking table, the shear stack and the wall-soil system, with the main area of interest laying on the wall itself and the soil mass in its vicinity, as well as the response of the free field. 4 LVDT transducers were used to measure the dynamic response and permanent displacements of the wall and 32 strain gauges were attached on the stem and the base of the wall, on three cross sections, to monitor the bending of the wall. The signal conditioning was made via appropriate amplifier and filter modules and data acquisitions frequency was set at 1,024 Hz. Additionally to the 57 data channels employed overall, a grid of coloured sand was used for monitoring the settlement of the backfill surface.

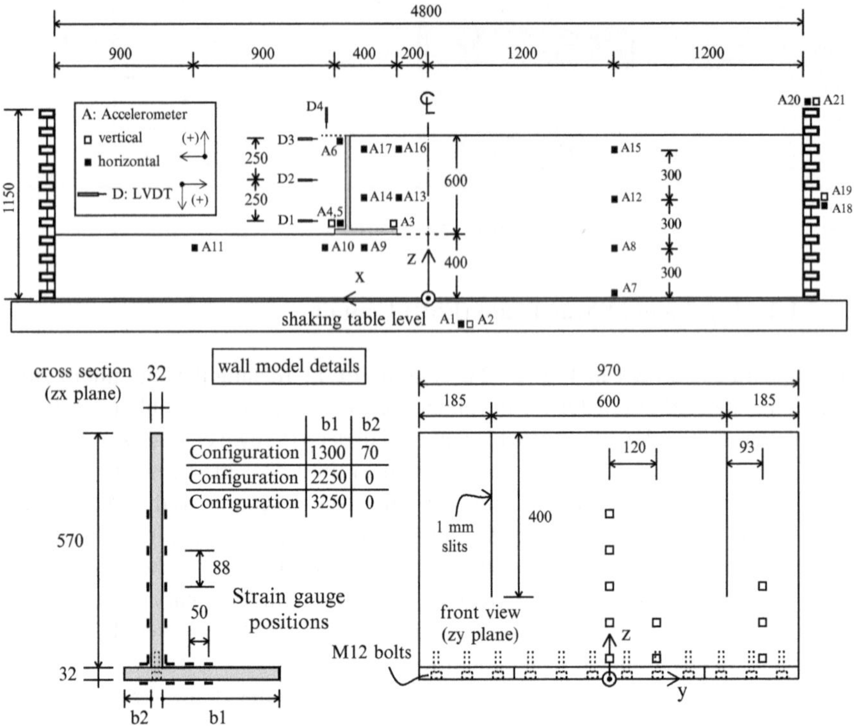

Fig. 27.3 Geometry and instrumentation of the shaking table model (dimensions in mm)

27.2.2 Shaking Table Model Setup

27.2.2.1 Model Geometry

The dimensions of the model are illustrated in Fig. 27.3, with the type and the positions of the instrumentation used. The model consists of an L-shaped retaining wall supporting a backfill of thickness 0.6 m and length five times its thickness, resting on a 0.4 m thick base soil layer (equal to the wall foundation width, B). The properties of the soil layers and detailed description of the instrumentation and the various wall configurations are provided in the below.

The retaining wall was constructed from Aluminium 5,083 plates of thickness 32 mm. The width of the wall stem is 970 mm. A central wall segment of 600 mm width was created by two 1 mm thick vertical slits penetrating 400 mm down into the wall. The location of the slits was 185 mm from each side of the wall stem. The base of the wall is subdivided into four 240 mm-wide aluminium segments that are each secured to the wall stem using three M12 bolts. Properties of the aluminium alloy are: unit weight $\gamma = 27$ kN/m^3, Young's modulus $E = 70$ GPa, Poisson's ratio $\nu = 0.3$.

Table 27.1 Soil properties

Soil layers	Voids ratio, e	Relative density, D_r	Unit weight, γ_d kN/m^3	Friction angle, φ [a]
Foundation	0.61	60 %	16.14	42°
Backfill	0.72	22 %	15.07	34°

[a]Estimated from Cavallaro et al. (2001)

Table 27.2 Pseudostatic critical accelerations and associated safety factors (SF)

Test configuration	Critical acceleration for $SF_{sliding} = 1$	$SF_{Bearing\ capacity}$ at critical sliding acceleration	Critical acceleration for $SF_{Bearing\ capacity} = 1$	$SF_{sliding}$ at critical bearing capacity acceleration
No 1	0.18g	7.45	0.35g	0.68
No 2	0.14g	1.46	0.17g	0.93
No 3	0.23g	0.44	0.17g	1.14

27.2.2.2 Soil Material Properties

The required soil configuration consists of a dense supporting layer and a medium dense backfill. The material proposed for both layers is Leighton Buzzard (LB) sand BS 881–131, Fraction B ($D_{50} = 0.82$ mm, $G_s = 2.64$ Mg/m^3, $e_{min} = 0.486$, $e_{max} = 0.78$). This particular soil has been used extensively in experimental research at Bristol and a wide set of strength and stiffness data is available (detailed references in Bhattacharya et al. 2012). The empirical correlation between peak friction angle φ and relative density D_r provided by the experimental work of Cavallaro et al. (2001) was used for a preliminary estimation of the soil strength properties.

The packing density for each layer, determined from sand mass and volume measurements during deposition, and the corresponding predictions for the peak friction angles strength are summarized in Table 27.1. The base deposit was formed by pouring sand in layers of 150–200 mm from a deposition height of 0.6 m and then densifying by shaking. After densification, the height of the layer was reduced to 390 mm. The top layer was formed by pouring sand in axisymmetric conditions close to the centre of the desired backfill region, without further densification. The pouring was carried out by keeping the fall height steady, about equal to 200 mm, to minimize the densification effect of the downward stream of sand.

27.2.2.3 Model Wall Configurations

Three different Configurations (#1, #2, #3) for the wall model presented in Fig. 27.3 were used to provide different response in sliding and rocking of the base. The characteristics of these configurations are shown in Table 27.2. After testing Configuration 1, the wall model was modified for Configuration 2. The wall heel was shortened by 50 mm and the toe was totally removed. In Configuration 3, the geometry of Configuration 2 was retained, after increasing the frictional resistance of the base interface from 23.5° to 28° (approximately equal to the critical state

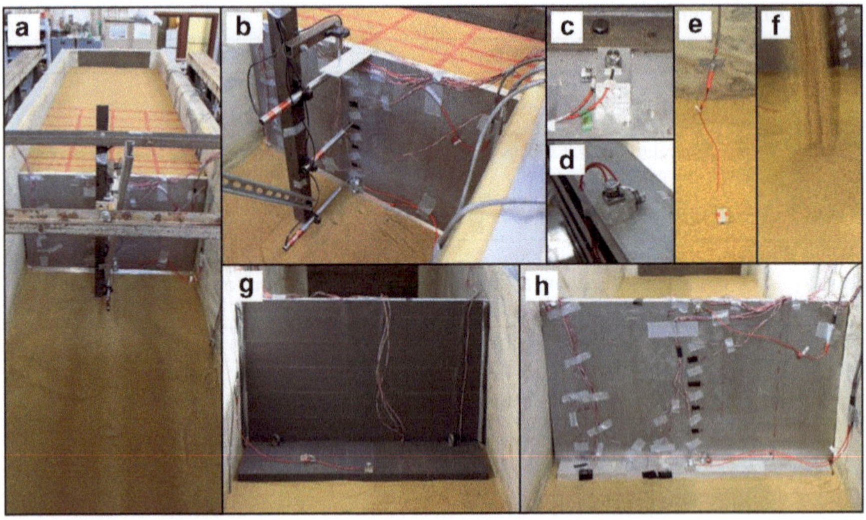

Fig. 27.4 Details of the experimental setup: (**a**) longitudinal aspect of the model, (**b**) wall face instrumentation – Config. 2, (**c**), (**d**) pairs of accelerometers on shaking table and upper ring, (**e**) backfill accelerometer, (**f**) sand pouring procedure, (**g**), (**h**) back and front view of the wall

angle), by pasting rough sandpaper. The interface friction angles were measured in-situ by means of static pull tests on the wall. The differences between these three Configurations, in terms of a pseudo-dynamic stability analysis according to EC7 (CEN 2003) are summarised in Table 27.2, ranging from a purely sliding-sensitive wall (Configuration 1), to a purely rotationally sensitive one (Configuration 3).

A set of characteristic pictures of model details and installation procedures are presented in Fig. 27.4.

27.2.3 Experimental Procedure

Each model Configuration was tested under the same dynamic excitation, described in detail in the following paragraphs. Two different input motions, harmonic and earthquake, were used in the form of sequential, increasing-amplitude time histories. The dynamic properties of the soil layers and the soil-wall interactive system were investigated by means of the white noise exploratory procedure described below. This kind of testing was repeated after every severe yielding of the system to track for significant changes in the dynamic response and properties.

27.2.3.1 White Noise Excitation

During white noise exploratory testing, a random noise signal of bandwidth *1–100 Hz* and *RMS* acceleration = 0.005g was employed. During each exploratory series test, and simultaneous data acquisition, system transmissibility was monitored using a two-channel spectrum analyser. The analyzer computes the frequency response function (FRF) between the input and the output signals of interest. Natural frequency and damping values for any resonances up to 40 Hz (i.e. within the seismic frequency range) were determined for well-defined resonances using the output of the analyser's curve fitting algorithm by means of a least-squares error technique. Additionally, transfer functions can be determined between any pair of data channels to get a clear view of the system dynamics.

27.2.3.2 Harmonic Excitation

This type of input acceleration was imposed by sinusoidal excitation consisting of 15 steady cycles. To smoothen out the transition between transient and steady-state response, the excitation comprises of a 5-cycle ramp up to full test level at the beginning of the excitation, and a 5-cycle ramp down to zero at the end. With reference to frequency and acceleration level, a set of five frequencies (4, 7, 13, 25 and 43 Hz) was used at low acceleration amplitude of 0.05g, to study the dynamic response of the system. The excitation frequency of 7 Hz was then selected for a series of harmonic excitations with increasing amplitude, until failure. The conditions of the excitation are considered to be essentially pseudostatic, as the above frequency is much smaller than the resonant frequencies of the system, with respect to both the free field and the soil-wall system.

27.2.3.3 Earthquake Excitation

Three earthquake records from the Italian and American database were selected for the earthquake testing. Specifically, the Tolmezzo record from the Friuli, 1976 earthquake, the Sturno record from Irpinia, 1980 and the Northridge record from Los Angeles, 1994. The authentic signals were scaled by a frequency scale factor of 5, derived from the scaling law $n^{0.75}$, which is valid for 1-g modelling (Muir Wood et al. 2002), assuming a geometrical scale factor of $n = 9$, corresponding to a prototype of 5.4 m high. The frequency-scaled signals were applied at low acceleration amplitude of 0.05g to measure the dynamic response of the model and then the Sturno record was selected for carrying out increasing amplitude dynamic testing, until failure in sliding or rocking of the retaining wall.

27.3 Experimental Results and Discussion

Some characteristic experimental measurements are presented in Figs. 27.5, 27.6, 27.7, 27.8 and 27.9, for the observed failure mechanisms, accelerations, dynamic and permanent displacements and bending moments, organized in sets of graphs suitable for direct comparisons. Most of the results presented herein mainly relate to

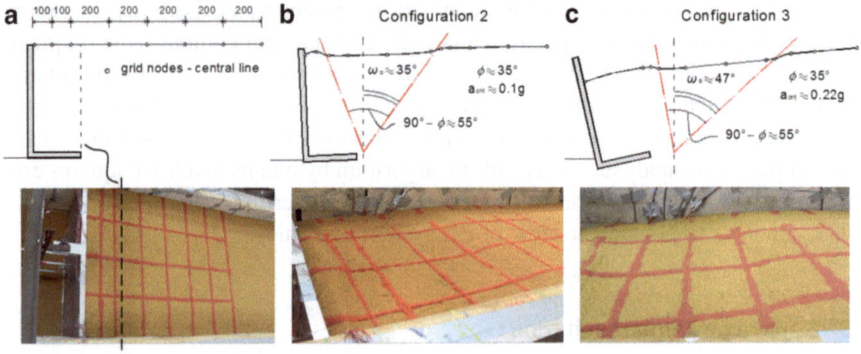

Fig. 27.5 Backfill surface settlement distribution at failure: (**a**) initial grid geometry – dimensions in mm, (**b**),(**c**) settlement distributions for Configurations 2 and 3, respectively

Configurations 1 and 3, for they exhibit yielding near similar acceleration conditions but in different modes. Limited results are presented for Configuration 2, which is significantly weaker compared to the others, failing almost simultaneously in foundation sliding and rotation (i.e., bearing capacity) mode, thus is less important for comparison reasons.

In Fig. 27.5, measured settlement profiles at the state of failure are plotted together with the assumed failure mechanisms for Configurations 2 and 3. The following are worthy of note: *First,* these failure mechanisms were observed only for transient earthquake loading, whereas in the case of harmonic excitation, the settlement profile could not clearly reveal the emergence of the main failure planes, as it had more-or-less a smooth parabolic shape. This can be explained in view of a non-uniform settlement and deformation mechanism and stronger dynamic effects imposed by earthquake loading. Although there is actually not a "rigid block" response in the retained soil mass, the experimental findings show that the earthquake excitation induces a more uniform acceleration distribution within the retained soil mass than the harmonic excitation, which corresponds to a more uniform stress field, as assumed in the pseudo-static analyses. *Second,* the assumed failure mechanisms confirm the estimations based on the material properties, the stability analysis and the yield accelerations presented in Tables 27.1 and 27.2 respectively.

The measurements of system displacement for all tested configurations are summarized in Fig. 27.6. The total, cumulative settlement and rotation of the wall, for each series of sequential input motions are presented in Fig. 27.6a, b, and the incremental displacements for each input motion are presented in Fig. 27.6c–e, indicating different behavior of the wall models under the same input. The measurements confirm the predictions for the expected failure modes and the levels of critical acceleration. The sliding failure is clearly visible in Configuration 1, as is the bearing capacity failure in Configuration 3. Configuration 2 although designed to be weaker in sliding, also exhibits significant rotational deformations caused by the high eccentricities induced by the seismic thrust. Rotational deformations are also observed in Configuration 1 for high acceleration

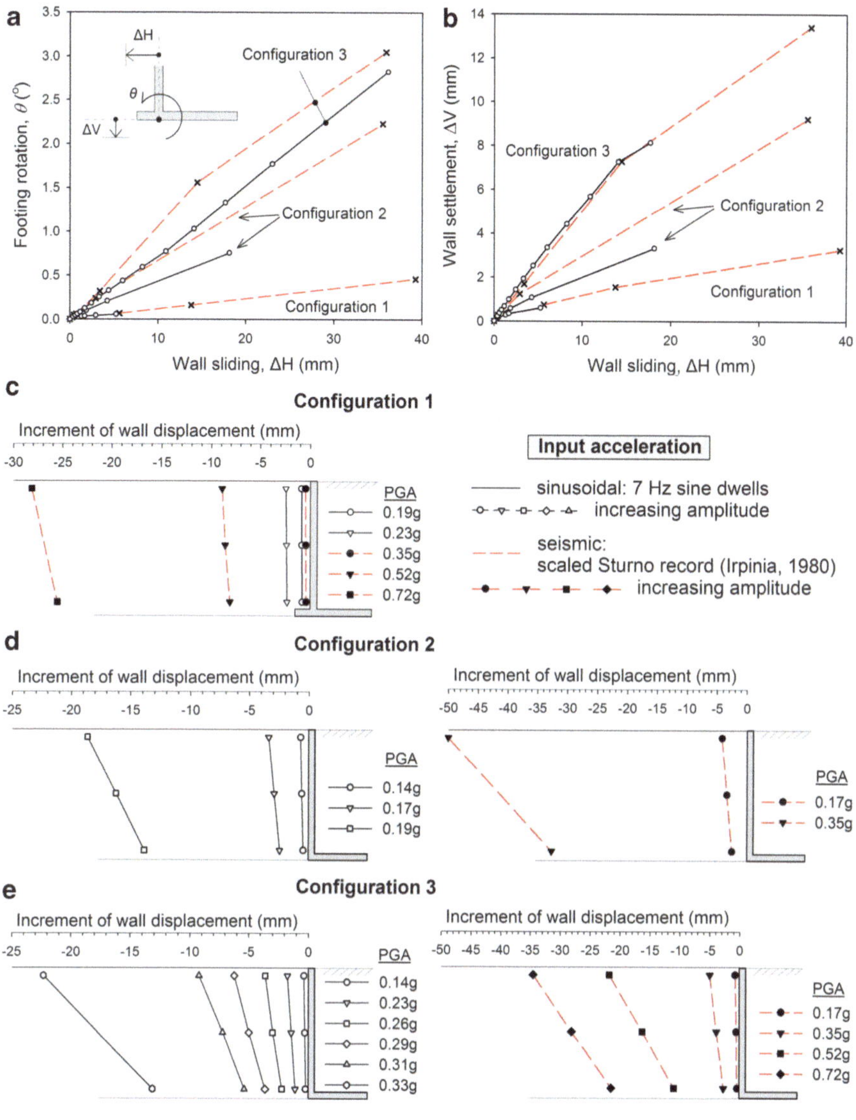

Fig. 27.6 Measurements of wall displacement and rotation for all configurations and various base excitations. (**a**) Cumulative footing rotation versus sliding (LVDT-D1), (**b**) Cumulative footing settlement (LVDT-D4) versus sliding, (**c**), (**d**) and (**e**) Incremental wall displacement (LVDTs D1-D2-D3) for configurations 1, 2 and 3, respectively

levels, revealing that walls resting on a compliant base exhibits local bearing capacity failure near the toe, due to high compressive stress concentration. This observation elucidates the importance of properly designing retaining structures to avoid developing significant rotational response.

Results from Configurations 1 and 3 under harmonic excitation are plotted together in Fig. 27.7.

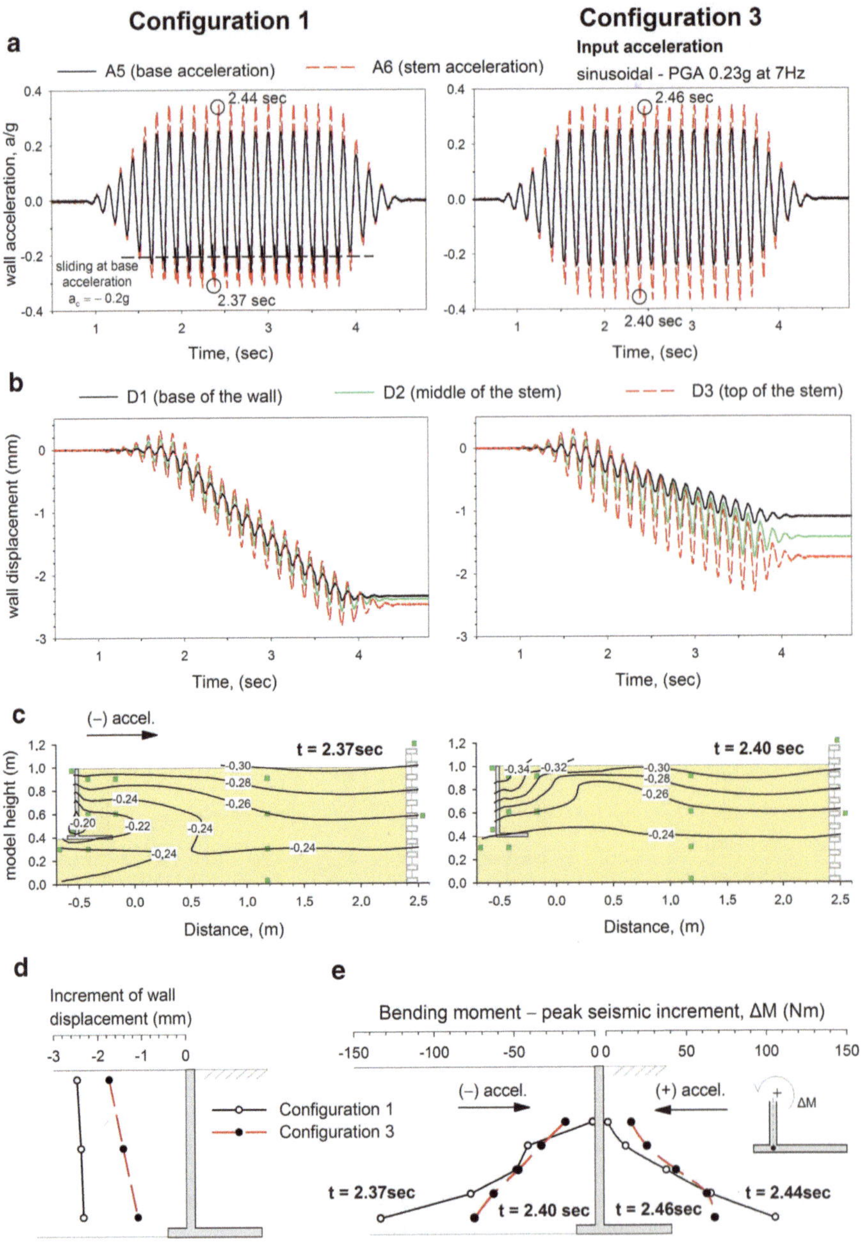

Fig. 27.7 Comparison of typical experimental results for Configurations 1 and 3 under harmonic-sinusoidal excitation: (**a**) measured wall accelerations, (**b**) corresponding wall displacements, (**c**) negative acceleration distribution (maximum inertial forces towards the wall), (**d**) increment of wall displacement (LVDTs D1-D2-D3) and (**e**) peak seismic increment of bending moment for positive and negative acceleration

The following can be observed: *First* the response of each Configuration is as expected. A translational response mode is evident in Configuration 1 and a rocking one in Configuration 3, respectively. Sliding discontinuities are obvious on the acceleration time histories of Configuration 1 (Fig. 27.7a), at a critical acceleration slightly higher than that of Table 27.2. Note that the translational yield acceleration is not steady, but always increases after every successive yielding, as even a small rotation causes penetration of the wall toe into the foundation soil thus increasing passive resistance. On the other hand, Configuration 3 starts rotating at initiation of yielding, without any evidence of sliding discontinuities on the recorded accelerogram. *Second*, in both cases the wall stem appears to have an amplified response, mainly because of foundation rocking and secondarily of pure bending of the stem. Naturally, this is more evident in Configuration 3. *Third*, both models exhibit a consistent, repeatable, behavior with respect to yielding.

The same results for earthquake loading on Configurations 1 and 3 are presented in Fig. 27.8. In this case, the input motion contains higher effective peak accelerations, but the number of important strong cycles (half cycle pulses) is only three. The sliding failure is again clearly visible in Configuration 1, as is the bearing capacity failure in Configuration 3 caused by the high eccentricities induced by seismic thrust. An important notice about the failure modes arising from the combination of the two comparisons is that the bearing capacity failure is most affected by the input acceleration level, whereas the behavior of pure sliding mechanisms is mainly controlled by the time interval of the strong motion, as known from sliding block theory. Accordingly, rotational mechanisms appear to be more critical under strong earthquakes, even though they are strong enough against sliding (Figs. 27.7d and 27.8d). Moreover, some rotational deformation is also observed in Configuration 1 for high acceleration level, revealing that any wall resting on a compliant base exhibits local bearing capacity failure near the toe, due to concentration of high compressive stresses. This observation elucidates the importance of properly designing retaining structures to avoid development of significant rotational response.

From the acceleration distributions of Figs. 27.7c and 27.8c, it can be observed that the earthquake loading results to conditions which are closer to the assumptions of pseudostatic analysis, as a soil mass moving together with the wall is evident, especially for the rotational mode of Configuration 3. Contrary to earthquake excitation, wall and soil under harmonic loading appear to respond in a quite different way. At last, the peak seismic increment of bending moments is compared in Figs. 27.7e and 27.8e. A noteworthy observation is that the earth pressure on the wall stem increases when the system moves towards the backfill, that is for an acceleration not critical for overall stability. On the other hand, at yielding acceleration of the system, earth pressure on the stem is minimum. This is in agreement with the findings of the analysis presented by Green et al. (2008) on a full scale numerical model and the experimental and numerical results of Al Atik and Sitar (2010). Comparing Configurations 1 and 3, it can be clearly identified that rotational modes induce lower earth pressure on the wall but different distribution leading to higher point of application of the thrust.

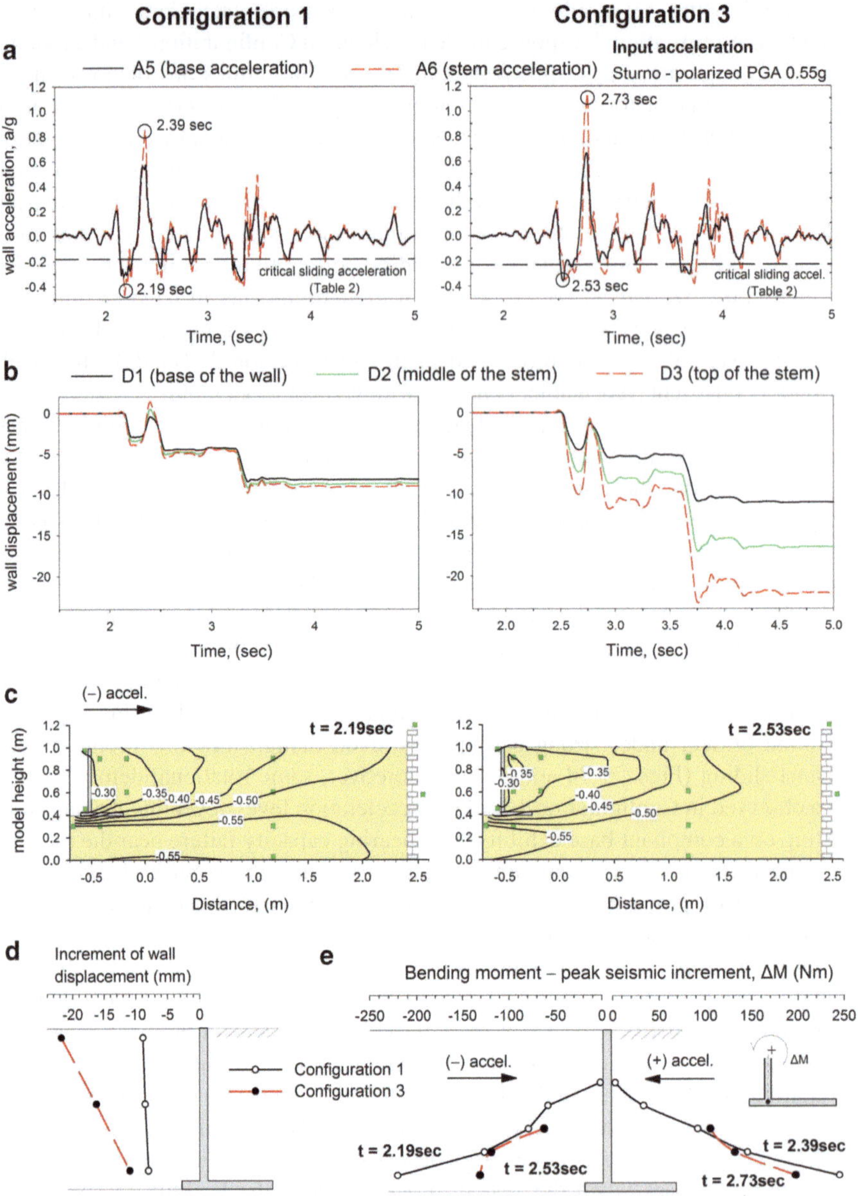

Fig. 27.8 Comparison of typical experimental results for Configurations 1 and 3 under seismic excitation: (**a**) measured wall accelerations, (**b**) corresponding wall displacements, (**c**) negative acceleration distribution (maximum inertial forces towards the wall), (**d**) increment of wall displacement (LVDTs D1-D2-D3) and (**e**) peak seismic increment of bending moment for positive and negative acceleration

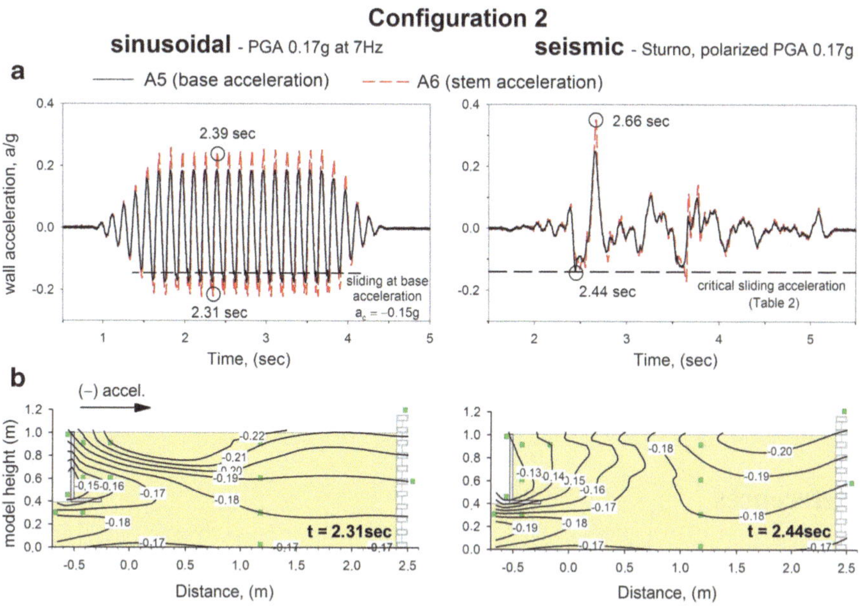

Fig. 27.9 Typical experimental results for Configuration 2: (**a**) measured wall accelerations for harmonic sinusoidal and seismic excitation, (**b**) negative acceleration distribution (maximum inertial forces towards the wall)

Finally, the response of Configuration 2 to harmonic and earthquake excitation with PGA = 0.17g presented in Fig. 27.9, exhibits a similar behavior to both Configurations 1 and 3, resulting to simultaneous sliding and rotational failure. As seen from both the acceleration time histories (Fig. 27.9a) and the failure mechanisms (Fig. 27.9b) the sliding failure mode prevails, which is consistent to the critical accelerations estimated in Table 27.2.

27.4 Conclusions

A series of shaking table tests on scaled models of cantilever retaining walls were conducted in the BLADE laboratory at the University of Bristol. The initial motivation of this experimental study was the validation of recent stress limit analysis solutions for the seismic design of this type of retaining structures (Evangelista et al. 2009, 2010; Kloukinas and Mylonakis 2011) in conjunction with the absence of any specific, relative regulations in established seismic codes, including EC8. Special issues related to stability design and response of walls founded on compliant base were also incorporated. Preliminary interpretation of the experimental findings confirms the predictions of the theoretical analysis, with reference to the failure mechanisms and the critical yield accelerations of the system. Pseudo-static stability analysis proves to behave adequately for both harmonic and seismic excitation, although important dynamic effects are evident in the first case, dealing with the response of the backfill and the wall stem. On the other hand, earthquake loading results to conditions which are closer to the assumptions of the pseudo-static analysis, notable the uniform distribution of the acceleration and the "rigid block" response of the backfill. Finally, the responses of the various wall configurations confirm the equivalent footing analysis of wall stability and highlight the importance of a proper design of walls founded on compliant base with respect to sliding and rocking.

Acknowledgments The research leading to these results has received funding from the European Union Seventh Framework Programme (FP7/2007–2013) for access to the Bristol Laboratory for Advanced Dynamics Engineering (BLADE), University of Bristol, UK under grant agreement n° 227887 [SERIES].

References

Al Atik L, Sitar N (2010) Seismic earth pressures on cantilever retaining structures. J Geotech Geoenviron Eng ASCE 136(10):1324–1333

Bhattacharya S, Lombardi D, Dihoru L, Dietz MS, Crewe AJ, Taylor CA (2012) Model container design for soil-structure interaction studies. In: Fardis MN, Rakicevic ZT (eds) Role of seismic testing facilities in performance-based earthquake engineering: SERIES workshop, vol 22, Geotechnical, geological and earthquake engineering. Springer, Dordrecht/New York

Cavallaro A, Maugeri M, Mazzarella R (2001) Static and dynamic properties of Leighton Buzzard sand from laboratory tests. In: Proceedings of 4th international conference on recent advanced in geotechnical earthquake engineering and soil dynamics and symposium in honour of Professor WD Liam Finn, San Diego, CA, 26–31 March 2001

CEN (2003) EN 1997–1:2003 Eurocode 7 geotechnical design, Part 1: general rules. European Committee for Standardization, Brussels

CEN (2004) EN 1998–1:2004 Eurocode 8 design of structures for earthquake resistance, Part 5: foundations, retaining structures and geotechnical aspects. European Committee for Standardization, Brussels

Crewe AJ, Lings ML, Taylor CA, Yeung AK, Andrighetto R (1995) Development of a large flexible shear stack for testing dry sand and simple direct foundations on a shaking table. In: Elnashai AS (ed) European seismic design practice. Balkema, Rotterdam

EAK 2000 (2003) Greek seismic code, Earthquake Planning and Protection Organization, Athens

Evangelista A, Scotto di Santolo A (2011) Dynamic active earth pressure on cantilever retaining walls. Comput Geotech 38(8):1041–1051

Evangelista A, Scotto di Santolo A, Simonelli AL (2009) Dynamic response of cantilever retaining walls. In: Proceedings of international conference on performance-based design in earthquake geotechnical engineering – from case history to practice – Tokyo, 15–18 June 2009

Evangelista A, Scotto di Santolo A, Simonelli AL (2010) Evaluation of pseudostatic active earth pressure coefficient of cantilever retaining walls. Soil Dyn Earthq Eng 30(11):1119–1128

Greco VR (1997) Stability of retaining walls against overturning. J Geotech Geoenviron Eng ASCE 123(8):778–780

Greco VR (2001) Active earth thrust on cantilever walls with short heel. Can Geotech J 38(2):401–409

Green RA, Olgun CG, Cameron WI (2008) Response and modeling of cantilever retaining walls subjected to seismic motions. Comput-Aided Civil Infrastruct Eng 23(4):309–322

Huntington W (1957) Earth pressures and retaining walls. Wiley, New York

Kloukinas P, Mylonakis G (2011) Rankine Solution for seismic earth pressures on L-shaped retaining walls. In: 5 ICEGE, Santiago, Chile, 10–13 January 2011
Lancellotta R (2007) Lower-bound approach for seismic passive earth resistance. Geotechnique 57(3):319–321
Muir Wood D, Crewe A, Taylor CA (2002) Shaking table testing of geotechnical models. UPMG-Int J Phys Model Geotech 2:01–13
Mylonakis G, Kloukinas P, Papantonopoulos C (2007) An alternative to the Mononobe–Okabe equations for seismic earth pressures. Soil Dyn Earthq Eng 27(10):957–969
NTC (2008) Italian building code, DM 14 Jan, GU n 29, 4 Feb, n 30
Rankine WJM (1857) On the stability of loose earth. Phil Trans Roy Soc Lond 147:9–27

Index

A
ABAQUS, 96, 101, 102, 108, 109, 113
Assessment
damage, 19, 21, 23, 32, 48, 56
earthquake-risk, 2
rapid, 15–34
safety, 17, 39–41, 246

B
Beam-column joint, 189, 191, 192, 194, 196–199, 201, 221–241, 243–258
Bond
model, 205, 211–218
slip, 218, 356, 361–366, 375
Building
full-scale RC, 201
masonry, 16, 17, 26–34, 156, 161, 168
multi-storey, 441, 442, 447–452
RC, 16–25, 47–59, 262, 278
school, 37–44, 246

C
Cantilever walls, 478, 480
Centrifuge
modeling, 408–412, 425–439
tests, 402–404, 408, 410, 411, 419, 427, 428, 432, 433, 435, 439
Concrete damage plasticity (CDP), 97
Confinement, 197, 204, 205, 209, 211, 213, 216–218, 245, 247, 249–251, 258, 323, 356, 363
Consecutive earthquakes, 174, 185
Cyclic load/-ing, 90, 96, 102, 157, 172, 179, 183, 207, 222, 231, 233, 252, 268, 354, 361, 366, 367, 369, 372, 375, 384, 444

D
Damage
assessment, 19, 21, 23, 32, 48, 56
evidence, 41–44
Dowels, 263–266, 278, 282–287, 289, 290, 294–296, 298–300, 304–306, 308–310, 324–326

E
Earth pressure, 390, 394, 397–399, 402–405, 479, 489
Earthquake
excitation, 41, 119, 185, 188, 189, 340, 457, 463, 485, 486, 491
risk assessment, 2
simulation, 333, 335, 426, 428, 434, 436, 438
Educational activities, 387
Emergency intervention, 65, 66
Energy, 408, 410, 415, 417, 418, 420, 423
dissipation, 129, 159, 188, 207, 244, 246, 252, 254, 256, 258, 269, 284, 305, 316, 351–353, 418
Existing structures, 56, 57, 116–118, 123, 124, 246, 303–327

A. Ilki and M.N. Fardis (eds.), *Seismic Evaluation and Rehabilitation of Structures*, Geotechnical, Geological and Earthquake Engineering 26,
DOI 10.1007/978-3-319-00458-7,

F
Failure mode(s), 40, 96, 105–108, 111, 207, 208, 214, 233, 246, 252–254, 256, 258, 283, 284, 296, 299, 304, 305, 361, 365, 372, 478, 486, 489, 491
Fibre reinforced polymer (FRP), 156, 203–218, 222, 279, 299, 300, 304
  carbon fibre reinforced polymer (CFRP), 250–252, 256, 275, 276, 289, 290, 292, 310, 311, 315, 323, 326
  strengthening, 245–250, 252, 254, 256, 258
Finite element
  nonlinear analysis, 222
  simulation, 96, 101, 102, 108, 111, 112, 229, 238, 239, 241
Fragility, 32, 33, 47–49, 56–59
  curves, 62–65, 67, 70, 74, 75
Free-field response, 465–470
Frequency drop, 85–89
Full scale, 155–168, 205, 217, 246, 248, 304–306, 326, 330, 442
  RC building, 201

H
Historic monuments, 115–153
Hybrid control, 329–340

I
Infill
  brick, 171, 172
  RC, 281–300, 303–327
Internal forces, 394, 398, 404

K
Kinematic interaction, 457, 458, 465, 467, 471, 472

L
Liquefaction, 425–439
  mitigation, 380, 382–385
  soil, 441–452
Load-bearing capacity, 122, 124, 127, 128, 141, 148, 149
Low-strength concrete, 221–241

M
Masonry, 4, 156, 157, 159–161, 163, 165, 166, 168, 246, 247, 283
  building, 16, 17, 26–34
  unreinforced, 77–93
  wall, 95–113, 172, 185
Model
  bond, 204, 211–218
  plastic damage, 95–113
Model code, 211–213
Multi-axial hybrid fabric, 156, 157

N
Natural stone, 155–168
Nonlinear
  dynamic analysis, 53, 56, 59, 122–124, 128, 139, 141, 268, 271
  finite element analysis, 222

O
Optimum damping, 330, 335
Outreach activities, 381, 386, 387

P
Pile
  bending, 458, 465, 473
  group, 457, 459, 465
  RC, 445
  steel sheet pile wall (SSPW), 442, 447, 449–452
Plastic damage model, 95–113
Post-tensioned metal straps, 187–202
Probabilistic scenarios, 65, 67–68, 70–74
Probability, 38–44, 48, 50, 54
Pseudo-dynamic tests, 176, 179, 183, 281–300, 313, 341–353

R
RC
  building, 16–25, 47–59, 187–202, 222, 251, 262, 278, 299
  infill, 281–300, 303–327
  pile, 445
  wall, 261–279, 281, 284, 305, 361, 372, 373
Reconstruction, 124, 129–153, 246
Rehabilitation, 66, 75, 156, 185
Retaining structures, 487, 489, 491
Retrofit/-ting, 155–168, 205, 281, 288, 299
  seismic, 246, 261, 262, 279, 303–327
Risk
  prioritization, 19, 34
  reduction, 62

seismic, 16, 17, 33, 34, 38, 44, 63, 66, 75, 185, 387
Rocking, 85, 86, 89–92, 105, 106, 111, 113, 262, 268, 279, 390, 394, 401, 403, 405, 407–423, 483, 485, 489, 492

S
Sand, 390–393, 403, 404, 411, 412, 414–423, 459, 461, 462, 472, 481, 483, 484
partially saturated, 381–386
Seismic
action, 80, 86, 90, 92, 246, 248, 249, 252, 332, 336, 478, 480
behavior, 77–93
performance, 16, 18, 21, 26, 32, 38, 178, 189, 199–201, 221–241, 258, 271, 276, 279, 355, 456
rehabilitation, 261–279, 299
response, 101, 132, 262, 341–353, 366, 372, 376, 390, 441, 457
retrofit/-ting, 246, 261, 262, 279, 303–327
risk, 16, 17, 33, 34, 38, 44, 63, 66, 75, 185, 387
scenarios, 66, 68
strengthening, 115–153, 171
Semi-active control, 330
Settlement, 17, 32, 394, 396–399, 402, 405, 408, 413–421, 423, 435, 438–439, 456, 481, 486, 487
Shaking table, 82, 155–168, 380–383, 385, 386, 457, 459, 461, 464–466, 470, 472
test/-ing, 122, 123, 153, 187–202, 480, 491
Shallow foundations, 407–423, 425–439
Shear
shear flexure interaction, 372–375
transfer, 285, 443
Sliding, 97, 105, 106, 111, 113, 164, 177, 178, 184, 282, 283, 296, 300, 304, 305, 359, 369, 374, 403, 405, 407–423, 478, 480, 483–487, 489, 491, 492
Soft-first-story, 341–353
Soil
liquefaction, 441–452
soil-pile-structure interaction (SPSI), 457, 464
soil-structure-interaction, 390, 408, 411, 417, 423
soil-wall interaction, 480
Soundproofing devices, 78–80
Square tunnels, 389–405
Stability control, 441–452, 480
Steel sheet pile wall (SSPW), 442, 447–452
Strengthening
FRP, 245, 246, 248–250, 254, 256–258
seismic, 115–153
Structural
damage, 38, 44, 48, 51, 54, 56, 59, 63, 171, 201, 341, 342, 426, 441, 447, 451, 452
safety, 2, 3

T
Test/-ing
centrifuge, 392, 402–404, 408, 410, 411, 419, 423, 427, 428, 432, 433, 435, 439
pseudo-dynamic, 176, 179, 183, 281–300, 313, 341–353
shaking table, 122, 123, 153, 187–202, 480, 491
Textile-Reinforced Mortar (TRM), 203–218

U
Uplifting footing, 262, 267–270, 274

V
Vulnerability assessment, 12, 39

MIX
Papier aus verantwortungsvollen Quellen
Paper from responsible sources
FSC® C105338

If you have any concerns about our products,
you can contact us on
ProductSafety@springernature.com

In case Publisher is established outside the EU,
the EU authorized representative is:
Springer Nature Customer Service Center GmbH
Europaplatz 3, 69115 Heidelberg, Germany

Printed by Libri Plureos GmbH
in Hamburg, Germany